VOLUME SIXTEEN

# Developments in Precambrian Geology

# NEOPROTEROZOIC-CAMBRIAN TECTONICS, GLOBAL CHANGE AND EVOLUTION: A FOCUS ON SOUTH WESTERN GONDWANA

DEVELOPMENTS IN PRECAMBRIAN GEOLOGY

Advisory Editor Kent Condie

Further titles in this series


1. B.F. WINDLEY and S.M. NAQVI (Editors)
   Archaean Geochemistry
2. D.R. HUNTER (Editor)
   Precambrian of the Southern Hemisphere
3. K.C. CONDIE
   Archean Greenstone Belts
4. A. KRÖNER (Editor)
   Precambrian Plate Tectonics
5. Y.P. MEL'NIK
   Precambrian Banded Iron-formations. Physicochemical Conditions of Formation
6. A.F. TRENDALL and R.C. MORRIS (Editors)
   Iron-Formation: Facts and Problems
7. B. NAGY, R. WEBER, J.C. GUERRERO and M. SCHIDLOWSKI (Editors)
   Developments and Interactions of the Precambrian Atmosphere, Lithosphere and Biosphere
8. S.M. NAQVI (Editor)
   Precambrian Continental Crust and Its Economic Resources
9. D.V. RUNDQVIST and F.P. MITROFANOV (Editors)
   Precambrian Geology of the USSR
10. K.C. CONDIE (Editor)
    Proterozoic Crustal Evolution
11. K.C. CONDIE (Editor)
    Archean Crustal Evolution
12. P.G. ERIKSSON, W. ALTERMANN, D.R. NELSON, W.U. MUELLER and O. CATUNEANU (Editors)
    The Precambrian Earth: Tempos and Events
13. T.M. KUSKY (Editor)
    Precambrian Ophiolites and Related Rocks
14. M. LEHTINEN, P.A. NURMI and O.J. RÄMÖ (Editors)
    Precambrian Geology of Finland: Key to the Evolution of the Fennoscandian Shield
15. MARTIN J. VAN KRANENDONK, R. HUGH SMITHIES and VICKIE C. BENNETT (Editors)
    Earth's Oldest Rocks


Photograph of the world's second largest canyon, the Fish River Canyon in southern Namibia. A marked angular unconformity separates high-grade metamorphic basement rocks of the 1.0 Ga Namaqua-Natal Belt from the overlying, undeformed strata of the latest Ediacaran/early Cambrian Nama Group; photo by H.E. Frimmel.

VOLUME SIXTEEN

DEVELOPMENTS IN PRECAMBRIAN GEOLOGY

# NEOPROTEROZOIC-CAMBRIAN TECTONICS, GLOBAL CHANGE AND EVOLUTION: A FOCUS ON SOUTH WESTERN GONDWANA

*Editors*

CLAUDIO GAUCHER
*Departamento de Geología, Instituto de Ciencias Geológicas, Facultad de Ciencias, Iguá 4225–11400 Montevideo, Uruguay*

ALCIDES N. SIAL
*NEG-LABISE, Departamento de Geologia, Universidade Federal de Pernambuco, C.P. 7852, Recife 50670-000, Brazil*

GALEN P. HALVERSON
*Geology & Geophysics, School of Earth & Environmental Sciences, University of Adelaide, North Terrace, Adelaide, SA 5005, Australia*

HARTWIG E. FRIMMEL
*Department of Geological Sciences, University of Cape Town, Rondebosch 7701, South Africa and Geodynamics & Geomaterials Research Division, University of Würzburg, Am Hubland, D-97074 Würzburg, Germany*

ELSEVIER

Amsterdam • Boston • Heidelberg • London • New York • Oxford
Paris • San Diego • San Francisco • Singapore • Sydney • Tokyo

Elsevier
Radarweg 29, PO Box 211, 1000 AE Amsterdam, The Netherlands
Linacre House, Jordan Hill, Oxford OX2 8DP, UK

First edition 2010



**British Library Cataloguing in Publication Data**
A catalogue record for this book is available from the British Library



**Library of Congress Cataloging-in-Publication Data**
A catalog record for this book is available from the Library of Congress

ISBN: 978-0-444-53249-7
ISSN: 0166-2635

For information on all Elsevier publications
visit our website at books.elsevier.com

Printed and bound in The Netherlands

10 11 12 13 14    10 9 8 7 6 5 4 3 2 1

# Contents

## PART 4 Neoproterozoic-Cambrian Evolution of the Rio de la Plata Palaeocontinent

## PART 6 Neoproterozoic-Cambrian Evolution of The Pampean Orogen



## PART 7 Microcontinents and Suspect Terranes in SW Gondwana



# SECTION III: Synthesis and Global Implications

## PART 8 Palaeogeographic Evolution from Rodinia to Gondwana

## PART 9 Neoproterozoic-Cambrian Biota



## PART 10 Neoproterozoic-Cambrian Biogeochemical Evolution



## PART 11 Neoproterozoic-Cambrian Palaeoclimatic Events

# Contributors

**F.G. Aceñolaza**
Instituto Superior de Correlación Geológica (INSUGEO), Miguel Lillo 205, 4000 San Miguel de Tucumán, Argentina (insugeo@csnat.unt.edu.ar)

**G.F. Aceñolaza**
Instituto Superior de Correlación Geológica (INSUGEO), Miguel Lillo 205, 4000 San Miguel de Tucumán, Argentina (acecha@webmail.unt.edu.ar)

**C.J.S. De Alvarenga**
Instituto de Geociências, Universidade de Brasília, Campus Universitário, Asa Norte, Brasília, DF 70910-900, Brazil (alva1@unb.br)

**M. Babinski**
Centro de Pesquisas Geocronológicas, Instituto de Geociências, Universidade de São Paulo, Rua do Lago 562, CEP 05508-080, São Paulo, SP, Brazil (babinski@usp.br)

**M.A.S. Basei**
Centro de Pesquisas Geocronológicas, Instituto de Geociências, Universidade de São Paulo, Rua do Lago 562, CEP 05508-080, São Paulo, SP, Brazil (baseimas@usp.br)

**G.H. Blanco**
Department of Geology, University of Johannesburg, PO Box 524, Auckland Park 2006, 13 Johannesburg, South Africa (blancogonzalo2@hotmail.com)

**P.C. Boggiani**
Instituto de Geociências, Universidade de São Paulo, Rua do Lago, 562, São Paulo, SP 05508-080, Brazil (boggiani@usp.br)

**J. Bossi**
Cátedra de Geología, Facultad de Agronomía, Garzón 780, Montevideo, Uruguay (jbossi@adinet.com.uy)

**B.B. de Brito Neves**
Instituto de Geociências, Universidade de São Paulo, Rua do Lago, 562, 05508-080 São Paulo, SP, Brazil (bbleybn@usp.br)

**N.M. Chumakov**
Geological Institute, Russian Academy of Sciences, Moscow 119017, Pyzhevskiy 7, Russia (chumakov@ginras.ru)

**C. Cingolani**
Centro de Investigaciones Geológicas (CIG), Universidad Nacional de La Plata-CONICET, calle 1 n. 644, 1900 La Plata, Argentina (ccingola@cig.museo.unlp.edu.ar)

**A.S. Collins**
Geology & Geophysics, School of Earth & Environmental Sciences, University of Adelaide, North Terrace, Adelaide, SA 5005, Australia (alan.collins@adelaide.edu.au)

**E.L. Dantas**
Instituto de Geociências, Universidade de Brasília, Campus Universitário, Asa Norte, Brasília, DF 70910-900, Brazil (elton@unb.br)

**M.A. Dardenne**
Instituto de Geociências, Universidade de Brasília, Campus Universitário, Asa Norte, Brasília, DF 70910-900, Brazil (marceldardenne@yahoo.fr)

**C.O. Drukas**
Centro de Pesquisas Geocronológicas, Instituto de Geociências, Universidade de São Paulo, Rua do Lago, 562, 05508-080 São Paulo, SP, Brazil

**T.R. Fairchild**
Instituto de Geociências, Universidade de São Paulo, Rua do Lago, 562, 05508-080 São Paulo, SP, Brazil (trfairch@hotmail.com)

**V.P. Ferreira**
NEG-LABISE, Departamento de Geologia, Universidade Federal de Pernambuco, C.P. 7852, Recife, PE 50670-000, Brazil (valderez@ufpe.br)

**M. Figueiredo**
Instituto de Geociências, Universidade de São Paulo, Rua do Lago, 562, 05508-080 São Paulo, SP, Brazil (mimff@hotmail.com)

**H.E. Frimmel**
Department of Geological Sciences, University of Cape Town, Rondebosch 7700, South Africa, and Geodynamics & Geomaterials Research Division, University of Würzburg, Am Hubland, D-97074 Würzburg, Germany (hartwig.frimmel@uni-wuerzburg.de)

**C. Gaucher**
Departamento de Geología, Instituto de Ciencias Geológicas, Facultad de Ciencias, Iguá 4225, 11400 Montevideo, Uruguay (gaucher@chasque.net)

**G.J.B. Germs**
University of Johannesburg, Private Bag X607, 6620 Oudtshoorn, South Africa (gagerms@global.co.za)

**L. Gómez Peral**
Centro de Investigaciones Geológicas (CIG), Universidad Nacional de La Plata-CONICET, calle 1 n. 644, 1900 La Plata, Argentina (lperal@cig.museo.unlp.edu.ar)

**G.P. Halverson**
Geology & Geophysics, School of Earth & Environmental Sciences, University of Adelaide, North Terrace, Adelaide, SA 5005, Australia (galen.halverson@adelaide.edu.au)

**M.T. Hurtgen**
Department of Earth and Planetary Sciences, Northwestern University, 1850 Campus Drive, Evanston, IL 60208, USA (matt@earth.northwestern.edu)

**A.J. Kaufman**
Departments of Geology and ESSIC, University of Maryland, College Park, MD 20742-4211, USA (kaufman@geol.umd.edu)

**M.S. Mantovani**
Instituto de Astronomia, Geofísica e Ciências Atmosféricas, Universidade de São Paulo, Rua do Matão 1226, São Paulo, SP 05508-090, Brazil (mantovani@usp.br)

**R.McG. Miller**
PO Box 11222, Windhoek, Namibia (rmmiller@mweb.com.na)

**A. Misi**
Centro de Pesquisa em Geofísica e Geologia, Instituto de Geociências, Universidade Federal da Bahia, Campus da Federação, 40170-290 Salvador-BA, Brazil (misi@ufba.br)

**A. Nutman**
Research School of Earth Sciences, The Australian National University, Canberra, ACT 0200, Australia; Beijing SHRIMP Centre, Chinese Academy of Geological Sciences, 26 Baiwanzhuang Road, Beijing 100037, China (nutman@bjshrimp.cn)

**C.R. Passarelli**
Centro de Pesquisas Geocronológicas, Instituto de Geociências, Universidade de São Paulo, Rua do Lago, 562, 05508-080 São Paulo, SP, Brazil (crpass@usp.br)

**A. Pedreira**
CPRM-Serviço Geológico do Brasil, Av. Ulysses Guimarães 2862, Salvador, Bahia 41213-000, Brazil (apedreira@sa.cprm.gov.br)

**A.C. Pedrosa-Soares**
CPMTC-IGC-Universidade Federal de Minas Gerais, CEP 31270-901, Belo Horizonte, MG, Brazil (pedrosa@igc.ufmg.br)

**M.M. Pimentel**
Instituto de Geociências, Universidade de Brasília, Campus Universitário, Asa Norte, Brasília, DF 70910-900, Brazil (marcio@unb.br)

**D.G. Poiré**
Centro de Investigaciones Geológicas (CIG), Universidad Nacional de La Plata-CONICET, calle 1 n. 644, 1900 La Plata, Argentina (poire@cig.museo.unlp.edu.ar)

**S.M. Porter**
Department of Earth Science, University of California at Santa Barbara, Santa Barbara, CA 93106, USA (porter@geol.ucsb.edu)

**R.V. Santos**
Instituto de Geociências, Universidade de Brasília, Campus Universitário, Asa Norte, Brasília, DF 70910-900, Brazil (rventura@unb.br)

**A.N. Sial**
NEG-LABISE, Departamento de Geologia, Universidade Federal de Pernambuco, C.P. 7852, Recife 50670-000, Brazil (sial@ufpe.br)

**O. Siga Jr.**
Centro de Pesquisas Geocronológicas, Instituto de Geociências, Universidade de São Paulo, Rua do Lago, 562, 05508-080 São Paulo, SP, Brazil (osigajr@usp.br)

**M. Egydio-Silva**
Instituto de Geociências, Universidade de São Paulo, Rua do Lago, 562, 05508-080 São Paulo, SP, Brazil (megydios@usp.br)

**M.A. da Silva Filho**[†]
In Memoriam (NEG-LABISE, Departamento de Geologia, Universidade Federal de Pernambuco, C.P. 7852, Recife 50670-000, Brazil)

**P. Sprechmann**
Departamento de Geología, Instituto de Ciencias Geológicas, Facultad de Ciencias, Iguá 4225, 11400 Montevideo, Uruguay (sprechma@chasque.net)

**A. Toselli**
Instituto Superior de Correlación Geológica (INSUGEO), Miguel Lillo 205, 4000 San Miguel de Tucumán, Argentina (ajtoselli@infovia.com.ar)

**A. Uhlein**
Universidade Federal de Minas Gerais, IGC-DEGEL-CPMTC, Belo Horizonte, MG 31270-901, Brazil (auhlein@uai.com.br)

**T.M. Will**
Geodynamics & Geomaterials Research Division, University of Würzburg, Am Hubland, D-97074 Würzburg, Germany (thomas.will@uni-wuerzburg.de)

# Preface

The 2nd Symposium on the Vendian-Cambrian of southwestern Gondwana was held in Montevideo in 2002, and associated field trips were carried out in Uruguay and Argentina with the sponsorship of UNESCO. It was during this conference that the idea of presenting a proposal to the IGCP (then International Geological Correlation Programme, now International Geoscience Programme) crystallised. The proposal was approved, becoming IGCP 478 project, entitled 'Neoproterozoic-Early Palaeozoic events in southwestern Gondwana'. Six consecutive meetings were successfully held: 2003 in Cape Town (South Africa), 2004 in São Paulo (Brazil), 2005 in Windhoek (Namibia), 2006 in Punta del Este (Uruguay), 2007 in Stellenbosch (South Africa) and 2008 during the 33rd International Geological Congress in Oslo, Norway. Field trips were also carried out to the Gariep Belt and Nama Group (2003), the Paraguay Belt (2004), the Damara Belt (2005), the Nico Pérez Terrane in Uruguay, Tandilia and Puncoviscana Formation in Argentina (2006) and the Saldania Belt (2007).

IGCP 478 has been a tremendous success based on a number of truly intercontinental research efforts. A wealth of new data emanated from the various IGCP-related field trips and from new collaborations between research groups across the continents. Most of this new information has been published in numerous individual research articles, including three thematic issues of international journals. Not surprisingly, our understanding of the geodynamic, palaeoenvironmental, climatic and biotic evolution of southwestern Gondwana has changed significantly over the past decade, partly because of the progress achieved in the course of IGCP 478, and also because of other research efforts, including those of thematically related IGCPs 493 (on the rise and fall of the Vendian biota) and 512 (on Neoproterozoic ice ages).

The Neoproterozoic was an eventful Era. These events comprise the demise of the supercontinent Rodinia and the formation of Gondwana, arguably the most intense orogenies in Earth's history, from the end of the Grenvillian to the Brasiliano/Pan-African, the most severe glaciations thus far recognised, the advent of animals and the first appearance of skeletons. Thus the Neoproterozoic spans a time that is of particular interest across multiple disciplines. Those continental areas that comprised the southwestern part of the supercontinent Gondwana at the end of the Neoproterozoic hold a series of keys to our understanding of the fundamental processes that shaped the Neoproterozoic Earth. This book presents an up-to-date review of our current understanding of the events that shaped the various parts of southwestern Gondwana as exposed in a number of Brasiliano/Pan-African tectonic belts and their corresponding forelands in South America and southwestern Africa. With 45 authors having contributed to this book, it represents the most comprehensive review of its kind and for the first time compares available information from South America and Southern Africa under the same cover.

The structure of the book follows a logical flow, starting with an introduction, followed by the systematic treatment of the various Neoproterozoic to Cambrian tectono-stratigraphic units of southwestern Africa and South America, and finally providing a synthesis of the presented data in a supercontinental and global context. Section 1 provides an updated and global overview of the main issues of Neoproterozoic research with focus on partly highly controversial aspects of biogeochemical and palaeoclimatic events. Section 2 describes the Neoproterozoic-Early Palaeozoic geology and events as recorded in the various parts of southwestern Gondwana, from the Amazonian, the São Francisco and the Río de la Plata cratons in South America to the Pan-African belts and forelands of southwestern Africa (Damara Orogen *sensu lato*). The Pampean Terrane and other Neoproterozoic microcontinents and suspected terranes are also covered. Several chapters that comprise Section 3 synthesise the implications of the findings presented in the previous section on our understanding of the interplay and relationships between tectonic events, palaeogeographic evolution, radiation and extinction of life forms as well as biogeochemical and palaeoclimatic events. Events recorded in other parts of the world are also discussed in order to complement the data from southwestern Gondwana and put them into a global context.

In spite of the topics dealt with in the various sections being all interrelated, each chapter and paper can be read and understood in isolation. For the sake of avoiding unnecessary and lengthy repetitions, a single comprehensive bibliography is provided at the end, covering all citations throughout the entire book, and several figures are referred to in more than one chapter. Each chapter is headed by a brief summary that helps the reader to readily navigate through the large amount of information presented in the book. The multitude of contributing authors invariably led to differences in style and conventions in the submitted manuscripts. This said, we endeavoured to adopt a consistent style, including a UK spelling convention. Ages of rocks or events are given in either 'Ma' or 'Ga', representing $10^6$ or $10^9$ years, respectively. Time spans unrelated to a specific datum are given as 'myr' (million years).

## ACKNOWLEDGEMENTS

This book project started with an invitation by Kent Condie, the series editor of 'Developments in Precambrian Geology' to one of us (H. Frimmel) to compile a book on a Neoproterozoic theme. Quickly this idea was taken up and it was only logical to ask Claudio Gaucher, the chief mastermind behind IGCP 478, to take over the leading role in this project. We are grateful to Kent Condie for having supported the idea of using this book series as platform to present a synthesis of the results achieved within IGCP 478, but also for his continuing support and encouragement throughout the various stages of book production. Constructive and helpful reviews of the original book proposal were provided by anonymous reviewers.

Technical assistance at Elsevier, especially by Linda Versteeg, Tirza van Daalen and Anita Koch, is gratefully acknowledged. Drafting was done by the authors or other staff members at the authors' institutions.

All contributions that make up this book were reviewed by internationally renowned experts. The nature of the book required some manuscripts to be long and full of detailed information, which made the reviewing of them a particularly time-consuming effort. We are especially grateful to a number of individuals who did not hesitate in sharing their expertise and time and made important contributions to the success of this book by providing critical, constructive and in some cases thought-provoking reviews. They are listed below:

Benjamin Bley de Brito Neves (Universidade São Paulo)
Carlos Cingolani (CIG, Universidad Nacional de La Plata)
Reinhardt Adolfo Fuck (Universidade de Brasilia)
D. García-Bellido (Universidad Complutense, Madrid)
Hartwig E. Frimmel (Universität Würzburg)
Claudio Gaucher (Universidad de la República, Uruguay)
Gerard J.B. Germs (University Johannesburg)
Kathleen Grey (Geological Survey of Western Australia)
Juan Carlos Gutiérrez-Marco (Universidad Complutense, Madrid)
Galen P. Halverson (University of Adelaide)
Léo A. Hartmann (Universidade Federal Rio Grande do Sul)
Walter D. Mooney (U.S. Geological Survey, Menlo Park)
Martin Okrusch (Universität Würzburg)
Marcio Martins Pimentel (Universidade de Brasilia)
Claudio Riccomini (Universidade São Paulo)
Renata Schmitt (Universidade Estadual Rio de Janeiro)
Graham Shields (Westfälische-Wilhelms Universität, Münster)
Alcides N. Sial (Universidade Federal de Pernambuco)
Peter Sprechmann (Universidad de la República, Uruguay)
Tony Tankard (Tankard Enterprises, Calgary)
William Randall Van Schmus (University of Kansas, Lawrence)
Detlef Walde (Universidade de Brasilia)
Sebastian Willman (Uppsala Universitet)
Shuhai Xiao (Virginia Polytechnic Institute and State University, Blacksburg)

Furthermore, special thanks goes to the International Union of Geological Sciences (IUGS) and UNESCO for 'seed funding' related to IGCP 478, which enabled fruitful scientific exchange. Jorge Ellis (UNESCO) is thanked for his continued support and for encouraging us to submit a proposal to the IGCP programme.

# SECTION I: INTRODUCTION

PART 1 Introduction

CHAPTER 1

# The Neoproterozoic and Cambrian: A Time of Upheavals, Extremes and Innovations ☆

Claudio Gaucher[1], Alcides N. Sial[2], Galen P. Halverson[3] *and* Hartwig E. Frimmel[4]

## Contents



## 1.1. The Neoproterozoic World

The Neoproterozoic Era (1,000–542 Ma) encompasses an eventful period in Earth history, comparable in length to the Phanerozoic Eon. Among the most notable events in the Neoproterozoic are the final amalgamation and demise of the oldest well-documented supercontinent (i.e. 'Rodinia': McMenamin and McMenamin, 1990; 'Palaeopangaea': Piper, 1982), the fusion of its cratonic pieces into Gondwana through an immense network of orogenic events, the most severe glaciations in Earth history ('Snowball Earth', Kirschvink, 1992a; Hoffman et al., 1998), large oscillations in the carbon and sulphur isotope composition of seawater in Earth history, the advent of animals, the first skeletonized organisms, the oldest evidence of predation and the colonization of the infaunal niche ('agronomic revolution': Seilacher, 1999). The large and ever increasing number of publications dealing with these and other aspects of the Neoproterozoic Earth suggests that in the years to come the list of extraordinary events and our understanding of the underlying mechanisms that drove them is likely to grow. The Neoproterozoic was without doubt a time of extremes, of evolution and innovation, which culminated in the birth of a habitable environment. This chapter aims to provide an updated summary of Neoproterozoic research and the intricate relationships between the solid Earth, the ocean–atmosphere, global climate and organismic evolution.

## 1.2. Tectonic Upheaval and Geodynamic Twists: The Tale of Two Supercontinents

The Neoproterozoic Era is bracketed by the end of the Grenvillian orogeny (1,300–900 Ma) and the Brasiliano-Pan-African system of orogenies (ca. 650–520 Ma, Figure 1.1): a full supercontinental cycle. On the basis of U-Pb detrital zircon ages found in major river mouths in four continents, Rino et al. (2008) argued that the Grenvillian and Brasiliano-Pan-African were the most intense orogenic events in Earth history. A legacy of the

☆ Gaucher, C., Sial, A.N., Halverson, G.P., Frimmel, H.E. 2009. The Neoproterozoic and Cambrian: a time of upheavals, extremes, and innovations. In: Gaucher, C., Sial, A.N., Halverson, G.P., Frimmel, H.E. (Eds): Neoproterozoic-Cambrian Tectonics, Global Change and Evolution: a focus on southwestern Gondwana. Developments in Precambrian Geology, 16, Elsevier, pp. 3–11.

[1] Departamento de Geología, Instituto de Ciencias Geológicas, Facultad de Ciencias Iguá 4225, 11400 Montevideo, Uruguay.
[2] NEG-LABISE, Departamento de Geologia, Universidade Federal de Pernambuco, C.P. 7852, Recife 50670-000, Brazil.
[3] Geology and Geophysics, School of Earth and Environmental Sciences, University of Adelaide, North Terrace, Adelaide, SA 5005, Australia.
[4] Department of Geological Sciences, University of Cape Town, Rondebosch 7700, South Africa.

*Developments in Precambrian Geology*, Volume 16
ISSN 0166-2635, DOI 10.1016/S0166-2635(09)01601-6

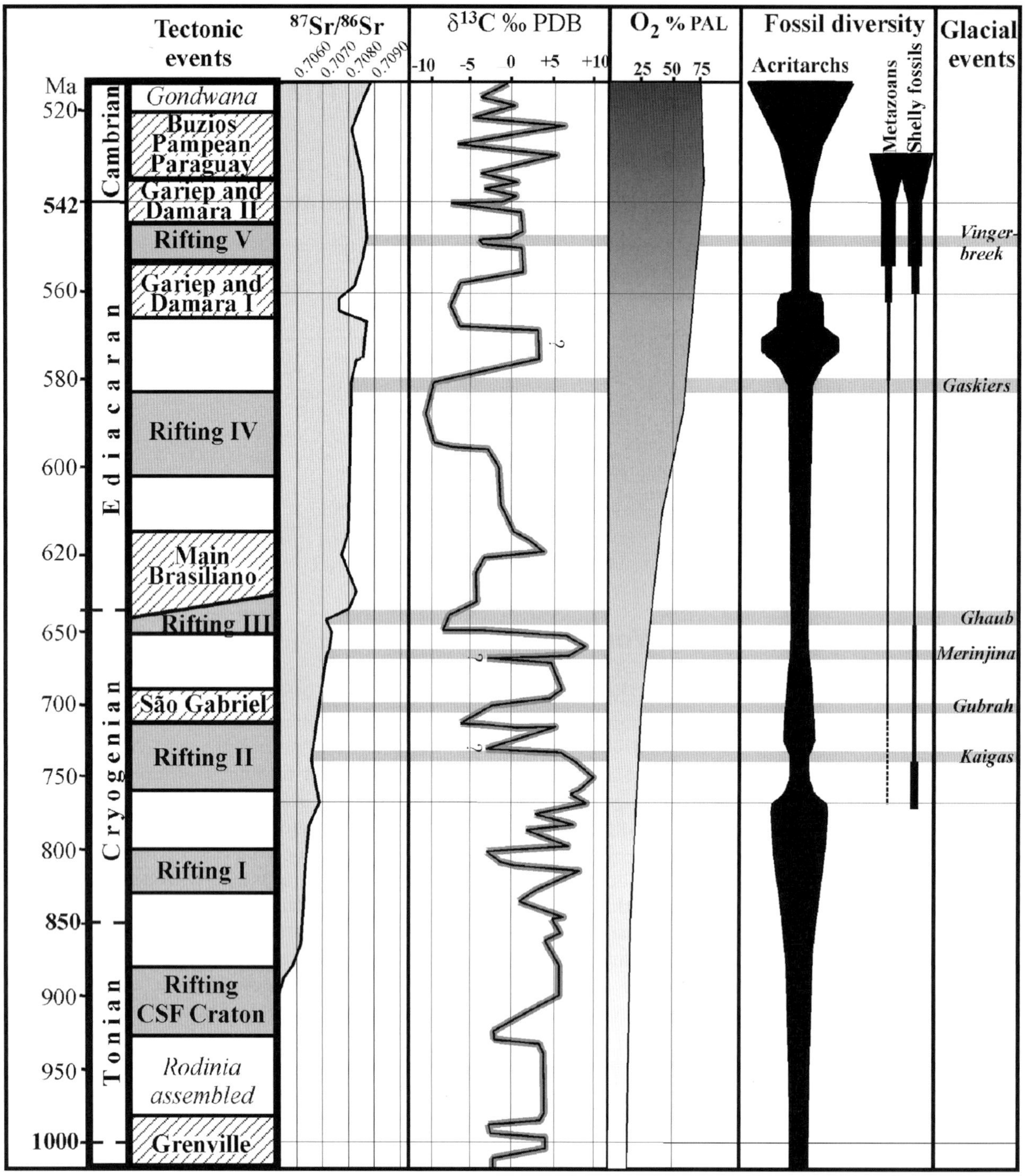


**Figure 1.1** Synthesis of Neoproterozoic chronostratigraphy, tectonic and glacial events, isotopic composition of seawater, atmospheric oxygenation and main extinction and radiation events. Note change in timescale at 635 Ma. $^{87}Sr/^{86}Sr$ curve mainly according to Halverson et al. (2007a); $\delta^{13}C$ curve from Halverson et al. (2005), Maloof et al. (2005) and Kah et al. (1999) with modifications to accommodate data from SW-Gondwana. Oxygenation curve according to Holland (2006). Rifting CSF Craton: Rifting in the Congo-São Francisco Craton (see Chapter 8 and references therein). Break-up of Rodinia (Li et al., 2008b; Chapter 8 and references therein): Rifting I: Greater India; Rifting II: South China, Kalahari Craton, Australia, East Antarctica; Rifting III: Amazonia and Río de la Plata cratons; Rifting IV: Baltica, northern Laurentia, Amazonia and Río de la Plata cratons; Rifting V: Opening of the Iapetus Ocean. Orogenic events shown are mainly referred to Gondwanan blocks (see Chapter 8 and references therein). Acritarch diversity (width in the Cambrian equals 75 species): see Chapter 9.1 and references therein. Shelly fossils and metazoan diversity: see Chapter 9.2.

break-up of Rodinia was a peak in passive margin abundance in the middle Neoproterozoic-Early Cambrian (Bradley, 2008), which accounts for the ample Neoproterozoic sedimentary record. Thus, it is clear that the Neoproterozoic was a tectonically dynamic period in Earth history; undoubtedly, these tectonic upheavals affected the concomitant evolution of the biosphere and dramatic fluctuations in seawater chemistry and climate (Section 1.3).

The early break-up of Rodinia appears to have been associated with one or more superplumes (Li et al., 1999, 2003, 2008b; Frimmel et al., 2001b; Puffer, 2002; Wang et al., 2007). The earliest superplume, temporally correlated with the initial break-up of Rodinia and situated beneath South China, India, East Antarctica and Australia, culminated ca. 825 Ma and may have been distinct from a subsequent peak in plume magmatism at ca. 780 Ma (Li et al., 2003). Whether these plumes triggered break-up or were a consequence of break-up remains a matter of debate. Mantle avalanches (Condie, 1998, 2000; Li et al., 2008b) or 'slab graveyards' (Rino et al., 2008), generated by the subduction of cold oceanic lithosphere at the margins of supercontinents, are often cited as possible mechanisms of plume inception (see Chapter 8). The shielding effect of supercontinents has been also invoked as a likely mechanism (Evans, 2003b). In combination, the global scale reorganizations of mass in the mantle may have been sufficient to drive large-scale, rapid true polar wander (Evans, 2003b). While such TPW events have been invoked as either a possible triggering mechanisms for Cryogenian glaciation or a means of rotating high latitude glacial deposits into the low latitudes (Kirschvink et al., 1997; Li et al., 2004), no data establishing this link has yet been established. The evidence for ca. 800 Ma TPW is derived from two unrelated sources (a) a compilation of radiometrically constrained palaeomagnetic data from mid-Neoproterozoic plume-related volcanics (Li et al., 2004) that suggests a single TPW event and (b) palaeomagnetic data from the mid-Neoproterozoic Akademikerbreen Group in Svalbard which suggests a pair of TWP events related to major shifts in the global carbon cycle (Maloof et al., 2006), but well in advance of the earliest evidence there for glaciation. Thus, independent data sets and theory are converging to imply at least one major TPW event sometime in the mid-Neoproterozoic. While this hypothesis is not yet fully tested and its potential connection to Cryogenian glaciation remains tenuous, it is a compelling idea that offers a potential means of elucidating the role of palaeogeography in the Earth system.

The break-up of Rodinia lasted >250 myr, with the final dispersion of Rodinia – the opening of the Iapetus Ocean – not beginning until ca. 600 Ma (Figure 1.1; Pisarevsky et al., 2008; see Chapter 8), after the nucleus of Gondwana had already begun to assemble (Collins and Pisarevsky, 2005). This protracted break-up of Rodinia is manifested in the seawater strontium isotope record, where a gradual rise from 0.7053 to 0.7072, related to erosion of old continental crust in the interior or Rodinia, characterizes the middle Neoproterozoic (Halverson et al., 2007a; Figure 1.1). A subsequent peak in $^{87}Sr/^{86}Sr$ of 0.7093 in the middle Cambrian (Montañez et al., 1996), the highest in Earth's history (see Section 1.3), probably resulted from the unroofing of the extensive Brasiliano-Pan-African system of orogens (Figure 1.1).

As will be discussed below and in the ensuing chapters, the combination of supercontinental break-up and subsequent amalgamation of Gondwana likely played a central role in the climatic, geochemical and biospheric evolution of the Neoproterozoic-Cambrian world.

## 1.3. The Ocean and Atmosphere: Oxygenation and Anomalies

The ocean and atmosphere are intimately linked through gaseous exchange at the air–water interface and weathering of the continents. Ocean chemistry and the redox state of the ocean–atmosphere changed markedly in the Neoproterozoic, as manifested in the reprise of banded iron formation (BIF), after a gap of nearly 1 Gyr encompassing the whole Mesoproterozoic (e.g. Beukes and Klein, 1992; Klein and Beukes, 1992; Holland, 2006). Canfield (1998) suggested, based on a compilation of sulphur isotope data on sedimentary sulphides, that the late Palaeoproterozoic-Mesoproterozoic deep ocean was euxinic (i.e. sulphidic and anoxic) thus inhibiting deposition of BIF through iron sequestration by sulphides. Although this hypothesis has been challenged by Holland (2006), who envisioned the gap in BIF deposition to be due to a non-stratified, mildly oxidized ocean devoid of $Fe^{2+}$, it has also been widely supported by varied data sets. Molybdenum concentrations (Scott et al., 2008), Mo isotope compositions (Arnold et al., 2004), Fe speciation data, sulphides S isotope compositions (Shen et al., 2002) and biomarker data (Brocks et al., 2005; Brocks and Schaeffer, 2008) in late Palaeoproterozoic-Mesoproterozoic black shales all support this view of common, if not prevailing euxinic conditions in the deep ocean, analogous to the modern Black Sea. Anbar and Knoll (2002) and Anbar (2008) elaborated on the biogeochemical consequences (such as nutrient abundances) of the Canfield (1998) model of a tripartite evolution of Precambrian ocean chemistry.

Given that evidence is converging to indicate that the Neoproterozoic represents a transition between a dominantly euxinic to oxygenated deep ocean, two key questions remain: (a) what was the nature of that transition and (b) how did it relate to the biospheric evolution that culminated in the Cambrian explosion? While many authors have argued for at least periodically sulphidic deep oceans in the Neoproterozoic (e.g. Hurtgen et al., 2005, 2006), Canfield et al. (2008) recently proposed instead, based on a large compilation of iron-speciation data, that at least during the latter half of the Neoproterozoic, the deep oceans were dominantly Fe-rich. The demise of sulphidic oceans could be a consequence, not of a return to the very low oxygen conditions that typified the Archaean, but rather a shortfall in the continental sulphur inventory, which limited the supply of sulphate to the oceans (Canfield, 2004). The well-known occurrence of BIF in glacial successions, such as the Rapitan Group of the northwest Canada (Klein and Beukes, 1993), the Sturtian of South Australia (Preiss, 1987), the Chuos (Beukes, 1973) and Numees (Frimmel et al., 2002) formations of Namibia and the Jacadigo Group and Puga Formation of the southern Paraguay belt in SW Brazil (Graf et al., 1994; Piacentini et al., 2007) clearly indicate widespread anoxic, non-sulphidic deep oceans. BIF and manganese formations unrelated to glacial deposits also occur, for example, in the Ediacaran Arroyo del Soldado Group of Uruguay (Gaucher, 2000; Gaucher et al., 2004b), and the Penganga Group of India (Mukhopayay et al., 1997; Holland, 2006), although the age of the latter is poorly constrained and may in fact be late Mesoproterozoic (Basu et al., 2008). While the iron-speciation data of Canfield et al. (2008) illustrate that the Neoproterozoic-Cambrian redox transition did not involve a simple switch from sulphidic to oxygenated deep oceans, the absence of persistent BIF deposition throughout the latest Neoproterozoic suggests that transitional ocean chemistry was not simply analogous to the iron-rich Archaean-Palaeoproterozoic deep oceans.

Fluctuations in the ocean redox state, in particular spanning the Cryogenian glaciations (see Chapter 10), appear likely. A number of authors argue that oceans were chemically stratified, with deep anoxic waters overlain by a thin, oxygenated water layer within the photic zone, where photosynthesis would provide the source of oxygen (Calver, 2000; Gaucher, 2000; Gaucher et al., 2004b; Shen et al., 2005). But it is becoming increasingly clear that the deep oceans became at least partially oxygenated some time in the middle Ediacaran Period, perhaps shortly after the Gaskiers glaciation (Fike et al., 2006; Canfield et al., 2007; Halverson and Hurtgen, 2007; McFadden et al., 2008).

Likely not coincidentally, the first Ediacaran fossils appear in deep water sediments in Newfoundland shortly after the Gaskiers glacials (Narbonne and Gehling, 2003). However, microfossils in South China suggest that the first Metazoa may have in fact evolved much earlier, in the earliest Ediacaran Period (Yin et al., 2007; Zhu et al., 2007). Since the redox state of the oceans has a profound impact on the biosphere, resolving the precise timing and nature of the Ediacaran oxygenation event and possible earlier fluctuations in the oxidation state of the oceans is required to clarify the still uncertain relationships between eukaryotic diversification and environmental change during the Neoproterozoic and Cambrian.

The carbon isotope record of marine carbonates ($\delta^{13}C_{carb}$) is exceptional within the scope of Earth's history. Neoproterozoic carbonates are characterized by typically highly positive $\delta^{13}C_{carb}$ values, with an average of approximately +5‰ (but with values commonly as high as +8‰ and rarely up to +12‰), punctuated by negative excursions that in some instances, drop below mantle values (−6‰; Halverson et al., 2005). Assuming a steady-state carbon cycle and constant, small fractionation between marine carbonate and seawater-dissolved inorganic carbon (DIC), the fraction of total carbon buried as organic matter ($f_{org}$) can be defined using the lever rule (e.g. Kump and Arthur, 1999) as

$$f_{org} = \frac{\delta^{13}C_{carb} + 6}{\Delta_p} \tag{1}$$

where $\Delta_p$ is the average net fractionation between contemporaneously produced carbonate and organic matter, essentially due to photosynthetic processes. For the Neoproterozoic, $\Delta_p$ is approximately 30‰ (Hayes et al., 1999) but in particular basins and at certain times, it appears to be highly variable (e.g. Fike et al., 2006). Assuming steady state, no drastic anomaly in the average value of mantle carbon during the Neoproterozoic, and an insignificant isotopic fractionation associated with carbonation of the oceanic crust (e.g. Hayes and Waldbauer, 2006), the lever rule requires $f_{org} \approx 0.4-0.5$ through most of the Neoproterozoic, plummeting during short periods to values as low as 0.1. If we consider that $f_{org} = 0.18$ in the present oceans (Hayes et al., 1999), the organic carbon burial rate during most of the Neoproterozoic Era must have been two to three times higher than today for most of the Neoproterozoic, decreasing to nearly half that value during glaciations.

This observation can be explained in terms of high rates of primary productivity, efficient organic carbon burial, or a combination of the two factors. In favour of the enhanced bioproductivity hypothesis (e.g. Kaufman et al., 1997; Gaucher, 2000; Gaucher et al., 2004b) are low-diversity, high-abundance acritarch assemblages interpreted to reflect recurrent eutrophic conditions in the Neoproterozoic (e.g. Gaucher, 2000, 2007; Gaucher et al., 2004b; Nagy et al., 2009). Enhanced bioproductivity has been also invoked as a mechanism of peak oil generation, black shale deposition and onset of climate cooling in the Phanerozoic (e.g. Cretaceous: Caldeira and Rampino, 1991; Barnes, 1999 and references therein). On the other hand, the alternative hypothesis of an unusually high organic carbon burial efficiency is supported by an above-average area of passive margins in the Neoproterozoic (Bradley, 2008), many of which were clustered in the low latitudes (Kirschvink, 1992a; Evans, 2000; Li et al., 2008b) where primary productivity is concentrated. In either case, it is likely that the high average $\delta^{13}C$ of marine carbonates was directly linked to the unique tectonic history of the Neoproterozoic.

Negative $\delta^{13}C$ anomalies are the second hallmark of the Neoproterozoic-Cambrian carbon isotope record (Figure 1.1). While such anomalies can also be explained by steady-state solutions (most easily a drop in $f_{org}$; Kump and Arthur, 1999), $\delta^{13}C$ values well below the canonical mantle value of $-6‰$ and, in places, apparently sustained for hundreds of thousands to millions of years (well in excess of the residence time of DIC in the oceans), strongly suggest non-steady state solutions. It should be noted that such extraordinary anomalies, such as the so-called Shuram (or Wonoka) anomaly in the middle Ediacaran, in which $\delta^{13}C$ drops as low as $-12‰$ and remained $<-6‰$ for at least a million years, if not much longer (e.g. Le Guerroué et al., 2006), invariably solicit suggestions of secondary origins (e.g. Kennedy et al., 2001; Bristow and Kennedy, 2008). Such scepticism is necessary in light of such highly unusual carbon isotope values. In the case of the ca. 580 Ma Shuram anomaly, however, virtually the identical carbon isotope pattern is reproduced in thick sections (usually several hundreds of metres; Halverson, 2006) on at least eight separate cratons. In the case of Siberia, the anomaly occurs in limestone preserving high ($>3{,}000$ ppm) Sr concentrations (Pokrovskii et al., 2006). Not only would a diagenetic mechanism require an extremely massive transfer of sedimentary organic carbon into replacement carbonate minerals, but also to produce such a similar carbon isotope pattern through secondary processes in so many geographically separated locations seems even more non-actualistic than prevailing models to explain such anomalies in terms of carbon isotope composition of seawater.

If indeed primary, then seemingly the only explanation for the Shuram anomaly is a massive addition of carbon derived from a reduced (i.e. $^{13}C$-depleted) carbon reservoir to the oceanic DIC reservoir. As discussed in more detail in Chapter 10, this could in theory be accomplished through release of methane from sedimentary clathrates or leakage of methane in a sulphate-poor ocean (e.g. Schrag et al., 2002), continental weathering of sedimentary organic matter (Workman et al., 2002) or the partial oxidation of a very large dissolved organic carbon reservoir in the oceans (Rothman et al., 2003). While the latter hypothesis is gaining traction (Fike et al., 2006; McFadden et al., 2008), it does not explain some of the earlier Neoproterozoic $\delta^{13}C$ anomalies (see Chapter 10) and no firm mechanism has been established to account for the unusual build up of DOC at this time. Another quandary regarding the Shuram anomaly is the huge demand of oxidants required to generate it (Bristow and Kennedy, 2008) at a time apparently immediately preceding and perhaps overlapping the first evidence for oxygenation of the deep ocean (Canfield et al., 2007) and first appearance of the Ediacaran biota (Narbonne and Gehling, 2003). The jury is still out on the driver, timing, timescale and significance of this extreme carbon isotope anomaly.

Even if the smoking gun for the Shuram anomaly can be identified, at least three and maybe four other prominent negative $\delta^{13}C$ anomalies, for which no consensus mechanism has been identified, occur in the middle Neoproterozoic. Like the Shuram anomaly, which appears to initiate prior to the middle Ediacaran Gaskiers glaciation (Halverson et al., 2005), negative anomalies presage two Cryogenian glaciations (Halverson et al., 2005; McCay et al., 2006). This consistent pattern of downturns in $\delta^{13}C$ preceding glaciation indicates a robust link between major perturbations to the global carbon cycle and icehouse events, begging the question of whether all such negative anomalies coupled to global cooling were driven by the same mechanism. But in this case, how do we explain negative anomalies that do not coincide with glaciation? Clearly, we have not yet decrypted the Neoproterozoic carbon isotope code.

## 1.4. Climatic Extremes and Conundrums

Neoproterozoic glacigenic sediments are widespread across the globe, so much so that both Douglas Mawson and Briand Harland recognized, over half a century ago, that these *infraCambrian* ice ages must have been unusually severe. Despite the episodic challenges to the glacial origin (e.g. Schermerhorn, 1974; Eyles and Januszczak, 2004; Direen and Jago, 2008) of many of the diamictites and associated deposits typically classified as glaciogenic, the prevailing view is that there were multiple, widespread glaciations. Establishing the age and global correlation of these glacial deposits, on the other hand, remains a topic of considerable disagreement and debate. It is now clear that there were three or more separate glaciations: at least two during the Cryogenian, and at least one (ca. 580 Ma; Thompson and Bowring, 2000; Bowring et al., 2003b; Calver et al., 2004) during the Ediacaran Period. But whereas new radiometric ages appear to be validating a synchronous end to the end-Cryogenian (i.e. Marinoan or Elatina) glaciation at 635 Ma (Hoffmann et al., 2004; Condon et al., 2005; Zhang et al., 2008b), the timing of earlier glaciations is becoming increasingly convoluted, with ages on alleged middle Cryogenian glacial deposits spanning nearly 100 myrs, from ca. 745 Ma in southern Africa (e.g. Kaigas and Chuos formations, Frimmel et al., 1996c; Key et al., 2001) to ca. 658 Ma from the *sensu stricto* Sturtian glacials (Merinjina Formation) of South Australia (Fanning, 2006), with a host of ages in between (Figure 1.1; for a recent review of these ages, see Allen and Etienne, 2008 and Xu et al., 2009). Slightly younger Re-Os ages from the Tapley Hill Formation and other equivalent interglacial sediments in Australia (Kendall et al., 2007, 2006) appear to reinforce a relative young age for the Sturtian glaciation, but imply a very short duration for the interglacial succession if the base of the type Ediacaran section is truly 635 Ma. These young ages for the interglacial sediments also conflict with a single U-Th-total Pb age of 680±23 Ma from an authigenic monazite in the Enorama Shale, that overlies the Tapley Hill Formation (Mahan et al., 2007). In summary, controversy remains over the ages and correlation of the earlier Cryogenian glaciations and new ages seem to raise as many new questions as they resolve.

All reliable palaeomagnetic determinations thus far obtained for cratons glaciated during the Cryogenian Period[1] indicate middle to low palaeolatitudes (e.g. Evans, 2000; Hoffman and Schrag, 2002). Since many of these syn-glacial deposits, including the famous tidal rhythmites of the Elatina Formation in South Australia (e.g. Williams, 1996) were deposited in the oceans, it is thus clear that glaciers reached sea level in the tropics, in many places transgressing carbonate platforms – a scenario that has not arisen since the Neoproterozoic. The superposed, thick carbonates that were deposited in many cases with little or no evidence of a depositional hiatus or reworking of the uppermost glacial deposits (Williams, 1979; Fairchild, 1993; Kennedy, 1996; Hoffman and Schrag, 2002) complete the picture of extreme climate switches that epitomize the Neoproterozoic Earth system. These cap carbonates invariably preserve negative carbon isotope excursions, which illustrate unequivocally that the most severe glacial events in Earth's history were intimately linked to perturbations to the global carbon cycle (Kaufman et al., 1997). Hypotheses abound for the origin of the cap carbonates and their affiliated negative $\delta^{13}C$ signatures (Kaufman and Knoll, 1995; Kennedy et al., 2001; Shields, 2005) and both are pillars of the snowball Earth hypothesis (Hoffman et al., 1998), but are not easily explained in other models that attempt to account for low-latitude glacial deposits, such as the high obliquity hypothesis (e.g. Williams, 1975, 1993) and the collapse of orbiting ice rings into the atmosphere (Sheldon, 1984). An alternative attempt to challenge the global nature of the glaciations, the 'zipper-rift' hypothesis (Eyles, 2008), which explains the Neoproterozoic glacial deposits as products of glaciated rift shoulders in the low latitudes, cannot account for the many low-latitude and extensive, end-Cryogenian glacial deposits that are found on tectonically stable, thermally subsiding basins, such as in northern Namibia, South Australia and northwestern Canada.

The so-called cap carbonates are extraordinary by any measure, hosting a suite of unusual sedimentary and geochemical features (Hoffman and Schrag, 2002) that cannot be easily explained by any modern analogue. For example, the 'cap dolostones' and associated barite and aragonite seafloor cements forming the base of the 635 Ma post-glacial transgressive sequence (Hoffman et al., 2007) contain isotopic signatures (e.g. boron isotopes in carbonate and triple oxygen isotopes in sulphate; Kasemann et al., 2005; Bao et al., 2008) indicative of very high $pCO_2$ in the aftermath of glaciation, while mega-oscillation ripples suggest sustained, intense winds whose shear could penetrate 200–400 m water depth (Allen and Hoffman, 2005). The barite cements and other physical and chemical features of the cap carbonate are also consistent with a highly physically and chemically stratified ocean, a predictable consequence of rapid melting of an immense volume of continental ice (e.g. Shields, 2005; Hoffman et al., 2007). However, the timescale of the meltback from snowball glaciation is a matter of serious debate (Hoffman et al., 2007). While it has commonly been assumed that continental ice sheets would melt in just a few thousand years, based on the large positive albedo feedback and assumed high $pCO_2$ (e.g. Hyde et al., 2000), palaeomagnetic data showing multiple magnetic reversals in coeval cap dolostones in the Paraguay Belt, Brazil (Trindade et al., 2003) and South Australia (Raub and Evans, 2006) challenge this assumption. If these are true reversals contemporaneous with deposition of the cap dolostone, then they imply a timescale for post-glacial melting on the order of hundreds of thousands to millions of years, inconsistent with catastrophic deglaciation as envisaged in the snowball Earth hypothesis.

Unfortunately, radiometric dating seems unlikely to resolve this debate any time soon, for not only would it be remarkable to find volcanic horizons at the right levels of a cap dolostone, but even the highest resolution zircon U-Pb ages are unlikely to resolve the difference between a few thousands of years and few hundred thousand years (Bowring and Schmitz, 2003). Nor has geochronology proved decisive in testing the time span of the presumed snowball Earth events, which are predicted to last approximately 10 myrs (Hoffman et al., 1998). While some attempts to constrain the time span of the Neoproterozoic glaciations have proven consistent with the snowball hypothesis, such as subsidence analysis (Hoffman et al., 1998) and measurement of Ir anomalies in cap carbonates (Bodiselitsch et al., 2005), these methods are indirect and not as convincing as direct radiometric ages.

[1] The Cryogenian Period has not yet been formally defined, but its upper boundary is already set to the base of the Ediacaran Period (ca. 635 Ma) and will surely span all prior Neoproterozoic glaciations.

This difficulty in precisely dating the time span of Neoproterozoic glaciations and using the sedimentary record to test the snowball hypothesis is also due to the fact that this record is fundamentally incomplete. Thus, even where there are thick glacial successions, determining their timing of deposition relative to the history of the glacial event is difficult (Hoffman, 2005). The documentation of thick and complex glacial successions (e.g. Christie-Blick et al., 1999; McMechan, 2000; Allen and Etienne, 2008) alone does not contradict the snowball Earth hypothesis, for coupled climate-ice sheet modelling suggests that even under full snowball conditions, thick, wet-based glaciers would grow on the continents within a few hundreds of thousands of years of the initial freeze over (Donnadieu et al., 2003). Furthermore, a large part of the glacial record may have been deposited during either the initial, pre-snowball phases or just prior to total meltback (Hoffman, 2005). Thus, it is likely that the physical processes responsible for the Neoproterozoic glacial sedimentary record would not be radically different than those operating in present glacial environments.

More crucial to the snowball Earth hypothesis is evidence of persistent open water during the glaciation, if it can be demonstrated that this open water existed in the low latitudes at any time other than the end of the glaciation. This question is not simply an academic debate over whether the Earth experienced a 'hard' or 'soft' (i.e. 'Slushball Earth') global glaciation, for even as little as $10^4$ km$^2$ of open water in the tropics might be sufficient to prevent sufficient $CO_2$ to build up in the atmosphere to overcome the high albedo of a virtually ice covered globe due to removal of carbon alteration of oceanic crust (Le Hir et al., 2008) – this assuming less than 0.2 bar of $CO_2$ is sufficient to initiate melting, which has been questioned by Pierrehumbert (2004). Evidence for such open water may be found in the form of oscillation ripples in the upper part of multiple cycles in the late Cryogenian Fiq Formation in Oman (Leather et al., 2002; Allen and Etienne, 2008). Others have argued for extensive open water based on computer climate simulations (Hyde et al., 2000; Peltier et al., 2007), the survival of several fossil taxa through glacial events (Corsetti et al., 2003; Moczydłowska, 2008a), biomarkers (Olcott et al., 2005) and the theoretical effects of global ice cover on phytoplankton and the rest of the marine biota (Gaidos et al., 1999; Corsetti et al., 2006).

This list of papers in support of or opposition to the snowball Earth hypothesis is too long to review in full in this chapter. However, a pair of recent papers by Bao et al. (2008, 2009) establishes convincingly that at least the end-Cryogenian glaciation was an exceptional event. Specifically, they have documented large negative $\Delta^{17}O$ anomalies in sulphate, which are most readily explained by extremely high $pCO_2$, both during (12,500–80,000 ppm) and after (> 10,000 ppm) the glaciation. High $pCO_2$ in the aftermath of glaciation is consistent with, but not unique to the snowball Earth hypothesis, but no other current hypothesis can rationalize such high $pCO_2$ during glaciation.

## 1.5. Permissive Ecology and Neoproterozoic-Cambrian Climate

Neoproterozoic glaciations may have played a key role in the diversification of eukaryotes and especially metazoans (Knoll and Walter, 1992). In the concept of 'permissive ecology' (Knoll, 2003b), Neoproterozoic ice ages provide both a negative and a positive influence on organisms. On the negative side, harsh glacial conditions, followed by extremely hot climates were an ecological bottleneck, clearly associated with acritarch extinction events (Vidal and Knoll, 1983; Vidal and Moczydłowska-Vidal, 1997; Moczydłowska, 2008a; see Chapter 9.1). Indeed, a compilation of acritarch diversity through the Neoproterozoic and Cambrian (Figure 1.1; Knoll et al., 2006, see Chapter 9.1) reveals a conspicuous dearth of acritarchs spanning at least two Cryogenian glaciations.

On the other hand, ice ages also foment biological diversification by providing vast ecospace to be colonized when ice retreats and climate stabilizes (Knoll, 2003b). This effect is presumably evidenced by the diversification of vendobionts and metazoans in the aftermath of the Gaskiers glacial event (Narbonne and Gehling, 2003). Perhaps even more impressive is the apparent diversification of acanthomorphic acritarchs (Zhou et al., 2007; Figure 1.1). In the latter case, however, an alternative hypothesis linking the Acraman impact in Australia to acritarch diversification cannot be ruled out (Grey et al., 2003; Grey, 2005; see Chapter 9.1), but in any case, still links diversification with environmental catastrophe. Knoll (2003b) further hypothesized that similar evolutionary radiations of eukaryotes did not take place in the aftermath of the older Cryogenian glaciations because there was not enough oxygen to sustain complex eukaryote physiology. A review of the possible relationships and feedbacks between Neoproterozoic ice ages and the biosphere has been provided by McMenamin (2004).

Decreasing amplitudes and increasing frequency of $\delta^{13}C$ oscillations characterize the Neoproterozoic-Cambrian transition (Brasier and Sukhov, 1998; Shields and Veizer, 2002; Maloof et al., 2005; Chapter 10) and coincide with a shift to a prevailing warm climate (Gaucher et al., 2007b). A warm Cambrian climate is indicated by apparently high $pCO_2$ at this time (e.g. Berner, 2004; Bao et al., 2008) and consistent with cyanobacterial and algal calcification events (Riding, 1994) and skeletal carbonate mineralogy (Zhuravlev and Wood, 2008). Change from a highly unstable Neoproterozoic global climate into a relatively stable and warm Cambrian environment has been ascribed to the progressive exhaustion of nutrients and establishment of a Cambrian superoligotrophic ocean (Martin, 1996; Barnes, 1999). From a permissive ecology point of view, both climate warming and climate stabilization removed a pre-existing stress factor, likely enabling the explosive diversification of metazoans. However, as we will discuss below, other, non-ecological factors were paramount in triggering the 'Cambrian explosion'.

The Cambrian explosion was preceded by a Neoproterozoic evolutionary radiation. Indeed, we may say that the Cambrian explosion had a long Neoproterozoic fuse. At the dawn of the Neoproterozoic, the oceans were already inhabited by eukaryotes that had reached macroscopic size and a multicellular grade of organization (Butterfield and Rainbird, 1998; Butterfield, 2000; Porter, 2004a; Knoll et al., 2006). Eukaryotic phytoplankton (the presumed affinity of most acritarchs) underwent two major periods of diversification in the Neoproterozoic (Figure 1.1), the first in the Tonian-early Cryogenian and the second (i.e. the ECAP: Ediacaran Complex Acanthomorph Palynoflora: Grey et al., 2003) in the early-middle Ediacaran. The Tonian-Cryogenian assemblage suffered a severe mass extinction in the Cryogenian, presumably related to glaciation. Acritarch diversity then remained low throughout the remainder of the Cryogenian, but a few taxa survived through to the Ediacaran Period (Moczydłowska, 2008a). The appearance of the ECAP be it in the middle Ediacaran, as suggested by the Australian record (Grey et al., 2003) or in the early Ediacaran, as suggested by the South China record (Zhu et al., 2007) was a truly explosive event (Vidal and Moczydłowska-Vidal, 1997; Knoll et al., 2006; Moczydłowska, 2008a; see Chapter 9.1). The ECAP suffered arguably the most severe Neoproterozoic mass extinction in the middle Ediacaran, with 75–90% of all species becoming extinct (Vidal and Moczydłowska-Vidal, 1997; see Chapter 9.1).

Judging from their fossil record, protozoans were also a Neoproterozoic innovation. The oldest fossil protozoans are the so-called vase-shaped microfossils (VSM), which appear in the pre-glacial Cryogenian (ca. 800 Ma) and have been convincingly classified as testate amoebae (Vidal, 1976; Porter et al., 2003). Agglutinated foraminifers (*Titanotheca coimbrae*) have been described from the late Ediacaran Arroyo del Soldado Group in Uruguay (Gaucher and Sprechmann, 1999; see Chapters 4.3 and 9.2), showing that amoeboid protozoans continued to diversify in the later Neoproterozoic. Ciliates have been recently described from the Doushantuo Formation in China (Ediacaran, 580 Ma), adding to the known diversity of Neoproterozoic Protozoa (Li et al., 2007). The advent and diversification of protozoans shows that heterotrophy reached a new level of complexity in the Neoproterozoic, although its origins may be considerably older (Bengtson, 2002). The enhanced primary productivity characteristic of Neoproterozoic oceans, leading to acritarch blooms (Gaucher, 2000, 2007; Gaucher et al., 2004b), may have promoted the diversification of eukaryotic predators/grazers (Porter and Knoll, 2000). The first evidence of macrophagous predation among metazoans and their implications will be dealt with below.

## 1.6. Emergence of Metazoa: A New World Order

Arguably the most important Neoproterozoic biotic innovation was the advent of metazoans. Among the *bona fide* metazoan fossils occurring in Neoproterozoic successions are sponges (*Palaeophragmodictya*: Gehling and Rigby, 1996; Brasier et al., 1997), cnidarians (*Namacalathus*: Grotzinger et al., 2000; *Vendoconularia*: Ivantsov and Fedonkin, 2002; *Corumbella*: Babcock et al., 2005; *Quadratitubus* and *Ramitubus*: Liu et al., 2008), molluscs (*Kimberella*: Fedonkin and Waggoner, 1997; Fedonkin et al., 2007b), arthropods (*Parvancorina*: Lin et al., 2006) and possibly annelids (*Cloudina*: Hua et al., 2005). The interpretation of the problematic fossil *Otavia* in middle Cryogenian rocks in Namibia as a sponge (Brain et al., 2001, 2003, 2008) and the occurrence of biomarkers typical of sponges in equivalent rocks in Oman (Love et al., 2009) suggest that the Porifera may have evolved significantly earlier and prior to the final global glaciation. The evolutionary picture of the synoptic palaeontological record closely resembles molecular clock estimates of key divergences of early Metazoa, the ages of which have been significantly reduced in recent years by the use of more precise calibration methods (i.e. based on invertebrates rather than vertebrates; Peterson and Butterfield, 2005). In their recent compilation of genetic data by Peterson et al. (2008), while Metazoa are rooted in the Cryogenian (with, e.g., the calcisponges diverging by ca. 760 Ma), bilaterian diversification was concentrated in the Ediacaran Period.

The evolution of the Vendobionta (Seilacher, 2007b and references therein) likely represents a separate radiation of organisms comparable to the Cambrian radiation of metazoans (Shen et al., 2008a). According to Shen et al. (2008a), the 'Avalon explosion' represents an independent, failed experiment with an evolutionary pattern similar to that of the Cambrian explosion. The affinities of the Vendobionta remain elusive, and recent hypotheses even deny their multicellular nature, suggesting that they represent giant foraminifera (xenophyophores, Seilacher, 2007b). Many, if not most, discoidal forms may actually represent microbial colonies, as argued by Grazhdankin and Gerdes (2007).

The advent of skeletons was a major Neoproterozoic innovation (Figure 1.1), which had many important repercussions: (a) it significantly changed the surface cycles of several elements, such as Ca, Si, P and C; (b) it enabled the eukaryotes to become reef builders and take over the reef ecosystem, a role previously monopolized by prokaryotes (cyanobacteria); and (c) it set the stage for a 'Darwinian arms race' (Vermeij, 1989), which fully developed in the Cambrian. Neoproterozoic shelly fossils include carbonate skeletons as in *Cloudina* (Germs, 1972a), *Namacalathus* (Grotzinger et al., 2000) and *Namapoikia* (Wood et al., 2002), siliceous scales as in the Tindir biota (Allison and Hilgert, 1986; McMenamin, 2004) and testate amoebae (Porter and Knoll, 2000), siliceous sponge spicules (Brasier et al., 1997), possibly phosphatic skeletons as in *Waltheria* (Gaucher and Sprechmann, 1999) and agglutinated protists like *Titanotheca* (Gaucher and Sprechmann, 1999; see also Chapter 9.2).

What may have triggered the advent of skeletons in the Neoproterozoic? A hint is provided by predatorial borings in *Cloudina* shells (Bengtson and Zhao, 1992; Hua et al., 2003) and in unmineralized tubular fossils (*Sinocyclocyclicus*, Liu et al., 2008), clearly showing that macrophagous predators were extant and food webs were likely complex by late Ediacaran times (Bengtson, 2002). The 'Darwinian arms race' or 'trophic escalation' frequently invoked as one of the main drivers of the Cambrian explosion (Vermeij, 1989; Gaucher and Sprechmann, 1999, Dzik, 2007; Chapter 9.2) was already underway in the late Neoproterozoic. The gap of ca. 40 myr between the earliest evidence of macrophagous predation and the Cambrian explosion may be explained by an ecological 'inhibitor', such as climatic instability and oxygen deficiency, typical of the Cryogenian and Ediacaran. Removal of these ecological barriers may have been a key factor leading to the Cambrian evolutionary radiation.

The response of eukaryotes to predation has been recently summarized by Jim Gehling as the 'Dig, Defend or Depart' (or Die!) strategy. Benthic organisms that could not build a mineralized skeleton ('Defend') may have become planktic ('Depart') to escape selective pressure (Bengtson, 2002). Butterfield (1997) argued that the advent of spiny acritarchs (acanthomorphs) may have followed this strategy. The third possible anti-predatory response was to dig for shelter, a behaviour aptly referred to as the 'Verdun Syndrome' by Dzik (2007). This author argued that the vast majority of latest Ediacaran and earliest Cambrian infaunal trace fossils represent shelters of animals feeding above the sediment surface, highlighting the protective function of the burrows. Finally, the 'agronomic revolution' (Seilacher, 1996, 1999), whereby animals were able to exploit the resources contained in the sediment, initiated near the Ediacaran/Cambrian boundary and radically changed the benthic ecosystem by the recycling of sediments and nutrients and the grazing of biomats, with obvious implications for the carbon cycle (McMenamin, 2004; Chapter 9.3). The Precambrian 'matgrounds' were irreversibly succeeded by Phanerozoic 'mixgrounds' (Seilacher, 1999). The final decline of stromatolites in the late Neoproterozoic may also be related to grazing pressure (see discussion in Bengtson, 2002). The agronomic revolution opened up a whole new set of ecological niches, which must have contributed to the Cambrian explosion.

## 1.7. Conclusions

The Neoproterozoic stands out as a period of innovations and upheavals. The extreme palaeoclimatic, palaeoceanographic and biotic events that characterize the Neoproterozoic Era may ultimately be a result of its unusual tectonic history. The final accretion and subsequent break-up of Rodinia, followed by the amalgamation of Gondwana were probably paramount in influencing Earth's surface environments, although the precise mechanisms remain controversial. Deep-Earth processes, such as mantle avalanches and superplumes, may have been the unheralded engines of a dynamic Neoproterozoic tectonic regime. Rapidly evolving palaeogeography may have in turn contributed to biogeochemical and climatic oscillations, which themselves were likely inextricably linked to biospheric evolution and ultimately the Cambrian explosion. Thus, Neoproterozoic-Cambrian Earth history is a case in point of the complexity and intrigue of the interactions between the deep Earth, the lithosphere, the oceans, the atmosphere and the biosphere. Without a doubt, continued vigorous research and debate about this fascinating Era will yield ever greater insight into the behaviour of the Earth system.

# SECTION II:
# NEOPROTEROZOIC-CAMBRIAN EVENTS IN SW GONDWANA

## PART 2 Neoproterozoic-Cambrian Evolution of the Amazonian Palaeocontinent

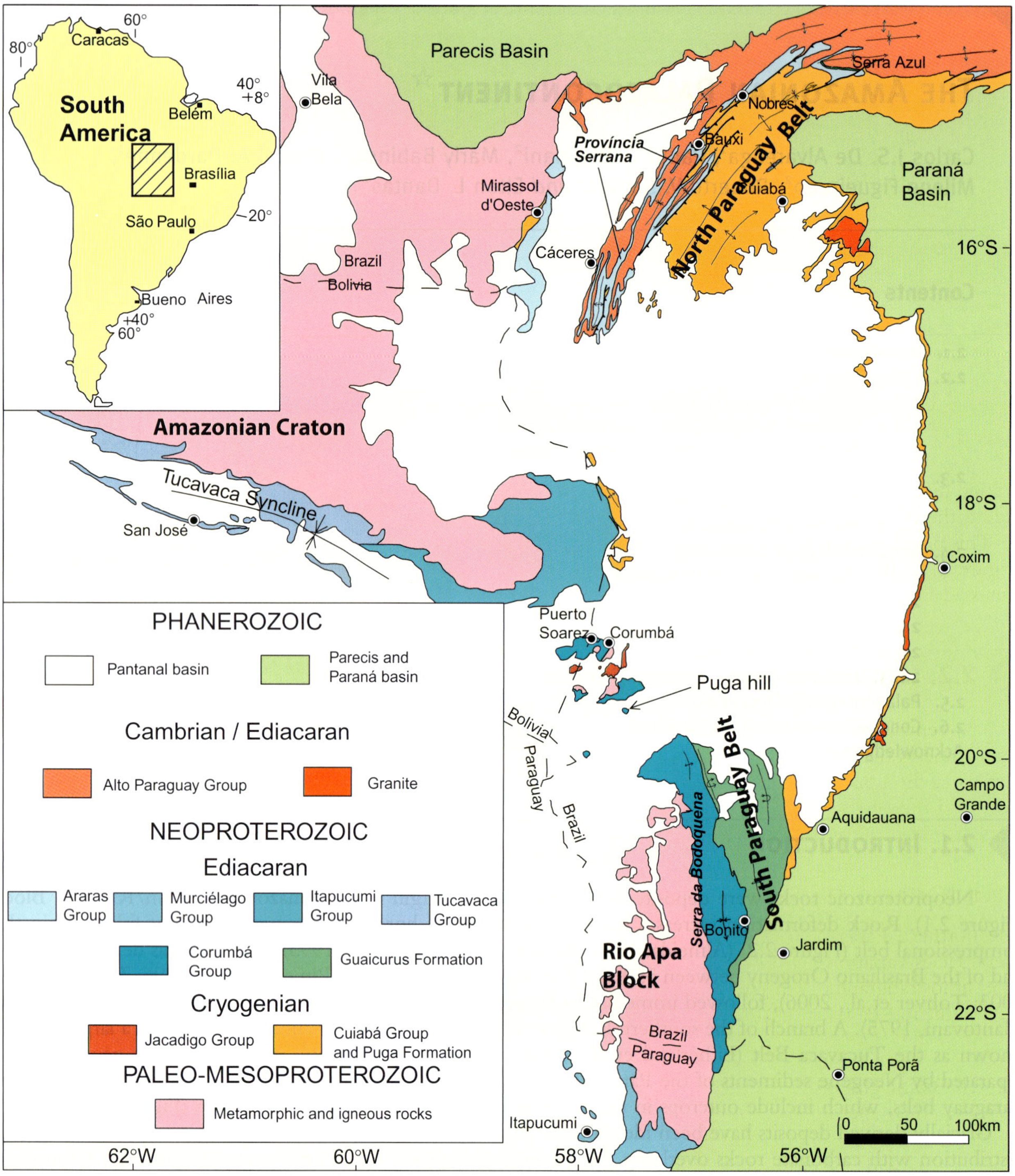


**Figure 2.1** Geological map of the Paraguay Belt on the southeastern part of the Amazonian Craton (modified from Schobbenhaus et al., 1981; Litherland et al., 1986; Boggiani, 1998).

Sedimentological and facies studies of biostratigraphy, palaeomagnetism and isotopic data of the Neoproterozoic rocks at the edge of the Amazonian/Rio Apa Craton have increased over the past decade and an improved picture of the relationships between their sections is emerging in the context of post-glacial carbonate facies in the Corumbá and Araras groups (Boggiani et al., 2003; Nogueira et al., 2003, 2007a; Misi et al., 2007). Palaeontological data for the Corumbá Group include new genera of organic-walled microfossils (Gaucher et al., 2003) in addition to *Cloudina lucianoi*, *Bavlinella faveolata* and scyphozoan *Corumbella werneri* (Hahn et al., 1982; Walde et al., 1982; Zaine and Fairchild, 1985, 1987). Various workers have presented composite $\delta^{13}C$, $\delta^{18}O$ and $^{87}Sr/^{86}Sr$ records for the carbonate rocks of the Corumbá and Araras groups (Boggiani, 1998; Boggiani et al., 2003; Gaucher et al., 2003; Nogueira et al., 2003, 2007a; Alvarenga et al., 2004, 2008; Font et al., 2006), although

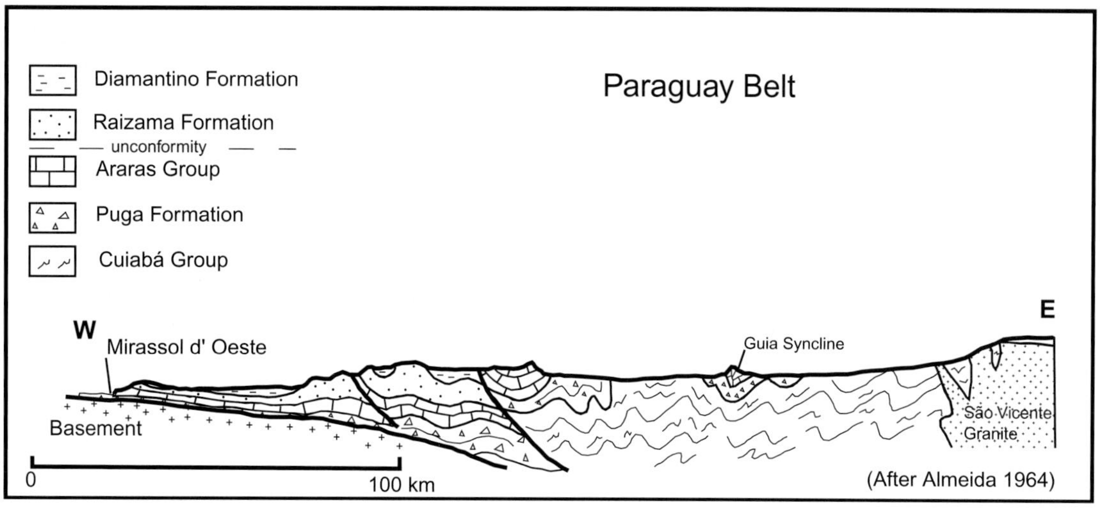


**Figure 2.2** Schematic section spanning the border of the Amazonian Craton and the Paraguay Belt (from Almeida, 1964).

striking differences are observed between these two units. The first palaeomagnetic data on a cap carbonate from the northern Paraguay Belt were reported by Trindade et al. (2003).

This paper presents a review of the stratigraphic units in the northern and southern Paraguay belts in Brazil (Figure 2.1). Biostratigraphy, and isotopic, palaeomagnetic and radiometric data available in the literature are discussed in order to establish a chronostratigraphic correlation.

## 2.2. Lithostratigraphy

In central-western Brazil, Neoproterozoic strata are preserved on the margins of the Amazonian/Rio Apa Craton and in the Paraguay Belt. Neoproterozoic sedimentary rocks crop out in different areas from the Província Serrana in the northern Paraguay Belt to the Serra da Bodoquena in the southern Paraguay Belt, passing by the Corumbá region and extending northwest into the Tucavaca Belt (Figure 2.1). The stratigraphies of these main areas of exposures are summarised in this paper.

### 2.2.1. Northern Paraguay Belt

The northern Paraguay Belt comprises, from base to top, four major lithostratigraphic sequences that exhibit lateral facies variations, mainly between the western border and the central part of the basin (Figure 2.3). The older lithostratigraphic units represent a passive margin and the younger ones represent a foreland basin related to the Brasiliano—Pan-African Orogeny (Almeida, 1974).

#### 2.2.1.1. Basal glacially influenced sequence

The basal glacially influenced sequence occurs in the cratonic cover (Puga Formation) as well as within the metasedimentary fold belt (FB, Cuiabá Group; Alvarenga and Trompette, 1992; Alvarenga et al., 2004). A correlation with the global Marinoan-Ghaub glaciation, dated at 635 Ma (Hoffmann et al., 2004; Condon et al., 2005), has been suggested for this glacial sequence in the northern Paraguay Belt (Nogueira et al., 2003; Alvarenga et al., 2004, 2008; Allen and Hoffman, 2005), despite the lack of geochronological data. The Puga Formation consists of diamictite associated with conglomerate, sandstone, siltstone and shale. The widespread occurrence of diamictite indicates a lateral transition from thinner coarse-grained beds close to the Amazonian Craton (Puga Formation) to thick fine-grained facies in the east (Cuiabá Group). The sedimentation model proposed for the glacially influenced unit involves three main glacial depositional systems: platformal, slope and outer slope (Alvarenga and Trompette, 1992). The platformal depositional systems, which cover the cratonic domain in the western inner shelf, were reworked by gravity flows on the outer shelf (Figure 2.3). The deposits in the inner shelf show alternating dominant massive diamictite, sandstones and fine-grained sediments with few dropstones. In the outer shelf, an association of massive diamictite, stratified diamictite and fine-grained sediments progressively replaces the dominant massive diamictite of the inner shelf. Glaciomarine sediments reworked by gravity flows and related to submarine fan deposits are associated with the slope depositional system. Diamictite, conglomerate and sandstone intercalations with occasional inverse and/or normal grading occur in the deeper parts of

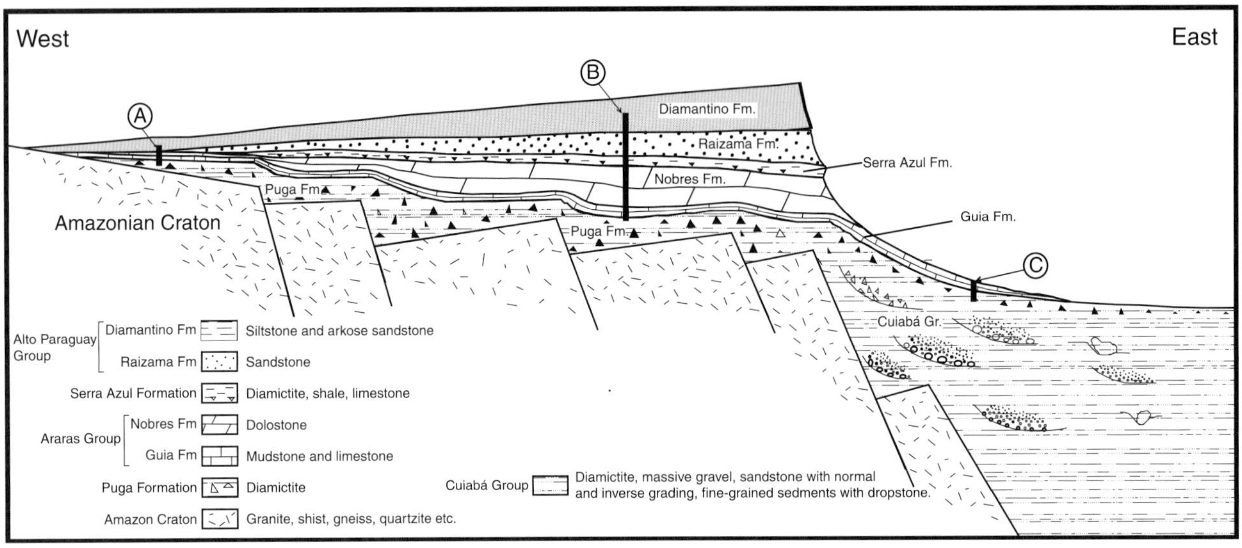


**Figure 2.3** Schematic stratigraphic cross-section of the Neoproterozoic depositional sequences along the southeastern edge of the Amazonian Craton, between 15° and 17° S latitude (modified from Alvarenga et al., 2008). A, B and C are the located sections illustrated in Figure 2.4.

the fan (Alvarenga and Trompette, 1992). Sandstone and siltstone intercalations represent inter-channel deposits formed by turbidity currents. Deposition on the outer slope system was dominated by fine-grained sediments (phyllite and metasiltstone) related to low-density turbidity currents, in which a decrease of glacial influence is indicated by the presence of a few isolated clasts or dropstones (Alvarenga and Trompette, 1992). This model of basin filling suggests that the Amazonian Craton was the main source area of the sediments, which were later reworked during the glacial event, forming submarine channels and turbidites on the slope and outer slope (Alvarenga and Trompette, 1992). Palaeomagnetic studies on the cap carbonate covering the Puga Formation on the Amazonian Craton indicate glacial sedimentation at low latitudes (Trindade et al., 2003).

### 2.2.1.2. Araras Group

The Araras Group (carbonate sequence) on the southeastern border of the Amazonian Craton overlies the Puga Formation and reaches a thickness of about 100–150 m on the western border (inner shelf) of the basin. Towards the east, this carbonate succession can reach up to 1,300 m in thickness (Luz and Abreu Filho, 1978; Alvarenga et al., 2004) in the middle shelf domain, and continuing to deeper sequences mud-rich limestone and laminated metasiltstone successions can be found towards the slope depositional system of the basin (Figures 2.3 and 2.4). The Araras Group can be subdivided into three formations.

The lowermost Mirassol d'Oeste Formation, which includes 18–30 m of the cap carbonate, overlies the Puga Formation (Nogueira et al., 2003; Alvarenga et al., 2008). This cap carbonate shows variable degrees of recrystallisation and contains stromatolites, tube-like structures, breccias, giant wave ripples (Nogueira et al., 2003; Alvarenga et al., 2004, 2008; Allen and Hoffman, 2005) and fan-like crystals that have been interpreted as aragonite pseudomorphs (Faulstich, 2005; Alvarenga et al., 2008).

The middle Guia Formation consists of laminated limestone, mud-limestone and mudstones up to 250 m thick in the middle shelf domain (Figure 2.4).

The upper part of the Araras Group is represented by the Nobres Formation, and it comprises shallow-water breccia, grainstone and packstone dolostone; towards the top of the formation sandstone intercalations about 1 m thick are common (Almeida, 1964; Alvarenga et al., 2004). The lower part of the Nobres Formation was recently subdivided into the Serra do Quilombo Formation (Nogueira et al., 2007a), but its extension was not confirmed for the whole belt and in this paper, the original denomination of the Nobres Formation (Hennies, 1966) is used.

Sedimentary rocks of these three formations reflect a shallower facies at the western edge of the basin (cratonic area) and progressive deeper water facies to the more distal platform in the east (fold belt).

### 2.2.1.3. Serra Azul Formation

The Serra Azul Formation includes discontinuous outcrops of diamictite and siltstone deposited on top of post-Marinoan-Ghaub carbonates of the Araras Group (Figures 2.3 and 2.4), and it represents the record of a second Neoproterozoic glaciation in the Paraguay Belt (Figueiredo et al., 2004, 2008; Alvarenga et al., 2007).

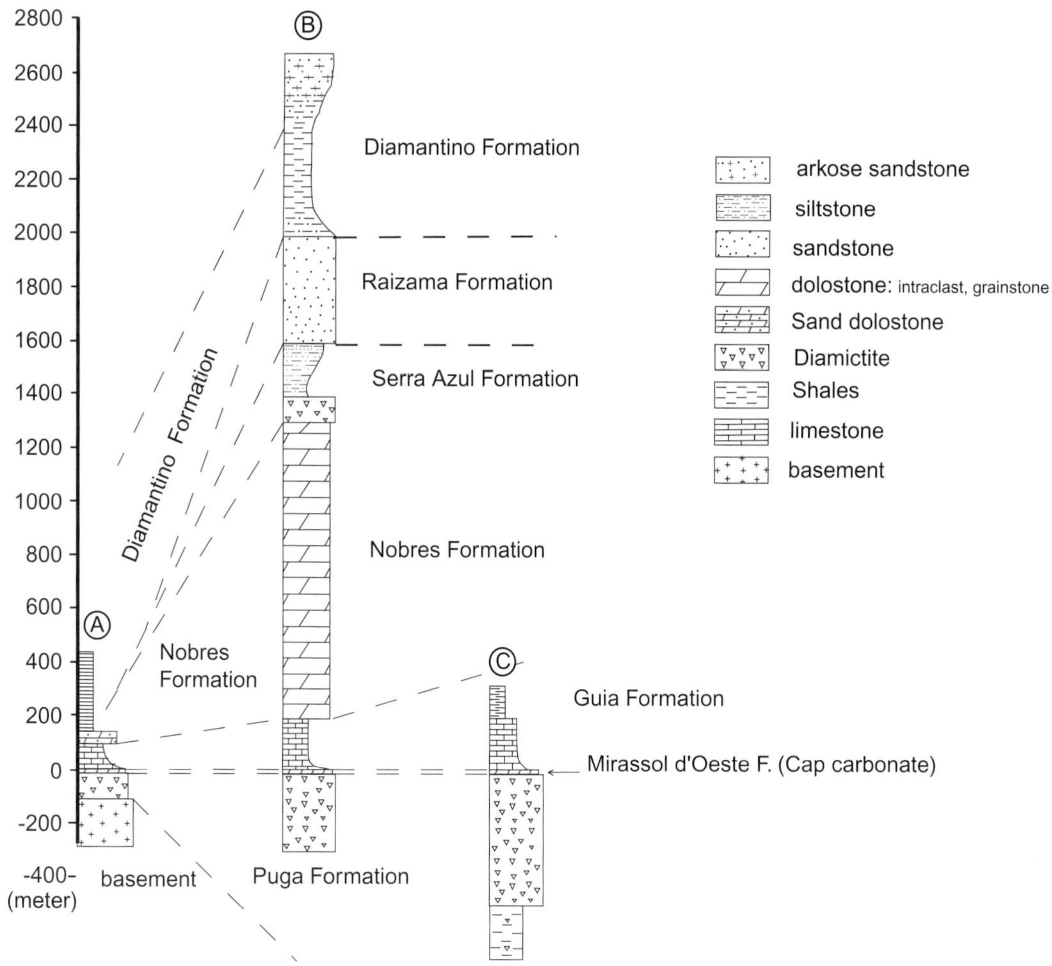


**Figure 2.4** Measured stratigraphic sections of the Neoproterozoic units on the northern Paraguay Belt, between 15° and 17° S latitude. A, B and C are located in Figure 2.3.

The basal unit comprises approximately 70 m of massive reddish diamictite with an abundant clayey–silty matrix that becomes yellowish and stratified in the last 60 cm (Figueiredo et al., 2008). The clasts are composed of sandstones, arkoses, cherts, quartz, quartzites, metaconglomerates, carbonates, claystones, diabases, granites, gneisses, basalts and rhyolites. The size of these clasts ranges from granule to boulder, varying from highly angular to well rounded. Some clasts are striated, faceted and polished, while boulders are bullet shaped. Their poorly sorted characteristics associated with a large compositional variety of clasts and an abundance of striated clasts suggests a glacial influence on the deposition of the diamictites.

The diamictites are sharply overlain by a 25-m thick succession of reddish laminated siltstone with sparse granules. This unit grades into a 150-m thick rhythmite succession whose basal part is composed of a clayey–silty rhythmite intercalated with episodic sandstone layers (5–10 cm) with hummocky cross-stratification in the Serra Azul region (Alvarenga et al., 2007; Figueiredo et al., 2008). The upper part of the rhythmite has more contribution of sand, grading into a heterolithic succession that becomes coarser and thicker towards the top of the section (Figueiredo et al., 2008). A limestone layer has been found in the clayey–silty rhythmite in the Nobres region (Figueiredo et al., 2008).

This glacial unit has been related to the Gaskiers glaciation (Figueiredo et al., 2004; Figueiredo, 2006; Alvarenga et al., 2007; Figueiredo et al., 2008) with an age of ca. 580 Ma (Bowring et al., 2003b). The unit has not yet been found in the southern Paraguay Belt.

### 2.2.1.4. Upper siliciclastic unit

The upper siliciclastic unit, known as the Alto Paraguay Group (Almeida, 1964), consists of two siliciclastic formations represented by sandstones and pebble conglomerates at the base (Raizama Formation), and red shale, siltstone and arkoses at the top (Diamantino Formation). The contact between these two units is transitional and consists of sandstone, shale and siltstone intercalations (Alvarenga et al., 2000).

predominantly siliciclastic units of the Corumbá Group (Cadieus and Cerradinho formations) lie under the Bocaina and Tamengo formations.

### 2.2.3.1. Corumbá Group

The Corumbá Group was first subdivided into the Cerradinho, Bocaina, Tamengo and Guaicurus formations (Almeida, 1965). Later, the Cadieus Formation was included as the lower unit of this group (Gaucher et al., 2003). The deposition of the Corumbá Group was initiated by crustal stretching (rift phase), including a mixed assemblage of clastic and carbonate rocks (Cadieus and Cerradinho formations) that unconformably overlies the Rio Apa Craton (Gaucher et al., 2003). The succeeding rocks include carbonate, marl and shale sediments of a drift sequence represented by the Bocaina, Tamengo and Guaicurus formations (Figures 2.5 and 2.7).

### 2.2.3.2. Cadieus and Cerradinho Formations

The Cadieus Formation is represented by conglomerate and arkose sandstone deposited in a proximal alluvial fan that grades to arkose sandstone, shale and grainstone of the Cerradinho Formation in a distal alluvial fan (Boggiani et al., 1993; Gaucher et al., 2003).

### 2.2.3.3. Bocaina Formation

The Bocaina Formation comprises shallow marine carbonates along hundreds of kilometres with various lithofacies, including grainstone, packstone, mudstone and stromatolites (Figure 2.8). In the Serra da Bodoquena, phosphatic dolostones were found at the top of this unit.

### 2.2.3.4. Tamengo Formation

The Tamengo Formation comprises up to 100 m of dark, organic-rich limestone-shale rhythmite with occasional grainstone limestone. Fine-grained facies in the Serra da Bodoquena region were deposited in deeper waters than those from the Corumbá region (Boggiani, 1998), which could explain the absence of the metazoan fossils *Cloudina* and *Corumbella.*

### 2.2.3.5. Guaicurus Formation

The Guaicurus Formation was first described by Almeida (1965) to name the thick shale sequence on top of the Corumbá Group that extends to the eastern part of Serra da Bodoquena and along the Miranda River valley (Figure 2.7).

The Tamengo and Guaicurus formations comprise a transgressive-highstand sequence that drowned the carbonate shelf of the Tamengo Formation; there is no equivalent of this unit in the northern Paraguay Belt.

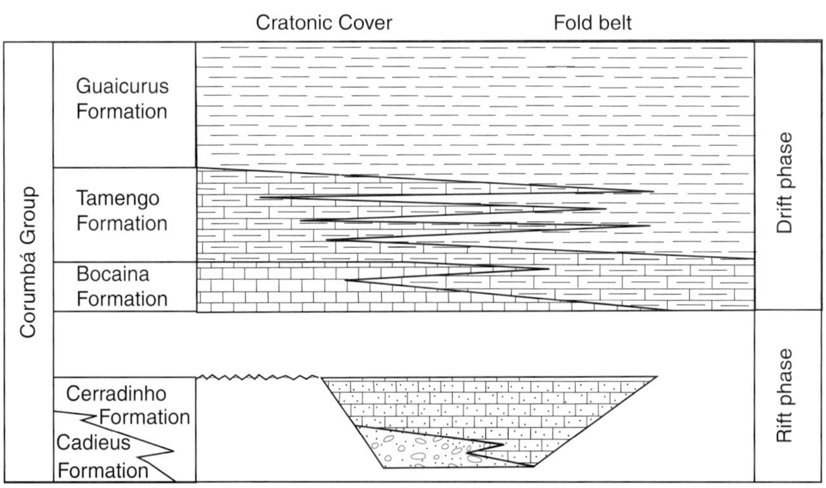


**Figure 2.7** Stratigraphic chart of the Neoproterozoic on the Serra da Bodoquena (modified from Boggiani and Alvarenga, 2004).

**Figure 2.8** Stromatolites in dolostones of the Bocaina Formation, Serra da Bodoquena.

## 2.3. Chemostratigraphy

The chemostratigraphic data available from the carbonate successions of the Araras and Corumbá groups consist of 300 carbon isotope and a few tens of Sr isotope data that were presented in recent papers (Boggiani et al., 2003; Gaucher et al., 2003; Nogueira et al., 2003, 2007a; Pinho et al., 2003; Alvarenga et al., 2004, 2008; Font et al., 2006; Riccomini et al., 2007).

### 2.3.1. Northern Paraguay Belt

#### 2.3.1.1. Araras Group

The Araras Group reaches a thickness of about 100–150 m over the Amazon Craton (border of the basin) and up to 1,300 m thick in the Paraguay fold belt (outer shelf and slope). The underlying Puga Formation is composed predominantly of diamictites, which are continuous for 300 km along the northern Paraguay Belt, between Cáceres, Bauxi, Nobres and Serra Azul (Figure 2.1). A Marinoan-Ghaub age for the Puga Formation has been suggested (Nogueira et al., 2003, 2007a; Alvarenga et al., 2004, 2008) based on similar cap carbonate and isotope data from other Neoproterozoic basins worldwide (Alvarenga et al., 2004; Allen and Hoffman, 2005; Nogueira et al., 2007a).

#### 2.3.1.2. Mirassol d'Oeste Formation

The basal Mirassol d'Oeste Formation is a 20-m thick sequence of transgressive cap dolomite deposited on top of diamictites over the Amazon Craton (Nogueira et al., 2003, 2007a; Alvarenga et al., 2004; Font et al., 2006) and in the FB (Alvarenga et al., 2008) (Figure 2.4). Cap dolomites over the Amazonian Craton show consistently negative $\delta^{13}C$ values ranging between −10.5‰ and −3.0‰ ($n = 96$), and $\delta^{18}O$ values between −8.4‰ and −1.9‰ ($n = 96$) (Figure 2.9). In the FB, the cap carbonate from a drill core shows $\delta^{13}C$ values ranging from −4.8‰ to −1.7‰ ($n = 7$) and $\delta^{18}O$ values between −8.0‰ and −5.1‰ ($n = 7$; Figure 2.9). A comparative carbon isotope stratigraphy between cap carbonate deposited over the Amazonian Craton and in the FB shows slightly higher $\delta^{13}C$ values in the FB (Alvarenga et al., 2008). The high and variable $^{87}Sr/^{86}Sr$ ratios (0.70852–0.71435) of the cap carbonate associated with their low Sr content (44–96 ppm) and high Mn/Sr ratios (>18) suggest that primary isotopic values were modified by post-depositional diagenetic alteration.

#### 2.3.1.3. Guia Formation

Limestones and mud-limestones (Guia Formation) deposited on top of the cap carbonate overlying the Amazonian Craton show $\delta^{13}C$ values between −5.4‰ and −2.7‰ ($n = 10$; Alvarenga et al., 2004, 2008; Nogueira et al., 2007a), while in the FB (mid shelf to foreslope basin) $\delta^{13}C$ values range from −5.3‰ to −0.1‰

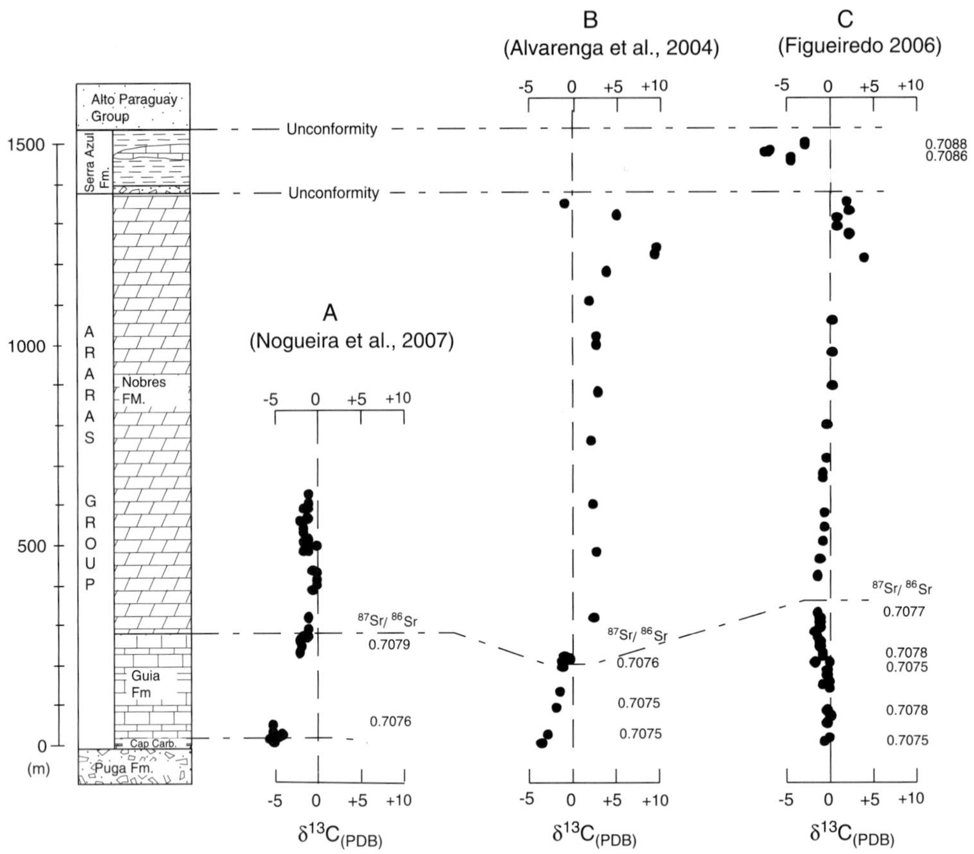


**Figure 2.9** Stratigraphic section and carbon isotopic profiles for the Araras Group and limestone of the Serra Azul Formation on the northern Paraguay Belt, between 15° and 17° S latitude. (A) $\delta^{13}C$ and $^{87}Sr/^{86}Sr$ from Nogueira et al. (2003, 2007a); (B) $\delta^{13}C$ and $^{87}Sr/^{86}Sr$ from Alvarenga et al. (2004); and (C) $\delta^{13}C$ and $^{87}Sr/^{86}Sr$ from Figueiredo (2006).

($n = 105$; Alvarenga et al., 2004, 2008; Nogueira et al., 2007a; Figueiredo, 2006; Figure 2.9). Oxygen isotope values for the same samples range between −8.3‰ and −5.8‰ ($n = 10$; Alvarenga et al., 2004, 2008; Nogueira et al., 2007a) over the Amazonian Craton, while in the FB, $\delta^{18}O$ values range from −10.6‰ to −6.3‰ ($n = 105$; Alvarenga et al., 2004, 2008; Nogueira et al., 2007a). These limestones and mud-limestones with high Sr concentrations (750−4,351 ppm) and low Mn/Sr ratios (0.01−0.20) have $^{87}Sr/^{86}Sr$ ratios between 0.70763 and 0.70780 ($n = 9$; Alvarenga et al., 2004, 2008) and probably represent the original Sr isotopic composition of the seawater. The limestone samples from the foreslope basin at Nossa Senhora da Guia Quarry show $\delta^{13}C$ values ranging from −5.3‰ to −0.8‰ and $\delta^{18}O$ values between −13.5‰ and −7.1‰ (Alvarenga et al., 2004).

### 2.3.1.4. Nobres Formation

The dolostones of the Nobres Formation are better exposed in the FB, and rocks of the Bauxi-Nobres region in the north and from the Cáceres area in the south of the FB present different isotopic pattern (Figure 2.9). In the Bauxi-Nobres region, $\delta^{13}C$ values range from −1.0‰ to +9.6‰ ($n = 45$), and $\delta^{18}O$ values range between −4.7‰ and −0.1‰ ($n = 45$; Pinho et al., 2003; Alvarenga et al., 2004; Figueiredo, 2006). In the Cáceres region, $\delta^{13}C$ values range from −2.2‰ to +0.2‰ ($n = 41$), while $\delta^{18}O$ values range between −6.7‰ and −3.4‰ ($n = 41$; Pinho et al., 2003; Nogueira et al., 2007a).

### 2.3.1.5. Serra Azul Formation

The isotope geochemistry data of the Serra Azul Formation include data in a limestone lens (about 12 m thick) in rhythmite from the upper part of the formation. $\delta^{13}C$ values range from −7.5‰ to −0.8‰, and $\delta^{18}O$ values from −7.8‰ to −5.5‰ ($n = 20$; Figueiredo and Babinski, 2008; Figueiredo, 2006; Figure 2.9). Nine samples of these limestones with Sr concentrations between 714 and 885 ppm and Mn/Sr ratios below 0.2 yielded $^{87}Sr/^{86}Sr$ ratios between 0.70863 and 0.70882 (Figueiredo and Babinski, 2008; Figueiredo, 2006).

## 2.3.2. Southern Paraguay Belt

### 2.3.2.1. Corumbá Group

The isotopic values for the carbonate rocks (Corumbá Group) in the southern Paraguay Belt were obtained for the cap carbonate overlying the diamictites at the Puga Hill and for the limestones of the Tamengo Formation in the Corumbá region (Boggiani, 1998; Boggiani et al., 2003; Gaucher et al., 2003).

### 2.3.2.2. Cap carbonate at Puga Hill

The cap carbonate described at Puga Hill consists of alternating pinkish and grey laminated limestone (rhythmite), with alternating peloidal and microcrystalline carbonate for ca. the first 12 m. Throughout this section, the values of $\delta^{13}C_{PDB}$ are homogeneous and are around −5‰ ($n = 9$) (Boggiani et al., 2003; Misi et al., 2007). The cap carbonate at Puga Hill is essentially calcium carbonate (MgO between 0.98% and 1.12%, $n = 9$), and the unique sedimentary feature of the cap carbonate is the rhythmic lamination.

### 2.3.2.3. Tamengo Formation

The Tamengo Formation occurs in Corumbá and along the eastern part of Serra da Bodoquena. The isotopic data for the limestones of the Tamengo Formation were obtained around Corumbá (Figure 2.1) in the Laginha, Saladeiro and Corcal quarries. Limestones of the lower part of the Tamengo Formation record a negative $\delta^{13}C$ excursion (−3.3‰ to −2.5‰), and in the direction of the top section, the $\delta^{13}C$ values vary between −1.3‰ and −0.3‰, following a positive $\delta^{13}C$ excursion (up to 5.8‰) to the upper section of the formation (Figure 2.10; Boggiani, 1998; Boggiani et al., 2003; Gaucher et al., 2003; Misi et al., 2007). At Serra da Bodoquena, about 200 km southeast of Corumbá, the limestones correlated to the Tamengo Formation show constant $\delta^{13}C$ values around 3‰ (Boggiani, 1998).

The $^{87}Sr/^{86}Sr$ ratios in the carbonates of the Tamengo Formation for both of these areas are between 0.7084 and 0.7085 (Boggiani, 1998; Babinski et al., 2008a).

## 2.4. Geochronological Constraints

Radiometric ages on rocks from the Paraguay Belt and related cover over the Amazonian Craton consist of Sm-Nd and Rb-Sr data in some siliciclastic rocks from the Puga Formation, the Cuiabá Group, the Araras Group

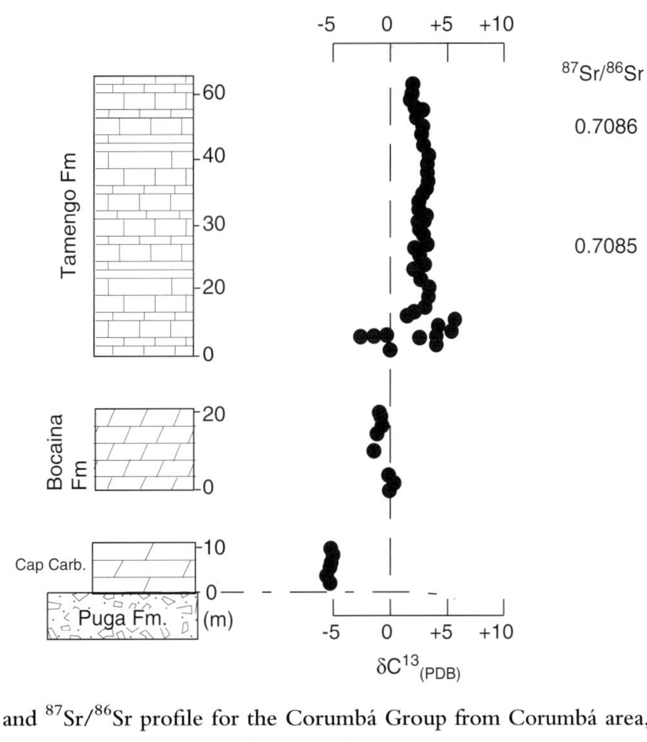


**Figure 2.10** Composite $\delta^{13}C$ and $^{87}Sr/^{86}Sr$ profile for the Corumbá Group from Corumbá area, Puga Hill and Serra da Bodoquena.

and the Alto Paraguay Group, and Pb/Pb analysis on cap carbonates from the Mirassol d'Oeste Formation. These geochronological data were obtained in rocks from the northern Paraguay Belt and are presented here according to the lithostratigraphic units described earlier.

### 2.4.1. Glacially influenced sequence

The Nd isotope signature of the metasedimentary rocks of this sequence was obtained in phyllites, quartzites and diamictites by Dantas et al. (2009).

The glacially influenced units (Cuiabá Group and Puga Formation) are dominated by fine-grained sediments (phyllites), sandstones, conglomerates and diamictites that occur in the upper part of this unit. Twenty-four analyses on this unit show $\varepsilon_{Nd}$ ($T = 600$) values ranging from $-12$ to $-7$ and Sm-Nd $T_{DM}$ model ages (DePaolo, 1981) of 1.6–2.2 Ga. The phyllites are the most common lithology and have a large interval of $\varepsilon_{Nd}$ ($T = 600$) values (between $-12.0$ and $-7.4$) as well as $T_{DM}$ model ages (between 1.7 and 2.2 Ga). The diamictites also have variable $T_{DM}$ model ages (1.6–2.1 Ga), but the $\varepsilon_{Nd}$ ($T = 600$) values are quite homogeneous, between $-7.0$ and 8.5. These $T_{DM}$ model ages suggest that rocks from the Amazonian Craton could be the source of the sediments, mainly the Ventuari-Tapajós and Rio Negro-Juruena provinces (Tassinari and Macambira, 1999).

### 2.4.2. Araras Group

The only direct dating available for these rocks was obtained on cap carbonates from the Mirassol d'Oeste Formation, at the base of the Araras Group, deposited on the cratonic area. Many stable isotopic studies have been done in this classic outcrop (Nogueira et al., 2003; Alvarenga et al., 2004; Font et al., 2006), and the good preservation of the samples allowed us to obtain good results. Twelve carbonate samples, collected along a 40-m thick section, were analysed for Pb isotopes, some in duplicate. The $^{206}Pb/^{204}Pb$ and $^{207}Pb/^{204}Pb$ isotope ratios from 21 analyses range from 18.47 to 65.18 and 15.70 to 18.55, respectively. These ratios show good alignment on the Pb isochron diagram and yielded a Pb-Pb isochron age of $633 \pm 25$ Ma (95% confidence), which was interpreted as the depositional age of these rocks (Babinski et al., unpublished data), and which confirms the Marinoan-Ghaub age of this succession previously suggested based on C and Sr isotope data.

The Nd isotope signature and $T_{DM}$ model ages have been determined in seven carbonates samples from the Araras Group. The Sm-Nd $T_{DM}$ model ages vary from 1.7 to 2.1 Ga, and $\varepsilon_{Nd}$ ($T = 600$) values from $-8.2$ to $-11$ (Dantas et al., 2009). Among these samples, the cap carbonates presented the less negative $\varepsilon_{Nd}$ ($T = 600$) values, which are closer to values reported for the matrix of the diamictites. Model ages of the limestones and the intercalated carbonate siltstones of the Guia Formation are similar. The more negative $\varepsilon_{Nd}$ ($T = 600$) values ($-11.3$ and $-11.4$) and $T_{DM}$ model ages of 2.0 Ga agree with the stratigraphic interval of the upper part of the Nobres Formation, which is related to the incursion of clasts (quartz and feldspar) in the carbonate succession. These $T_{DM}$ model ages suggest that the Amazonian Craton is the source area of the sediments.

### 2.4.3. Serra Azul Formation

Some indirect geochronological constraints are available for the Serra Azul diamictites deposited on top of the Araras Group carbonates. This stratigraphic position allows us to suggest that the diamictites of the Serra Azul Formation correspond to the Gaskiers glacial event, dated at 580 Ma (Bowring et al., 2003b). Geochronological data available for the Serra Azul Fm. include K-Ar ages of ca. 730 and 1,200 Ma obtained on volcanic clasts recovered from the diamictites, and Sm-Nd $T_{DM}$ model ages of 2.0 and 1.6 Ga obtained on the matrix of the diamictites and pelites of the Serra Azul Formation shows variations in sedimentary provenance (Figueiredo and Babinski, 2008). The $\varepsilon_{Nd}$ ($T = 600$) values range from $-6.7$ to $-9.5$. These Sm-Nd $T_{DM}$ ages suggest that the mixture of rocks from the Ventuari-Tapajós, Rio Negro-Juruena, and even the younger Rondonian-San Ignácio and Sunsás provinces (Tassinari and Macambira, 1999) could be the sources of sediments to the Serra Azul deposits. The source area of the younger volcanic clasts (ca. 730 Ma), however, has not yet been identified in the region and could be related to the rift phase of the Paraguay Belt during the earlier stages of Rodinia break-up.

### 2.4.4. Alto Paraguay Group

Geochronological data are available for siliciclastic rocks of the Raizama and Diamantino formations.

Shales from the lower part of the Diamantino Formation yielded an Rb-Sr isochron age of $568 \pm 20$ Ma, which was interpreted as the age of diagenesis of these sediments (Bonhomme et al., 1982; Cordani et al., 1985).

Sm-Nd $T_{DM}$ model ages were also determined on seven samples of siliciclastic rocks from the upper unit of the Paraguay Belt, ranging from 1.5 to 1.8 Ga (Dantas et al., 2009). The $\varepsilon_{Nd}$ ($T = 600$) values are not so negative and vary from −4.9 to −9.7. These Sm-Nd values confirm the presence of a younger source of sediments filling the upper part of the basin, as confirmed by the 730 Ma volcanic clasts found in the diamictites of the Serra Azul Formation (Figueiredo and Babinski, 2008).

### 2.4.5. Corumbá Group – Tamengo Formation

Limestones in the Corcal Quarry close to Corumbá have abundant *Cloudina* and few intercalations of volcanic tuffs. Euhedral crystals of one of the ash beds yielded a $^{238}U/^{206}Pb$ average age of $543 \pm 3$ Ma ($n = 17$; 95% confidence) (Babinski et al., 2008a). The $^{87}Sr/^{86}Sr$ ratios cluster at 0.7084−0.7085, which agrees with the Ediacaran age.

## 2.5. Palaeomagnetic Constraints

The first palaeomagnetic data on the Neoproterozoic cap carbonate of the Puga Formation were obtained from 19 samples in the Terconi Quarry in Mirassol d'Oeste, MT (Trindade et al., 2003). These palaeomagnetic data indicate five polarity reversals along the first 20 m of the cap dolostones, suggesting a primary magnetisation and low palaeolatitudes (22°+6/−5) for the cap dolostone (Trindade et al., 2003). The palaeomagnetic reversals imply slow sedimentation (millions of years), contrary to the snowball Earth hypothesis.

## 2.6. Conclusions and Regional Analysis

The lithostratigraphy of the Neoproterozoic in the Paraguay Belt is summarised in the columnar sections from the northern and southern Paraguay belts (Figure 2.11). There is no consensus about the lithostratigraphic correlation between the north and the south sections of the Paraguay Belt. This will be one of the focal points of the following discussion, including an integrated discussion of the lithostratigraphy, glacial episodes, palaeontology, chemostratigraphy and ages of the rocks.

Glacial episodes have been found in both sectors of the Paraguay Belt, although the evidence is more convincing in the northern belt. The Neoproterozoic record of the northern Paraguay Belt contains two glacial events: an older one of Marinoan-Ghaub glacial age (630 Ma) represented by the Puga Formation and the Cuiabá Group rocks, and a younger one, probably related to the Gaskiers glaciation (580 Ma) recorded by the Serra Azul Formation (Figure 2.11A). The Puga Formation and the correlated Cuiabá Group are interpreted as being related to the Marinoan-Ghaub glaciation because of the cap-carbonate lithofacies, palaeomagnetic data, and associated $\delta^{13}C$ and $^{87}Sr/^{86}Sr$ isotope trends (Nogueira et al., 2003; Trindade et al., 2003; Alvarenga et al., 2004, 2008; Font et al., 2006). Striated clasts and dropstones are direct evidence for the glaciation during the sedimentation of the Puga Formation and the Cuiabá Group, where the latter involves glacially influenced turbidite sedimentation

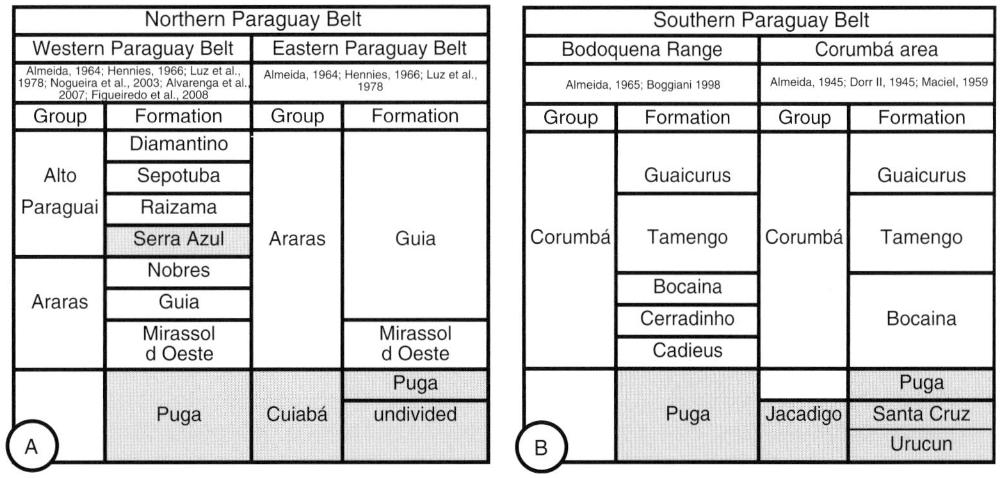


**Figure 2.11** Neoproterozoic stratigraphic nomenclature in the Paraguay Belt.

(Alvarenga and Trompette, 1992). The Puga diamictites have a well-developed, post-glacial cap carbonate (Nogueira et al., 2003; Font et al., 2006), whereas the younger Serra Azul diamictite has no identified cap carbonate, and siliciclastic mudstone rock lies directly above it (Alvarenga et al., 2007; Figueiredo et al., 2008). The cap carbonate of the Araras Group, named the Mirassol d'Oeste Formation, has the same structures (tube, wave ripple and fan-like) as those found in the cap carbonate in the Otavi Group, Namibia (Hoffman and Schrag, 2002; Nogueira et al., 2003; Allen and Hoffman, 2005). The main difference between the Serra Azul and the Puga glaciogenic units is the absence of a transgressive carbonate succession overlying the diamictites of the Serra Azul Formation.

In the southern Paraguay Belt, the rocks interpreted as deposited during a glacial period are the Jacadigo Group and the Puga Formation. The presence of boulders in the iron formation of the Jacadigo Group has been used to suggest a glacial origin for these rocks (Walde et al., 1981; Hoppe et al., 1987; Urban et al., 1992). These granite pebbles and boulders generally do not cross-cut the underlying bedding, even the biggest ones, which are larger than 1 $m^3$, and the bedding deformation around the boulders occurs associated with compaction and is more pronounced above the boulders than beneath them (Trompette et al., 1998). Trompette et al. (1998) thus argued against glacial origin of these deposits, suggesting instead reworked gravitational flows. The diamictites of the Puga Formation in the southern Paraguay Belt do not show evidence of striated clasts or dropstones, but their widespread areal distribution, underlying a carbonate succession (Araras and Corumbá groups) is strong evidence for their stratigraphic correlation to other glacial deposits of Marinoan-Ghaub age ($\sim$635 Ma) in the Paraguay Belt.

Another difference between the stratigraphy of the northern and the southern Paraguay belts is the occurrence of limestones and shales from the Tamengo and Guaicurus formations, both with microfossil contents (Gaucher et al., 2003). The Tamengo Formation is a transgressive phase at the top of the Corumbá Group, and it is not represented on the top of the Araras Group, which was covered by the younger glacial rocks of the Serra Azul Formation (Gaskiers glacial age). The depositional age of the Tamengo Formation has been considered Ediacaran (Hahn et al., 1982; Walde et al., 1982; Zaine and Fairchild, 1985; Gaucher et al., 2003), and it is now well constrained by precise SHRIMP ages on zircons extracted from volcanic ashes, yielding an age of $543 \pm 3$ Ma (Babinski et al., 2008a).

Regarding the composite $\delta^{13}C$ data, an important negative incursion is associated with the cap-carbonate succession in the Araras Group (Nogueira et al., 2003, 2007a; Alvarenga et al., 2004, 2008; Font et al., 2006). The negative incursion is more important in the first 20 m of the cap dolostone than in the limestone-shale rhythmite of the Guia Formation. These negative $\delta^{13}C$ excursions have also been found in the lower Corumbá Group (Puga Hill), and they demonstrate how such values are important for correlation between different sections (Boggiani et al., 2003). The positive $\delta^{13}C$ excursion in the Araras Group is seen in the Bauxi and Nobres areas, and it is probably related to the influence of high organic content.

In the northern Paraguay Belt, stratigraphic and isotopic studies have focused on the glacial and post-glacial deposits linked to the Marinoan-Ghaub glaciation, while in the southern Paraguay Belt, stratigraphic sections and isotopic profiles remain incomplete, precluding a full reconstruction of the evolution of the Paraguay Belt.

## *ACKNOWLEDGEMENTS*

This chapter is a contribution to the International Geological Correlation Programme (IGCP) Project 478 'Neoproterozoic-Early Palaeozoic Events in SW-Gondwana'. In October 2004, a field trip was organised in the Paraguay Belt within the framework of IGCP Project 478 led by C. Gaucher and G. Germs. We thank the enthusiastic participation on this field trip from Gerard Germs, Thomas R. Fairchild, Francisco E.S. Pinho, J.C. Möller, Hans J. Hofmann, Daniel Poiré, Nicolas Beukes, Gonzalo Blanco, Jens Gutzmer, Stefan Schroeder, Andreas Braun, Ginaldo A.C. Campanha, Lucas Warren, Daniel Oliveira, Paulo Tibana, Karina Pamoukaghlian, Jane Nobre and Mario Brito. We thank Detlef Walde, Claudio Gaucher and Alcides N. Sial for constructive reviews.

# PART 3 Neoproterozoic-Cambrian Evolution of the São Francisco Palaeocontinent

CHAPTER 3

# The São Francisco Palaeocontinent☆

A.N. Sial[1], M.A. Dardenne[2], A. Misi[3], A.J. Pedreira[4], C. Gaucher[5], V.P. Ferreira[1], M.A. Silva Filho[1], A. Uhlein[6], A.C. Pedrosa-Soares[6], R.V. Santos[2], M. Egydio-Silva[7], M. Babinski[7], C.J.S. Alvarenga[2], T.R. Fairchild[7] *and* M.M. Pimentel[2]

## Contents



## 3.1. Introduction

The São Francisco Craton (Almeida, 1977) and its counterpart in Africa, the Congo Craton (Trompette, 1994), represent sectors of a Neoproterozoic palaeocontinent preserved from the Brasiliano-Pan African orogeny as recorded in their marginal belts. The limits between the São Francisco Craton and the surrounding belts (Brasília, Araçuaí, Rio Pardo, Sergipano, Riacho do Pontal and Rio Preto belts; Figure 3.1) are marked by intense folding and overthrusting. In the interior of the craton, the Neoproterozoic sedimentary cover is only gently deformed or horizontal and preserved in isolated basins with distinct names: the São Francisco, Irecê and Una-Utinga basins. The São Francisco basin includes basal assemblages of Palaeo- and Mesoproterozoic age (e.g. Martins-Neto and Alkmim, 2001). However, in this chapter, we restrict to the discussion to the Neoproterozoic sedimentary assemblages comprising the São Francisco Supergroup (Alkmim and Martins-Neto, 2001; Martins-Neto and Alkmim, 2001), which is characterised by a glaciogenic unit at the base Jequitaí Formation and an argillaceous–calcareous–arkosic unit (Bambuí Group) at the top.

The Neoproterozoic evolution of the São Francisco Craton, including the Brasiliano marginal belts and the sedimentary cratonic cover, is reviewed in this chapter. The lithostratigraphy, Neoproterozoic-aged mineralisation (lead, zinc, fluorine, barium and phosphates) and available geochronological and stable isotope data are also presented.

☆Sial, A.N., Dardenne, M.A., Misi, A., Pedreira, A.J., Gaucher, C., Ferreira, V.P., Silva Filho, M.A., Uhlein, A., Pedrosa-Soares, A.C., Santos, R.V., Egydio-Silva, M., Babinski, M., Alvarenga, C.J.S., Fairchild,T.R., Pimentel, M.M. 2009. The São Francisco Palaeocontinent. In: Gaucher, C., Sial, A.N., Halverson, G.P., Frimmel, H.E. (Eds): Neoproterozoic-Cambrian Tectonics, Global Change and Evolution: a focus on southwestern Gondwana. Developments in Precambrian Geology, 16, Elsevier, pp. 31–69.

[1] NEG-LABISE, Department of Geology, Federal University of Pernambuco, C.P. 7852, Recife, PE 50670-000, Brazil.
[2] Institute of Geosciences, University of Brasília, Brasília-DF 70910-900, Brazil.
[3] Institute of Geosciences, Federal University of Bahia, Salvador, BA, Brazil.
[4] Geological Survey of Brazil (CPRM), Av. Ulysses Guimarães, 2862, Salvador, BA 41213-000, Brazil.
[5] Departamento de Geología, Facultad de Ciencias, Iguá 4225, 11400 Montevideo, Uruguay.
[6] Federal University of Minas Gerais, IGC-DEGEL-CPMTC, Belo Horizonte, MG 31270-901, Brazil.
[7] Institute of Geosciences, University of São Paulo, Rua do Lago 562, São Paulo, SP, 05508-080, Brazil.

*Developments in Precambrian Geology*, Volume 16
ISSN 0166-2635, DOI 10.1016/S0166-2635(09)01603-X

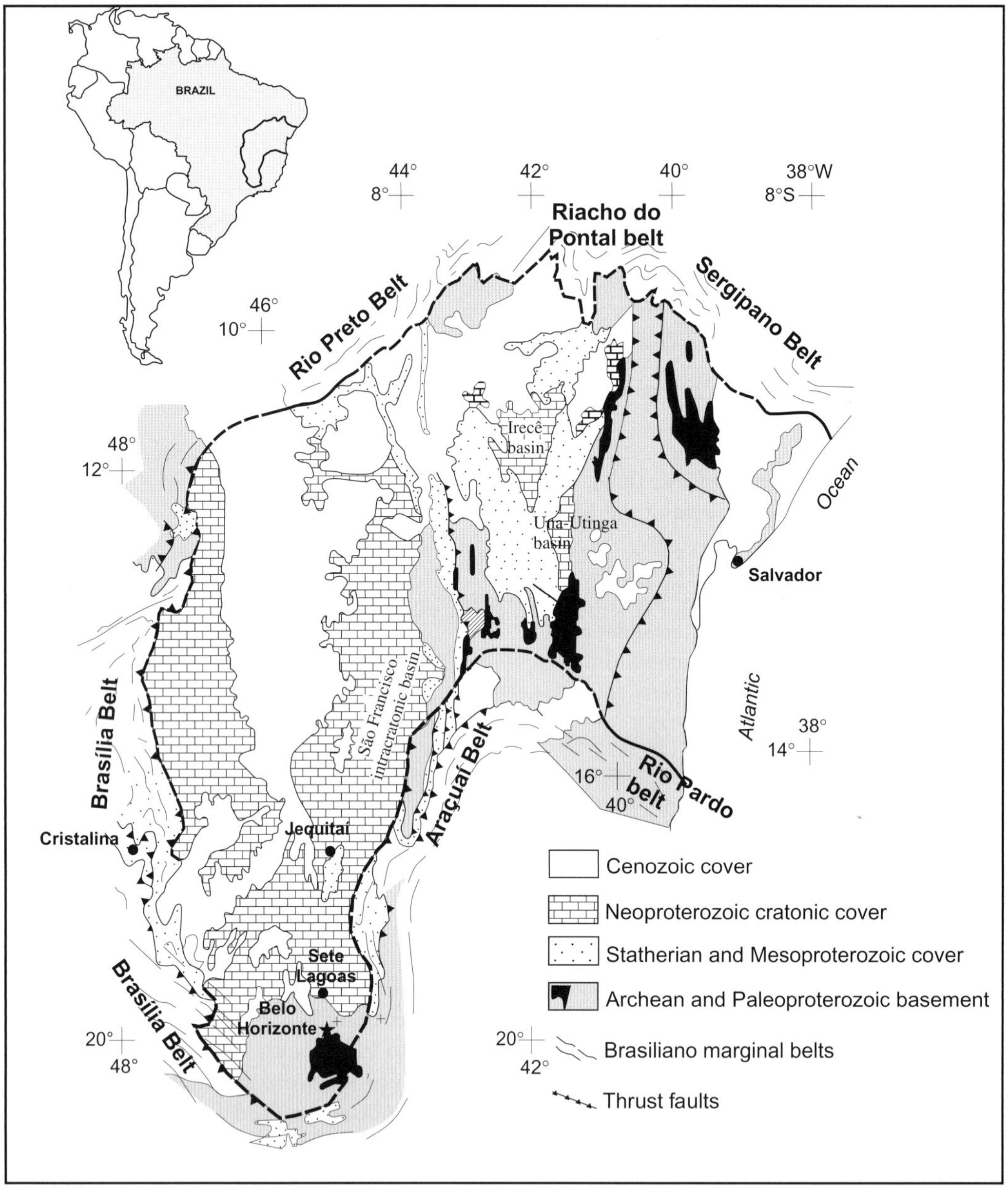


**Figure 3.1** Simplified geological map of the São Francisco Craton showing the extent of Neoproterozoic cover (modified from Misi et al., 2005a,b).

It is well known that the late Neoproterozoic was marked by significant changes in Earth's climate. Clues to these environmental perturbations are recorded in glacial deposits, sometimes formed in equatorial latitudes (Sohl et al., 1999) and often found overlain by dolostones or limestones (so-called cap carbonates; e.g. Kennedy et al., 1998). These strong climatic variations seem to have occurred between 750 and 580 Ma, during at least three distinct glacial epochs, commonly termed Sturtian, Marinoan and Gaskiers glacial events (Halverson et al., 2005). In this chapter, the geological record of Neoproterozoic glacial events and corresponding stable isotopic data for related cap carbonates in the São Francisco Craton are also discussed.

## 3.2. Synthesis of the Lithostratigraphic Units

### 3.2.1. The Neoproterozoic cover of the São Francisco Craton

#### 3.2.1.1. The Bambuí Group

Over the São Francisco Craton, the Bambuí Group (São Francisco basin; Figure 3.1) is the most important Neoproterozoic sedimentary unit, covering large cratonic areas and occupying, in central Brazil, segments of the eastern side of the Brasilia belt. It overlies the Paranoá Group from which it is separated by an unconformity filled

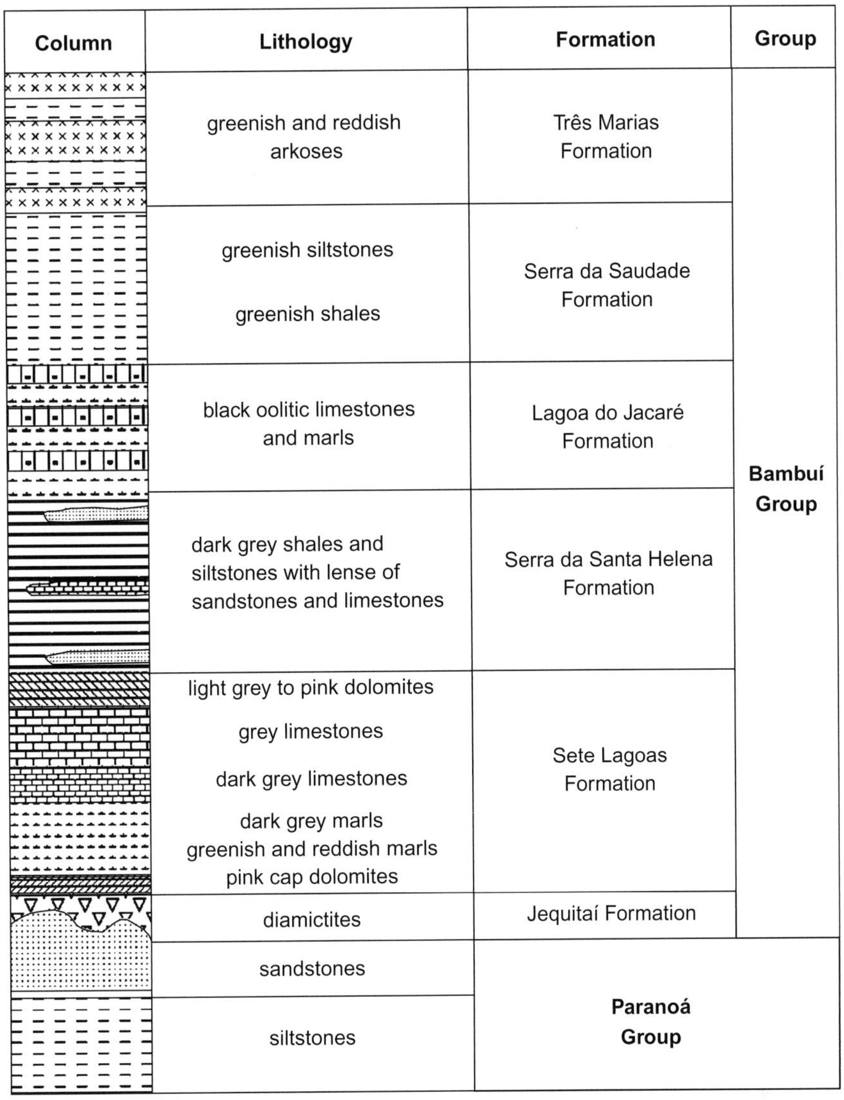


**Figure 3.2** Lithostratigraphic column of the Bambui Group (Dardenne, 2007a).

by glacial diamictites of the Jequitaí Formation (Figure 3.2). The Paranoá sedimentary rocks consist mostly of mature siliciclastic sediments such as quartzites, with intercalations of metasiltstones and minor lenses of limestones and dolostones. This group has been divided into nine lithostratigraphic units (Faria, 1985), beginning with a paraconglomerate, followed by transgressive and regressive siliciclastic-dominated cycles, and ending with pelites and dolostones containing *Conophyton metula* Kirichenko stromatolites (Cloud and Dardenne, 1973; Cloud and Moeri, 1973; Dardenne et al., 1976; Dardenne, 1979). Available geochronological and microfossil data for the Paranoá Group point to an age of 1,170–950 Ma and a source region in the Palaeoproterozoic sialic basement of the craton, suggesting sedimentation on a passive margin (Guimarães, 1997; Pimentel et al., 1999).

The Jequitaí Formation (Figure 3.3a and b), at the base of the Bambuí Group, consists of clast-supported, greenish or grey diamictites (quartzite, granite, gneiss, limestone, dolostone and siltstone) with clay-rich matrix, sometimes marly, and minor siltstone and sandstone lenses. This unit crops out discontinuously in central Brazil and overlies both the Paranoá rocks and the basement (Dardenne, 1978a, 1979, 2000; Karfunkel and Hoppe, 1988; Uhlein, 1991, 2004; Uhlein et al., 1999) and represents syn-glacial deposition over a wide area of the São Francisco Craton. Locally, a thin horizon of a light-grey to pink cap dolostone overlies the glaciogenic diamictites.

Deposits of the Jequitaí Formation, probably Sturtian in age ($\sim$750 Ma), are well exposed at the margins of Serra do Cabral and Agua Fria areas in the southeastern portion of the São Francisco Craton. This formation, up to 150 m thick, overlies quartzites of the Palaeo-Mesoproterozoic Espinhaço Supergroup with a slight unconformity in the Serra do Cabral area. The tops of these quartzites show subglacial erosion structures such as grooved and striated pavements with striae-oriented ENE-WSW. The Jequitaí Formation consists of massive and stratified diamictites with granules, pebbles and boulders of gneisses, granites, quartzites and carbonates (Uhlein et al., 2004). The diamictites are massive in their lower part and in their upper part exhibit alternating clast-rich and

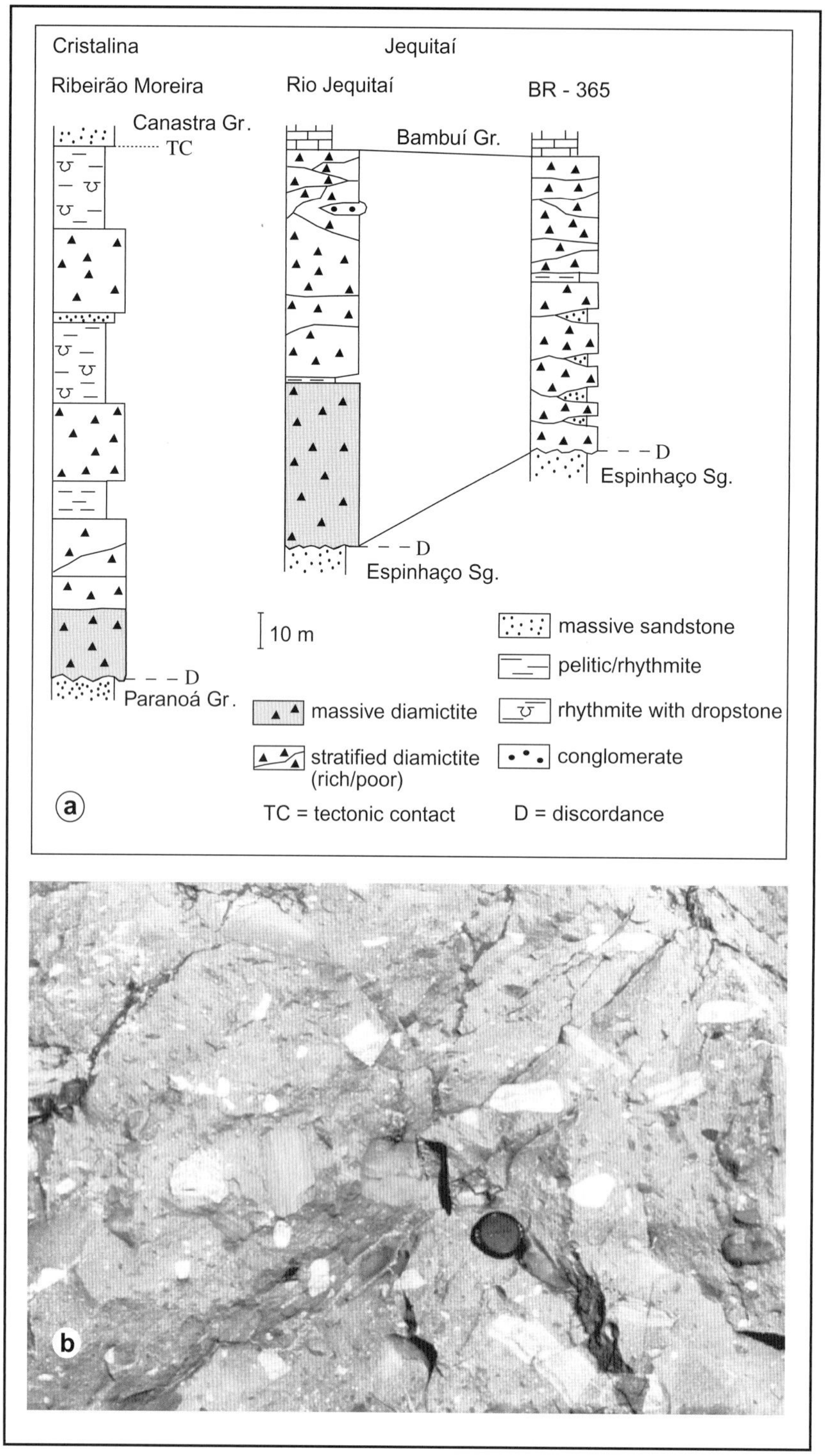


**Figure 3.3** (a) Lithofacies variations of the Jequitaí Formation near Serra do Cabral, Minas Gerais, and Cristalina, Goias (Uhlein et al., 1999); (b) diamictite of the Jequitaí Formation at Serra da Água Fria, Minas Gerais.

clast-poor beds (Figure 3.3). This formation also contains lenticular sandstones, siltstones and a few laminated siltstone–mudstone intercalations. Clast-poor diamictites overlying striated pavement suggest two phases of the glacial event. The first phase, probably on the continent, produced ice erosion (striated pavement) and the second one a clast-poor diamictite deposited by gravity flows during ice retreat and sea-level rise.

The Jequitaí Formation, also named the Carrancas Formation, is correlative to the Bebedouro Formation that underlies carbonates of the Salitre Formation (Una Group) in the northeastern part of the São Francisco cratonic area. In the Onça do Pitangui map sheet, in the state of Minas Gerais, layers of unmetamorphosed varvites (Figure 3.4) of the Carrancas Formation form a 30 m-thick deposit with dropstones (informally named the Moema sequence by Rocha-Campos et al., 2007) towards the top of the Jequitaí Formation.

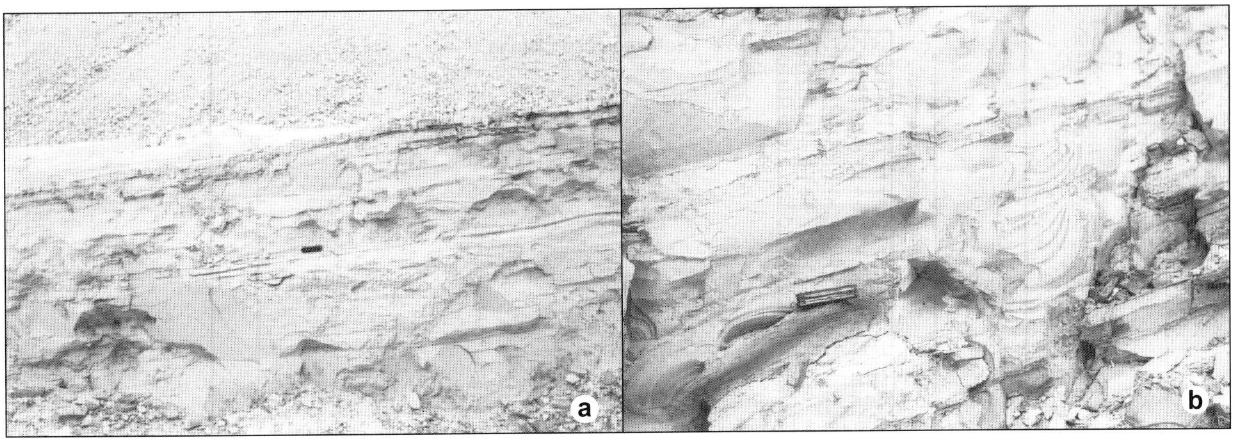


**Figure 3.4** (a and b) Unmetamorphosed rhythmite deposit of the Carrancas Formation. Locally, dropstones are observed in this deposit. Onça do Pitangui map sheet, Minas Gerais, eastern Brazil.

The post-glacial transgression flooded the craton, during which time pelitic and carbonate sediments of the lower Bambuí Group were deposited (Figure 3.2) over at least 300,000 $km^2$ (Dardenne, 1978a,b, 1979, 2000; Misi and Kyle, 1994). This sedimentary association, which follows the Jequitaí glaciation, is repeated in three regressive megacycles. Each of these megacycles begins with a regional marine transgression associated with basin subsidence, the evidence for which is seen in deep pelitic marine facies, passing upwards into shallow platform, subtidal and supratidal facies.

From base to top, these megacycles are arranged as follows. *Megacycle I* is pelitic-calcareous, corresponding to the Sete Lagoas Formation, forming a coarsening-upward sequence with calcilutites in the basal portion and passing into limestones and dolostones at the top. At the base, characteristic pink dolomitic calcilutites with green argillaceous films are generally observed. They pass progressively to dark-grey and black-laminated calcilutites. The limestones, generally dark-grey to black and microcrystalline, are well stratified and homogeneous, but lenticular in large scale. To the top, carbonate facies are predominantly dolomitic, distinguished by intraclasts and oolites. The first cycle ends with an extensive subaerial exposure marked by tepee structures, moulds of sulphate nodules, dissolution breccias, and laminated and columnar stromatolites (Dardenne, 1978a,b, 1979; Dardenne and Freitas-Silva, 1999).

*Megacycle II* begins with the Serra de Santa Helena Formation, which is dominantly pelitic at the base and signals sudden but broad subsidence of the basin. The pelites are followed by dark-grey platformal limestones of the Lagoa do Jacaré Formation, which were deposited in an environment dominated by storms and tidal currents.

*Megacycle III* is pelitic to sandy and represented by the pelitic Serra da Saudade Formation at the base, deposited in deep-platform environment with episodic influence of storms, and by the Três Marias Formation at the top, dominated by arkose and deposited in a shallow-platform environment influenced by storm currents with episodic tidal to supratidal facies. The pelites are distinctly greenish in colour, containing abundant detrital mica grains in bedding planes, while the greenish or reddish arkoses are fine-grained and rich in plagioclase and mica. Some contain volcanic lithic fragments derived from the erosion of a Brasiliano magmatic arc to the east and are regarded as a foreland basin molasse.

Along the São Francisco River, between the cities of Januária, Itacarambi, Montalvânia and Manga, well-developed and continuous carbonate horizons (Figure 3.5) have been named the Januária and Nhandutiba formations (Dardenne, 1978a,b). The former represents a typical coarsening-upward regressive megacycle, consisting, from bottom to top, of dark-grey calcilutites, dark-grey to grey limestones with lamellar breccias or oolites, pink saccharoidal dolarenites and beige lithographic dolostones. Evidence of emergence is sometimes observed between saccharoidal dolarenites and lithographic dolostones, indicating discontinuity or unconformity in the upper portion of the Januária Formation. This unconformity underlines the impermeable level of lithographic dolostone that has controlled the percolation of basinal fluids responsible for development of secondary dolomitisation, formation of saccharoidal dolarenite and development of Mississippi valley-type Pb-Zn-F mineralisation.

The carbonate and pelite facies are considerably thicker in the region between the Januária–Itacarambi cratonic area and Vazante (Figure 3.6), reflecting regional subsidence induced by differentiate movement along N-S faults during deposition, generating significant westward deepening of the basin (Alvarenga and Dardenne, 1978; Dardenne, 1978a,b, 1979, 2000).

The cap dolostone at the base of the Sete Lagoas Formation overlying Carrancas diamictites has been dated at ca. $740 \pm 22$ Ma (whole-rock Pb-Pb isochron on carbonates; Babinski et al., 2007) and this age, along with C and Sr isotope data for carbonates of the Bambuí Group (Santos et al., 2000, 2004; Vieira et al., 2007), supports a Sturtian age for the Carrancas glaciation and a Cryogenian age for at least the base of the Bambuí Group.

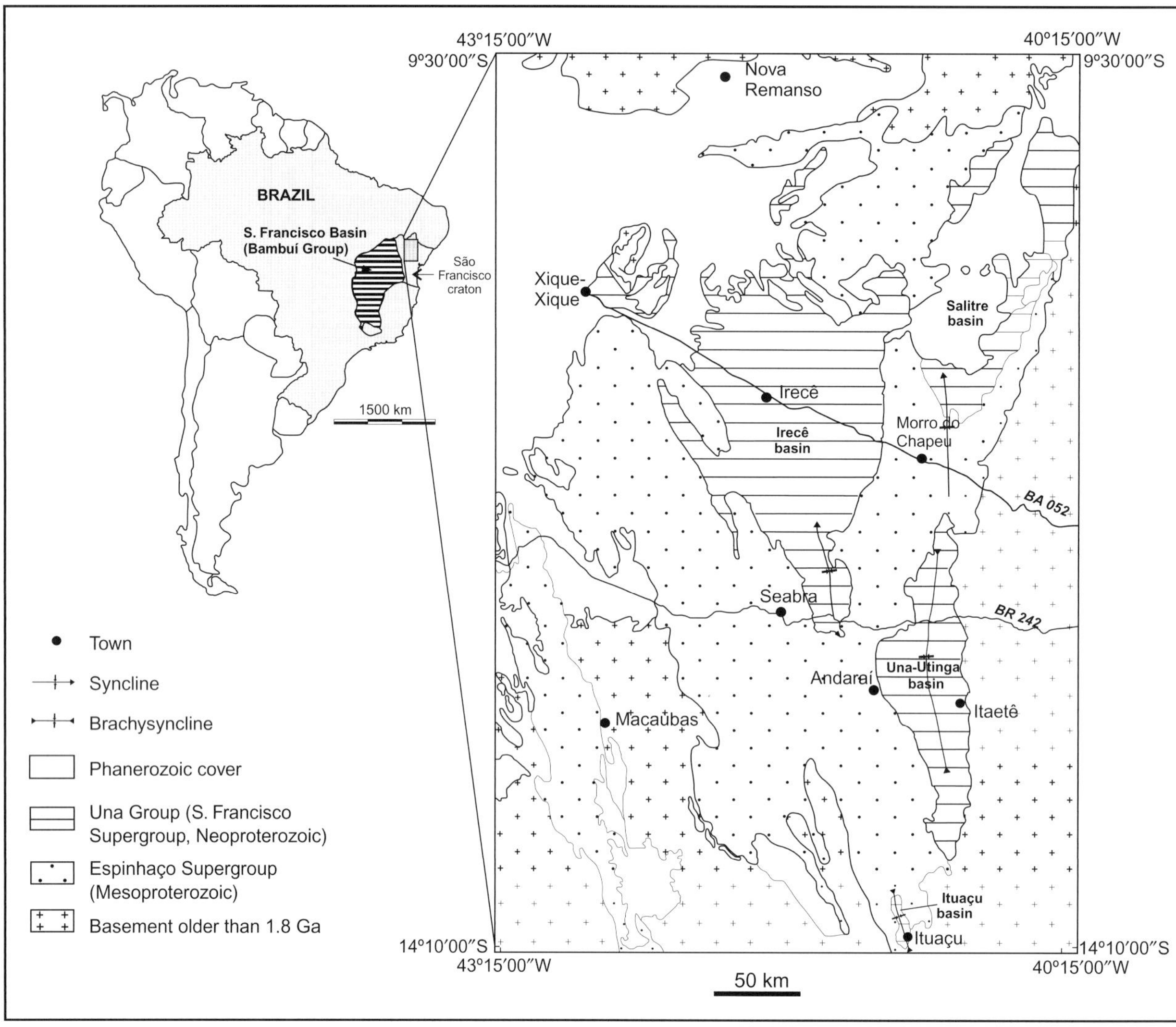


**Figure 3.7** Simplified geological map showing the Neoproterozoic "sub-basins" of the Una Group in the central area of Bahia state, Brazil (modified from Guimarães, 1996).

## 3.2.2. The surrounding belts

### 3.2.2.1. The Brasilia belt

In the western margin of the São Francisco Craton, a thick Meso-Neoproterozoic sedimentary/metasedimentary pile forms the Brasilia belt (Araí, Paranoá, Serra da Mesa, Araxá, Ibiá, Vazante, Canastra and Bambui) that extends for more than 1,000 km (Figure 3.9) and is part of a passive margin association. These rocks are mostly undeformed and unmetamorphosed over the craton and increasingly deformed and metamorphosed westward, reaching amphibolite and granulite facies conditions in the central part of the belt (Dardenne, 1978a,b, 2000; Fuck et al., 1994; Pimentel et al., 2000; Piuzana et al., 2003). The evolution of the deformation and metamorphism reflects the vergence of this belt towards the São Francisco Craton. The tectonic zonation proposed by Costa and Angeiras (1971) was reformulated by Fuck et al. (1994) who identified an internal zone in the west, a cratonic zone in the east and an external zone in between.

A WNW-ESE lineament at about the latitude of Brasília allows a subdivision of the Brasília belt into two segments with distinct geotectonic histories. In the northern segment, sedimentary units were metamorphosed to greenschist facies or are unmetamorphosed and the well-preserved stratigraphy allows the reconstitution of the palaeogeography and depositional systems. In the area to the north of Brasília, where sequences of the Paranoá and Araí groups occur, the compressional tectonics is expressed in right lateral transcurrent faults and thrusts that locally have affected the sedimentary cover.

The southern segment exhibits distinct tectonic features when compared to the northern one. Deformation and metamorphism are very intense and have obliterated stratigraphic relationships between the various units. The Araxá, Canastra, Ibiá and Vazante groups are involved in a complex, imbricated system of nappes and thrusts

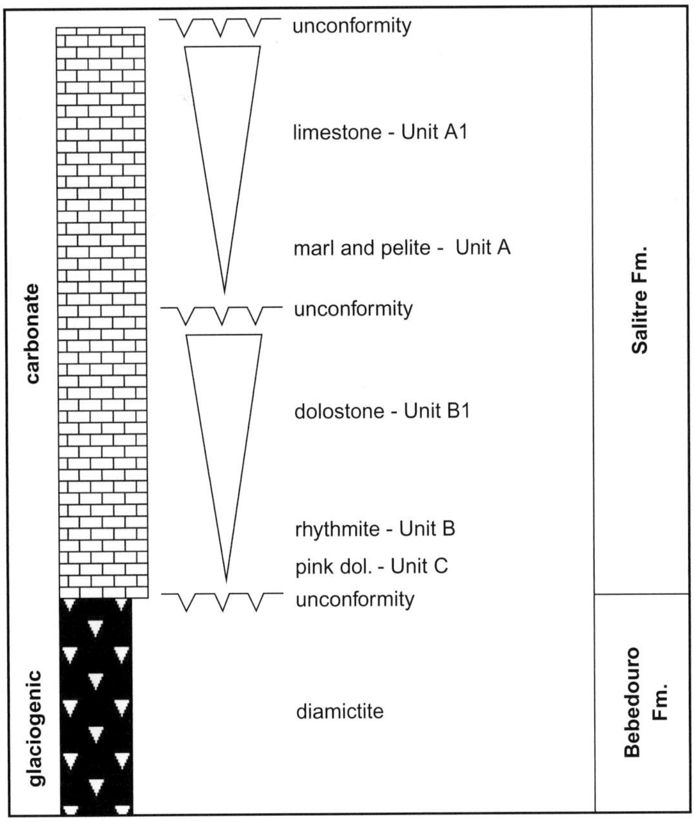


**Figure 3.8** Composite stratigraphic cartoon showing the two megasequences of the Una Group (modified from Misi et al., 2007).

**Table 3.1** Correlations between the Una and Bambui groups at the formational level.

| Group | Bambuí (Formations) | Una (Formations) |
|---|---|---|
| Lithotypes | | |
| Arkose, siltstone | Três Marias | – |
| Siltstone, pelites | Serra da Saudade | – |
| Black oolitic limestone | Lagoa do Jacaré | Unit A1 |
| Marl, shales | Serra Santa Helena | Unit A |
| Dolostone with tepee | Sete Lagoas | Unit B1 |
| Laminated limestone | | Unit B |
| Pink dolostone | | Unit C |
| Diamictite, arkose, pelitic rocks | Jequitaí, Carrancas | Bebedouro |

indicating a large amount of tectonic transport, on the order of tens to hundreds of kilometres. The contacts between the various involved assemblages correspond to low-angle shear zones, with sheath folds and lateral ramps. The external zone is characterised by a well-developed thin-skinned system of folding and overthrusting from west to east, with increasing deformation and metamorphism towards the internal zone to the west. From east to west, it is possible to recognise successive rock assemblages belonging to the Bambuí, Vazante and Ibiá-Araxá groups.

Lithostratigraphic units of the Bambuí Group are easily identified on the flanks of anticlines and synclines, frequently faulted, with N-S to N20E trending axes, as in the Formosa and Bezerra areas to the east of Brasília. In the core of anticlines, one observes sandstones, arkoses, rhythmites and stromatolitic dolomites of the Paranoá Group, overlain in erosional unconformity by Bambuí sedimentary rocks of the Jequitaí, Sete Lagoas, Serra de Santa Helena, Lagoa do Jacaré, Serra da Saudade and Três Marias formations (Figure 3.2).

In the Cristalina area (Figure 3.1), diamictites of the Jequitaí Formation occur in the flanks of a large brachianticline, disconformably overlying quartzites of the Paranoá Group (Faria, 1985; Cukrov et al., 2004; Uhlein et al., 2004). In this area, phyllites of the Canastra Group overthrust Jequitaí diamictites.

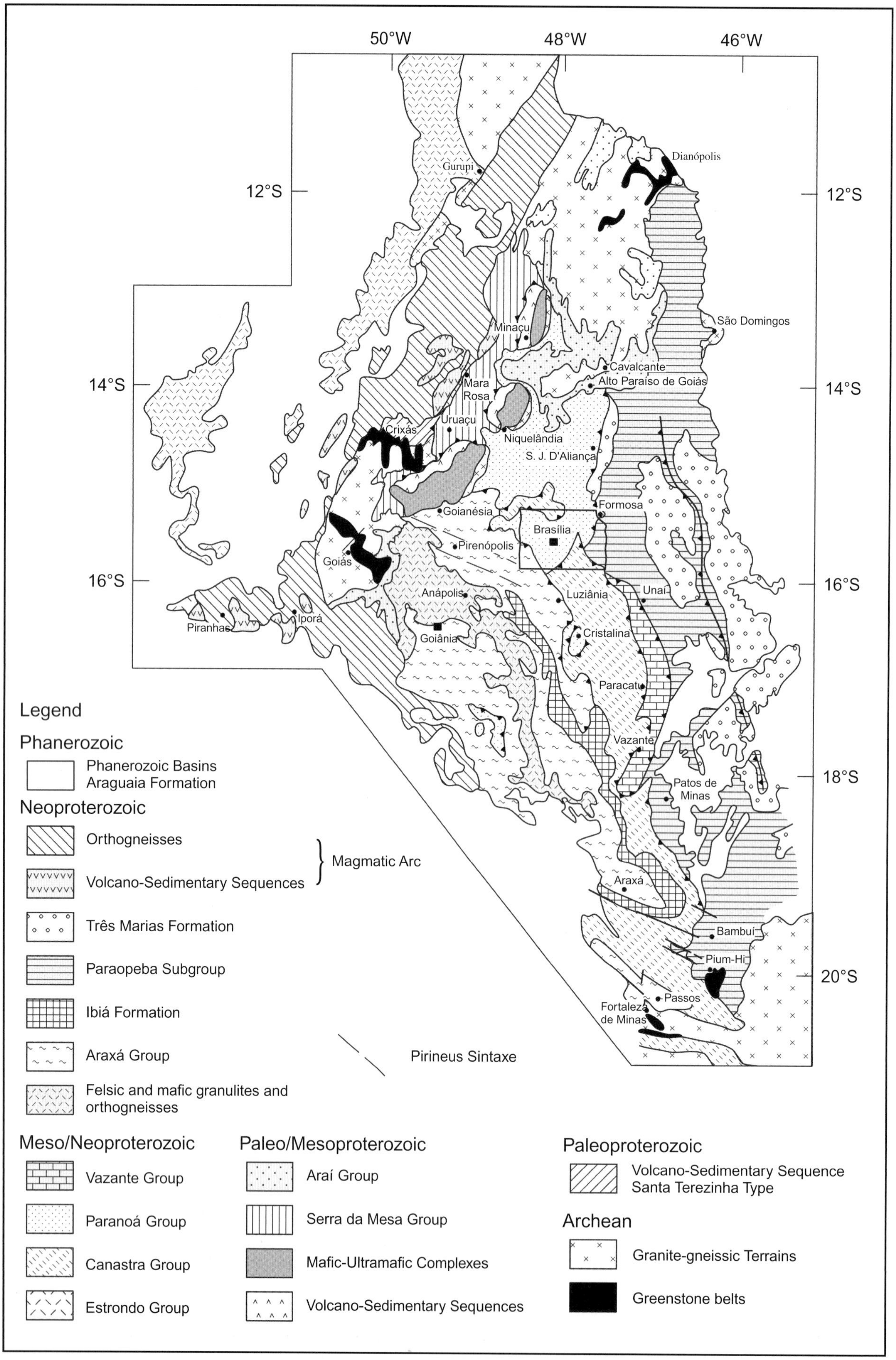


**Figure 3.9** Simplified geological map of the Brasilia belt showing the Palaeo-Meso-Neoproterozoic units, according to Dardenne (2000), Pimentel et al. (2001) and Valeriano et al. (2004). Pb-Zn deposits: (1) Vazante; (2) Morro Agudo; (3) Fagundes; (4) Ambrosia (Cunha et al., 2007).

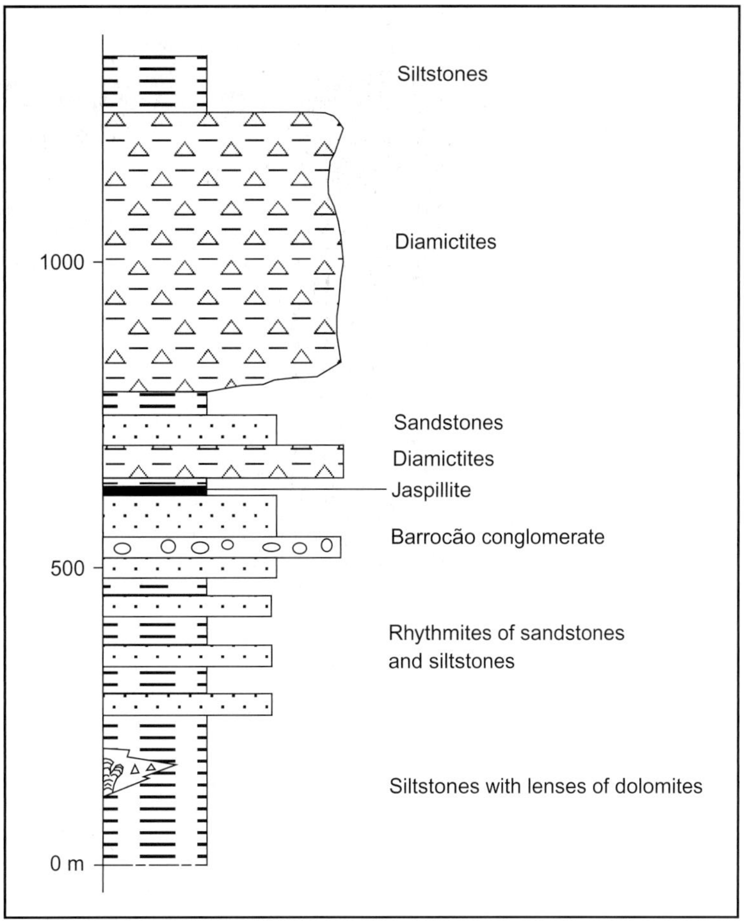


**Figure 3.10** Schematic geological column of the Lagoa Formation (Seer et al., 1989; Baptista, 2004).

In the southwestern portion of the São Francisco Craton, unusual conglomeratic facies occur in the Lagoa Formosa, Samburá and Carmo do Rio Claro formations of the upper Bambuí Group. The Lagoa Formosa Formation (Seer, 2001) crops out for over 300 km between the towns of Tiros and Presidente Olegário (Serra do Abaeté Mountains), where it shows little deformation with only a weak schistosity indicating northeastern vergence. This assemblage (Figure 3.10) probably correlates with the Serra da Saudade Formation and comprises two interbedded sequences: (i) stratified and massive diamictites (500–1,000 m thick) and (ii) a well-stratified 1,000–2,000 m-thick sequence of fine-grained diamictites, siltstones, sandstones, greywackes, jaspillites and limestones (Uhlein et al., 2004). Pebbles and boulders are predominantly of intrabasinal arkoses and siltstones floating in a greenish grey pelitic to sandy-pelitic matrix. Some clasts are composed of limestone, chert, quartzite and granite-gneiss. These paraconglomerates are interpreted as gravitational debris flows originated by regional uplift of the internal zone of the Brasília belt. This interpretation is supported by a northeasterly decrease in the size of clasts in the diamictites from boulders to granules. At the base of the Lagoa Formosa Formation, thin horizons of phosphorite are intercalated with green pelites, with abundant potassium-rich illite, and known as the Verdete facies. Carbonate beds (calcirudites, calcarenites and columnar stromatolites up to 10 m thick) are interbedded with siltstones in the lower portion of the unit and may correlate with either the top of the Lagoa do Jacaré Formation (Dardenne, 1979) or the base of the Lagoa Formosa Formation (Lima and Uhlein, 2005).

The Samburá Formation (Branco, 1957) is the name given to various minor occurrences of conglomerates, generally isolated within pelites of the upper portion of the Bambuí Group, showing similar characteristics. Most of these conglomerates are clast-supported and composed mainly of granitic, rhyolitic, arkosic and quartzitic pebbles. They are interpreted to have been deposited in eastward-tapering alluvial fans (Figure 3.11) (Castro and Dardenne, 1996).

The Carmo do Rio Claro Formation near the Furnas dam, described by Heilbron et al. (1987), is a combination of conglomerates, diamictites, arkoses and siltstones that correlate with the Samburá Formation, as well as slates, phyllites and marbles. The diamictites, which contain rounded gneiss, metapelite and metasandstone

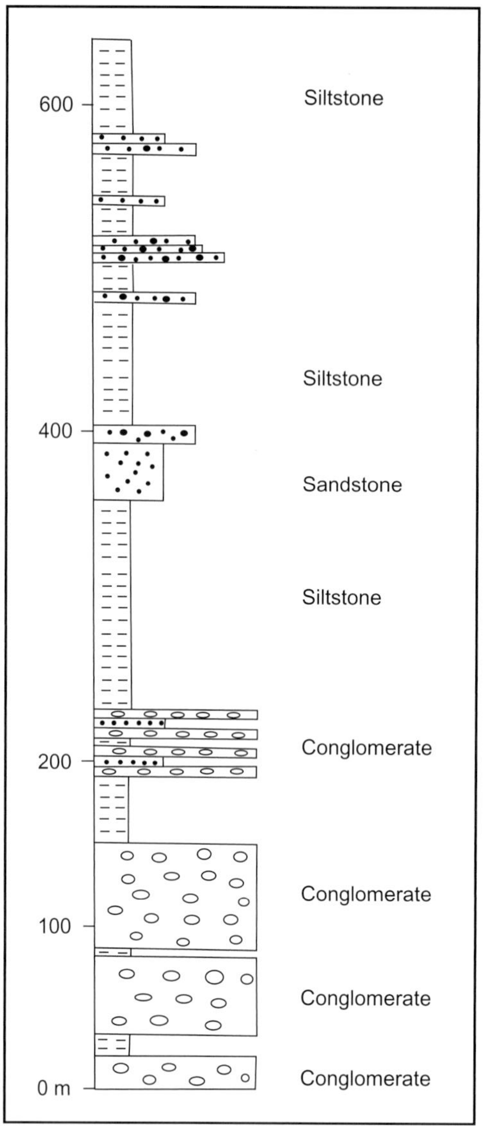


**Figure 3.11** Schematic geological column of the Samburá Formation near Samburá hydroelectric dam (after Castro and Dardenne, 1996).

pebbles, are considered to be derived from the west. The sequence is overthrust by rocks of Canastra and Araxá groups, which form the Passos Nappe. The Carmo do Rio Claro Formation is stratigraphically and genetically equivalent to the Lagoa Formosa Formation.

The Vazante Group metasediments are a thick sequence of marine pelites and dolostones that crop out for over 300 km along a north–south trend in northwestern Minas Gerais state, spanning the cities of Coromandel, Lagamar, Vazante, Paracatu and Unaí. Misi et al. (2007) have argued that the Vazante and Bambui groups are correlative, based on equivalent Sr isotope signatures in both successions ($\sim$0.7074) and the interpretation of seismic profiles by Romeiro-Silva and Zalan (2005). This controversial argument conflicts with the contrasting C isotope records between the two units and new Re-Os ages (unpublished) of organic shales of the Serra do Garrote, Serra do Poço Verde and Serra da Lapa formations that imply a late Mesoproterozoic age (A.J. Kaufman and K. Azmy, oral communications).

The Vazante Group is subdivided into seven formations (Figure 3.12), from base to top: Santo Antônio do Bonito, Rocinha, Lagamar, Serra do Garrote, Serra do Poço Verde, Morro do Calcario and Lapa formations (Dardenne, 2001).

The Santo Antônio do Bonito Formation (Souza, 1997), considered to be the basal formation, consists of beds of white quartzites, sometimes conglomeratic, intercalated with slates. In the Santo Antônio do Bonito and Santo Inacio rivers, this formation is characterised by the presence of diamictite horizons within the quartzites that contain limestone, dolostone, metasiltstone and granitic pebbles floating in a pelitic matrix, which is locally phosphatic. Larger concentrations of phosphate are found in the slaty facies and in phospharenite layers, rich in

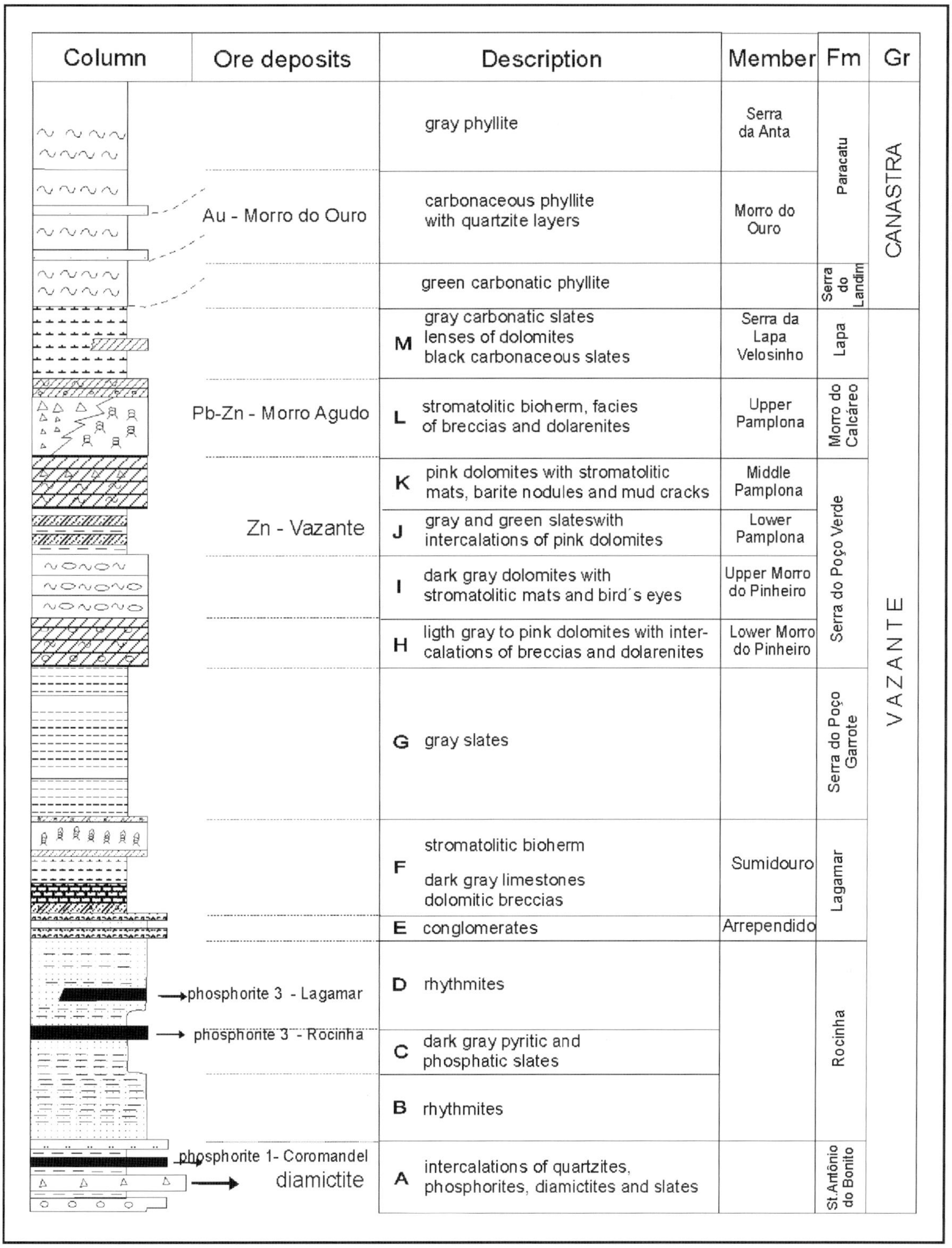


**Figure 3.12** Lithostratigraphic column of the Vazante Group (Dardenne, 2007b).

intraclasts and pellets (Phosphorite 1). The diamictites represent debris flows deposited in relatively deep waters in a glacial environment.

The Rocinha Formation (Dardenne et al., 1997) is composed of a basal rhythmic sandy and pelitic sequence that is gradational with the underlying Santo Antônio do Bonito Formation. The upper Rocinha Formation is a thick package of regularly intercalated slates and metasiltstones. It passes upward into dark-grey marly and pyritic slates, with thin phosphatic laminations that transition into the intraclast- and pellet-rich phospharenites (Phosphorite 2), forming the Rocinha phosphate deposit. In the upper part of the formation, rhythmic sediments (quartzites and siltstones) contain the Lagamar phosphate deposit composed almost entirely of phospharenites (Phosphorite 3).

The Lagamar Formation is a psammite–pelite–carbonate unit represented in its basal portion by alternations of conglomerate, quartzite, metasiltstone and slate. The conglomerates, ascribed to the Arrependido Member, consist of a framework of quartzite, metasiltstone and dark-grey limestones clasts. The psammites are overlain by dolomitic intraformational breccias passing upward into dark-grey, well-stratified limestones with intercalations of lamellar breccias followed by stromatolitic dolostones, which comprise beige to pale pink bioherms composed of microbialaminated dolostones, oncolitic dolarenites and dolorudites, and columnar stromatolites with convex and conical laminations classified as *C. metula* Kirichenko and *Jacutophyton* type (Dardenne et al., 1976). Laterally and vertically, these bioherms interfinger with marly metasiltstones and slates.

The Serra do Garrote Formation (Madalosso and Valle, 1978; Madalosso, 1980), which is about 400 m thick, comprises thick, dark-grey to greenish grey slates, sometimes rhythmic, carbonaceous and pyritic, with fine-grained quartzite intercalations. The Serra do Poço Verde Formation (Dardenne, 1979) is dominantly dolomitic and is divided into four members (Rigobello et al., 1988), described successively from the base to the top. The 500 m-thick Lower Morro do Pinheiro Member is composed of light-grey- and/or pink-laminated dolostones with cyanobacteria mats, intercalated with oncolitic dolarenites and intraformational breccias and lenses of dolostones with columnar stromatolites. The Upper Morro do Pinheiro Member is composed of medium- to dark-grey microbialaminated dolostones with common birdseye fenestrae, intercalated with layers of dolarenites, lamellar breccias and carbonaceous shales (thickness from 300 to 500 m). The occurrence of *Conophyton cylindricus* Maslov or *C. metula* Kirichenko (Moeri, 1972; Cloud and Dardenne, 1973) near the Cabeludo village is probably related to this unit. The Lower Pamplona Member is 100–200 m thick and composed of grey, green and purple siltstones intercalated with micritic and microbialaminated dolostones with small lenses of fine-grained to conglomeratic sandstones. The ~400 m-thick Middle Pamplona Member is composed of light-grey to pink microbialaminated dolostones with locally developed mud cracks and nodular barite, intercalated with beds of dolarenite, lamellar breccia and columnar stromatolites, and with lenses of shales. The 200–300 m-thick Upper Pamplona Member is characterised by the presence of light-grey to pink stromatolitic dolostones constituting biostromes and bioherms with convex-laminated columns and associated with oolitic and oncolitic dolarenites and dolorudites. In the region of Morro Agudo, Paracatu and Unaí, this unit corresponds to the Morro do Calcario Formation, which is here >900 m thick and composed mainly of dolorudites derived from reworked but still partially preserved stromatolitic bioherms and associated oolitic and oncolitic dolarenites. In this region, the Morro do Calcario and Serra do Poço Verde formations cannot be easily subdivided as they are in the Vazante region.

The Lapa Formation constitutes the upper portion of the Vazante Group described in the region of Paracatu as Serra do Velosinho and Serra da Lapa members. These are represented by carbonaceous phyllites, marly metasiltstones, dolomitic lenses and quartzite layers. The dolomitic lenses show various facies, including microbialaminated dolostones, columnar stromatolites and intraformational breccias, interfinering with the pelitic sequence that regionally overlies the dominantly dolomitic formations of Morro do Calcario and Serra do Poço Verde. In the Unaí region, the Lapa Formation comprises a rhythmic turbidite sequence of siltstones and pelite, locally with lithic sandstones and conglomerates intercalated with dark-grey slates. The clasts are fragments of phyllites, fine-grained quartzites and cherts.

The presence of *C. cylindricus* Maslov and *C. metula* Kirichenko in the dolomitic bioherms near Cabeludo and Lagamar implies a correlation with the Paranoá Group (Dardenne, 2000). Carbon and Sr isotope signatures for dolomitic rocks of these groups are fairly distinct from signatures for the Bambuí Group carbonates and are more consistent with a Mesoproterozoic age (Santos et al., 2000, 2004). Consequently, the glaciogenic diamictite of the Santo Antônio do Bonito and the overlying sedimentary sequence of Vazante Group have an unknown age, but are inferred to be Kaigas-equivalent (c. 750 Ma) or older.

Recently, various workers (Olcott et al., 2005; Azmy et al., 2005) have proposed the occurrence of a regional glacial event spanning most of the Upper Pamplona Member and Lapa Formation in the Vazante area, and Morro do Calcario, Mocambo and Lapa Formations near The Morro Agudo mine. This contention is based on the presence of a layer of brecciated rocks with dropstones, interpreted as possible glacial diamictites with a negative $\delta^{13}C$ excursion in overlying dolostones. These breccias may be alternatively interpreted as intraformational dolomitic breccias, possibly of debris flow origin (Dardenne, 2007b). If confirmed, the glacial episode recorded in the Lapa Formation would be related to the Sturtian glaciation, represented regionally by the Jequitaí Formation. This suggests that the Vazante Group constitutes a stratigraphic equivalent of the upper portion of the Paranoá Group, deposited in a passive margin setting, as previously proposed by Campos Neto (1984) and Dardenne (2007a,b).

#### 3.2.2.2. The Araçuai belt

The name Araçuaí was formerly applied only to the arcuate marginal belt located along the east and southeast edges of the São Francisco Craton (Almeida, 1977; Figure 3.1). The Araçuaí orogen was defined after the identification of several geotectonic components (e.g. ophiolites, pre-collisional magmatic arc and syn-collisional

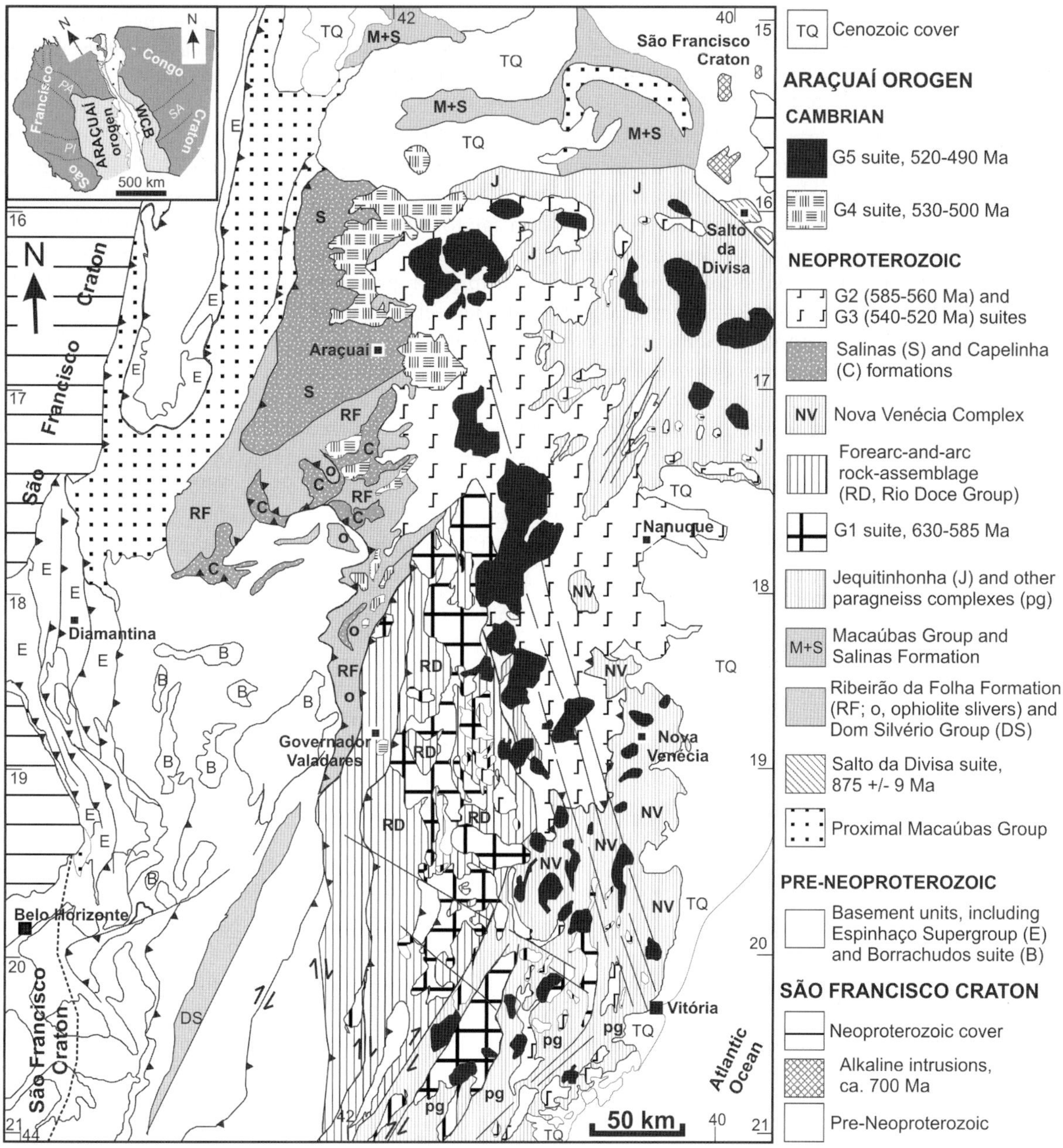


**Figure 3.13** Geological map of the Araçuaí orogen (modified from Pedrosa-Soares et al., 2008) and its location in relation to the São Francisco–Congo Craton (see inset, modified from Alkmim et al., 2006b: WCB, West-Congo belt; Neoproterozoic aulacogens: PA, Paramirim; PI, Pirapora; SA, Sangha).

granites) related to that marginal belt (Pedrosa-Soares et al., 2001, 2008). This orogen is 700 km long and up to 500 km wide and encompasses the entire region between the São Francisco Craton and the Atlantic continental margin, between 15°S and 21°S in eastern Brazil (Figure 3.13), including the southeastern limit of the São Francisco Craton. The Araçuaí orogen and its counterpart located in southwestern Africa (the West-Congo belt; Trompette, 1994; Tack et al., 2001) form a confined orogen evolved into an embayment carved into the São Francisco–Congo palaeocontinent (Pedrosa-Soares et al., 2001, 2008; Alkmim et al., 2006b).

The basement of the Araçuaí orogen, not described here, consists of Archaean to Mesoproterozoic units, including the rift-sag Espinhaço Supergroup and the related anorogenic Borrachudos suite (Uhlein et al., 1998b; Martins-Neto, 2000; Pedrosa-Soares and Wiedemann-Leonardos, 2000b; Noce et al., 2007).

The Araçuaí orogen comprises several Neoproterozoic stratigraphic units, such as the Macaúbas Group, which includes glacially influenced deposits and deep marine sedimentation associated with meta-mafic–ultramafic ophiolite slivers, tonalitic batholiths making up a pre-collisional magmatic arc (G1 suite), orogenic metasedimentary successions (e.g. Salinas Formation and Nova Venécia Complex) and a large sector dominated by syn-collisional granites (G2 suite). The belt is intruded by Cambrian-age post-collisional plutons (Figure 3.13).

Two magmatic episodes yield age constraints for the rift stage of the precursor basin of the Araçuaí orogen. The older age from a meta-dolerite dyke of Pedro Lessa, located ca. 50 km to the southeast of Diamantina, was dated at 906 ± 2 Ma (U-Pb, zircon and baddeleyite; Machado et al., 1989). The A-type granites of Salto da Divisa yielded a zircon U-Pb age of 875 ± 9 Ma (Silva et al., 2007).

The precursor rift basin was filled by units of the proximal Macaúbas Group (Figure 3.13). The oldest rift deposits consist of metamorphosed sandstones, arkoses, conglomerates and pelites that lack evidence of glaciation and represent fluvial sedimentation (Karfunkel and Hoppe, 1988; Noce et al., 1997; Martins, 2006). U-Pb ages of detrital zircon grains constrain the maximum sedimentation age for this early rift stage at around 900 Ma (Pedrosa-Soares et al., 2008). These pre-glacial deposits of the Macaúbas basin are covered by metamorphosed, generally massive diamictites with minor sandstones, pelites and rhythmites, representing glacio-terrestrial to gravitational glacio-marine sedimentation (Pedrosa-Soares et al., 1992; Uhlein et al., 1998b, 1999; Martins-Neto et al., 2001; Martins, 2006). This lower diamictite-rich unit was thrust over the Neoproterozoic carbonate-pelite cover of the São Francisco Craton and can be correlated with the glaciogenic Jequitaí Formation of the cratonic region (Karfunkel and Hoppe, 1988; Uhlein et al., 1999; Martins-Neto and Hercos, 2002). The eastern part of the proximal Macaúbas Group includes a thick pile of sediments with layered diamictites, pebble-bearing iron formations, graded sandstones and pelites with dropstones, representing glaciogenic debris flows and turbidites deposited in submarine fan environment (Pedrosa-Soares et al., 1992; Uhlein et al., 1999; Martins-Neto et al., 2001). This succession includes mafic volcanic rocks, metamorphosed to greenschist facies, with pillow structures and other features of subaqueous flows and is derived from tholeiitic basalt protoliths with a dominant within-plate signature (Gradim et al., 2005). Inherited zircon grains from older felsic rocks yielded U-Pb ages from Archaean to Late Mesoproterozoic (the youngest at 1,156 ± 21 Ma), with a whole-rock Sm-Nd $T_{DM}$ model age of ca. 1.5 Ga (Babinski et al., 2005). However, some samples with oceanic signatures and slightly positive $\varepsilon Nd_{(900\,Ma)}$ of +0.23, together with the inherited zircon grains, suggest a transitional mafic magma that migrated through a thinned continental crust. Accordingly, these greenschists provide strong evidence of volcanism in an extensional marine basin floored by thin continental crust, during the very late rift stage of the Macaúbas basin (Gradim et al., 2005; Pedrosa-Soares et al., 2008).

The upper succession of the proximal Macaúbas Group consists of interlayered sandstones and pelites, metamorphosed from the greenschist to lower amphibolite facies, representing the proximal passive margin sedimentation of the Macaúbas basin. The youngest detrital zircon grain from a sandstone layer constrains the maximum sedimentation age of this shelf succession to around 864 Ma (Pedrosa-Soares et al., 2000a, 2008).

The distal part of the Macaúbas Group is a passive margin sequence free of diamictite, known as the Ribeirão da Folha Formation (Figure 3.13). This unit was formerly considered to be part of the Salinas Formation (Pedrosa-Soares et al., 1992, 1998, 2001), but since the redefinition presented by Lima et al. (2002), it has been separated (Pedrosa-Soares et al., 2008). The Ribeirão da Folha Formation was metamorphosed progressively from west to east to the garnet, staurolite, kyanite and sillimanite zones (Pedrosa-Soares and Wiedemann-Leonardos, 2000b; Queiroga et al., 2006). The western part of the Ribeirão da Folha Formation comprises banded quartz-mica schists (fine-grained turbidites) with thin calc-silicate (marl) and marble lenses. The eastern Ribeirão da Folha Formation is associated with thrust slices of oceanic meta-mafic and meta-ultramafic rocks, constituting a tectonically dismembered ophiolite complex. This eastern succession is rich in peraluminous mica schists (pelagic pelites) with intercalations of sulphide-bearing cherts and diopsidites associated with massive sulphide bodies, oxide, silicate and sulphide-type banded iron formations, and graphite schists (Pedrosa-Soares et al., 1992, 1998, 2001; Pedrosa-Soares and Wiedemann-Leonardos, 2000b; Suita et al., 2004; Queiroga et al., 2006). The meta-mafic rocks associated with the Ribeirão da Folha Formation are coarse-grained massive ortho-amphibolites, representing gabbroic protoliths, and medium- to fine-grained banded ortho-amphibolites that seem to be derived from dolerites and basalts (Pedrosa-Soares et al., 1998, 2008; Suita et al., 2004). These ortho-amphibolites have a geochemical signature akin to modern ocean-floor mafic rocks and are comparable to similar rocks of other Neoproterozoic ophiolites (Pedrosa-Soares et al., 1998, 2001; Pedrosa-Soares and Wiedemann-Leonardos, 2000b; Suita et al., 2004). The U-Pb (LA-ICPMS) age of 660 ± 29 Ma obtained from zircon crystals of a plagiogranite vein suggests the epoch of magmatic crystallisation of at least part of the oceanic magmatic rocks (Queiroga et al., 2007).

The Jequitinhonha Complex comprises a thick paragneiss (kinzigitic) association, consisting of migmatised biotite gneisses, with variable contents of garnet, cordierite, sillimanite and graphite, thick intercalations of graphite-rich gneiss, and minor quartzite and calc-silicate rocks. The paragneiss protoliths comprise marine arkosic greywackes to peraluminous pelites, deposited under oxidising conditions, with horizons of carbonaceous black mud that were deposited in a restricted marine environment. The sedimentary protoliths are interpreted to have been deposited during the passive margin stage of the Macaúbas basin (Pedrosa-Soares and Wiedemann-Leonardos, 2000b).

The western external domain of the Araçuaí orogen, adjacent to the southeastern edge of the São Francisco Craton, is a west-verging fold-thrust belt, with increasing intensity of metamorphism from west to east.

The northern segment of the external tectonic domain bends to the east, outlining the northern curvature of the orogen, where it shows north-verging structures and metamorphic grade increasing from north to south (Pedrosa-Soares and Wiedemann-Leonardos, 2000b; Pedrosa-Soares et al., 2001, 2008; Uhlein, 2004; Alkmim et al., 2006b).

Along the western external domain, the major normal faults of the precursor rift basin, active during the Tonian, were reactivated as thrust faults during the contractional orogenic event (Uhlein et al., 1999; Martins-Neto and Hercos, 2002; Uhlein, 2004). The main deformational event, related to the syn-collisional stage, is characterised by zones with asymmetrical folds showing westward vergence separated by mylonitic belts of brittle–ductile shear zones. An east–southeast dipping S1 cleavage or mylonitic schistosity with down-dip stretching lineation and kinematic indicators such as S-C surfaces, patterns of deformed and boudinaged veins, asymmetric deformation around porphyroclasts and clasts, and sheath folds indicate a westward tectonic transport towards the São Francisco Craton (Uhlein, 1991; Pedrosa-Soares et al., 1992; Uhlein et al., 1995, 1998b; Cunningham et al., 1998; Alkmim et al., 2006b). Thrusts and reverse faults are defined by mylonites in which S-C fabrics or porphyroclast tails indicate top-up-to-the-west movement. Pebbles within Macaúbas diamictite layers were stretched with their long axis parallel to the E-W mineral lineation (Uhlein, 1991; Pedrosa-Soares et al., 1992). Rocks of the eastern proximal Macaúbas Group show asymmetric extensional crenulation cleavage ($S_2$) as the axial surface of $F_2$ folds. These structures show top-down-to-the-east tectonic transport and indicate a regional-scale normal-sense shear zone probably related to the extensional (gravitational) collapse of the Araçuaí orogen (Marshak et al., 2006).

The internal tectonic domain is the high-grade metamorphic core of the orogen, which includes the pre-collisional magmatic arc, paragneiss complexes and a huge anatectic zone dominated by syn-collisional S-type granites and post-collisional intrusions (Figure 3.13). The internal tectonic domain is generally west-verging, but the back-arc zone between 17°30′S and 20°S shows syn-collisional tectonic transport to the east (Pedrosa-Soares et al., 2001, 2006, 2008; Alkmim et al., 2006b; Vauchez et al., 2007). The southern sector of the Araçuaí orogen and its connection with the Ribeira orogen are marked by syn-collisional west-vergent thrusts cut by late, dextral, strike–slip shear zones (Cunningham et al., 1998; Heilbron et al., 2004; Peres et al., 2004; Alkmim et al., 2006b). High-grade metamorphism and partial melting are syn-kinematic with the regional foliation and took place at ca. 585–560 Ma. The internal domain shows widespread granitoids of the G1 and G2 suites (Figure 3.13). In general, G1 and G2 batholiths display high-temperature (600–700°C) solid-state foliation parallel to the magmatic orientation, both concordant with the regional foliation of the country rocks, suggesting synchronous deformation of the magmatic rocks and their host rocks (Uhlein et al., 1998a; Pedrosa-Soares et al., 2001, 2006; Vauchez et al., 2007).

The Araçuaí orogen records four evolutionary stages whose timing is constrained by U-Pb ages of the pre-collisional (ca. 630–585 Ma), syn-collisional (ca. 585–560 Ma), late collisional (ca. 560–530 Ma) and post-collisional (ca. 530–490 Ma) magmatic episodes (Pedrosa-Soares et al., 2001, 2008; Silva et al., 2005).

The pre-collisional stage (ca. 630–585 Ma) is mainly represented by the I-type G1 plutonic suite. This suite represents the roots of a magmatic arc formed in a continental active margin setting (Pedrosa-Soares and Wiedemann-Leonardos, 2000b; Nalini-Junior et al., 2000b; Pedrosa-Soares et al., 2001, 2008). It consists of deformed batholiths and stocks composed of tonalite, granodiorite and minor diorite, with dioritic to mafic enclaves. Chemical compositions from many samples of several G1 plutons outline a consistent metaluminous, calc-alkaline, volcanic arc signature (Pedrosa-Soares and Wiedemann-Leonardos, 2000b; Nalini-Junior et al., 2000b, 2005; Pedrosa-Soares et al., 2001; Campos et al., 2004; Martins et al., 2004b). U-Pb zircon and monazite ages constrain the evolution of the G1 suite to between ca. 630 and ca. 585 Ma (Noce et al., 2000; Nalini-Junior et al., 2000b; Whittington et al., 2001; Silva et al., 2005). The location of this magmatic arc relative to the zone with oceanic slivers suggests subduction to the east and a suture zone roughly located along the 42°W meridian, south of 17°S (Pedrosa-Soares et al., 1998, 2001, 2008).

The forearc-and-arc rock assemblage includes orogenic units that represent the volcano-sedimentary pile of the magmatic arc and/or contain sediments supplied by it (e.g. Rio Doce Group; Vieira, 2007), together with strongly deformed tonalitic bodies, thrust slices of basement rocks and probable passive margin deposits (Pedrosa-Soares et al., 2008).

The Nova Venécia Complex comprises pelite-rich sediments deposited in the back-arc basin of the Araçuaí orogen (Noce et al., 2004). This complex contains migmatised, high-grade paragneisses rich in biotite, garnet, cordierite and/or sillimanite, with lenses of calc-silicate rocks (Pedrosa-Soares et al., 2006, and references therein). U-Pb ages for detrital zircons constrain the maximum age of sedimentation to around 608 ± 18 Ma (revised age by Pedrosa-Soares et al., 2008, after data from Noce et al., 2004). High-grade metamorphism and partial melting are syn-kinematic with the regional foliation and took place at around 580–570 Ma, but the complex was also involved in late orogenic episodes of granite genesis (Noce et al., 2004; Munhá et al., 2005; Pedrosa-Soares et al., 2006).

To the northwest of the magmatic arc is found the Salinas Formation (formerly included in the Macaúbas Group; e.g. Pedrosa-Soares et al., 2001; Lima et al., 2002), a metamorphosed greywacke–pelite succession with intercalations of clast-supported conglomerates locally rich in clasts of volcanic rocks, and calc-silicate (meta-marl) lenses, named. U-Pb data for detrital zircons from greywacke samples suggest a maximum sedimentation age of 588±24 Ma for the Salinas Formation (revised age by Pedrosa-Soares et al., 2008, after data from Lima et al., 2002). Furthermore, zircons from cobbles of felsic volcanic rocks of the Salinas conglomerates yielded U-Pb magmatic ages around 630–600 Ma (Pedrosa-Soares et al., unpublished data). As the magmatic arc of the Araçuaí orogen is the most likely source for these volcanic clasts, as well as for the ca. 588 Ma detrital zircons, the Salinas Formation is considered to be an orogenic sedimentary pile (Lima et al., 2002; Santos, 2007; Pedrosa-Soares et al., 2008). The Salinas Formation records the regional deformation of the Araçuaí orogen, but in low-strain zones, well-preserved sedimentary features, such as convoluted bedding and intraformational breccias suggesting sedimentation in a tectonically active marine environment (Lima et al., 2002; Santos, 2007).

The Capelinha Formation is a metamorphosed sandstone succession, with intercalations of pelite and detrital iron formation and is also interpreted to be sourced from the adjacent orogen (Queiroga et al., 2006; Pedrosa-Soares et al., 2008).

The G2 suite represents the syn-collisional stage (ca. 585–560 Ma). It mainly consists of foliated, S-type, biotite±garnet±cordierite±sillimanite granites and two-mica granites, with common mylonitic features. Xenoliths and roof pendants of metasedimentary rocks in several stages of assimilation and of variable size are abundant (Celino et al., 2000; Nalini-Junior et al., 2000a,b; Pedrosa-Soares and Wiedemann-Leonardos, 2000b; Pedrosa-Soares et al., 2001, 2006; Campos et al., 2004). U-Pb ages suggest that the main episode of G2 granite generation took place at around 575–560 Ma (Pedrosa-Soares and Wiedemann-Leonardos, 2000b; Campos et al., 2004; Silva et al., 2005; Pedrosa-Soares et al., 2006), although some G2 granites crystallised at around 582 Ma (Nalini-Junior et al., 2000a). Large batholiths dominated by mylonitic garnet–biotite granites reveal well-preserved magmatic features with similar magmatic crystallisation ages (Pedrosa-Soares et al., 2006; Vauchez et al., 2007).

The late collisional stage apparently lasted from ca. 560 to 530 Ma, but these age limits are poorly constrained geochronologically. The S-type G3 suite is late to post-collisional. It consists of garnet- and/or cordierite-rich leucogranite, which generally occur as veins and small intrusions cutting G2 granites. Zircon and monazite U-Pb data from G3 leucogranite samples have yielded ages from ca. 550 to 520 Ma (Whittington et al., 2001; Campos et al., 2004; Silva et al., 2005; Pedrosa-Soares et al., 2006).

The post-collisional stage (530–490 Ma) is related to the gravitational collapse of the Araçuaí orogen (Pedrosa-Soares and Wiedemann-Leonardos, 2000b; Pedrosa-Soares et al., 2001). G4 and G5 are plutonic suites associated with this stage and both lack the regional foliation (Figure 3.13). The G4 suite includes relatively shallow (5–15 km), S-type granitic intrusions with variable contents of biotite, muscovite and/or garnet. They have circular surface outcrops and sometimes form clusters of amalgamated plutons with locally preserved pegmatoidal roofs (Pedrosa-Soares and Wiedemann-Leonardos, 2000b, and references therein). The G5 suite predominantly consists of I-type granitic intrusions, which may include charnockitic, enderbitic and/or mafic portions, and minor mafic bodies with subordinate granitic and/or charnockitic facies. Magma mingling and mixing features are very common. The granitic rocks are generally porphyritic to subporphyritic and have a high-K calc-alkaline signature (Wiedemann et al., 2002; Campos et al., 2004; Martins et al., 2004b).

The evolution of the precursor basin and orogenic stages of the Araçuaí orogen are presented in Table 3.2. This orogenic belt defines the southeastern limit for the São Francisco Craton, but it does not represent a common plate margin orogen. The Araçuaí orogen (together with its counterpart located in Africa, the West-Congo belt) is better characterised as a confined orogen because it developed inside an embayment (like a large gulf or a Red Sea-type basin) carved into the São Francisco–Congo palaeocontinent. Its precursor basin was not completely ensialic but was filled by continental rift (including glaciogenic) and passive margin sediments. The subsequent orogeny emplaced slivers of ophiolite and a pre-collisional magmatic arc, evidence that rifting proceeded to ocean-floor production (Pedrosa-Soares et al., 2001, 2008).

Regional structural studies provide explanations for the tectonic mechanisms underlying the different stages of the Araçuaí orogen (Uhlein, 2004; Alkmim et al., 2006b). The so-called "nutcracker tectonics" model explains this orogen as a series of kinematically distinct deformation stages: (1) early north-verging deformation; (2) development of the west-verging fold-thrust belt against the São Francisco Craton and an east-vergent zone in the back-arc region; (3) lateral escape to the south along dextral strike–slip faults; (4) late-stage gravitational collapse with development of extensional sites and intrusion of post-collisional plutons (Alkmim et al., 2006b).

### 3.2.2.3. The Rio Pardo basin

The Rio Pardo Group is a Neoproterozoic-Early Palaeozoic sedimentary fill of the Rio Pardo basin in the southeastern portion of the São Francisco Craton (Figure 3.14). The first reference to rocks of this group was

**Table 3.2** Evolution of the precursor basin and orogenic stages of the Araçuaí orogen, southeastern Brazil (age references in text).

| Age | Stage; environment and/or process | Stratigraphic unit and/or structural feature |
|---|---|---|
| <900 Ma | Early continental rift; pre-glacial, fluvial | Diamictite-free, lowermost proximal Macaúbas Group |
| ca. 875 Ma | Continental rift; pre-glacial, fluvial to marine | Salto da Divisa anorogenic suite; Diamictite-free proximal Macaúbas Group |
| <875 Ma | Late continental rift; mainly gravitational glacio-marine sedimentation and transitional mafic volcanism | Diamictite-rich and diamictite-turbidite sequences with transitional basalts of the proximal Macaúbas Group |
| <864 Ma | Proximal passive margin; continental shelf, marine transgression | Diamictite-free, sand-pelite unit of the uppermost proximal Macaúbas Group |
| ca. 816 Ma to ? | Distal passive margin and ocean spreading; continental slope to ocean floor | Ribeirão da Folha Formation, Dom Silvério Group and mafic-ultramafic oceanic rocks (ophiolite slivers) |
| > 630 Ma | Start of oceanic lithosphere subduction from west to east | Nucleation of north-verging structures (cf. nutcracker tectonics model; see text) |
| 630–585 Ma | Pre-collisional; continental magmatic arc and related basins | G1 suite, Rio Doce Group, Nova Venécia Complex and Salinas Formation |
| 585–560 Ma | Syn-collisional; fold-and-thrust tectonics, syn-kinematic metamorphism and partial melting | G2 suite and migmatization of other units; west- and east-verging thrust-and-folding with low-angle dip shear zones |
| 560–500 Ma | Late- to post-collisional; adiabatic post-kinematic S-type anatexis, partially coeval to lateral escape to the south | G3 and G4 suites; NNE to NE-trending, high-angle dip, dextral, strike-slip shear zones controlled the lateral escape |
| 520–490 Ma | Post-collisional; gravitational collapse, rising of mantelic magma and partial melting of deep crust | G5 suite; extensional sites along NE and NW trends of lineaments |

made in 1870 (Hartt, 1941) when conglomeratic schist and fine-grained sandstones were described and the impression of a plant resembling *Asterophyllites scutigera* Dawson (previously known from the Upper Devonian of St. John, New Brunswick, Canada) was found. The Baixo Rio Pardo rocks were thus assigned a Lower Palaeozoic age. The interval for deposition of rocks of this group is still poorly constrained; Karmann et al. (1989) suggested a depositional range from 1,100 to 600 Ma.

Until the middle 1960s, studies of this basin were at the reconnaissance level, and the stratigraphy was subdivided into the Rio Pardo and Salobro formations (Oliveira and Leonardos, 1940; Almeida, 1968). Later, the low-grade metasedimentary rocks of the basin were divided into six formations, formalised by Pedreira et al. (1969): the Panelinha, Camacan, Salobro, Água Preta, Serra do Paraíso and Santa Maria formations, with the Rio Pardo Formation, previously used by Almeida (1968), elevated to group status. The "layer cake" stratigraphy was somewhat modified in subsequent studies (Andrade and Nunes, 1974; Siqueira et al., 1978) that placed the Salobro Formation at the top of the group and introduced other minor changes. Karmann et al. (1989) once again revised the stratigraphy and the interpretation of the basin's tectonic evolution, from an aborted-rift, giving way to a thermally subsiding basin, followed by a foreland basin.

A stratigraphic chart for the Rio Pardo Basin and its basement is shown in Figure 3.15. The first column shows the ages (palaeontological or geochronological) of the distinct units; the second, third and fourth columns show the relationships among the lithostratigraphic units (group, subgroup and formations); in the central panel the same relationships are in a two-dimensional display, in a north–south section and, in the last two columns, the possible tectonic evolution and depositional environments for the units.

Rocks of the Rio Pardo Basin were deposited on Archaean basement (Moraes Filho and Lima, 2007) that consists of charnockites, enderbites and granulitic trondhjemites, as well as enderbitic-trondhjemitic orthogneisses and meta-gabbronorite (Ibicaraí Complex). These rocks were intruded by rocks of the Pau Brasil suite, which include charnockites, monzo-diorites, granites, quartz-monzonites and tonalities, the Anuri Syenite, and diabase dykes (Souto et al., 1972). The Itapetinga Complex is composed of syeno-granitic to tonalitic orthogneisses, intruded by anorogenic granitoids that crop out in the southern region of the basin.

The basal unit of the Rio Pardo Group, the Panelinha Formation, is observed as discontinuous outcrops in the northern, northwestern and western borders of the basin and its lower contact is non-conformable on the

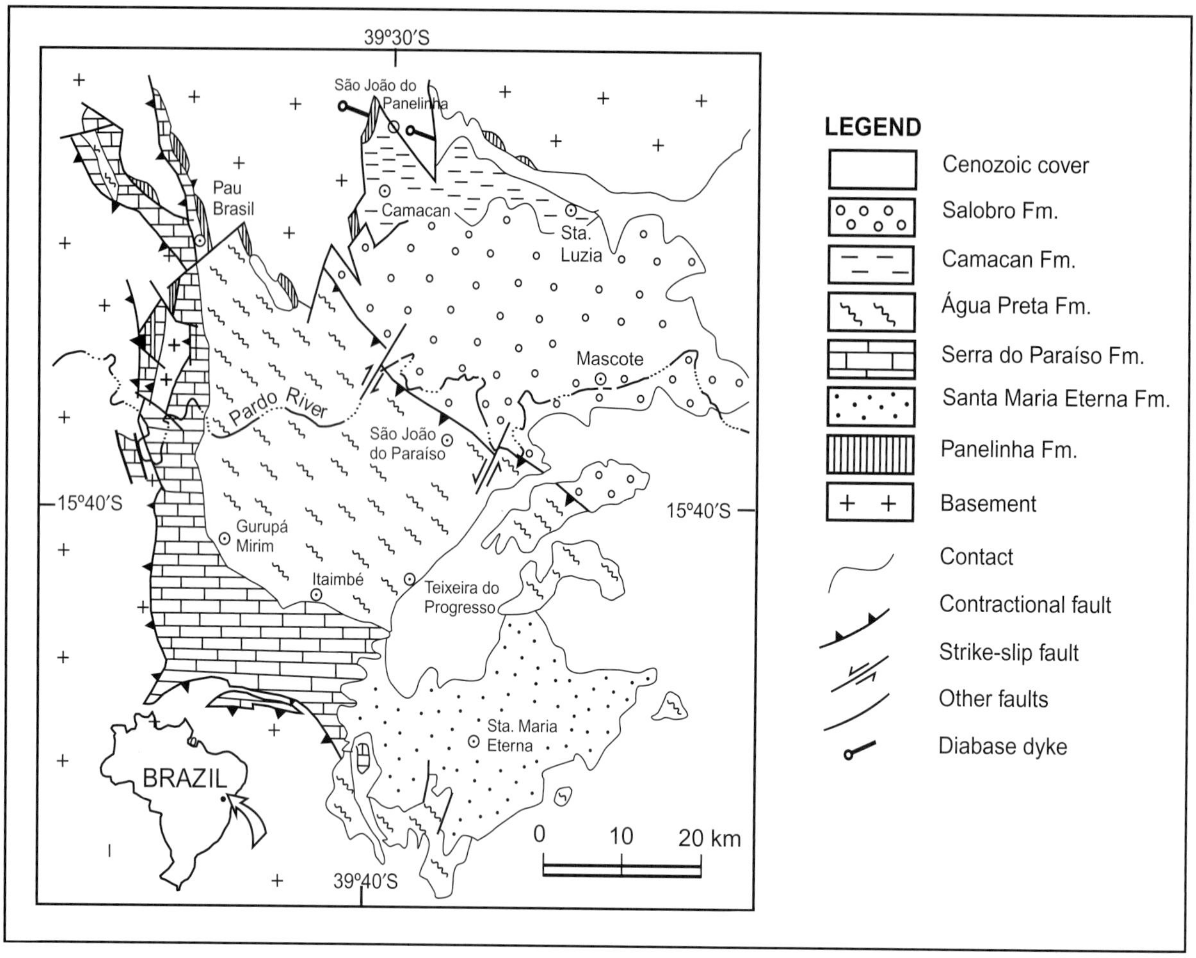


**Figure 3.14** Simplified geological map of the Rio Pardo basin, eastern São Francisco Craton (modified from Pedreira, 1999).

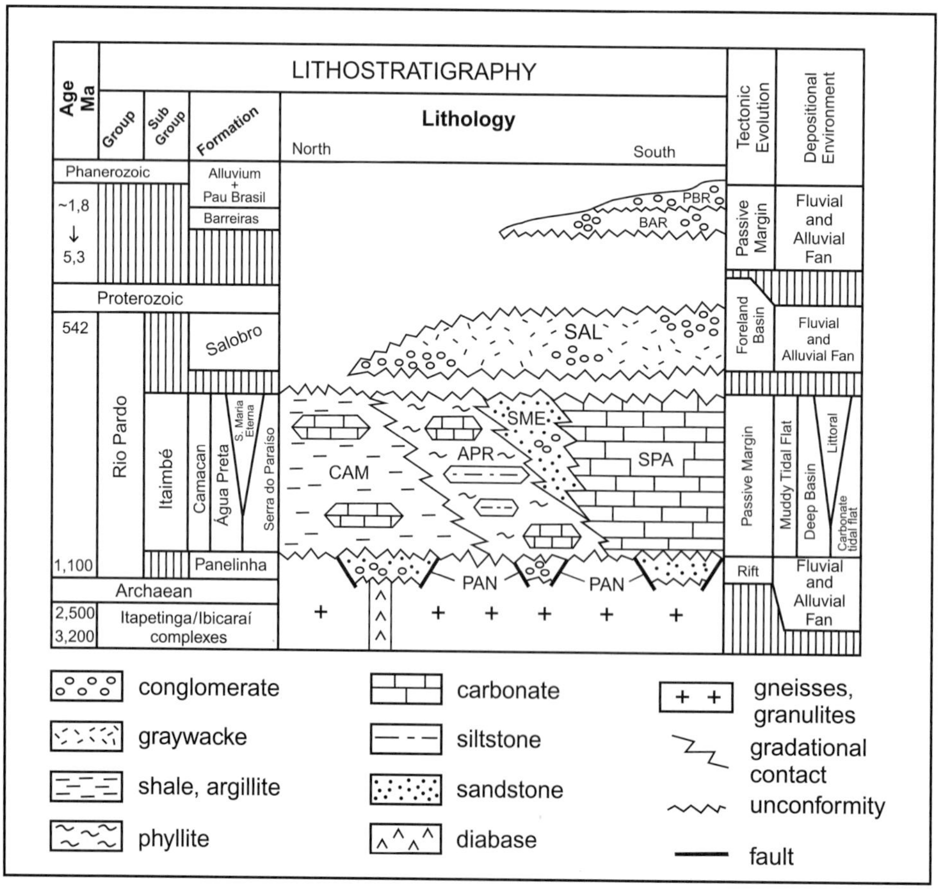


**Figure 3.15** Stratigraphic chart for the Rio Pardo Group (modified from Pedreira, 1999).

crystalline basement. Its upper contact is with sedimentary rocks of the Camacan, Água Preta, Serra do Paraíso and Salobro formations (Figure 3.14). The Panelinha Formation is about 200 m thick (Karmann et al., 1989) and its lower portion is composed of immature breccias with coarse-grained matrix, superposed by polymictic metaconglomerates with pebbles of basement rocks, including syenite gneisses (Pedreira et al., 1969). These coarse clastics grade upwards into massive and laminated metagreywackes with conglomerate intercalations and disseminated pyrite, followed by metarkoses with conglomeratic intercalations; the metarkoses show parallel lamination, cross-bedding and graded bedding.

The Itaimbé Subgroup (Figure 3.14) was proposed by Karmann et al. (1989) to encompass the Camacan, Água Preta, Santa Maria Eterna and Serra do Paraíso Formations, whose contacts are lateral and separated by possible unconformity from the Panelinha Formation and unconformably overlain by the Salobro Formation.

The Camacan Formation (Pedreira et al., 1969) crops out in the northern border of the basin, along a belt between the towns of Camacan and Santa Luzia, and has a thickness of 200 m. The northern limit of the belt is the São Pedro River and the southern one is found in the Chororão, Rochedo and Lapão ranges, in the domain of the Salobro Formation (Figure 3.14). The lower and upper contacts of the Camacan Formation are unconformable, respectively, with the Panelinha and Salobro formations. The Camacan Formation is composed of metasiltstones and slates, argillaceous metalimestones and dark-grey dolomites, pyrite-bearing metagreywackes and fine-grained sandstones. The base of the formation consists of limestones and dolostones with intraclasts, followed by a thick interval where fine-grained sandstones alternate with argillite, shale and slate. Approximately in the middle part of this formation, a bed of greywacke separates this lower succession from an upper one of similar composition. Costa Pinto (1977) suggested a transition from continental environment, as evidenced by evaporation of the base of this formation, to open and then restricted marine, during which time carbonates were deposited.

The Água Preta Formation (Pedreira et al., 1969) crops out in the central part of the basin, in an area limited by the town of Pau Brasil and the villages of Gurupá-Mirim, Itaimbé and Teixeira do Progresso, as well as in erosional windows, east and northeast of Teixeira do Progresso village. The northeastern contact of the Água Preta Formation with the Salobro Formation is represented by a contractional fault (Rio Pardo–Água Preta Fault; Karmann et al., 1989). The contact with the carbonate facies of the Serra do Paraíso Formation is interfingered. The minimum thickness of the Agua Preta Formation is of about 2,300 m.

In the Água Preta Formation, laminated phyllites and metasiltstones predominate, followed by micaceous and calcareous sandstones; subordinate argillaceous and dolomitic metalimestones and metadolostones also occur. This formation contains a few lenses of carbonate rocks southeast of the town of Pau Brasil. Costa Pinto (1977) analysed C and O isotopes in one of these lenses and found isotopic signatures similar to those in the Serra do Paraíso Formation.

The Santa Maria Eterna Formation (Lima et al., 1981) consists of massive dolomitic marbles with intercalations of grey metadolostone and micaceous quartzite. It crops out in the southeastern portion of the basin and is mostly overlain by the Neogene Barreiras Formation. It is in fault contact with the Água Preta Formation and interfingers with the Serra do Paraíso Formation (Karmann et al., 1989). Owing to the lack of continuous outcrops, its thickness has never been estimated.

The Serra do Paraíso Formation (Pedreira et al., 1969) makes up the southern and western borders of the basin. It is bound below by the basement and the Água Preta Formation. The former contact is generally in the form of reverse faults while the latter is gradational and interfingered. In the type-section, south of Itaimbé village, the formation is 1,575 m thick. North of the town of Pau Brasil, this formation consists of metadolomites and calcareous metadolomites. Tepee structures occur close to its base, indicating subaerial exposure. On the road between Pau Brasil and the Pardo River, it consists of micaceous metalimestones, metasiltstones and phyllites. In the region of the Serra do Paraíso–Gurupá Mirim, it is predominantly sand-carbonate metarhythmites, metalimestones and metarenites, while along the BR-101 road, it begins with calc-schists at the base, followed by alternating metalimestones, metadolomites and quartzites.

The Salobro Formation crops out in the northeastern sector of the Rio Pardo Basin, in an area of high relief formed by ranges such as the Chororão, Rochedo, Pacuipe, and Lapão. Its contact to the southwest with Água Preta Formation is a reverse fault, with tectonic transport to northeast (Rio Pardo–Água Preta Fault). The northern contact is a significant unconformity with the Camacan Formation. To northeast, the Salobro Formation is covered by the Barreiras Formation. Karmann et al. (1989) determined a thickness of about 5,000 m.

The Salobro Formation consists of immature metarenites, fine- and coarse-grained greywackes, metarkoses, metasiltstones, slates and metaconglomerates (both clast- and matrix-supported) rhythmically interstratified. Conglomerate lenses occur throughout this formation, the thickest being the Lapão conglomerate, which crops out in the northeastern sector of the formation, east of the town of Santa Luzia, and was formerly washed for diamonds. It is poorly sorted, with clasts whose diameter ranges from 1 to 50 cm; the clasts are from the basement (granulites, syenite gneisses), chert, metadolomite and metalimestone. The other conglomerate lenses have similar composition, with clasts of phyllites and fine-grained quartzites. The Serra do Lapão conglomerate, in the lower Salobro Formation, occurs south of the town of Santa Luzia and comprises metadolostone pebbles with a calcarenite matrix.

The source for the Salobro Formation sediments was represented by two areas of high topographic relief (granulites and syenites from the north and sedimentary rocks from the south). The burial must have been fast, as indicated by the immaturity of the sediments. The interfingering facies of these conglomerates occurred on tectonic ramps, as demonstrated elsewhere by Steidtmann and Schmitt (1988).

The sedimentation in the Rio Pardo Basin started with the deposition of alluvial fans in graben-type basins, limited by extensional faults. The alluvial fans were formed by the conglomerates and breccias of the Panelinha Formation (Figure 3.14). Distally, these fans grade to fluvial systems (Figure 3.15) with abandoned channels and swampy areas evidenced by horizons of conglomerate intercalated with pyritic greywackes in the upper part of the formation.

Rifting ceased before development of oceanic crust, and the Camacan, Água Preta, Serra do Paraíso and Santa Maria Eterna formations (Itaimbé Subgroup) were deposited on the Panelinha Formation under marine conditions during the subsequent thermal subsidence phase. In the deeper parts of the basin, turbidites of the Água Preta Formation were deposited. Carbonate lenses (Mutuns creek, Nancy, Gurupá Mirim and others) in gradational contact with these turbidites were probably deposited in shallower environments.

The Água Preta Formation grades into Camacan, Serra do Paraíso and Santa Maria Eterna formations towards the shallower parts of the basin. Evidence for shallow deposition is found in all of these units: ripple marks and features that indicate subaerial exposure, such as intraclasts and mud cracks, occur in the Camacan Formation; intraclasts and tepees are found in the Santa Maria Eterna Formation; and stromatolites at the base of the Serra do Paraíso Formation indicate deposition under inter- to subtidal conditions. These formations can be interpreted, respectively, to have been deposited in muddy tidal flat, littoral and carbonate platform environments (Figure 3.15). The balance between the siliciclastics influx and the shape of the coastline can account for the distinct facies within these formations. Facies variability is most pronounced in the Serra do Paraíso Formation, whose depositional environment grades from outer platform to supratidal. Delgado et al. (2003) interpreted these formations as deposited on a passive margin during a retrograding depositional cycle.

After lithification of the limestones, sands and clays, tectonic reactivation uplifted both the basement north of the basin and the sedimentary rocks south of the Rio Pardo–Água Preta Fault. Both are source areas for the sediments of the Salobro Formation, as indicated by the pebble composition of conglomerates in the unit.

This brief description of the Rio Pardo Group demonstrates that the sedimentary environment and tectonic setting during deposition of the Rio Pardo Group sediments evolved from an intracontinental rift through a phase of thermal subsidence and into a foreland basin.

### 3.2.2.4. The Sergipano belt

The Sergipano is a Brasiliano/Pan-African fold-and-thrust belt (Figures 3.11–3.13) located on the northeastern edge of the São Francisco Craton. The structure and lithology of this belt were compared to the Ndjolé Series of northern Gabon by Allard and Hurst (1969). Later, Cordani (1973) proposed a comparison with the sequences of Mbalmayo-Bengbis, Dja and Sembe Ouesso of southern Cameroon and the extreme north of the Congo. According to Trompette (1994), the Sergipano belt represents the Brazilian continuation of the Oubanguides and, together, they form a roughly E-W elongated mega-orogen more than 5,000 km long. More recently, the Sergipano belt has been correlated with the Yaoundé belt (Cameroon, Africa) by Oliveira et al. (2006) based on palaeocontinent reconstructions, lateral variation of rock units and tectonic histories.

This belt was formed by continental collision between the Congo–São Francisco Craton and the Pernambuco-Alagoas massif during the Brasiliano/Pan-African orogeny (Brito Neves et al., 1977). Initially, it was interpreted as a geosyncline (Humphrey and Allard, 1968; Silva Filho and Brito Neves, 1979) and later as a collage of lithostratigraphic domains (Davison and Santos, 1989; Silva Filho, 1998) or as a Neoproterozoic fold-and-thrust belt produced by inversion of a passive margin basin located at the northeastern edge of the ancient São Francisco plate (D'el Rey Silva, 1999).

The São Francisco Craton, to the south of the Sergipano belt, is a granite-greenstone belt sequence with high-grade terranes with ages ranging from 3,400 to 2,080 Ma (Oliveira et al., 2006). The Pernambuco-Alagoas massif, to the north of Sergipano belt, comprises Palaeoproterozoic to Mesoproterozoic high-grade gneisses and migmatites intruded by Neoproterozoic granitoids (Brito Neves et al., 1982; Silva Filho et al., 2002). The Sergipano belt consists, from north to south, of six lithostratigraphic domains (Figures 3.16–3.18) separated by major shear zones: Canindé, Poço Redondo, Marancó, Macururé, Vaza Barris and Estância (Santos and Souza, 1988; Davison and Santos, 1989; Silva Filho, 1998). The Macururé, Vaza Barris and Estância are composed mostly of metasedimentary rocks with metamorphic grade, varying from unmetamorphosed in the Estância domain, through greenschist grade in the Vaza Barris, to amphibolite facies in the Macururé domain.

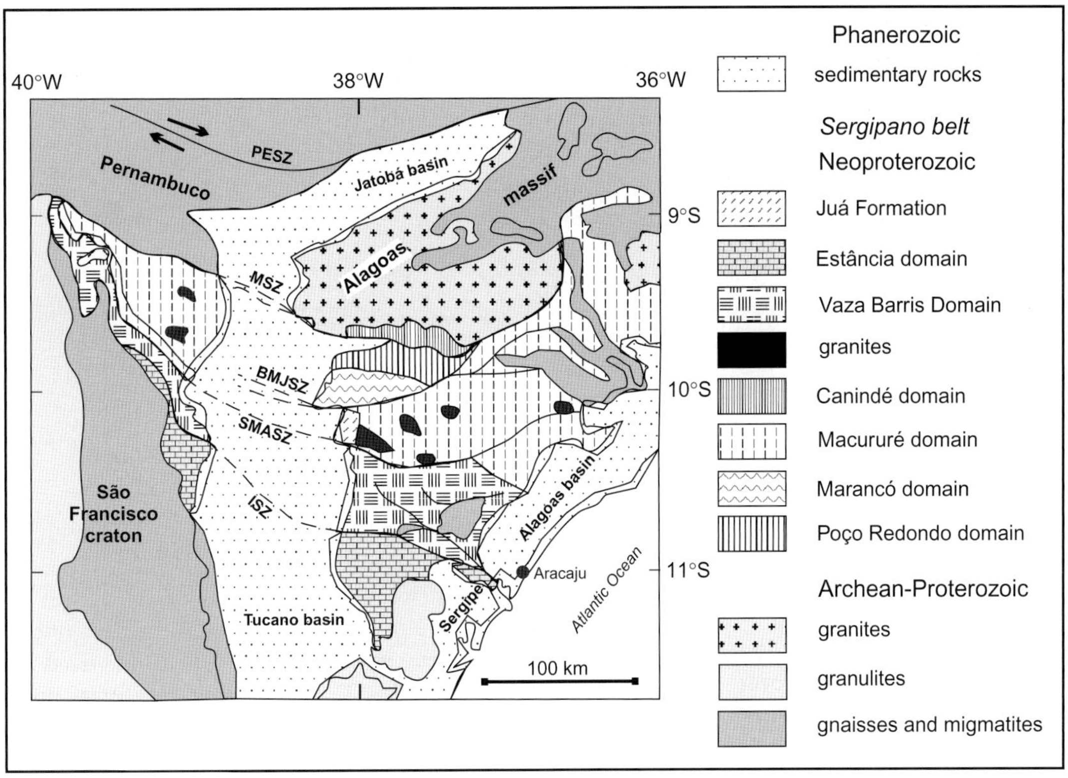


**Figure 3.16** Domains of the Sergipano belt (modified from D'el Rey Silva, 1995). MSZ, BMJSZ, SMASZ and ISZ represent, respectively, the Macururé, Belo Monte-Jeremoabo, São Miguel do Aleixo and Itaporanga shear zones.

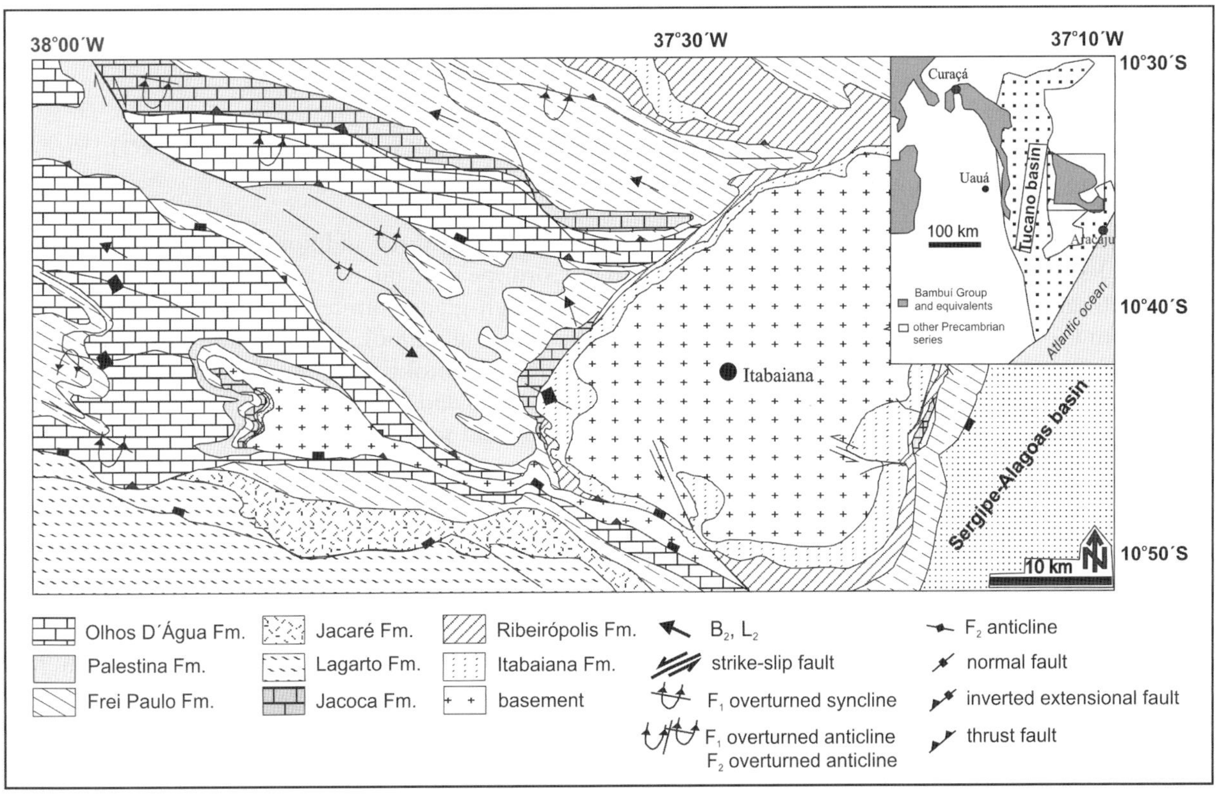


**Figure 3.17** Simplified geological map of the eastern portion of the Sergipano belt (modified from D'el Rey Silva, 1999).

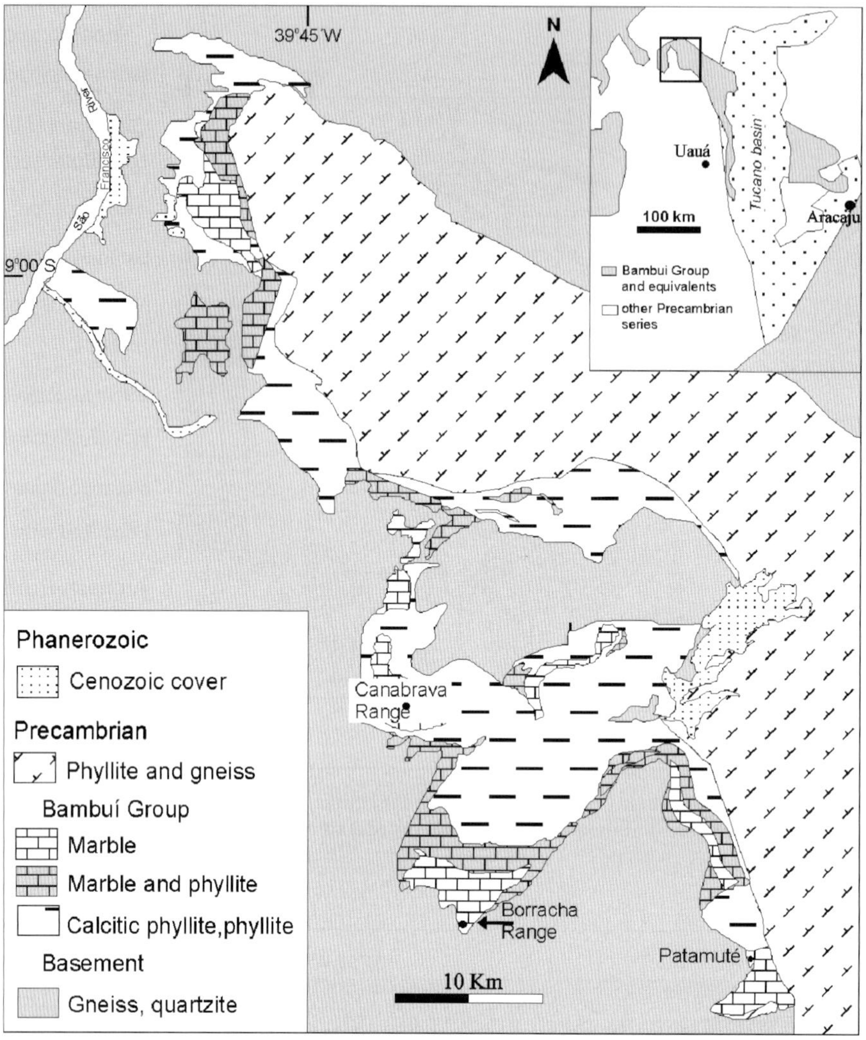


**Figure 3.18** Simplified geological map of the western portion of the Sergipano belt (modified from Jordan, 1973).

Silva Filho and Torres (2002) and Silva Filho et al. (2003) suggested three additional domains: Rio Coruripe, Viçosa and Pernambuco-Alagoas. It is possible that the Rio Coruripe domain, metamorphosed to granulite facies and later retrograded to amphibolite and greenschist grade, is a high-grade counterpart of the Macururé domain, judging from the lateral continuity of similar rock types (Oliveira et al., 2006). All these domains, except for the Estância and Vaza Barris, were intruded by Brasiliano/Pan-African granitoids. Additional descriptions of all of the domains in the Sergipano can be found in Davison and Santos (1989) and Santos et al. (1998), and a summary on the geochronological data, including recent geochronology, is found in Oliveira et al. (2006).

The Canindé domain (Figure 3.16) comprises (a) an elongated pink granite sheet (Garrote unit); (b) a metavolcano-sedimentary sequence (Novo Gosto unit) represented by fine-grained amphibolite, marble, graphite schist, micaschist and metagreywacke; (c) subvolcanic microgabbro-quartz diorite complex (Gentileza unit); (d) the Canindé gabbro-leucogabbro complex (gabbro, gabbronorite, peridotite and pegmatitic gabbro) and (e) granitic plutons (tonalites, granodiorites and quartz-syenites). The ages of the rocks in this domain range from about to 715 to 609 Ma (Silva Filho et al., 1997; Nascimento et al., 2005; Oliveira et al., 2006).

The Poço Redondo domain (Figure 3.16) is composed of biotite gneisses, migmatites and granitic intrusions, such as the Serra Negra augen-gneisses, the Sitios Novos granodiorite and leucogranite sheets. Grey gneisses from the palaeosome of migmatites yielded ages slightly older (960–980 Ma U-Pb SHRIMP ages; Carvalho et al., 2005) than the Serra Negra augen-gneisses (952 Ma U-Pb SHRIMP age; Carvalho et al., 2005). The Sitios Novos granodiorite has been dated at $651 \pm 11$ Ma (U-Pb TIMS; Carvalho et al., 2005). The Marancó domain is mainly formed of a metavolcano-sedimentary sequence (quartzite, conglomerate, micaschist, phyllites and lenses of andesite, dacite and quartz porphyry) with peridotite and amphibolite lenses (Figure 3.16). The presence of andesites and dacites, dated at about 603 Ma (Carvalho et al., 2005) and conformably interleaved with phyllites, and of granodiorite stocks dated at $595 \pm 11$ Ma (Silva Filho et al., 1997), points to an Ediacaran

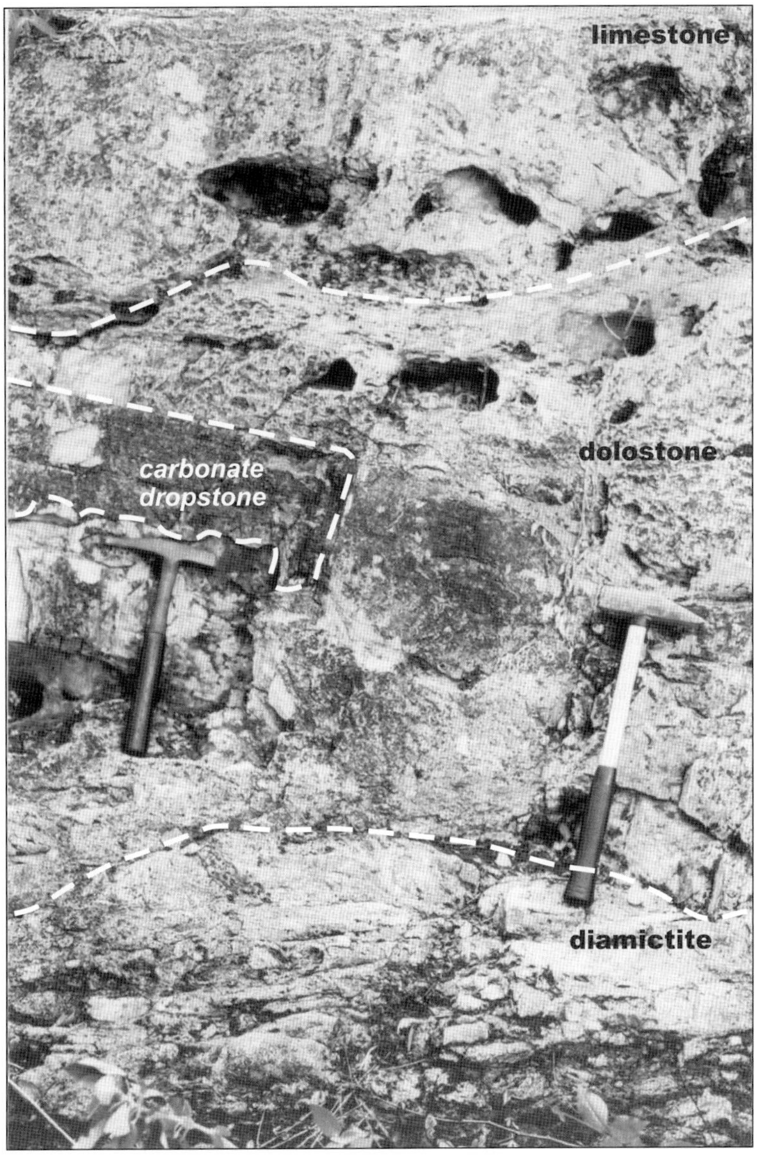


**Figure 3.19** Diamictite of the Juetê Formation, overlain by carbonates of the Acauã Formation, at Serra da Borracha, about 20 km west from Patamuté village, state of Bahia.

depositional age. The Macururé domain consists mainly of garnet micaschists and phyllites (Figure 3.16) with some quartzite and marble. These metasedimentary rocks were intruded by granitic plutons dated at about 625–630 Ma (Bueno et al., 2005; Long et al., 2005).

The Vaza Barris domain (Figure 3.16) shows two cycles of sedimentation, both with a continental to shallow-marine, basal siliciclastic megasequence, overlain by a carbonate sequence (D'el Rey Silva, 1995, 1999). The lower siliciclastic megasequence is represented by the Juetê (sandstone, diamictite; Figure 3.19), Ribeirópolis (silty phyllites, metagreywackes, pebbly phyllites, diamictites; Figure 3.20) and Itabaiana (conglomerate, quartzite, metasiltite) formations. The lower carbonate megasequence is known as the Jacoca Formation, which is stratigraphically equivalent to the Acauã Formation. These two megasequences form the Estância-Miaba Group. The upper siliciclastic megasequence (Simão Dias Group) is represented by the Lagarto-Palmares, the Jacaré and the Frei Paulo formations and is overlain by the Vaza Barris Group that comprises diamictites of the Palestina Formation and the upper carbonate megasequence, the Olhos D'Água Formation. All of these rocks spread continuously across the craton margin into the Sergipano belt where they occur around the Itabaiana and Simão Dias basement domes and are overlain by a metadiamictite and metacarbonate belonging to the upper megasequence. According to D'el Rey Silva (1999), both the basement and cover experience the same Neoproterozoic compressive deformation under subgreenschist facies.

The thick Olhos D'Água Formation is composed of a sequence of marbles and interbedded green, calcareous chlorite schists and silty phyllites; it overlies diamictites and pebbly metagreywackes of the Palestina Formation

**Figure 3.20** Diamictite of the Ribeiropolis Formation in sharp contact with dolostones of the Jacoca Formation, Capitão Farm, state of Sergipe. Dolostone shows wavy contact and some bedding is observed in the diamictite. Some clasts seem to be concentrated and salient at the top of diamictite, which suggests lithification and erosion prior to deposition of the dolostone.

**Figure 3.21** Pebbly metagreywacke of the Palestina Formation, sometimes with dropstones (about 12 km from Rosario village, state of Bahia).

(Figure 3.21). Marble beds are interbedded with blue to black, fine-grained metalimestone and grey metadolostone. The thick carbonates around the Simão Dias dome (Figure 3.17) pass upward into supertidal–intertidal facies with oolites, and wave-reworked structures indicating a near-shore environment (D'el Rey Silva, 1995). The Jacoca Formation (older) and Olhos D'Água Formation (younger) carbonates have been interpreted as cap carbonates by Sial et al. (2000a, 2006).

An acidic tuff within the Ribeiropolis diamictites in an outcrop on the Ribeiropolis-Gloria road, 7 km from Ribeiropolis town, was dated at around 730 Ma (zircon, U-Pb; B.B. Brito Neves, written communication), suggesting that the deposition of the overlying Jacoca carbonates happened right after the Sturtian glaciation event. The youngest U-Pb age of zircon grains (615 Ma) from the Frei Paulo metagreywackes above the Olhos D'Água carbonates constrains the deposition of Jacoca and Olhos D'Água to the interval 730–615 Ma.

The Estância domain (Figure 3.16) comprises, from base to top, sandstones and argillites of the Juetê Formation, dolostones and limestones of the Acauã Formation, sandstones and conglomerate lenses of the Largato Formation, and sandstone and minor conglomerate lenses of the Palmares Formation (Silva Filho and Brito Neves,

1979). Although D'el Rey-Silva (1999) assumed that deposition in the Estância and Vaza Barris domains occurred during two cycles of sedimentation on a passive continental margin of the ancient São Francisco plate (detritus sources from the São Francisco Craton), Oliveira et al. (2006) proposed instead deposition in foreland basins, based on detrital zircon populations with ages between 570 and 657 Ma and an inferred provenance from the Sergipano belt and Borborema province to the north.

### 3.2.2.5. The Riacho do Pontal and Rio Preto belts

The Riacho do Pontal and Rio Preto belts border the northern portion of the São Francisco Craton (Figure 3.1) and are part of the southern portion of the Borborema Province (Brito Neves, 1983; Brito Neves et al., 2000). The Riacho do Pontal belt consists of the southern frontal zone of the Borborema Province that overthrusts directly onto the basement of the São Francisco Craton and represents a subautochthonous metasedimentary sequence (Jardim de Sá et al., 1992; Oliveira, 1998a). The Rio Preto belt is located at the northwestern border of the São Francisco Craton and is interpreted as an intracontinental belt (Egydio-Silva, 1987; Egydio-Silva et al., 1990; Andrade Filho et al., 1994; Trompette, 1994).

The Riacho do Pontal belt stretches from the southern region of the states of Piaui and Pernambuco to the northern region of the state of Bahia. This belt can be divided into four zones or structural domains, from south to north: (a) cratonic domain, (b) southern zone, (c) central zone and (d) internal or northern zone (Figure 3.22). The cratonic domain consists of biotite-hornblende gneisses, migmatites, gneissic batholiths of granodioritic-tonalitic composition and of patches of metasedimentary rocks. The rocks in this domain are Archaean to Palaeoproterozoic in age, but affected by the Pan-African/Brasiliano cycle. The boundary between the Riacho do Pontal belt and the São Francisco Craton is somewhat unclear in this domain. A southward thrust overriding the craton was part of a Pan-African/Brasiliano tectonic event that also deformed carbonates of the Una Group (Alkmim et al., 1993; Brito Neves, 1983).

The southern zone is represented by a thrust-fold belt with southern vergence onto the São Francisco Craton (Angelim, 1988; Gomes, 1990; Jardim de Sá et al., 1992; Oliveira, 1998a,b). The rocks that form the Riacho do Pontal belt are variably referred as the Salgueiro or Casa Nova groups, depending on the location. These rocks are terrigenous-shelf metasedimentary rocks represented by garnet micaschists, quartzites, marbles and gneisses. Deep-sea, flysch-like metagreywackes with carbone and/or feldspathic clasts are regarded as turbidites. Syn-tectonic granitoids are interwoven into the metasedimentary rocks, as are foliated sheets of two-mica monzogranitic tonalites, known as Rajada orthogneisses. The syn-tectonic Serra da Boa Esperança granite and quartz-syenite yielded an age of $555 \pm 10$ Ma (Rb-Sr, Pb-Pb; Jardim de Sá et al., 1996). The WSW-ENE foliation shows NNW-SSE stretching lineation and sheath folds with axes subparallel to that lineation and resulted from the same event that has generated the nappes. The thrust deformation induced a decollement of the metasedimentary sequence over the cratonic massif (Jardim de Sá et al., 1992).

The central zone consists of metabasalts, amphibolites, chlorite-, talc-, tremolite- and actinolite-bearing rocks regarded as ultramafic intercalations in schists, and metachert. This zone contains the Monte Orebe and Brejo Seco mafic–ultramafic massifs with metasedimentary intercalations (Marimon, 1990; Moraes, 1992). The mafic rocks are Neoproterozoic tholeiitic ocean-floor basalts (Moraes, 1992), believed to be an ophiolitic complex.

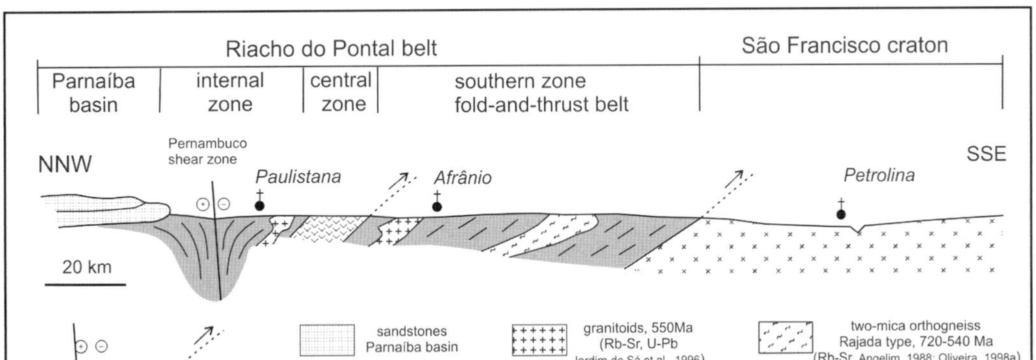


**Figure 3.22** Cross-section for the Riacho do Pontal belt that can be divided into four zones or structural domains, from south to north: (a) cratonic domain, (b) southern zone, (c) central zone and (d) internal or northern zone (Uhlein et al., 2008).

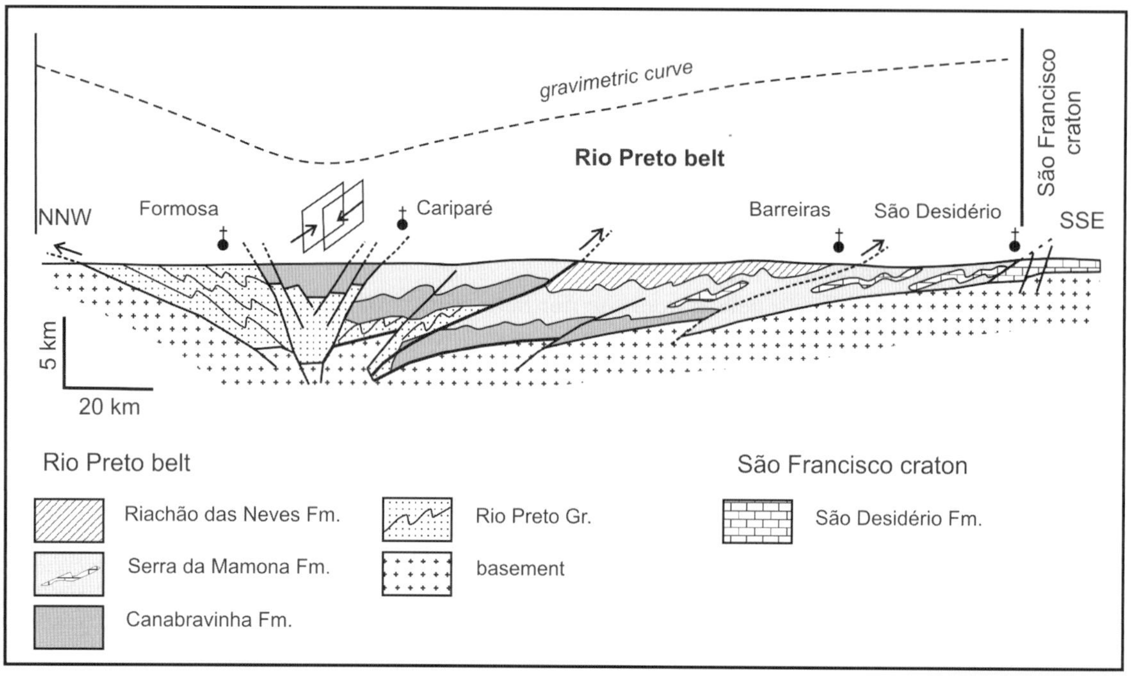


**Figure 3.23** Cross-section for the Rio Preto belt (Egydio-Silva, 1987).

The internal zone is bounded by ductile right-lateral shear zone and corresponds to a tectonically complex region where subvertical mylonitic foliation shows recumbent S structures with right-lateral movements. This zone consists of remnants of Brasiliano metasedimentary sequences, intrusive granitoids and slices from the basement that form stretched amygdaloidal bodies. The mylonitic zone is up to 1 km thick with proto-mylonites through ultramylonites (Vauchez and Egydio-Silva, 1992; Vauchez et al., 1995).

The Rio Preto belt is considered a marginal fold belt forming a westward extension of the Riacho do Pontal and Sergipano belts. This belt is of great structural and lithostratigraphic interest because it displays a transition between cratonic facies, in the south, to metasedimentary rocks at the central part, and gneisses of the basement towards the northern border. The metasedimentary sequences are equivalent to the São Francisco Supergroup in the state of Bahia.

From south to north, the Rio Preto belt can be subdivided into three structural units, each with distinct lithologic, structural and metamorphic characteristics, separated by disrupted structures (Figure 3.23). In the southern unit, in the portion of the São Francisco Craton and adjacent to this belt (the São Desidério region), the sequence begins with microcrystalline limestone with argillaceous intercalations, characterised by 50 m-thick horizontal layers, and shows no signs of metamorphism or deformation (cratonic domain). This unit was named the São Desidério Formation (Egydio-Silva, 1987; Egydio-Silva et al., 1989) and probably correlates with the Sete Lagoas Formation of the Bambuí Group in Minas Gerais (Figure 3.24).

To the north, cratonward from the fold belt, these limestones are overlain by clastic rocks with intercalations of marls and limestones. This sequence (Serra da Mamona Formation) has been correlated with the Santa Helena Formation (Egydio-Silva, 1987; Egydio-Silva et al., 1989). North of Barreiras, they are overlain by meta-arkoses, fine- to medium-grained quartzites, locally feldspathic, metagreywackes and rare carbonate intercalations. This sequence (Riachão das Neves Formation) is considered partially equivalent to the Três Marias Formation of the Bambuí Group (Egydio-Silva, 1987; Trompette, 1994).

In the central portion of the belt, the lower part of the succession is more complete with the occurrence of boulder diamictites, with clasts up to 50 cm in diameter of gneiss, quartzite, metasiltstone, marble and schist. Referred here as the Canabravinha Formation (Figure 3.25), this unit has been correlated to the Bebedouro Formation, although a glaciogenic origin remains unsubstantiated. Further north, in the region of Formosa do Rio Preto (Figure 3.23), the Rio Preto Group contains predominantly micaceous quartzites and schists with a few intercalations of amphibolites and itabirites. It has been correlated with the Chapada Diamantina Group (Inda and Barbosa, 1978).

The general structure of the Rio Preto belt is of an asymmetric fan, probably associated with late ENE-WSW shearing that also occurred in the Borborema Province. Structural analysis of this belt defined three structural units (Egydio-Silva, 1987; Egydio-Silva et al., 1990; Andrade Filho et al., 1994). The central and southern ones, well developed between the craton and the Cariparé thrust, show a single deformation phase with southern vergence

**Figure 3.24** Limestones of the São Desidério Formation near São Desidério, Rio Preto belt, probably correlated with the Sete Lagoas Formation of the Bambuí Group.

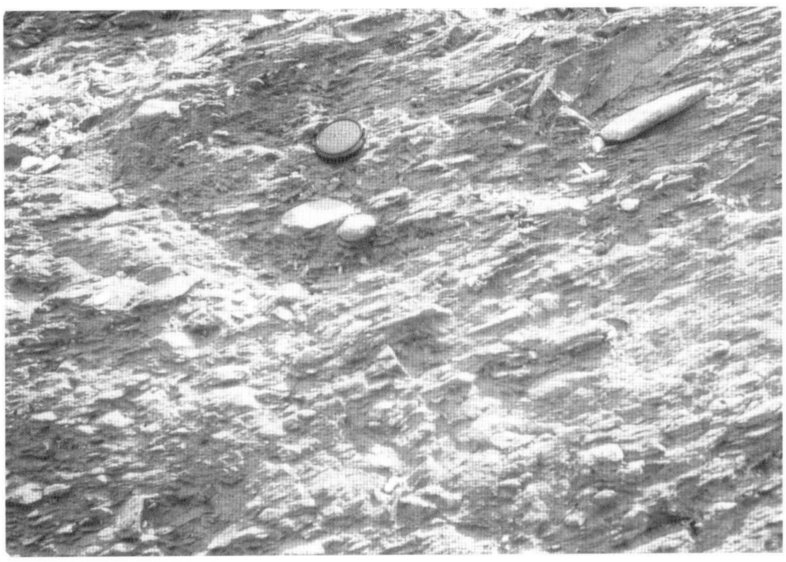

**Figure 3.25** Diamictites of the Canabravinha Formation, Rio Preto belt.

towards the São Francisco Craton. The northern unit is metamorphosed to the greenschist facies, is polyphase and is thrust northwards over the Piauí basement. The absence of granitic intrusions and rare volcanism favour an ensialic origin for this belt (Egydio-Silva, 1987; Trompette, 1994).

The Riacho do Pontal belt appears to have participated in the collision between the São Francisco Craton and the Pernambuco-Alagoas massif. The basin can be schematically reconstructed with terrigenous-shelf sediments and deep-sea sediments (turbidites) in a passive margin context. Mafic–ultramafic massifs (Monte Orebe and Brejo Seco) may be remnants of a narrow oceanic crust-floored basin (Oliveira, 1998a,b). The Brasiliano collisional deformation is complex, with thrusts towards the south over the São Francisco Craton associated with greenschist to amphibolite metamorphism (Jardim de Sá et al., 1992). In the internal zone, a continental-scale subvertical dextral ductile shear zone is the main structural feature.

The Rio Preto belt is a good example of an intracontinental fold belt whose central structure seems to have been controlled by the geometry of a Mesoproterozoic trough and by the intervention of late shears. The sedimentary basin, as reconstructed, is composite, with a northern trough of Mesoproterozoic age filled by the Rio Preto Group and juxtaposed against a younger subsiding southern basin. In this younger basin, a mostly clastic and thickened equivalent of the late Proterozoic Una Supergroup was deposited. The migration of the depocenter of the basins towards the south implies the intervention, before deposition of the Bebedouro Group, of epeirogeny uplifting northern highlands against which the Neoproterozoic basin was juxtaposed (Egydio-Silva, 1987; Trompette, 1994; Andrade Filho et al., 1994).

## 3.3. Isotope Chemostratigraphy

C and O isotope studies were performed across different sections of some sequences in the southwestern portion of the São Francisco Craton (Santos et al., 2000, 2004; Alvarenga et al., 2007). The most complete profile is at the Serra de São Domingos region, where over 700 m of Bambuí Group sediments was deposited atop siliciclastic rocks of the Paranoá Group. These two groups are separated by a karstic limestone breccia infilled with well-sorted sand grains that may be the lateral equivalent of Jequitaí Formation diamictites. In the base of this section, dolostones of the upper Paranoá Group have $\delta^{13}C$ values close to 0‰ (VPDB), similar to values observed in other areas where this unit is exposed. The Paranoá–Bambuí transition is characterised by a spike of low $\delta^{13}C$ values ($\sim -5$‰), followed by limestones with slightly positive values. The uppermost dolostones of the Conselheiro Mata Group (Espinhaço Supergroup) and dolostones of the Rio Pardo Grande Formation (lowermost portion of the Macaúbas Group) display $\delta^{13}C$ values around 0‰ (Figure 3.26; Santos et al., 2000), typical of Mesoproterozoic to early Neoproterozoic carbonates. The most remarkable aspect of these profiles is a positive $\delta^{13}C$ excursion in the upper Sete Lagoas Formation that continues through the Serra de Santa Helena and Lagoa do Jacaré Formation limestones (Figure 3.26). This excursion, identified across different profiles of the northern and southern reaches of the Bambuí basin and used to reconstruct the palaeogeography of the Bambui basin (Santos et al., 2004), is a robust stratigraphic marker related to a Neoproterozoic positive carbon isotope excursion recognised worldwide.

A sedimentological and C–O isotope study carried out by Vieira et al. (2007) in nine sections of the Sete Lagoas cap carbonate sequence at its classical outcropping area, in the southern tip of the São Francisco Craton, helped refine its stratigraphy. According to these authors, the Sete Lagoas Formation comprises two shallowing-upward megacycles, each cycle bound by a flooding surface and amalgamated with a third-order sequence boundary. The first megacycle is represented by deep-platform deposits with abundant crystal fans (aragonite pesudomorphs) that show negative $\delta^{13}C$ values ($-4.5$‰) that grade upward to storm wave and tide-influenced layers with values around 0‰. A second megacycle is characterised by a thick, mixed substorm wave-base succession deposited during another transgression that drowned the platform, with lime mudstone–pelite rhythmite that grades to crystalline limestone rich in organic matter with $\delta^{13}C$ values up to $+14$‰.

Vieira et al. (2007) proposed that the very high $\delta^{13}C$ values observed in the second megacycle, coupled with geochronological data, support correlation with post-Sturtian sequences. Some differences in the depositional record for the Sete Lagoas and other post-Sturtian units described in North America, Australia and Namibia are perhaps due to the deposition of the Sete Lagoas carbonates in shallower settings, preserving a thick record of storm- and wave-influenced sedimentation, not found elsewhere. Alternatively, these differences may be ascribed to diachronous deposition of the post-Sturtian cap carbonate sequences.

Chemostratigraphic studies of the Una Group have been carried out by Torquato and Misi (1977), Misi and Kyle (1994) and Misi and Veizer (1998). Sulphur, carbon and oxygen isotope analysis, as well as Sr isotope determinations, confirmed the evolution of the sedimentary sequences as proposed by Misi and Souto (1974) and by Misi (1979). In the study by Misi and Veizer (1998), samples for isotope analysis were collected along the eastern border of the Irecê "basin" covering the stratigraphic section of carbonate lithofacies from the laminated limestones of Unit B to the black oolitic limestones of Unit A1. Based on petrographic studies and trace element determinations, the best preserved samples (micrites, with Mn/Sr $<0.09$ and Sr $>300$ ppm) were selected for isotope analysis. For the best preserved samples, there is a progressive increase in $\delta^{13}C$ ($+0.3$‰ to $+8.7$‰), decline in $^{87}Sr/^{86}Sr$ (0.70789–0.70745) and a relatively uniform $\delta^{18}O$ signal ($-5$‰ to $-7.5$‰ VPDB) upsection (Figure 3.27).

Previous C and O isotope studies by Torquato and Misi (1977) of 63 whole-rock carbonate samples from the same stratigraphic section demonstrated the same increase in $\delta^{13}C$ values in the upper section (Unit A1, ca. $+8$‰) and showed significant negative shifts in the pinkish dolostones of Unit C, in the lower section (around $-6$‰; $n = 12$). Considering only the best preserved samples and considering the data of Torquato and Misi (1977) from the same section, it is possible to show the variation of the isotopic composition through the Una Group (Figure 3.27).

$^{87}Sr/^{86}Sr$ variation, when compared with the seawater evolution curves through the Neoproterozoic (Jacobsen and Kaufman, 1999), suggests an age of sedimentation between 750 and 600 Ma. $\delta^{13}C$ variation, with a negative shift at the base and a high positive excursion in the upper section, is compatible with that observed in the Bambuí Group (Misi et al., 2007).

A preliminary C and O isotope study on carbonates of the Rio Pardo Group was carried out by Costa Pinto (1977), but as the 120 samples involved lack of stratigraphic control, they are not used in our compilations. Costa

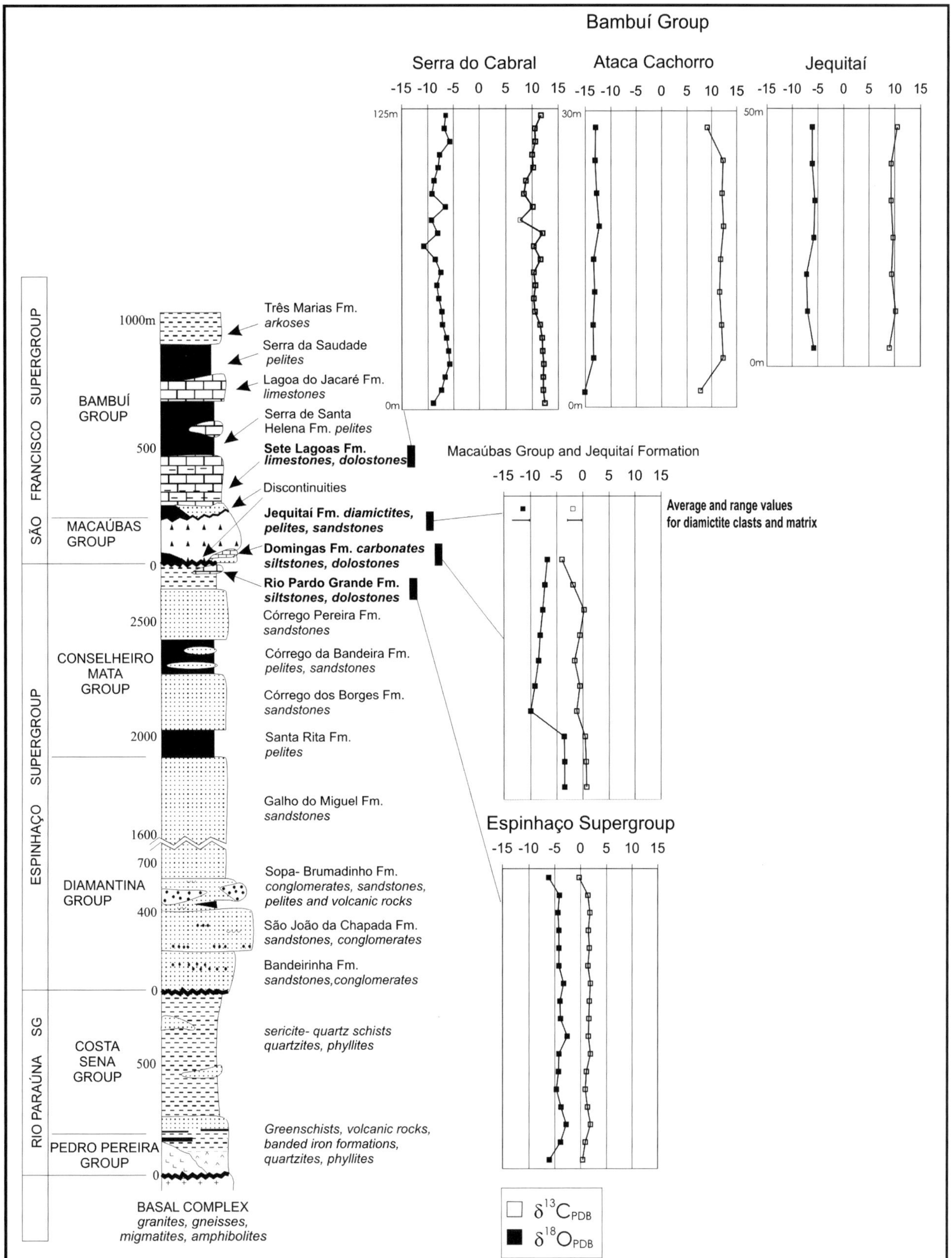


**Figure 3.26** C and O isotope profiles for carbonates of the Bambuí Group at Serra de São Domingos, southeastern portion of the São Francisco Craton and for carbonates of the Domingas Formation (Macaúbas Group) and Rio Pardo Grande Formation of the Espinhaço Supergroup (modified from Santos et al., 2004).

CHAPTER 4.1

# Extension and General Evolution of the Río de la Plata Craton ☆

Jorge Bossi[1] *and* Carlos Cingolani[2]

## Contents



## 4.1.1. Introduction

### 4.1.1.1. Origin, extension and boundaries of the Río de la Plata Cratonic region

Following the fragmentation of the Rodinia supercontinent, Archaean to Mesoproterozoic cratonic blocks were amalgamated to constitute the Gondwana continent during Neoproterozoic-Cambrian times (Brito Neves et al., 1999; Hoffman, 1999; Fuck et al., 2008). One of these continental blocks of western Gondwana is the Río de la Plata Craton (RPC) (Almeida et al., 1973, 1976, 2000; Cordani et al., 2000), which comprises part of southern Brazil, Paraguay, Uruguay and the central-eastern sector of Argentina (Figure 4.1.1). However, the main exposures of the RPC, are located within its eastern border and include the Asunción arch, the Taquarembó Block, and the Nico Pérez, Piedra Alta and Tandilia terranes (Figure 4.1.2). The Nico Pérez Terrane (NPT) is the oldest and includes Palaeoarchaean granite-greenstone sequences (Hartmann et al., 2001).

To the north, the RPC (Figure 4.1.2) is bounded by the southern Amazonian craton (Bolivia-Brazil), whereas to the east it is bounded by the allochthonous Cuchilla Dionisio suspect terrane (Uruguay). The RPC is limited towards west by the Pampia Terrane of Argentina (Ramos, 1988; Rapela et al., 2007). According to geochronological studies of basement rocks (Cingolani and Varela, 1973; Rapela et al., 2003) and based on at least two boreholes in the Claromecó basin (Lesta and Sylwan, 2005), it was deduced that the RPC was not part of the basement of the Sierra de la Ventana fold and thrust belt. It is important to note that low-grade metapelites occurring in the Punta Mogotes borehole have been correlated to similar rocks (Rocha Formation) of the Cuchilla Dionisio Terrane (CDT) in Uruguay (Cingolani and Bonhomme, 1982; Ramos, 1988), showing similar detrital zircon ages (Rapela et al., 2008).

☆ Bossi, J., Cingolani, C., 2009. Extension and general evolution of the Río de la Plata Craton. In: Gaucher, C., Sial, A.N., Halverson, G.P., Frimmel, H.E. (Eds.): Neoproterozoic-Cambrian tectonics, global change and evolution: a focus on southwestern Gondwana. Developments in Precambrian Geology, 16, Elsevier, pp. 73–85.

[1] Facultad de Agronomía, Cátedra de Geología, Garzón 780, Montevideo, Uruguay.
[2] Centro de Investigaciones Geológicas (CIG), Universidad Nacional de La Plata-CONICET, calle 1 n. 644, 1900-La Plata, Argentina.

*Developments in Precambrian Geology*, Volume 16
ISSN 0166-2635, DOI 10.1016/S0166-2635(09)01604-1

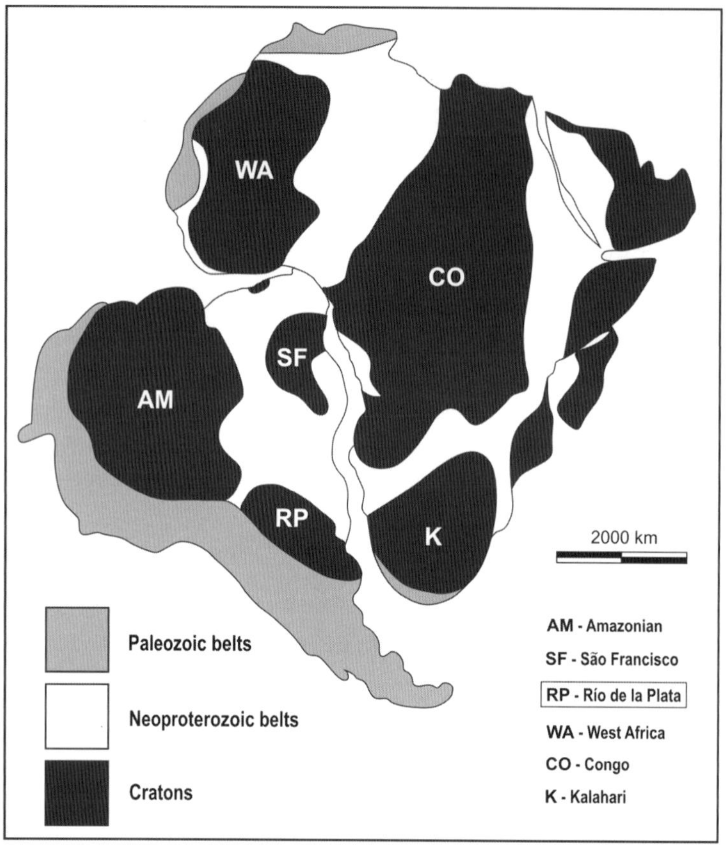


**Figure 4.1.1** Pre-drift configuration of Western Gondwana showing the location of the Río de la Plata Craton.

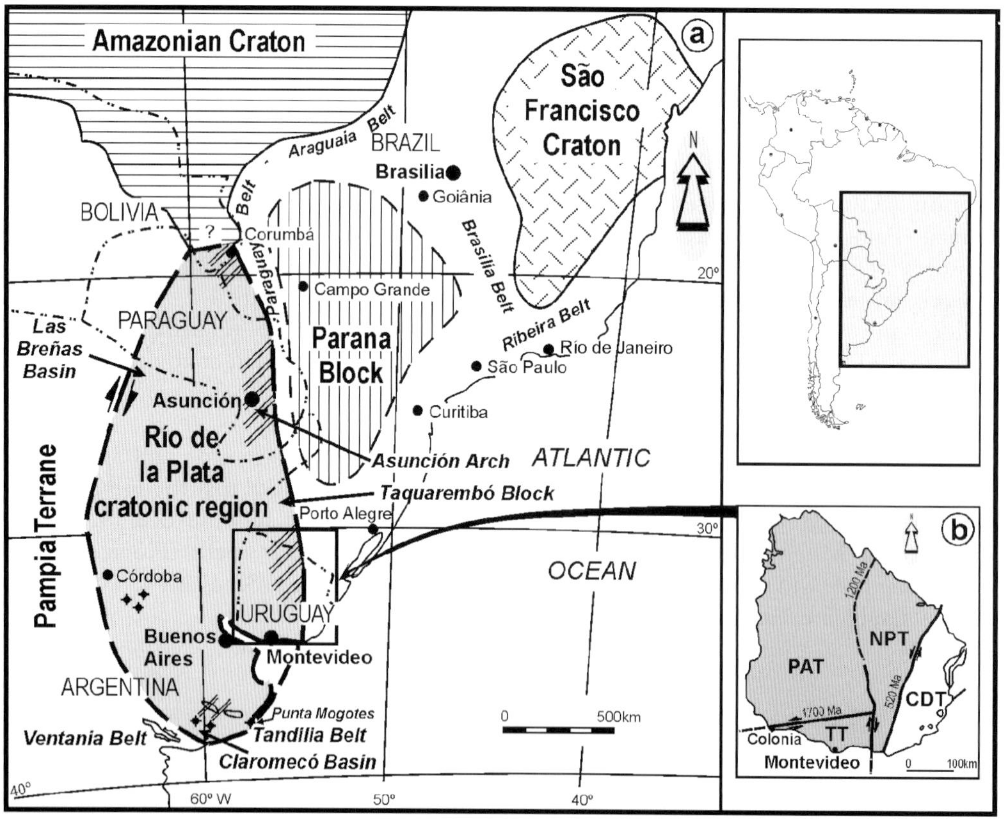


**Figure 4.1.2** (a) Southeastern South American sector showing the inferred limits of the Río de la Plata cratonic region (modified from Gaucher et al., 2003). (b) Detail of the Uruguayan basement, comprising the Piedra Alta (PAT), Tandilia (TT) and Nico Pérez (NPT) terranes as part of Río de la Plata cratonic region, and the exotic Cuchilla Dionisio Terrane (CDT).

The southwestern Pacific margin of Gondwana records a protracted terrane accretion history during Neoproterozoic-Palaeozoic times simplified as 'Terra Australis Orogen' by Cawood (2005).

### 4.1.1.2. Geophysical data

Extensive areas of the RPC are covered by Early Palaeozoic to Cenozoic sedimentary basins, therefore knowledge regarding its subsurface continuation is based on geophysical research and borehole data (Rapela et al., 2007 and references therein). Geophysical studies (Booker et al., 2004) indicate that the boundary between the RPC and the Early Palaeozoic Pampia Terrane could be a complex faulted margin (Ramos et al., 2002; von Gosen and Prozzi, 2005) as a continuation of the Transbrasiliano Shear Zone (Cordani et al., 2001) that was probably active as early as the Late Neoproterozoic. The recognition that longitudinal, dextral, ductile shear zones were associated with magmatism suggests transpressional movements.

Late Proterozoic to Early Palaeozoic palaeomagnetic poles were obtained from the Sierra de Ánimas rift-related magmatic complex (520 Ma), exposed in the Piriápolis area (Uruguay; Sanchez Bettucci and Rapalini, 2002) and at the Sierra de los Barrientos area of Tandilia, Argentina (Rapalini, 2006).

### 4.1.1.3. Neoproterozoic-Lower Palaeozoic sedimentary cover

The north-south extension of the RPC is based on correlation of similar remnants of Neoproterozoic-Lower Palaeozoic sedimentary covers. The main units are: Corumbá Group (Brazil), Itapucumí Group (Paraguay), Las Ventanas Formation and Arroyo del Soldado Group (Uruguay) and Sierras Bayas Group-Cerro Negro Formation (Argentina; Gaucher et al., 2003, 2005b, 2008a; Blanco and Gaucher, 2005; Poiré and Spalletti, 2005). It is noteworthy that westwards of the mentioned platform, the Neoproterozoic (Brasiliano Cycle) belts have not been neither directly observed nor detected using geophysical methods within the inferred Argentinian limits of the RPC.

### 4.1.1.4. Palaeocontinent concept

The 'Río de la Plata palaeocontinent' concept geographically coincides with the RPC as originally proposed by Almeida et al. (1976). On the other hand, Dalziel (1992) used the 'Plata Block' term for his palaeogeographic reconstructions. Other authors continued naming craton of all the regions with rock ages more than $2,000 \pm 100$ Ma that represent a foreland with respect to the Brasiliano mobile belt (500–700 Ma; Dom Feliciano Belt *sensu* Fragoso César et al., 1987; Basei et al., 2006; Oyhantçabal et al., 2006, 2007). New geological, palaeontological, geochemical and geochronological evidence suggest that the term 'palaeocontinent' is the best to be used in order to analyse all the geological information without any preconception, such as a 'stable craton with a mobile belt'.

## 4.1.2. Main Outcrop Areas

### 4.1.2.1. Southern Brazil

As shown in Figure 4.1.2, the southernmost portion of the Brazilian shield (Hartmann et al., 2002a; Gastal et al., 2005) comprises three main geotectonic domains: (1) the RPC represented by the Taquarembó block; (2) the Dom Feliciano Terrane or collisional belt (550–600 Ma) and (3) the São Gabriel arc (700–900 Ma). The Late Neoproterozoic magmatic events were contemporaneous with the beginning of a large foreland system or extensional event: the Camaquã basin active between 605 and 470 Ma (Babinski et al., 1996; Paim et al., 2000). The São Gabriel arc includes metamorphosed volcano-sedimentary sequences (Vacacaí Group) intruded by low-K calc-alkaline metagranitoids (Cambai Complex). All these rocks formed during the 700–900 Ma São Gabriel Orogeny (Gastal et al., 2005; Saalman et al., 2006; see Chapter 4.6), and are largely characterised by a depleted juvenile signature. The relationship between the RPC and the Camaquã basin, a thick Neoproterozoic to Ordovician volcano-sedimentary succession (Paim et al., 2000; Gastal and Lafon, 2001), will be discussed in detail below (see Chapter 4.6).

In southwestern Brazil (Mato Grosso do Sul state), remnants of a carbonate-siliciclastic platform cover bearing Ediacaran fossils are included in the Corumbá Group (Boggiani et al., 1993; Figure 4.1.2). All lines of evidence presented by Gaucher et al. (2003) indicate that this Group is coeval with the Arroyo del Soldado Group (Uruguay), showing similar depositional environments and eustatic sea-level changes as a part of the same shelf. These units were deposited in a passive, Atlantic-type continental margin (Boggiani et al., 1993; Boggiani, 1998;

Gaucher et al., 2003). The platform deepened to the east, with granitic-metamorphic source areas located to the west. The age of deformational events affecting the Corumbá Group are constrained around 540–490 Ma (see Chapter 2).

### 4.1.2.2. Paraguay

At the Asunción Arch (Figure 4.1.2), a high-grade Palaeoproterozoic basement complex is composed of orthogneisses, banded granulites, amphibolites and quartzites, which record reworking events at 620–500 Ma (Cordani et al., 2001). Ramos (1988) suggested the possibility that the Río de la Plata cratonic region could extend to the Asunción region. Predominantly carbonatic rocks of the Itapucumí Group were assigned to the Ediacaran on the basis of the occurrence of *Cloudina* (Boggiani and Gaucher, 2004), and correlated in part with the Corumbá Group (Boggiani, 1998). Due to the collision between the RPC and the ensialic Paraná-Paranapanema continental block located to the east, the Paraguay-Araguaia belt (ancient 'Brazilides Ocean') was formed and the Neoproterozoic-Cambrian sedimentary basin was closed and deformed.

### 4.1.2.3. Uruguay

The main outcrops of the Río de la Plata Cratonic region are located in Uruguay. It was considered tectonically homogeneous (Dalla Salda et al., 1988; Bossi and Navarro, 1991 and references therein), but detailed studies show that it is composed of three tectono-stratigraphic terranes named Piedra Alta, Tandilia and Nico Pérez, separated by first-order shear zones (Bossi and Campal, 1992; Bossi et al., 1993c, 2005). As shown in Figure 4.1.2b, the Piedra Alta Terrane (PAT) is separated from the Tandilia Terrane (TT) by the Colonia Shear Zone of ca. 1,700 Ma. Neoproterozoic, siliciclastic-carbonate sedimentary cover remnants (Piedras de Afilar Formation), unconformable overlying the Palaeoproterozoic basement, are preserved in the southern portion of this terrane. Pamoukaghlian et al. (2006) assigned this unit to the Neoproterozoic based on carbon isotope chemostratigraphy (see Chapters 4.2–4.4). Dolerites and the La Tuna Granite, probably of Lower Cambrian age, intrude the Piedras de Afilar Formation.

The suture between the NPT and the Piedra Alta and Tandilia terranes is the Sarandi del Yí lineament (SYL), recording dextral shear at 1,200–1,250 Ma. The age of the SYL of ca. 1,200 Ma was obtained by K-Ar on synkinematic muscovites from pegmatites crystallised along thrust planes (1,253 ± 32 Ma: Cingolani, in Bossi et al., 1998) and by $^{40}Ar$-$^{39}Ar$ on mafic dikes deformed by the shear zone, yielding ages between 1,370 and 1,170 Ma (Teixeira et al., 1999; see Chapter 4.6). The NPT is mainly composed of a Palaeoarchaean basement (granite-greenstone belt after Hartmann et al., 2001), Palaeoproterozoic orthogneisses and supracrustal volcano-sedimentary sequences (Bossi et al., 1965; Bossi and Navarro, 1991; Hartmann et al., 2001), intruded by the 1,784 ± 3 Ma Illescas rapakivi granite (Figure 4.1.3). A 'Grenvillian' metamorphic complex (Campal et al., 1995) occurs in the NPT. An important thrust separates the Grenvillian rocks from the Carapé tectonic slab (Bossi et al., 2007), which has been dated K-Ar on synkinematic muscovite at 572 ± 11 Ma (Cingolani, in Bossi and Campal, 1992). The eastern side of the RPC is limited by the sinistral Sierra Ballena megashear of ca. 530 Ma, separating the NPT from the allochthonous CDT (Figure 4.1.3). The suture between cratonic areas to the west and remnants of a magmatic arc to the east (CDT) are characterised by the absence of a continuous Neoproterozoic, Brasiliano-aged schist belt. These facts could indicate that the amalgamation of the CDT to NPT was due to lateral rather than orthogonal accretion (Basei et al., 2000, 2005; Bossi and Gaucher, 2004).

The interpretation that the NPT is a para-autochthonous terrane displaced to the north during Cambrian times was suggested (Gaucher, 2000; Bossi and Gaucher, 2004), related to a sinistral reactivation of the SYL. Main evidence for the north displacement of the NPT is the presence of Archaean detrital zircons in Neoproterozoic sediments of the Tandilia System (Gaucher et al., 2008b), presently 600 km to the south.

In this review of the Uruguayan part of the RPC we will discuss the following events, which still remain controversial or poorly studied: granitogenesis between 630 and 530 Ma; the compressive event represented by the Carapé tectonic slab; the significance of shear zones at ca. 525 Ma and the Sierra de Ánimas alkaline magmatism. Finally, the 'Lavalleja Group' will be discussed, because there are conflicting Neoproterozoic and Mesoproterozoic ages for the sedimentation and/or metamorphism of this unit.

#### 4.1.2.3.1. Neoproterozoic-Cambrian granitic magmatism in the RPC

The NPT records magmatic activity in the late Neoproterozoic and Cambrian that can be grouped in three main events at 660–630, 590–580 and 540–530 Ma. The geochemical signature, origin and evolution of the small granitic intrusions in NPT are not well known. Umpierre and Halpern (1971), Soliani (1986), Preciozzi et al. (1993) and Hartmann et al. (2002a) suggested, based in isotopic data, a relationship with a magmatic arc during the

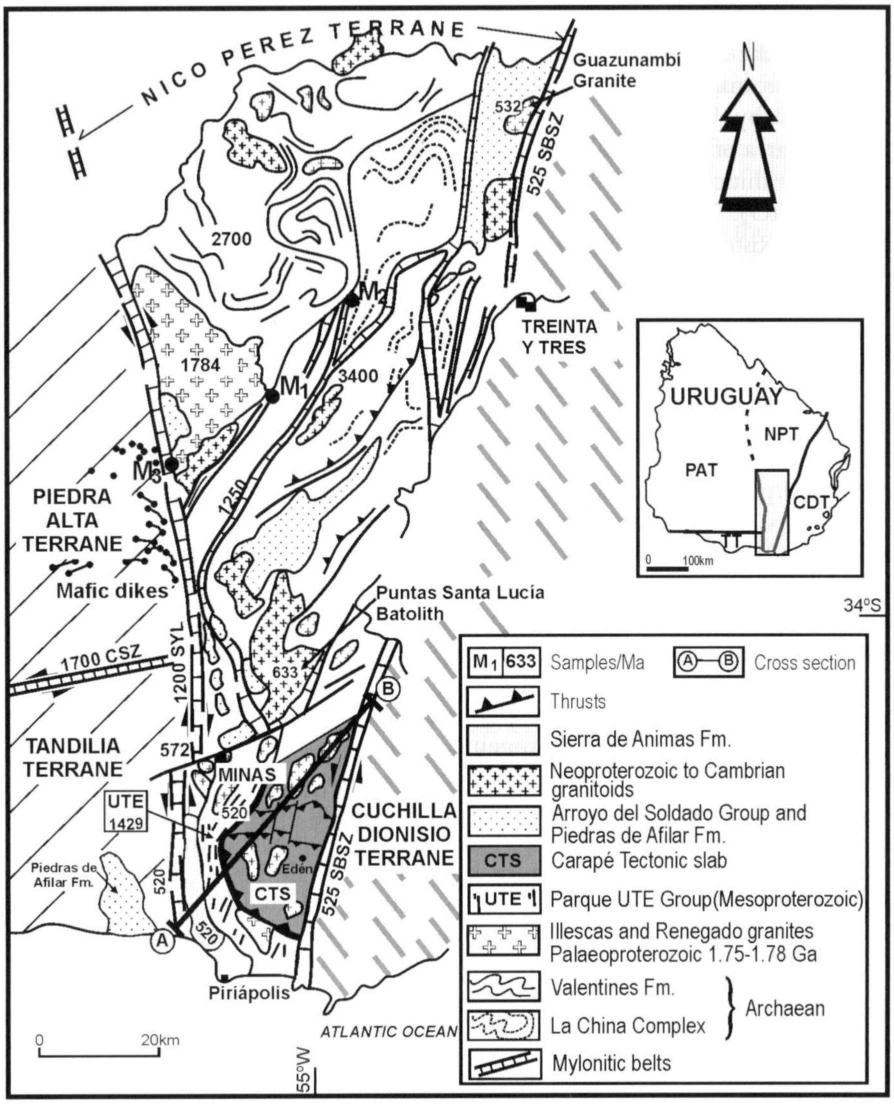


**Figure 4.1.3** Main geological units and boundaries of the Nico Pérez Terrane and neighbouring areas modified after Bossi and Gaucher (2004). CSZ: Colonia Shear Zone; SBSZ: Sierra Ballena Shear Zone; SYL: Sarandí del Yí lineament. Section AB: see Figure 4.1.4.

Brasiliano Cycle, with west-directed subduction and a long distance from the trench. Metamorphic rocks of the Brasiliano Cycle are lacking within the terrane, but a conspicuous late Ediacaran to basal Cambrian, folded sedimentary cover (Arroyo del Soldado Group) is preserved, which bears abundant fossils (Gaucher, 2000). It is clear that the Sierra Ballena Shear Zone (SBSZ) affected this sedimentary cover (Gaucher et al., 1998). Furthermore, the Arroyo del Soldado Group overlies with erosional unconformity the Puntas del Santa Lucía Granite (633 ± 12 Ma, U-Pb SHRIMP: Hartmann et al., 2002a) and the Mangacha Granite of 583 ± 7 Ma (U-Pb SIMS, Gaucher et al., 2008b). On the other hand, the Arroyo del Soldado Group is intruded by the Guazunambí Granite which yielded a Rb-Sr isochronic age of 532 ± 11 Ma (Kawashita et al., 1999a; Figure 4.1.3). K-Ar ages ranging from 532 ± 16 to 492 ± 14 Ma for pelites of the Arroyo del Soldado Group constrain the age of anchimetamorphism and cooling of the unit (Cingolani et al., 1990b; Gaucher, 2000). Thus, all lines of evidence suggest that between 580 and 540 Ma stable conditions prevailed at the eastern margin of the RPC. Gaucher et al. (2008a) argue that magmatism pre-dating the Arroyo del Soldado Group (580 Ma or older) is associated to extensional tectonics (see also Chapter 4.6). Magmatism post-dating the Arroyo del Soldado Group is clearly linked to tangential continental collision (Bossi and Gaucher, 2004). For a different interpretation of these events see Oyhantçabal et al. (2007). One apparent contradiction arises from K-Ar ages of 572 ± 11 Ma on synkinematic muscovites (Cingolani, in Bossi and Campal, 1992) associated to the emplacement of the Carapé tectonic slab. If this single age is taken at face value, it would indicate an important compressive event coeval with the transition from rifting to drifting in the Arroyo del Soldado basin, which is clearly not compatible (see below).

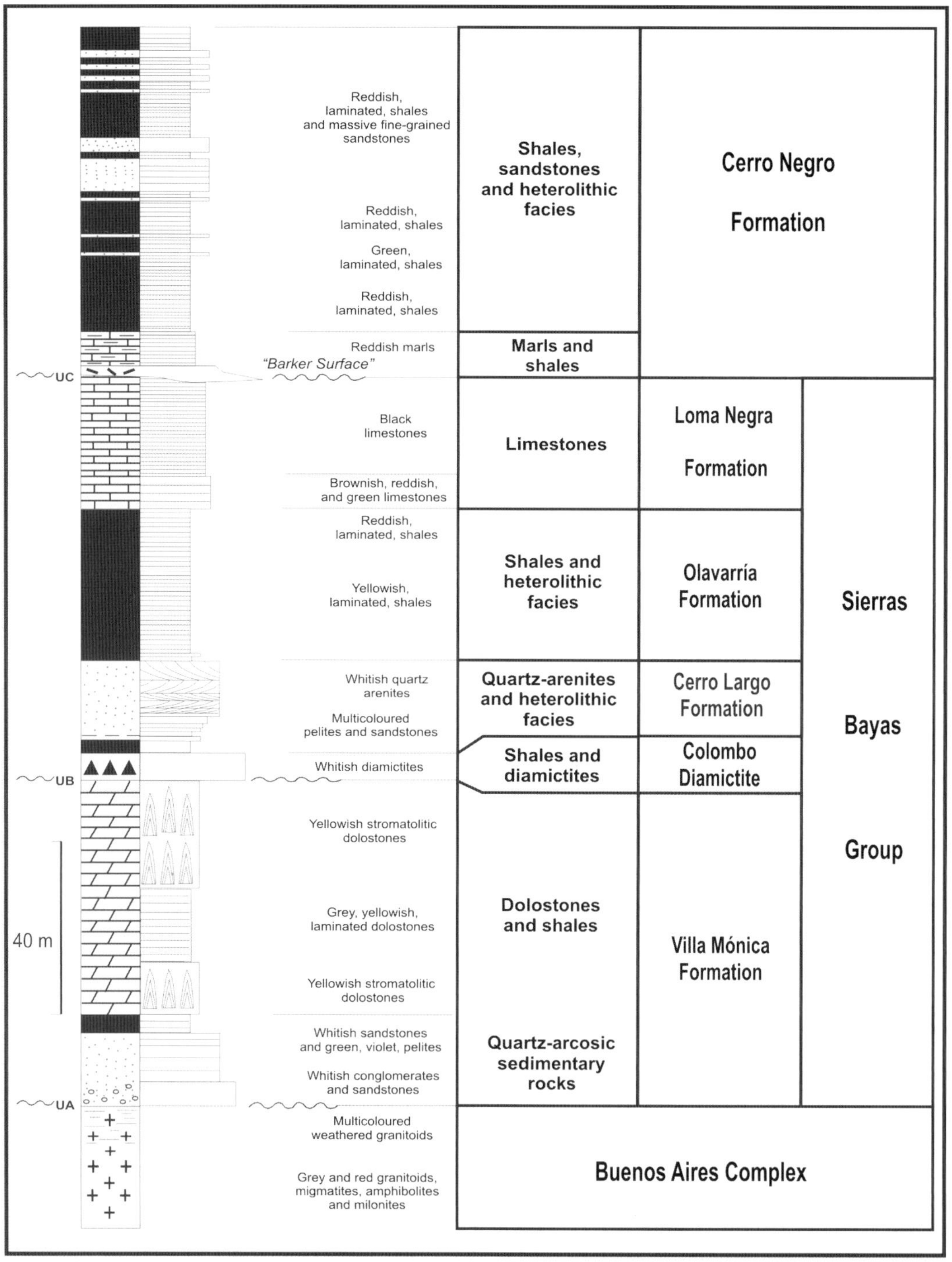


**Figure 4.2.1** Stratigraphic column of the Sierras Bayas Group and Cerro Negro Formation in the western part of Tandilia (Olavarría region).

From a lithostratigraphic standpoint (Figure 4.2.1) the Neoproterozoic sedimentary successions comprise the Villa Mónica Formation (Poiré, 1993), Cerro Largo Formation (Poiré, 1993), Las Aguilas Formation (Zalba, 1978), Olavarría Formation (Andreis et al., 1996) and Loma Negra Formation (Borrello, 1966), all of these grouped into the Sierras Bayas Group (Dalla Salda and Iñiguez, 1979; Poiré, 1993). The Cerro Negro Formation (Iñiguez and Zalba, 1974) overlies the Sierras Bayas Group with erosional unconformity and is of latest Neoproterozoic age. Finally, an Early Palaeozoic succession is represented by the Balcarce Formation (Dalla Salda and Iñiguez, 1979).

Between the crystalline basement and the sedimentary cover, arkosic and quartz-kaolinitic saprolites represent palaeoweathering surfaces (Zalba et al., 1992). Diamictites of the Sierra del Volcán Formation (Spalletti and del Valle, 1984) occur between the crystalline basement and the Balcarce Formation. Its age is still controversial and will be discussed below. Another unit of restricted areal extent is the sub-surface Punta Mogotes Formation, which was only recorded in the homonymous borehole at 400 m depth in the Mar del Plata area (Figure 4.1.7).

From a sequence-stratigraphic point of view, Iñiguez et al. (1989) distinguished three Neoproterozoic sequences, one Ediacaran-Cambrian sequence and a final Ordovician sequence, to which Poiré and Spalletti (2005) added one more Neoproterozoic sequence (Diamante sequence). Most of the four sedimentary sequences of the Sierras Bayas Group and Cerro Negro Formation are bounded by regional unconformities (Poiré, 1987, 1993, 2002, 2004; Leanza and Hugo, 1987; Barrio et al., 1991; Gómez Peral, 2008). Thus, six depositional sequences are currently recognised in Tandilia: the Neoproterozoic Tofoletti, Malegni, Diamante and Villa Fortabat sequences, the La Providencia sequence (latest Neoproterozoic-Cambrian) and the Ordovician-Silurian Batán sequence (Spalletti et al., 1996; Andreis and Zalba, 1998; Poiré et al., 2003b; Poiré and Spalletti, 2005). The confirmation of the stratigraphic position of the Las Aguilas Formation below the limestones of the Loma Negra Formation (Poiré and Spalletti, 2005), as first suggested by Leveratto and Marchese (1983), helps to clarify and simplify the stratigraphy of the sedimentary cover in Tandilia.

## 4.2.2.1. Sierras Bayas Group

The Neoproterozoic sedimentary cover of Tandilia in the Sierras Bayas range (close to Olavarría, Figure 4.1.7) comprises a 185 m-thick succession (Sierras Bayas Group, Figure 4.2.1) composed of three depositional sequences separated by regional unconformities (Poiré, 1987, 1993) and overlain by the latest Ediacaran Cerro Negro Formation.

### 4.2.2.1.1. Weathered crystalline basement

The oldest regional unconformity (UA, Figure 4.2.1) is developed between the crystalline basement and the sedimentary cover. An arkosic saprolite occurs there, representing a palaeoweathering surface (Poiré, 1987; Zalba et al., 1992). This altered level is 4 m thick and well exposed in the Villa Mónica Quarry (Olavarría area, Figure 4.1.7). It is composed of yellow, red, light green and grey, 'pseudostratified', unconsolidated saprolite. Mineralogically, it is made up of quartz, microcline, plagioclase, biotite, muscovite and abundant Fe-oxides and illite, thus reflecting basement rock mineralogy.

### 4.2.2.1.2. Villa Mónica Formation (Tofoletti sequence)

The Villa Mónica Formation, the oldest depositional sequence (Tofoletti sequence, 52–70 m thick), comprises two sedimentary facies associations: a siliciclastic member at the base and a carbonate member at the top (Figure 4.2.1). Whereas the basal member is composed of shallow-marine siliciclastic rocks (conglomerates, quartz-arenites, arkoses, diamictites and shales), the upper member is characterised by shallow-marine stromatolitic dolostones and shales. The sandstone unit is interpreted as marine sediments, which covered the weathered and peneplainised RPC during a transgressive event. The contact with the underlying palaeosols (saprolite) is well exposed in quarries in the Olavarría area (Poiré, 1993). Gómez Peral et al. (2005) and Poiré et al. (2005a) reported phosphates and hydrothermal pyrophyllite in shales of the upper part of the siliciclastic member.

The overlying dolostone member includes very conspicuous stromatolites representing a stromatolitic platform, and associated clastic, non-stromatolitic carbonates, probably deposited in an intertidal environment. Detailed studies have allowed to identify the following groups and forms (Poiré, 1987, 1989, 1990): *Colonella* fm., *Conophyton ?ressotti*, *Conophyton* fm., *Cryptozoon* fm., *Gongylina* fm., *Gymnosolem* fm., *Inzeria* fm., *Jacutophyton* fm., *Jurusonia nisvensis*, *Katavia* fm., *Kotuikania* fm., *Kussiella* fm., *Minjaria* fm., *Parmites* fm., *Parmites* cf. *cocrescens* and *Stratifera* fm. (see Chapter 4.3). Total organic carbon (TOC) content of dolostone samples from the Villa Mónica Formation average 0.8% (Gómez Peral et al., 2007). On the basis of stromatolite biostratigraphy, Poiré (1987) assigned dolostones of the Villa Mónica Formation to the Riphean (ca. 800–900 Ma). A Rb-Sr age from interbedded shales of 793±32 Ma was reported by Cingolani and Bonhomme (1982), but its significance remains controversial. Carbon isotope data were presented by Gómez Peral et al. (2007), and will be discussed in Chapter 4.4.

An unconformity (UB, Figure 4.2.1) separates the Tofoletti sequence (Villa Mónica Formation) from the overlying Malegni sequence (Cerro Largo Formation). The top of the dolostones is irregular, as clearly shown by its variable thickness (36–52 m). Above the unconformity UB, predominantly diamictitic deposits occur, which were recently named as Colombo Diamictites (Poiré and Gaucher, 2007, see below).

Diagenesis of the dolostones of the Villa Mónica Formation can be described in terms of two dolomitisation processes that affected these rocks, which probably were developed from low-magnesium calcite precursor (Gómez Peral, 2008). Three different cement generations were recognised: Stage 1, crystallisation of a dolosparitic cement occasionally showing crystals with iron-rich nuclei; Stage 2, quartz cement precipitation in form of euhedral crystals; and Stage 3, high-Mg-calcite cement made up of anhedral to euhedral macrosparitic crystals.

### 4.2.2.1.3. Colombo Diamictite

The Colombo Diamictite (8 m; Figure 4.2.1) is composed of whitish, massive mudstones and claystones bearing lonestones up to 2.5 m in diameter and showing convolute bedding (Figure 4.5.1F). Chert breccias and fine orthoconglomerate also occur. This sedimentary package has been considered as the base of the Cerro Largo Formation (Poiré, 1993) but its possible separation as a new unit developed between the Villa Mónica and Cerro Largo formations is being considered. Evidences of synsedimentary tectonism and its stratigraphic position suggest correlation of the Colombo diamictite with the rift-related Las Ventanas Formation in Uruguay (see Chapter 4.5).

### 4.2.2.1.4. Cerro Largo Formation (Malegni sequence)

The Cerro Largo Formation represents the second depositional sequence (Malegni sequence, 40 m thick). As discussed above, it is not clear whether the Colombo Diamictite belongs to this sequence or represents a separate unit. Finely bedded, varicoloured, glauconitic sandstones, heterolithic facies and cross-bedded quartz-arenites characterise the Cerro Largo Formation. Outcrops of the white, well-sorted quartz-arenites of the Cerro Largo Formation typically occur at the core of the main hills (up to 300 m high) of the Olavarría region. This coarsening- and shallowing-upward sequence represents the transition from subtidal nearshore to shoreface environments, suggesting progradational processes.

Trace fossils have been reported from sandstones of the Cerro Largo Formation (*Palaeophycus* isp. and *Didymaulichnus* isp.) but they are dubious and were re-interpreted as desiccation cracks associated to biomats (Porada and Bouougri, 2008; see Chapter 4.3).

The contact between the Cerro Largo and Olavarría formations (Malegni and Diamante sequences) in the Olavarría region is transitional, but in the Barker area there are breccias at the base of the younger sequence (Las Aguilas Formation).

### 4.2.2.1.5. Olavarría and Las Aguilas formations (Diamante sequence)

In the Olavarría region (Figure 4.1.7) quartz-arenites of the Cerro Largo Formation pass transitionally into siltstones and claystones of the Olavarría Formation (Andreis et al., 1996; Figure 4.2.1). This contact is well exposed at the Volquetes road cut and in the Cementos Avellaneda quarry.

The Olavarría Formation (maximum thickness: 37 m) consists of two sections, from base to top: yellowish heterolithic facies (27 m thick) and reddish claystones (8 m thick). Palaeoenvironmental interpretations suggest shallow-marine deposits in a transgressive systems tract (TST).

At the Cuchilla de las Aguilas (Barker area, Figure 4.1.7), three lithofacies occur in the Las Aguilas Formation, from base to top (Zalba et al., 1988; Andreis, 2003): silicified calcareous breccias with silicified oolites and peloids; reddish-to-whitish claystones; and a coarsening-upward, heterolithic (sandstone-shale) sequence. Sedimentary structures in the latter facies include plane-parallel lamination, ripples, hummocky cross bedding and synaeresis and desiccation cracks.

In the Diamante quarry (Barker area), the middle part of the unit comprises red claystones with high iron content (32–70% $Fe_2O_3$) up to 9 m in thickness (Alló, 2001), which could be correlated with other Late Neoproterozoic iron deposits, such as the Jacadigo Group in SW Brazil (Mato Grosso do Sul state), or BIF in the lower Arroyo del Soldado Group (ASG) (Gaucher, 2000; Gaucher et al., 2003, 2004c; Figure 4.2.5). No glacial deposits are associated to the ironstones of the Las Aguilas Formation.

### 4.2.2.1.6. Loma Negra Formation (Villa Fortabat sequence)

The Loma Negra Formation, the youngest depositional sequence of the Sierras Bayas Group (Villa Fortabat sequence, 40 m thick), is composed almost exclusively of red and black micritic limestones, deposited by suspension fall-out in open marine ramp and lagoonal environments.

This sequence can be divided from base to top in two members based on lithology: reddish micritic limestones (8 m thick) followed by black micritic limestones (up to 32 m thick). In these limestones, the following diagenetic processes were recognised by Gómez Peral et al. (2007): Stage 1: Recrystallisation of the calcitic precursor to non-planar micrite (1–10 μm), microsparite (10–25 μm) or sparite, preserving the existing microstructure (1–25 μm). Stage 2: Chemical dissolution generates porosity in form of veins and voids. Stage 3: Precipitation of calcite cements filling porosity. Stage 4: Irregular stylolites due to pressure–solution. Stage 5: Silicification along veins and as partial replacement of carbonate crystals.

On top of the Loma Negra Formation a regional unconformity occurs (Barrio et al., 1991), which has been recently named 'Barker Surface' (Poiré and Gaucher, 2007; Figure 4.2.1; Chapter 4.5). This surface has been related to a sea-level drop that exposed the Loma Negra shelf. Meteoric dissolution of carbonates led to an expressive karstic surface, in which residual clays and brecciated chert accumulated.

Gaucher et al. (2005b) reported for the first time the occurrence of *Cloudina* cf. *C. riemkeae* Germs (1972a) in the reddish micritic limestones (see Chapter 4.3). *Helminthopsis* isp. and probable medusa resting traces have been found in the reddish limestones (Poiré et al., 2003b).

No glacigenic sediments occur at the base of limestones of the Loma Negra Formation, and there is no sign of an erosional unconformity either (Poiré, 2002). Therefore, carbonates of the Loma Negra Formation do not represent 'cap carbonates' (Hoffman et al., 1998b). Cozzi et al. (2002) discussed the Snowball Earth hypothesis as exemplified by the Ediacaran Shuram Formation of Oman. The Sierras Bayas Group and its correlation with the ASG support a 'slushball' Earth scenario, at least for Ediacaran glacial events (Poiré, 2002; Gaucher et al., 2004c; see Chapter 4.5).

### 4.2.2.2. Cerro Negro Formation (La Providencia sequence)

The Cerro Negro Formation (La Providencia Depositional sequence) unconformably overlies the Loma Negra Formation, filling palaeovalleys and a karst relief. It is a 100–400 m-thick unit characterised by reddish, greenish or brown-olive claystones and heterolithic facies (fine-grained sandstone–claystone interbeds), mainly deposited in upper to lower intertidal conditions.

The lower part of the Cerro Negro Formation consists of reddish residual clays, micritic limestones and marls. Chert breccias and phosphatic rocks occur as karst infill (Leanza and Hugo, 1987; Barrio et al., 1991). The upper contact with the Balcarce Formation is not exposed.

The Cerro Negro Formation yielded abundant acritarchs mainly assigned to *Leiosphaeridia*, which are consistent with a latest Ediacaran age for the unit (Cingolani et al., 1991; Gaucher et al., 2005b, see Chapters 4.3 and 9.1).

### 4.2.2.3. Balcarce Formation (Early Palaeozoic)

The Balcarce Formation has been studied in detail by del Valle (1987a). It overlies the crystalline basement or the older sedimentary units (Cerro Negro Formation, Sierra del Volcán Diamictite and Punta Mogotes Formation).

The Balcarce Formation (100–400 m thick) is composed of white quartz-arenites and granule-conglomerates with subordinate levels of kaolinitic mudstones and quartz-pebble conglomerates. The geometry of the sandstone beds is sheet-like; most sedimentary bodies are bound by convex-upward surfaces, although some wide channel-like features are also present. Planar and tangential cross bedding is the dominant structure within sandstone bedsets, and large-scale sigmoidal bodies are common in most sections. Sheet-like and lenticular, sandstone–mudstone interbeds are commonly intercalated with sandstone beds. Trace fossils are abundant at the top-bedding surface of sandstones in sandstone–mudstone intercalations. The quarries in the Batán and Chapadmalal area (Figure 4.1.7) expose the stratigraphic architecture of the Balcarce Formation. Based on their contrasting geometry, two main groups can be defined in this siliciclastic succession: one group is characterised by a subhorizontal stacking pattern (aggradational geometry) and the other shows very well developed depositional clinoforms (progradational geometry).

Tidal processes are inferred for the cross-bedded sandstone facies (bars) and heterolithic facies. Large- to medium-scale, laterally persistent bodies of cross-bedded sandstones, exhibit rhythmic lateral variations in the thickness of foresets and in clay content due to spring and neap tide alternation. Clay drapes covering foresets and other depositional surfaces, herringbone cross bedding, opposite palaeocurrent trends in successive sedimentary bodies and the occurrence of reactivation surfaces also suggest tidal deposition. The migration and accretion of bidimensional sand bars seem to be controlled by highly asymmetrical time–velocity tidal currents. Subordinated, high-energy storm episodes are suggested by hummocky cross-bedded sandstones, sheet conglomerates armouring previous tidal sand bodies, and heavy mineral concentrations in the wavy sandstone laminae of heterolithic facies.

An epicontinental, shallow-marine open shelf is inferred for the Balcarce Formation in Tandilia. Most sedimentary facies were developed in the nearshore and inner shelf environments of a tide-dominated and storm-influenced platform. The internal anatomy of the clinoforms of the Balcarce Formation clearly shows that controls on sedimentation failed to remain uniform as progradation occurred. The interdigitation of interbar heterolithic deposits and subtidal sand bar deposits within this falling stage systems tract suggests that the shallow-marine system was affected by higher order oscillations in relative sea level.

The progradation to the south of the clinoforms reported by Poiré et al. (2003b) confirms that the margin of the Balcarce basin was located to the north of the Tandilia region (Teruggi, 1964; Dalla Salda and Iñiguez, 1979).

The Balcarce Formation contains abundant and diverse trace fossils (Poiré et al., 1984; del Valle, 1987b; Poiré and del Valle, 1996; Spalletti and Poiré, 2000; Seilacher et al., 2002; see Chapter 4.3). The age of the Balcarce Formation has been considered as early Palaeozoic. The unit lacks body fossils that could allow a more precise age assignment. Trace fossils broadly constrain the age between Cambrian and Silurian, but Seilacher et al. (2002) favour a Lower Silurian age based on the occurrence of *Cruziana ancora*. More recently, Rapela et al. (2007) reported U-Pb detrital zircon ages as young as 475–480 Ma (Early Ordovician) from the Balcarce Formation, suggesting a Late Ordovician to Lower Silurian age for the unit.

#### 4.2.2.4. Sierra del Volcán Diamictite

The Sierra del Volcán Diamictite (Spalletti and del Valle, 1984) is a thin sedimentary unit (4 m) cropping out between the crystalline basement and quartz-arenites of the Balcarce Formation at the homonymous hill. This sequence is composed of two diamictite facies (Spalletti and del Valle, 1984): (a) basal, brownish, coarse- to fine-grained sandstone, sandy shales and yellowish shales with scattered, often faceted dropstones of variable size. Dropstones are polihedrical, often in vertical position, affecting both ripple bedding and lamination structures. (b) Grey and whitish, muddy or sandy, massive diamictite (Figure 4.5.1G) bearing orthoquartzite and granitoid boulders up to 30 cm in diameter. They grade up-section into poorly stratified, finer diamictites with local syndepositional corrugations.

On the basis of grain-size studies and field occurrence, Spalletti and del Valle (1984) suggested that the basal sandstones were deposited in a marine, littoral environment, between the beach and the nearshore–offshore transition zone. The diamictites were inferred to be glaciomarine deposits developed near a peneplanised continent in the shallow platform of an epeiric sea.

Based on its stratigraphic position, the Sierra del Volcán Diamictite was originally considered to represent an Ediacaran glacial event (Spalletti and del Valle, 1984; Pazos et al. 2008). However, Poiré and Spalletti (2008) suggested a possible Ashgillian (Hirnantian) age for these glacial deposits on the basis of Lower Ordovician U-Pb ages of detrital zircon grains (van Staden et al., unpublished data) from the Sierra del Volcán Diamictite. Therefore, this glacial unit should be referred to the Hirnantian, which has a very important record in southern South America and South Africa.

The geochronological information provided by detrital zircons from the Sierra del Volcán Diamictite and the Balcarce Formation suggests that the kaolinite-rich Balcarce Formation could represent the onset of Silurian greenhouse conditions. The Lower Silurian age for the Balcarce Formation suggested by Seilacher et al. (2002) on the basis of trace fossils is consistent with this scenario.

### 4.2.3. Nico Pérez Terrane, Uruguay

The Nico Pérez Terrane of Uruguay represents the oldest crustal block belonging to the RPC. Its pre-Neoproterozoic basement is composed mainly by Archaean and Palaeoproterozoic units (Bossi and Ferrando, 2001; Hartmann et al., 2001). Mesoproterozoic volcano-sedimentary successions occur in the southern part of the terrane (see Chapter 4.6).

Thick Neoproterozoic sedimentary successions include the Mina Verdún Group (MVG) (Poiré et al., 2003a, 2005b), the Las Ventanas Formation (Midot, 1984) and the ASG (Gaucher et al., 1996; Gaucher, 2000).

The controversy regarding the age of the former Lavalleja or Fuente del Puma Group has been dealt within the previous chapter. It suffices here to say that its Neoproterozoic age has not been thoroughly demonstrated yet, and an important portion is actually Mesoproterozoic in age (Parque UTE Group: Bossi et al., 2008). All Neoproterozoic detrital zircon U-Pb SHRIMP ages reported by Basei et al. (2008) from this unit are moderately to highly discordant, except for only one grain that yielded 715 ± 26 Ma. On the other hand, the same sandstones are intruded by gabbros that yielded an U-Pb zircon age of 1,492 ± 4 Ma (Oyhantçabal et al., 2005). Thus, it is still uncertain if there are any Neoproterozoic rocks in the unit formerly known as Lavalleja Group.

#### 4.2.3.1. Mina Verdun Group

The MVG was erected by Poiré et al. (2003a, 2005b) to include a sedimentary succession exposed in the homonymous mine near Minas (Figure 4.2.2). The age of the unit is loosely constrained between Late Mesoproterozoic and Early Neoproterozoic (Tonian) by carbon isotope chemostratigraphy and field relationships (Gaucher et al., 2006, 2008a; see 4.4).

The original definition included four formations (Poiré et al., 2003a, 2005b), from base to top: (a) Don Mario Formation (black shales); (b) La Toma Formation (greenish marls); (c) El Calabozo Formation (limestones) and (d) Gibraltar Formation (massive dolostones, dolomitic marls and subordinate red limestones). It is worth noting that at the stratotype base and top are covered. Subsequently, Gaucher et al. (2004a) assigned carbonates exposed

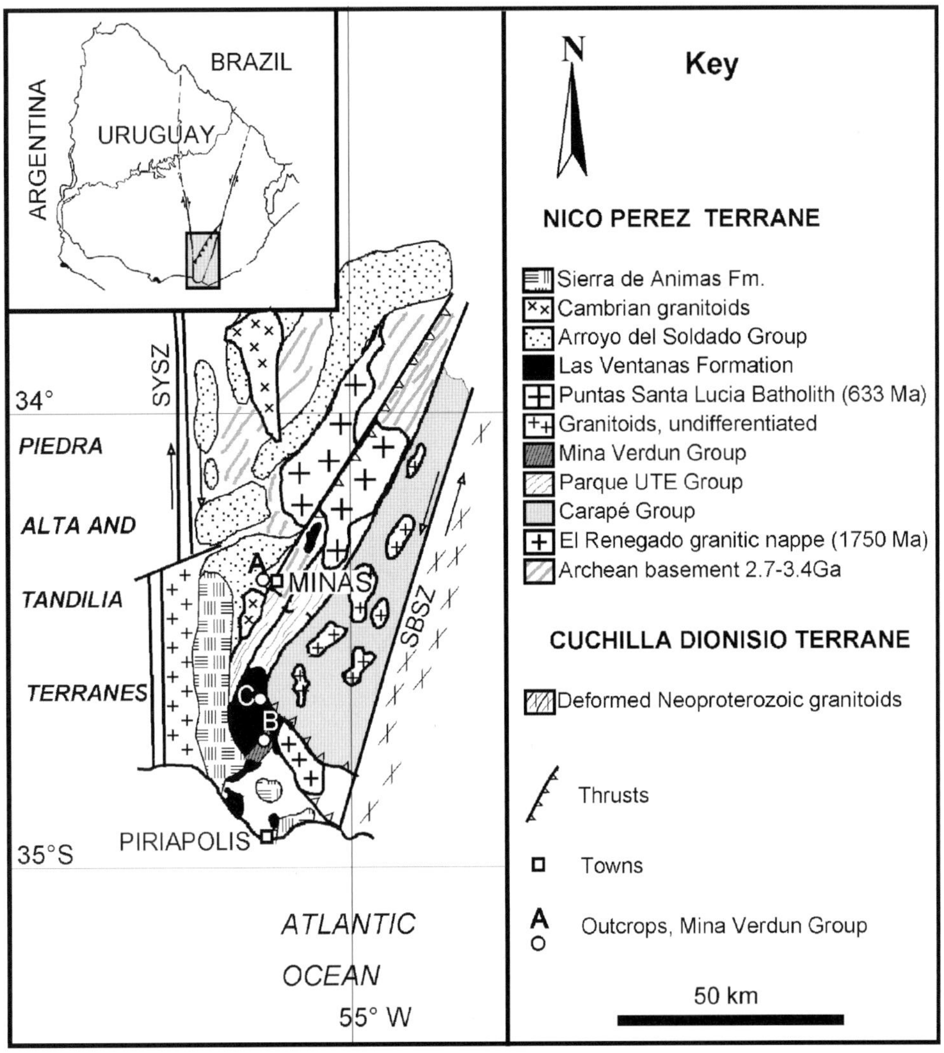


**Figure 4.2.2** Geological map of the southern Nico Pérez Terrane, showing outcrops of the Mina Verdún Group. SYSZ: Sarandí del Yí Shear Zone, SBSZ: Sierra Ballena Shear Zone. A: Mina Verdún Quarry; B: Burgueño Quarry; C: Paso del Molino section. Modified from Gaucher et al. (2008a).

in the Burgueño Quarry, 40 km to the south of Minas (Figure 4.2.2) to the MVG. There, acid volcanic and pyroclastic rocks are concordantly folded with the rest of the succession, and were grouped into the Cerro de las Víboras and Nueva Carrara formations, respectively (Gaucher et al., 2007a). Thus the currently accepted lithostratigraphy of the MVG comprises six formations, which will be described below in ascending order.

### 4.2.3.1.1. Cerro de las Víboras Formation

Its type area is located in the homonymous hill to the NW of Burgueño Quarry (Gaucher et al., 2007a; Figure 4.2.2). The Cerro de las Víboras Formation, 80 m in thickness, represents the base of the MVG there; its lower contact is not exposed. The unit is made up of foliated, greenish-grey metarhyolites with β-quartz phenocrysts and sericitic matrix. Lapilli-tuff intercalations showing black fiammes are common. Rhyolites are very resistant to weathering and erosion, resulting in the highest and steepest hills of the area. Up section, acid pyroclastic rocks are interbedded with pelites and sandstones of the lower Don Mario Formation, showing that the contact between both units is transitional (Figure 4.2.3).

### 4.2.3.1.2. Don Mario Formation (Poiré et al., 2003a)

It is composed of massive and laminated, organic-rich, black shales 40 m in thickness. At the base, arkosic sandstone and pyroclastic intercalations occur (Gaucher et al., 2007a). XRD analyses of shales show quartz, calcite, clay minerals (illite, chlorite, illite-smectite and chlorite-smectite interlayer) and feldspars. Their geochemistry

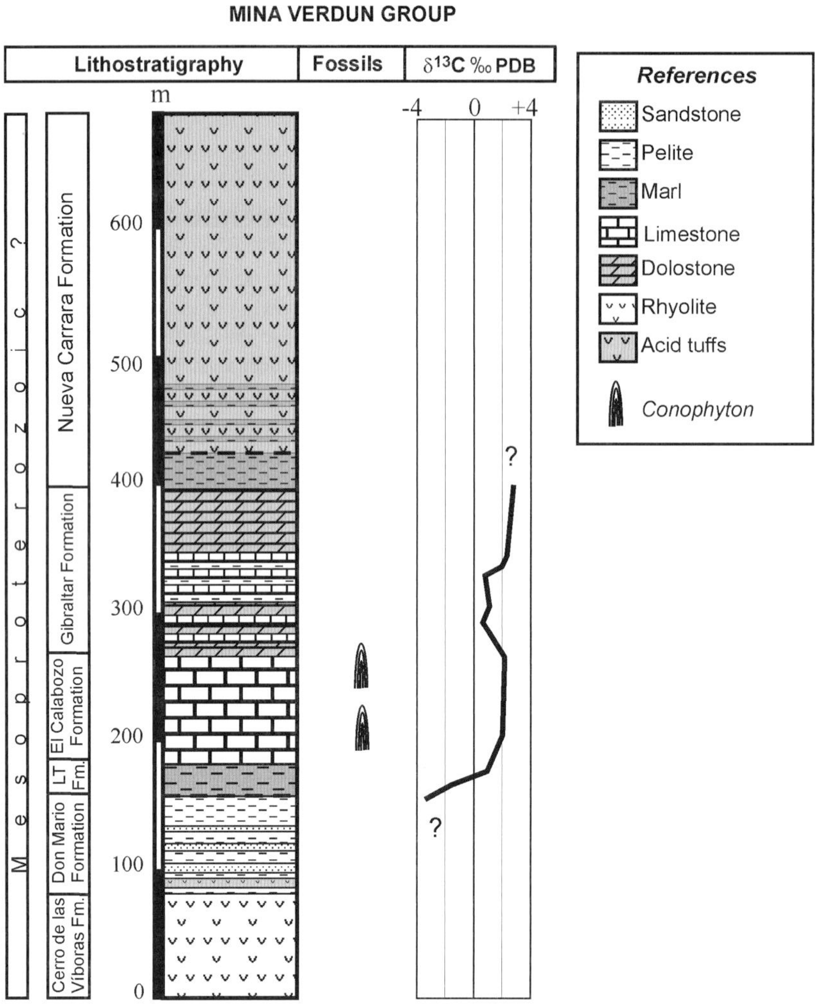


**Figure 4.2.3** Simplified stratigraphic column of the Mina Verdún Group, after Gaucher et al. (2007a). LTFm: La Toma Formation.

suggests a high degree of weathering and reworking in the source area, which was made up of differentiated upper continental crust (Poiré et al., 2005b). Anoxic conditions prevailed in the basin during deposition of the Don Mario Formation, leading to high TOC values and the occurrence of disseminated sulphides.

### 4.2.3.1.3. La Toma Formation

Dark green marls up to 15 m in thickness concordantly overly black shales of the Don Mario Formation (Poiré et al., 2003a; Figure 4.2.3), representing the onset of carbonate deposition. Marls are composed of up to 45% $SiO_2$ and 35% $CaCO_3$ (Poiré et al., 2005b). Carbonate minerals are represented by minute grains of both calcite and dolomite, the former being more abundant. Clay minerals are composed of illite and scarce chlorite. The unit is concordantly overlain by massive and stromatolitic limestones of the El Calabozo Formation.

### 4.2.3.1.4. El Calabozo Formation

This is probably the most characteristic and economically important unit of the MVG. It comprises massive, laminated and stromatolitic limestones typically showing coniform stromatolites (*Conophyton*), up to 170 m in thickness (Poiré et al., 2003a; Figure 4.2.3). Limestones are very pure, with up to 93% $CaCO_3$ at its stratotype or even 95% at Burgueño Quarry (Figure 4.2.2), making them an important raw material for the cement and lime industries. In thin section they show a homogeneous mosaic of micritic and subsparitic calcite with stylolites and corrugated microbial lamination. Subordinate quartz, feldspars and illite have been observed by XRD. Collapse breccias probably resulting from karstic phenomena also occur. Columnar stromatolites assigned to *Conophyton* fm. are by far the most abundant group. Scarce but well-preserved, dendroid, branching and digitate stromatolites were also observed (see Chapter 4.3). In the Burgueño Quarry, a pervasive decolouration of limestones has been explained by contact metamorphism of nearby granites, causing the reaction of organic carbon with carbonates and affecting its carbon-isotopic composition (Gaucher et al., 2006, 2007a; see Chapter 4.4).

#### 4.2.3.1.5. Gibraltar Formation (Poiré et al., 2003a)

An erosional unconformity separates the El Calabozo Formation from the overlying Gibraltar Formation (Figure 4.2.3), which has been interpreted as a karstic palaeorelief by Poiré et al. (2005b). This contact is also deformed due to boudinage of the more competent dolostones, as shown by Gaucher et al. (2007a). A 4 m-thick diagenetic breccia is often at the base of the Gibraltar Formation. The unit, 60 m in thickness, comprises the following lithologies from base to top: (a) black marls and shales with pink limestone intercalations, (b) pink and red limestones, often dolomitic and (c) massive, mainly light yellow or reddish dolostones, showing abundant carbonate veins. Dolostones are the most abundant and characteristic rocks of the Gibraltar Formation. They are in turn overlain by green marls at Burgueño Quarry, where the top of the unit is well exposed (Gaucher et al., 2007a).

The occurrence of stromatolites in the Gibraltar Formation was recently recognised in two drill-cores at its stratotype (Poiré et al., 2006). The strong dolomitisation of the Gibraltar Formation and its poor exposure do not allow a detailed palaeoenvironmental interpretation (Poiré et al., 2005b). However, the preservation of the stromatolites in cores is surprisingly good and can be used to determine environmental conditions. Gymnosolenid and few *Conophyton* stromatolites have been so far identified.

#### 4.2.3.1.6. Nueva Carrara Formation

The Gibraltar Formation transitionally passes into greenish tuffs of the Nueva Carrara Formation (Gaucher et al., 2007a; Figure 4.2.3). Green, chloritic marls occur at the base, and are often interbedded with green, acid lapilli-tuffs. Up section, 300 m lapilli-tuffs with subordinate metamarl intercalations occur. Tuffs show β-quartz and alkali feldspar phenocrysts and a matrix composed of chlorite and quartz. Pumice and fiammes made up of chlorite, muscovite and quartz occur, as well as accessory lithic and crystal fragments of dacite, microcline and perthitic feldspars.

In all outcrop areas, the MVG is overlain with angular and erosional unconformity by polymictic orthoconglomerates and red diamictites of the Las Ventanas Formation (Blanco and Gaucher, 2005; Gaucher et al., 2008a), of early Ediacaran age. Thus, the MVG may be considerably older than Early Ediacaran, probably of Mesoproterozoic age (Gaucher et al., 2006, 2008a; see Chapter 4.4). Both the MVG and Las Ventanas Formation are intruded by a subvolcanic, longitudinal (E-W) mafic dyke swarm that yielded a K-Ar whole rock datum of $485 \pm 13$ Ma considered as a minimum age (González et al., 2004). These dykes may belong to the $581 \pm 13$ Ma Nico Pérez mafic dyke swarm (see Chapter 4.6).

### 4.2.3.2. Las Ventanas Formation

The Las Ventanas Formation, erected and first mapped by Midot (1984), is composed of a thick, fining- and thinning-upward volcano-sedimentary sequence covering an area of ca. 120 km$^2$ of the southern Nico Pérez Terrane (Blanco and Gaucher, 2005; Figure 4.2.2). Polymictic orthoconglomerates predominate, passing into sandstones and siltstones up-section. The structure in its type area corresponds to a broad syncline (Cerro Las Ventanas Syncline) with an axis oriented S20W.

Masquelin and Sánchez Bettucci (1993) first correlated outcrops of Las Ventanas Formation with the predominantly conglomeratic Playa Hermosa Formation, which they considered Neoproterozoic or Cambrian in age. This correlation has been confirmed by Blanco and Gaucher (2005), which regarded the Playa Hermosa Formation as a deep-water equivalent of the Las Ventanas Formation. A detailed stratigraphic and sedimentologic analysis of the Playa Hermosa Formation is due to Pazos et al. (2003, 2008).

The Las Ventanas Formation is subdivided into the following members, from base to top (Blanco and Gaucher, 2005; Figure 4.2.4):

(a) La Rinconada Member, made up of basic volcanics, mainly basalts (44–52% $SiO_2$), volcanoclastic breccias and subordinate basic tuffs, reaching 500 m in thickness. Basalts are often vesicular and represent stacked flows. Volcanoclastic breccias (locally named 'Darwin breccia') are more common in the upper part of the unit.
(b) The Quebrada de Viera Member, up to 3,800 m in thickness, overlies the Rinconada Member with erosional unconformity. It is composed of polymictic, mostly clast-supported orthoconglomerates, grading to conglomerate-sandstone intercalations up section. Glacigenic, reddish and greenish diamictites occur only at the base of the unit (Figure 4.2.4), filling a karst surface developed on top of the MVG (Gaucher et al., 2008a;

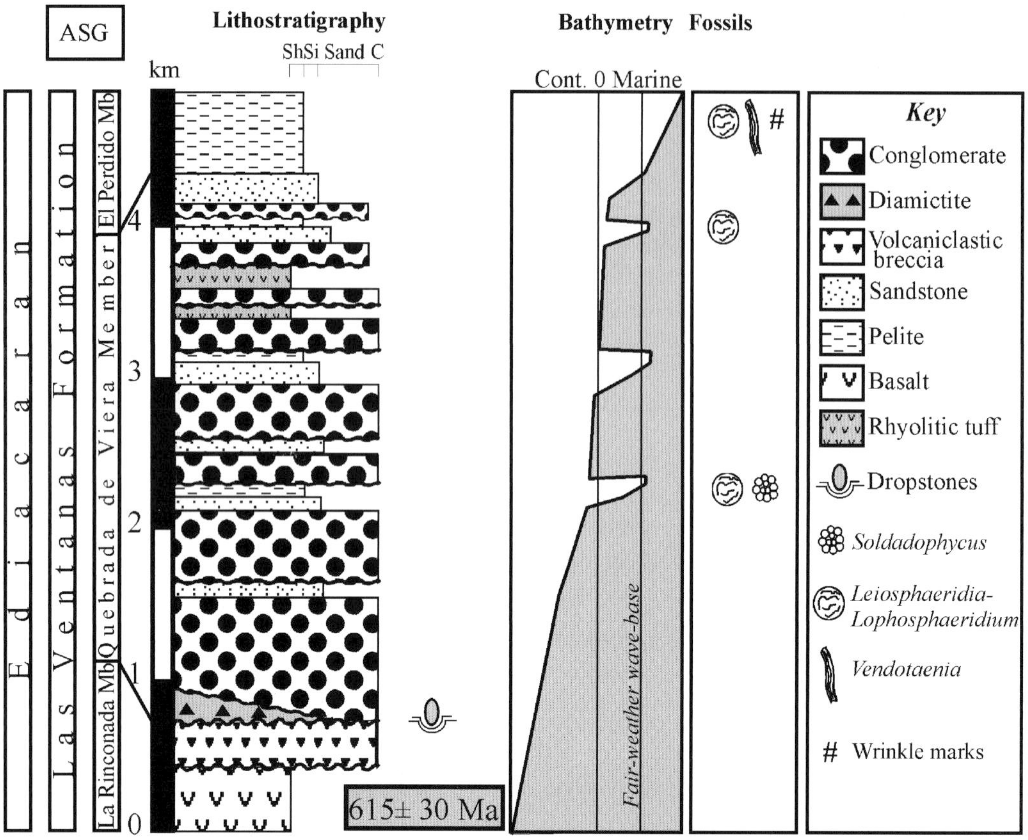


**Figure 4.2.4** Synthetic stratigraphic column of the Las Ventanas Formation, modified from Gaucher et al. (2008a). ASG: Arroyo del Soldado Group.

see Chapter 4.5). Lithoclast composition of orthoconglomerates shows provenance from a bimodal volcanic and granitic source area. Up to 37% basalt clasts in certain conglomerate levels suggest little chemical weathering and/or a steep palaeorelief. Cross bedding and normal grading are common sedimentary features of the conglomerates. Sandstones are texturally and mineralogically immature arkoses, often showing hummocky cross stratification at the top of this member (Blanco and Gaucher, 2005). Whereas basalt clasts are more common at the base, rhyolitic clasts and intercalated rhyolitic tuffs occur at the top.

(c) El Perdido Member concordantly overlies the Quebrada de Viera Member and is characterised by laminated siltstones and shales up to 600 m in thickness (Figure 4.2.4), showing reddish colours at the base and green and grey colours at the top (Blanco and Gaucher, 2005). Abundant pyrite at the top of this member indicates dysoxic or anoxic conditions.

Conglomerates of the Las Ventanas Formation represent alluvial fan and fan delta deposits (Pecoits et al., 2004; Blanco and Gaucher, 2005), on the basis of (Einsele, 2000): (1) predominance of unchannelised, conglomerate deposits; (2) occurrence of both stream deposits (predominant) and sedimentary gravity flows; (3) great thickness of conglomerates and (4) intercalation of conglomerates with marine siltstones and sandstones at the top of the Quebrada de Viera Member.

Thus, the Las Ventanas Formation represents a deepening-upward sequence (Figure 4.2.4), recording evolution from an alluvial fan-dominated environment to shallow-marine conditions with occasional storms. Sedimentary structures, thickness of conglomerate deposits, abundance of unstable lithoclasts and immaturity of sandstones point to a steep palaeorelief. Bimodal volcanism evolves from mainly basic (basalts) at the base to mainly acid (rhyolitic) at the top. An extensional geotectonic setting, possibly a rift, was postulated by Blanco and Gaucher (2005) for the Las Ventanas Formation (see Chapter 4.6). In favour of this hypothesis are the bimodal, synsedimentary volcanism, strong palaeorelief, great thickness of alluvial fan conglomerates and the evolution from continental to open marine environments.

Acritarchs and other organic-walled microfossils occurring in palynological macerations of siltstones of Las Ventanas Formation (Quebrada de Viera and El Perdido members) were described and illustrated by Blanco and Gaucher (2005) and Gaucher et al. (2008a, see Chapter 4.3).

As noted already by Blanco and Gaucher (2005), the lithostratigraphic column presented by Pecoits (2003), although correct in its lower part, duplicates the stratigraphy of the Las Ventanas Formation by the failure to recognise the structure, which is an open syncline. Grading in the lower half of the column presented by Pecoits (2003) and Pecoits et al. (2008) is normal, and turns to reverse in the upper half. Likewise, clast composition is symmetrical, the upper part being a specular image of the lower. This is easily explained by tectonic duplication, as mentioned by Blanco and Gaucher (2005).

The recent proposal of the 'Maldonado Group', made up of the Las Ventanas and San Carlos formations (Pecoits et al., 2008), also deserves consideration here. The Las Ventanas and San Carlos formations crop out in different tectonostratigraphic terranes (Nico Pérez Terrane and Cuchilla Dionisio Terrane, respectively) and are separated by the crustal-scale Sierra Ballena Shear Zone, comprising up to 10 km-thick ultramylonites. Both units were never observed in contact. U-Pb ages of detrital zircons, Sm-Nd isotopic signature of granitoids and its general geologic evolution show that the Cuchilla Dionisio Terrane ( = Punta del Este Terrane) is an exotic block of African affinity (Bossi and Gaucher, 2004; Basei et al., 2005, 2008; Gaucher et al., 2008b; see Chapter 4.6). According to the International Stratigraphic Guide, a group is 'a sequence of two or more *contiguous or associated* formations with significant and diagnostic lithologic properties in common' (Salvador, 1994, p. 35). None of these criteria apply to the San Carlos and Las Ventanas formations; therefore the Maldonado Group is not a valid lithostratigraphic unit.

## 4.2.3.3. Arroyo del Soldado Group

The ASG is a key unit for unravelling the Ediacaran to Cambrian evolution of southwestern Gondwana because of its position at the boundary between the African and South American cratons and also due to its fossiliferous nature. It was originally described from restricted areas in the north-western and central Nico Pérez Terrane (Gaucher et al., 1996). The ASG occurs over an area in excess of 30,000 km$^2$ (Figures 4.1.3 and 4.2.2), with a maximum NS dimension of 320 km and EW dimension of 130 km. Outcrops of the ASG cover an area of ca. 3,000 km$^2$ (Gaucher, 2000). The main geological features (lithostratigraphy, sedimentology, deformation) were described in detail by Gaucher (2000), therefore only the lithostratigraphic framework and recently acquired data will be discussed here.

### 4.2.3.3.1. Lithostratigraphy and sedimentary environments

The ASG is a platform succession unconformably overlying a mainly Archaean to Neoproterozoic basement in the Nico Pérez Terrane, representing the drift phase of the oceanic basin opened during rifting represented by the Las Ventanas Formation (Blanco and Gaucher, 2005; Gaucher et al., 2008a). The group is characterised by an alternation of predominantly siliciclastic (Yerbal, Barriga Negra, Cerro Espuelitas and Cerros San Francisco formations) and thick carbonate units (Polanco and Cerro Victoria formations), which extend laterally over hundreds of kilometres with little facies variability. These formations, in ascending order, are characterised as follows (Gaucher, 2000; Figure 4.2.5):

(a) The *Yerbal Formation* is a fining-upward, mainly siliciclastic unit more than 1,500 m thick. Sandstones occur at the base, siltstones dominate up section, and BIF, dolostone and chert intercalations occur at the top (Gaucher et al., 2004c).

(b) The *Polanco Formation* is composed of ~900 m of bluish-grey, pure limestones, limestone–dolostone rhythmites and, more rarely, pure dolostones (Gaucher, 2000; Gaucher et al., 2004c). It is the most geographically extensive unit of the ASG.

(c) The *Barriga Negra Formation* is composed of 1,500 m of conglomerates and breccias. Whereas carbonate breccias directly derived from the underlying Polanco Formation characterise the base, an increasing proportion of extra-basinal (basement) clasts occur up section (Gaucher, 2000). At the top, chloritic and glauconitic sandstones and conglomerates are interbedded with siltstones containing marine acritarchs (Gaucher, 2000; Figure 4.2.5). Thus, depositional environments of the Barriga Negra Formation are transitional at the base, continental in the middle and again marine at the top, recording a regression and shelf exposure followed by renewed flooding. The unit wedges out towards the east, showing that the deeper areas of the shelf were not exposed during the regression.

(d) The *Cerro Espuelitas Formation* is 1,200 m thick and composed of intercalated dark shales and chemical sediments, such as oxide-facies BIF and chert (Gaucher, 2000).

(e) The ~300 m-thick *Cerros San Francisco Formation* is composed of mature quartz-arenites and subarkoses (Montaña and Sprechmann, 1993; Gaucher, 2000). It rests with erosional unconformity on shales of the Cerro Espuelitas Formation or directly on the pre-Ediacaran basement.

(f) The *Cerro Victoria Formation* is 400 m thick and comprises stromatolitic dolostones with trace fossils indicative of Cambrian age (Sprechmann et al., 2004; Gaucher et al., 2007c).

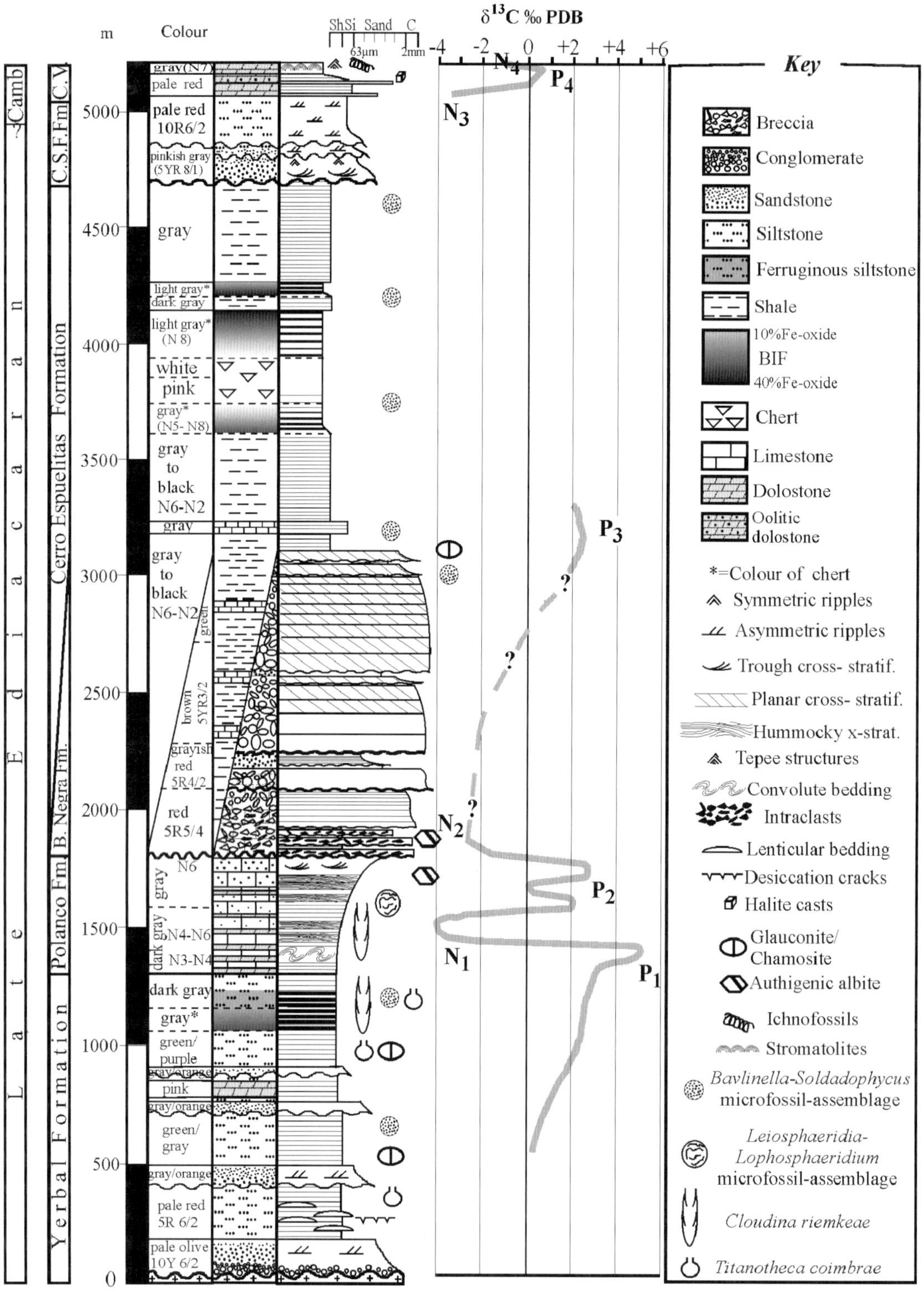


**Figure 4.2.5** Simplified stratigraphic column of the Arroyo del Soldado Group, modified from Gaucher (2000). CSF Fm: Cerros San Francisco Formation. CV: Cerro Victoria Formation. A more detailed discussion of the bio- and chemostratigraphic data is given in Chapters 4.3 and 4.4, respectively.

Volcanic, volcanoclastic and pyroclastic rocks do not occur in the ASG. Carbonates were deposited on pericontinental, storm-dominated carbonate ramps. Siliciclastic units were deposited in a basin with a gentle palaeoslope, resulting in sandstones being texturally and mineralogically mature quartz-arenites or subarkoses (Gaucher, 2000). Palaeocurrent analyses, thicknesses and facies distribution suggest that the basin opened to the east and south. Together, these features indicate a stable, Atlantic-type continental margin as the geotectonic setting of the Arroyo del Soldado basin (Gaucher, 2000), which has been recently confirmed by detrital zircon U-Pb ages of Arroyo del Soldado sandstones (Gaucher et al., 2008b; Blanco et al., 2009; see Chapter 4.6).

This platform was probably the depositional setting of the correlative Sierras Bayas Group of Argentina (Gaucher et al., 2005b) and the Corumbá Group of Brazil (Gaucher et al., 2003; Figure 4.1.2).

In a recent paper, Pecoits et al. (2008) confirm the stratigraphic scheme put forward by Gaucher (2000), but correlate the Cerros San Francisco Formation with the Piedras de Afilar Formation, thereby synonimizing both units as Piedras de Afilar Formation. However, a number of reasons militate against this correlation (Gaucher et al., 2008b), which will be discussed in Section 4.2.4. Correlations of the ASG with other units of the RPC will be dealt with in Chapters 4.3 and 4.4.

#### 4.2.3.3.2. Age constraints

The age of the ASG is radiochronologically constrained by: (a) a maximum U-Pb SHRIMP age of 633 ± 12 Ma for the Puntas del Santa Lucía pluton (Hartmann et al., 2002a), which the ASG unconformably overlies; (b) a better maximum age constraint is provided by an U-Pb SIMS zircon age of 583 ± 7 Ma for the Arroyo Mangacha Granite, which is overlain by sandstones of the Cerros San Francisco Formation (Gaucher et al., 2008b); (c) a detrital zircon U-Pb LA-ICP MS age of 566 ± 8 Ma for sandstones of the Barriga Negra Formation (Blanco et al., 2009) and (d) a minimum Rb-Sr isochron age of 532 ± 11 Ma for the Guazunambí Granite (Kawashita et al., 1999a), which intrudes the ASG. In addition, K-Ar ages ranging between 532 ± 16 and 492 ± 14 Ma have been reported for the recrystallisation of pelites within the group (Cingolani et al., 1990b; Gaucher, 2000). Thus, deposition of the ASG is bracketed between 583 and 532 Ma (Late Ediacaran-Lower Cambrian), with the Barriga Negra and younger formations being deposited after 566 ± 8 Ma.

Biostratigraphic data further allow to assign the lower and middle ASG to the Late Ediacaran, and the upper ASG to the lowermost Cambrian. Most significant is the occurrence in the lower-middle ASG of an assemblage of shelly fossils including *Cloudina riemkeae* Germs (1972a) and *Titanotheca coimbrae* Gaucher and Sprechmann (1999), as well as a low-diversity acritarch assemblage dominated by *Bavlinella faveolata*, *Soldadophycus* and *Leiosphaeridia* (Gaucher et al., 1996, 1998, 2004b; Gaucher, 2000; Figure 4.2.5, see Chapter 4.3). Carbon and strontium isotope chemostratigraphy supports a Late Ediacaran age for the lower-middle ASG (Gaucher et al., 2004c,d, 2007c). Substantial evidence supports a Lower Cambrian correlation for the Cerro Victoria Formation. Large ichnofossils representing complex infaunal burrow systems, classified by Sprechmann et al. (2004) as *Thalassinoides* isp., are common in the formation. Carbon-isotopic curves for the Cerro Victoria Formation (Gaucher et al., 2007c) closely resemble Lower Cambrian (lower Nemakyt-Daldyn or Fortunian) $\delta^{13}C$ global curves between 542 and 535 Ma (Derry et al., 1994; Montañez et al., 2000; Maloof et al., 2005), characterised by low-amplitude, high-frequency secular variations (see Chapter 4.4).

## 4.2.4. Piedras de Afilar Formation

The Piedras de Afilar Formation crops out in a restricted area of the recently recognised Tandilia Terrane (Bossi et al., 2005; Figures 4.1.3 and 4.1.6) in southern Uruguay (Jones, 1956; Bossi et al., 1998; Spoturno et al., 2005). It is more than 700 m thick and made up of a basal sandstone succession that passes upwards into shales and then carbonates at the top (Figure 4.2.6). The sedimentary succession, as described by Coronel et al. (1982), is composed of, in ascending order: 350 m of predominantly quartzitic sandstones and shale interbeds with cross stratification, ripple marks and flute casts common in the sandstones; 350 m of laminated siltstones; and fine-grained, intensively folded limestones. Dolerite sills intrude and significantly alter the stratigraphic succession at many levels (Figure 4.2.6). According to Pecoits et al. (2008), the Piedras de Afilar Formation comprises, at its stratotype, 430 m coarse to very fine, cross-bedded sandstones, followed by 150 m of laminated siltstones and capped by 25 m of greenish and grey limestones (Figure 4.2.6).

The basement of the Piedras de Afilar Formation is made up of amphibolite-facies metasedimentary rocks of the Montevideo Formation (Pando Belt), as well as the Soca rapakivi granite (Oyhantçabal et al., 1998). The intrusion yielded an U-Pb SHRIMP zircon age of 2,056 ± 6 Ma (Santos et al., 2003). The Piedras de Afilar Formation rests with erosional and angular unconformity on these basement units, thus representing an unmetamorphosed sedimentary cover of the RPC.

### 4.2.4.1. Age and correlations

A maximum age constraint is provided by the youngest detrital zircon U-Pb ages of 1,008 ± 13 and 1,006 ± 31 Ma reported by Gaucher et al. (2008b) for the Piedras de Afilar Formation. Pamoukaghlian et al. (2006) reported $\delta^{13}C$ values between +5.05 and +5.80‰ PDB for carbonates in the upper Piedras de Afilar Formation (Figure 4.2.6). $\delta^{13}C$ values between +5 and +6‰ PDB are more often encountered in Neoproterozoic carbonate

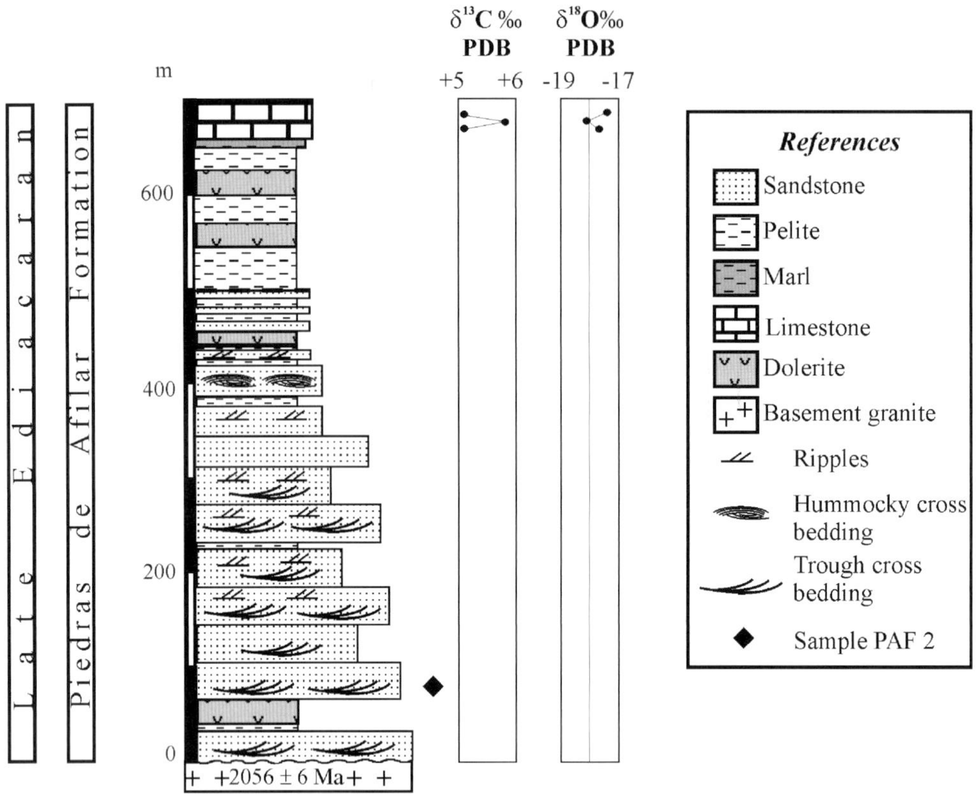


**Figure 4.2.6** Simplified stratigraphic column of the Piedras de Afilar Formation, based on Pamoukaghlian et al. (2006), Coronel et al. (1982), Gaucher et al. (2008b) and Pecoits et al. (2008). Basement ages according to Santos et al. (2003). U-Pb detrital zircon ages for sample PAF 2 were reported by Gaucher et al. (2008b, see Chapter 4.6).

successions, especially in the Cryogenian and Ediacaran (Halverson et al., 2005; see Chapter 10). Cambrian carbonates, on the other hand, are characterised by $\delta^{13}C$ values between −4 and +2.5‰ PDB (Montañez et al., 2000). Therefore, the data reported by Pamoukaghlian et al. (2006) suggest a Cryogenian or Ediacaran age for the Piedras de Afilar Formation.

Pecoits et al. (2008) suggest correlation of the Piedras de Afilar and Cerros San Francisco formations on lithostratigraphic grounds alone. Four independent lines of evidence militate against this correlation: (a) $\delta^{13}C$ values of carbonates of the upper Piedras de Afilar Formation (Pamoukaghlian et al., 2006) are much higher than $\delta^{13}C$ of carbonates overlying the Cerros San Francisco Formation (Cerro Victoria Formation), which range between −3 and +0.6‰ PDB (Gaucher et al., 2007c; Figure 4.2.5); (b) stromatolitic carbonates of the Cerro Victoria Formation are distinctively different from marls and clastic carbonates of the upper Piedras de Afilar Formation; (c) source areas of both units were very different; the Piedras de Afilar Formation being derived mainly from Meso- and Palaeoproterozoic sources, and the Cerros San Francisco Formation from Archaean and Palaeoproterozoic rocks (Blanco et al., 2007, 2009; Gaucher et al., 2008b; see Chapter 4.6) and (d) the poorly preserved acritarchs reported by Pecoits et al. (2008) for the Piedras de Afilar Formation actually favour correlation with the Yerbal Formation and not with the Cerros San Francisco Formation, as pointed out by Pamoukaghlian et al. (2006).

The available data point to a correlation of the Piedras de Afilar Formation with the Yerbal Formation (ASG) and the Cerro Largo-Olavarría formations (Sierras Bayas Group), because: (a) they are all overlain by carbonates characterised by positive $\delta^{13}C$ values up to 5.5‰ PDB (Gaucher et al., 2004c; Pamoukaghlian et al., 2006; Gómez Peral et al., 2007; see Chapter 4.4); (b) they represent deepening-upward sequences, with quartz-arenites at the base and pelites at the top; (c) in all cases provenance is dominated by Palaeoproterozoic detritus and a significant (but variable) proportion of Mesoproterozoic grains (Gaucher et al., 2008b; Blanco et al., 2009; see Chapter 4.6) and (d) the depositional environment of the three units was an Atlantic-type shelf deepening to the east and south. According to this correlation, the Piedras de Afilar Formation represents a shallower lateral equivalent of the Yerbal Formation.

Regardless the preferred correlation of the Piedras de Afilar Formation, the latter should be maintained as a separate unit. The formation has its own, distinctive lithologic characteristics, its outcrops are separated from the possible correlates by the crustal-scale Sarandí del Yí Shear Zone, and the name is well established in the literature since Jones (1956). More specifically, the International Stratigraphic Guide states that '... in case of subtle lateral transition, where boundaries have to be arbitrary, the use of one name in a broad sense may be preferable to the use of two or more names' (Salvador, 1994, p. 40). Considering that important facies changes (not 'a subtle transition') and a large shear zone occur between the Piedras de Afilar Formation (not an 'arbitrary boundary') and its possible correlates, it becomes clear that the name should be maintained.

If the correlation of the unit with the Yerbal Formation is confirmed, as seems the case, the Piedras de Afilar Formation may be included in the ASG as a lateral equivalent of the Yerbal Formation in the Tandilia Terrane. This is at odds with the proposal of Pecoits et al. (2008), which suggested to apply the name 'Piedras de Afilar Formation' also to the outcrops assigned to the Cerros San Francisco Formation in the Nico Pérez Terrane.

## *ACKNOWLEDGEMENTS*

This study has been partially funded by CSIC, Uruguay (projects: 'Estratigrafía de secuencias carbonáticas del Neoproterozoico del Terreno Nico Pérez' and 'Estratigrafía, petrografía, edad y distribución del Grupo Mina Verdún'). Insightful reviews by Léo A. Hartmann and Peter Sprechmann helped improve an earlier draft. We are indebted to José M. Canalicchio (Cementos Avellaneda, SA) and Flavio García Repetto (Cementos Artigas, SA) for providing access to quarries and logistical support.

CHAPTER 4.3

# Biostratigraphy☆

Claudio Gaucher[1] and Daniel Poiré[2]

## Contents



## 4.3.1. Introduction

Neoproterozoic stromatolites were first recognised in the Río de la Plata Craton (RPC) by González Bonorino (1954) in the Sierras Bayas Group (Argentina). The first body fossils were also described from this unit by Fairchild (1978) and Pothe de Baldis et al. (1983), consisting of organic-walled microfossils (acritarchs). In Uruguay, Montaña and Sprechmann (1993) were the first to recognise Cambrian stromatolites in the Cerro Victoria Formation. The first report of Neoproterozoic body fossils (acritarchs) in Uruguay is due to Gaucher and Schipilov (1994) and Gaucher et al. (1996). Since these first reports, around a dozen papers appeared that deal with Neoproterozoic and Cambrian fossils in the RPC. The sedimentary successions deposited on and around the RPC record the existence of a diversified biota in Ediacaran times, including cyanobacteria, acritarchs, shelly fossils, stromatolites, trace fossils and – possibly – rare members of the Ediacara fauna.

## 4.3.2. Tonian and Cryogenian Fossil Record

Due to the fact that rifting of the RPC from Rodinia begun around 650 Ma (see Chapter 4.6), the Early Neoproterozoic sedimentary record is very poor and controversial. Two units of potentially lower Neoproterozoic age are fossiliferous, namely the Mina Verdún Group in Uruguay and the Villa Mónica Formation (Sierras Bayas Group) in Argentina.

The Mina Verdún Group is a marine succession occurring in the Nico Pérez Terrane (Figures 4.2.2–4.2.3). Limestones of the Calabozo Formation contain an almost monospecific assemblage of stromatolites, dominated by *Conophyton* (Poiré et al., 2003a, 2005b; Gaucher et al., 2004a, 2006, 2007a). They consist of unbranching, subcylindrical columnar stromatolites with typical conical laminae (Figure 4.3.1A–B). Their profile is angulate to geniculate, and their cross section circular to oblong (Poiré et al., 2005b). Column height reaches 70 cm, and

☆Gaucher, C., Poiré, D., 2009. Biostratigraphy. Neoproterozoic-Cambrian evolution of the Río de la Plata Palaeocontinent. In: Gaucher, C., Sial, A.N., Halverson, G.P., Frimmel, H.E. (Eds.): Neoproterozoic-Cambrian tectonics, global change and evolution: a focus on southwestern Gondwana. Developments in Precambrian Geology, 16, Elsevier, pp. 103–114.

[1] Departamento de Geología, Instituto de Ciencias Geológicas, Facultad de Ciencias Iguá 4225, 11400 Montevideo, Uruguay.
[2] Centro de Investigaciones Geológicas (CIG), Universidad Nacional de La Plata-CONICET, calle 1 n. 644, 1900-La Plata, Argentina.

Developments in Precambrian Geology, Volume 16
ISSN 0166-2635, DOI 10.1016/S0166-2635(09)01606-5

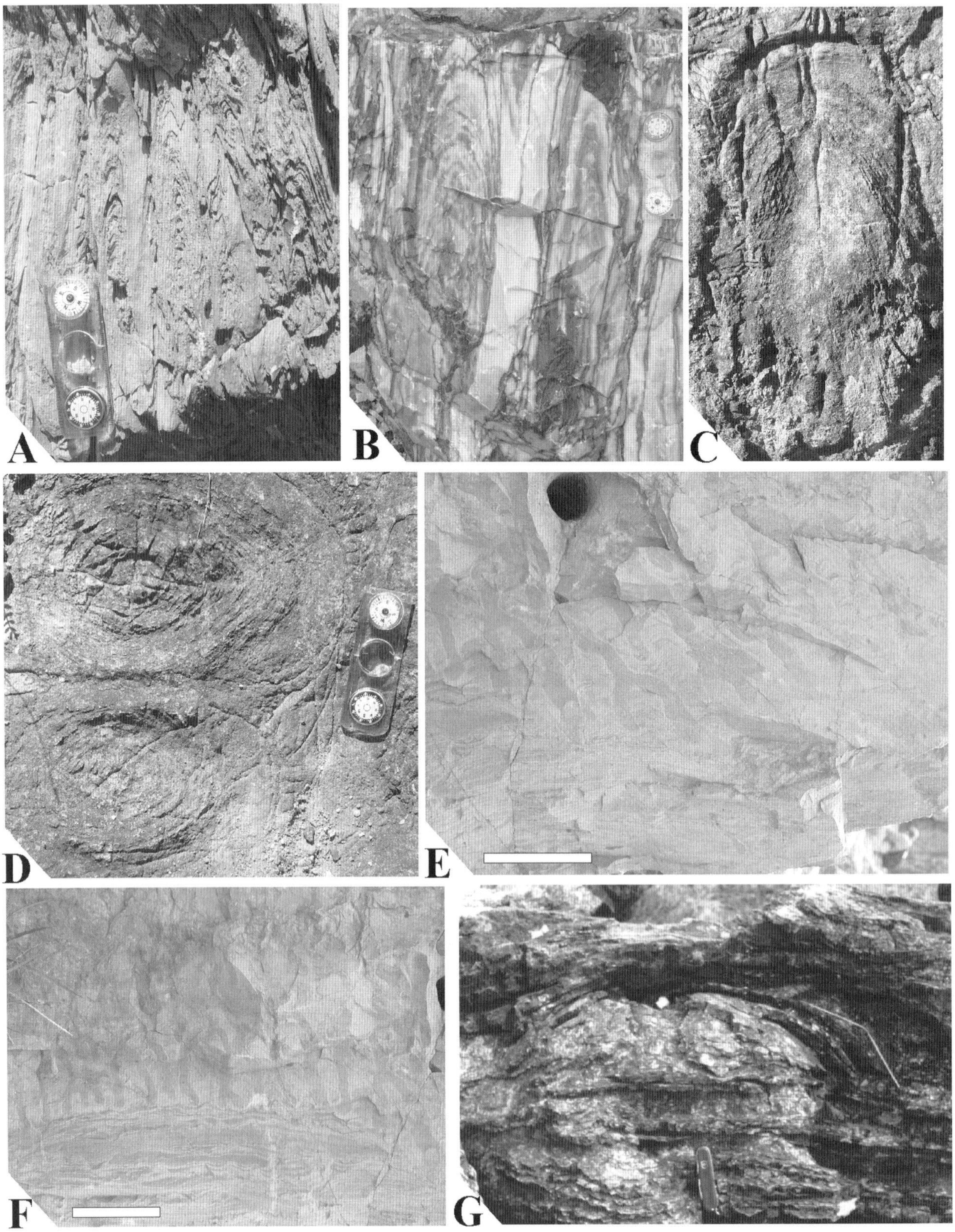


**Figure 4.3.1** Stromatolites of the Mina Verdún (A, B), Sierras Bayas (C–F) and Arroyo del Soldado (G) groups. (A) *Conophyton* from the Mina Verdún Group at its stratotype. (B) Larger, adjacent *Conophyton* columns from the Mina Verdún Group at Burgueño Quarry. (C, D) *Conophyton* from the Villa Mónica Formation at Piedra Amarilla Quarry, in longitudinal (C) and cross section (D). Height of column in C: 21 cm. (E, F) Gymnosolenids developed on top of stratiform stromatolites, Villa Mónica Formation, same locality as C, D. (G) Large, domical stromatolite of the Cerro Victoria Formation at its stratotype, from Gaucher (2000). Scale is 8 cm long in A, B, D; 10 cm in E, F and 9 cm in G.

diameter 25 cm. Rare occurrences of digitate columnar stromatolites were reported from the Mina Verdún Group by Poiré et al. (2005b, 2006).

A minimum age of the Mina Verdún Group is provided by the unconformably overlying Las Ventanas Formation, of Early Ediacaran age (Blanco and Gaucher, 2005; Gaucher et al., 2008a; see below). Carbon isotopic data presented by Gaucher et al. (2006, 2007a) show low-amplitude $\delta^{13}C$ oscillations and a plateau around 2‰ Pee Dee Belemnite (PDB), thus consistent with a Late Mesoproterozoic to Tonian age for the Mina Verdún Group (Kah et al., 1999; see Chapter 4.4). Other *Conophyton*-dominated stromatolite assemblages include: (1) the Itaiacoca Group (Sallun Filho and Fairchild, 2005) of SE Brazil (Ribeira Belt), with a minimum U-Pb TIMS age of $934 \pm 36$ Ma (Siga Jr. et al., 2006); (2) the Paranoá Group of central Brazil (Brasília Belt), possibly deposited between 1,200 and 900 Ma (Sallun Filho and Fairchild, 2005), and showing a C-isotope curve similar to the Mina Verdún Group (Santos et al., 2000) and (3) the Uluksan and Nunatsiaq groups of the basal Bylot Supergroup (Canada), dated by Pb-Pb method at 1,200 Ma (Sherman et al., 2002). According to Walter (1994), the period between 1,350 and 1,050 Ma marks the highest abundance and diversity of coniform stromatolites. Thus, it is still uncertain if the Mina Verdún Group was deposited in the Late Mesoproterozoic or Early Neoproterozoic. U-Pb dating of interbedded rhyolites and tuffs described by Gaucher et al. (2007a) might help clear this issue.

The Villa Mónica Formation (lower Sierras Bayas Group) hosts a diverse stromatolite assemblage described in detail by Poiré (1987). Stromatolites were grouped into 16 morphotypes by Poiré (1987, 1993), namely (Figure 4.3.1C–F) *Colonella* fm., *Conophyton resotti*, *Conophyton* fm., *Cryptozoon* fm., *Gongylina* fm., *Gymnosolen* fm., *Inzeria* fm., *Jacutophyton* fm., *Jurusonia nisvensis*, *Katavia* fm., *Kotuikania* fm., *Kussiella* fm., *Minjaria* fm., *Parmites* fm., *Parmites* cf. *cocrescens* and *Stratifera* fm. (Poiré, 1987, 1993).

Acritarchs assigned to the species *Leiosphaeridia minutissima*, *Leiosphaeridia tenuissima* and *Synsphaeridium* sp. were described from shales of the Villa Mónica Formation by Gaucher et al. (2005b; Figure 4.3.2I). The assemblage is characterised by low-diversity, dominance of leiosphaerids and large size of *L. tenuissima*, reaching 450 μm (Figure 4.3.2I).

Whereas stromatolite biostratigraphy (Poiré, 1987) and Rb-Sr ages of pelites between $793 \pm 32$ Ma and $805 \pm 28$ Ma (Bonhomme and Cingolani, 1980) point to a Tonian or Early Cryogenian age, the acritarch assemblage described by Gaucher et al. (2005b) favors an Early Ediacaran age.

## 4.3.3. Ediacaran Fossil Record

The Ediacaran is well represented in the RPC by a number of sedimentary successions, namely the Las Ventanas Formation, Arroyo del Soldado Group and Sierras Bayas Group–Cerro Negro Formation (see Chapter 4.2). The fossil content of each of these units will be discussed below.

### 4.3.3.1. Las Ventanas Formation

This unit represents a taphrogenic succession associated to rifting of the RPC from Rodinia (see Chapter 4.6). Sedimentary environments range from continental alluvial fans at the base to fan deltas in the middle part and open marine shales at the top. Acritarchs were described by Blanco and Gaucher (2004, 2005) and Gaucher et al. (2008a) from the upper Las Ventanas Formation (Quebrada de Viera and El Perdido members). The assemblage is leiosphaerid dominated and includes the following species (Figure 4.3.2F–H): *L. tenuissima*, *L. minutissima*, *Lophosphaeridium* sp., *Soldadophycus bossii*, *Soldadophycus major* and *Soldadophycus* sp. The vendotaenid *Vendotaenia antiqua* also occurs (Gaucher et al., 2008a). The microflora is characterised by relatively low diversity, dominance of *L. tenuissima*, absence of acanthomorphic acritarchs and large size of sphaeromorphs, reaching 400 μm in diameter. A number of species are shared with acritarch assemblages preserved in the overlying Arroyo del Soldado Group. Differences between both assemblages include the occurrence of abundant *Bavlinella faveolata* and smaller size of leiosphaerids in the Arroyo del Soldado Group (Figure 4.3.6). The assemblage occurring in the Las Ventanas Formation is assigned to the Early Ediacaran Leiosphere Palynoflora (EELP) (Grey, 2005; see Chapter 9.1), which possibly spans the interval between the base of the Ediacaran (end of the Marinoan Glaciation) and the termination of the Gaskiers Glaciation (582 Ma) or the Acraman impact event (Grey, 2005). An Early Ediacaran age is also supported by a K-Ar age of $615 \pm 30$ Ma for basalts (Sanchez Bettucci and Linares, 1996) of the lower Las Ventanas Formation (La Rinconada Member) and an Ar-Ar age of $579 \pm 1.5$ Ma for the Pan de Azúcar Pluton, which intrudes the Las Ventanas Formation (Oyhantçabal et al., 2007).

### 4.3.3.2. Lower and middle Arroyo del Soldado Group

The fossil record of this unit is the richest and – thus far – most intensively studied in the whole RPC. Whereas the Yerbal, Polanco, Cerro Espuelitas and Barriga Negra formations have been assigned to the Ediacaran, the

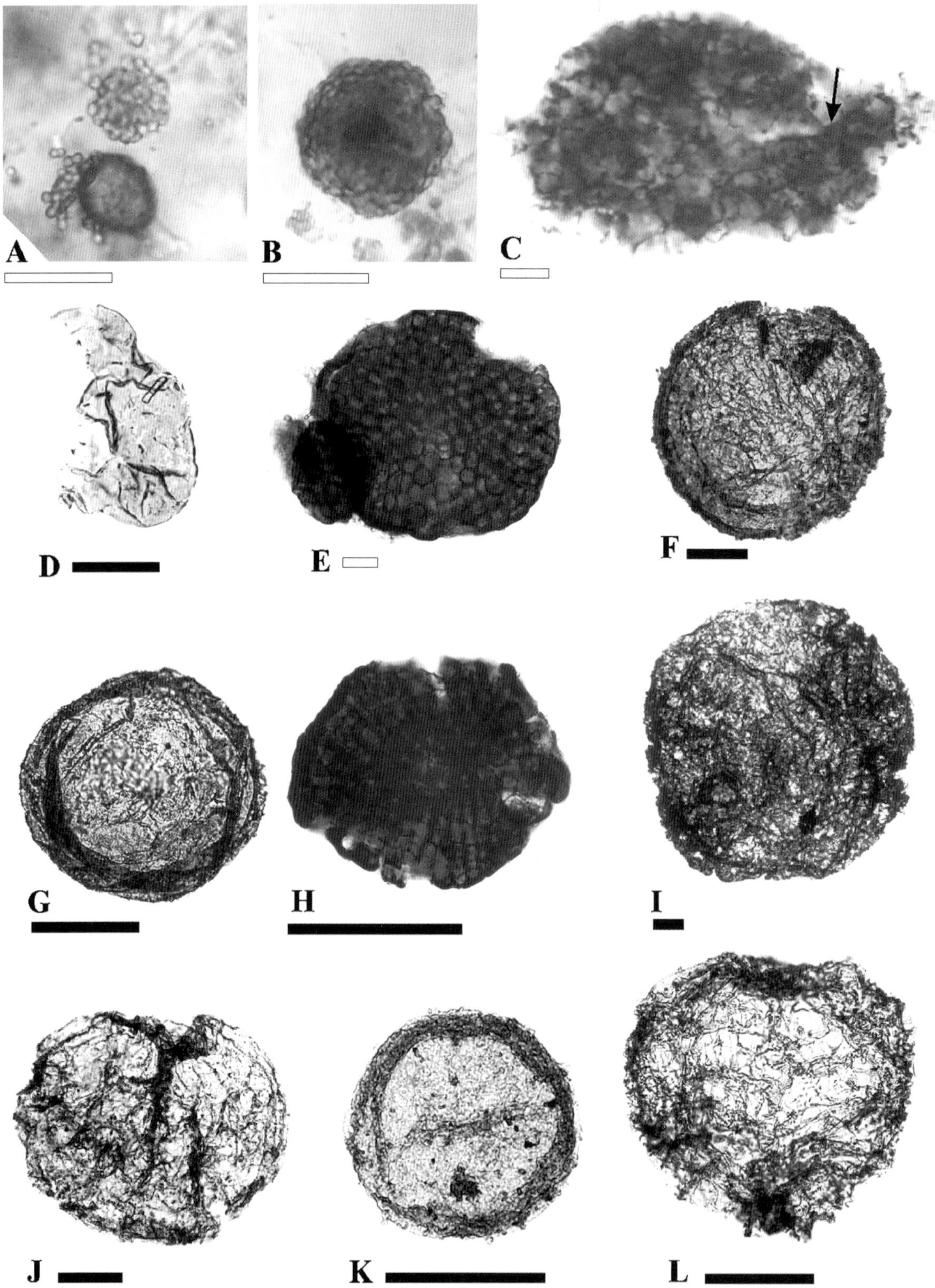


**Figure 4.3.2** Organic-walled microfossils of the Arroyo del Soldado (A–E) and Sierras Bayas (I–K) groups, and Las Ventanas (F–H) and Cerro Negro (L) formations. Specimens isolated by means of acid maceration otherwise indicated. (A, B) *Bavlinella faveolata* in thin sections from the uppermost Barriga Negra (A) and Yerbal (B) formations. Note in A the sheathed specimen releasing baeocytes (from Gaucher, 2000). (C) *Soldadophycus* cf. *S. major* Gaucher (2000) in a thin section of chert from the uppermost Yerbal Formation, La Salvaje Anticline. Note transition from sphaeroidal into filamentous cells (arrowed). (D) *Leiosphaeridia tenuissima*, Polanco Formation (from Gaucher, 2000). (E) *Soldadophycus bossii*, saucer-shaped colony with "daughter" colony, Yerbal Formation (from Gaucher et al., 2004b). (F) *L. tenuissima*, Las Ventanas Formation (from Gaucher et al., 2008a). (G) *Lophosphaeridium* sp. showing prominent concentric folds and small verrucae, Las Ventanas Formation (from Gaucher et al., 2008a). (H) *Soldadophycus* sp. saucer-shaped colony with radial rows of cells (Las Ventanas Formation; Gaucher et al., 2008a). (I) Large *L. tenuissima*, Villa Mónica Formation (from Gaucher et al., 2005b). (J, K) *L. tenuissima*, Cerro Largo Formation (from Gaucher et al., 2005b). (L) *L. tenuissima*, Cerro Negro Formation (Gaucher et al., 2005b). Note tenuous, strongly plicated vesicle wall (same as in F and J). Hollow scale bars represent 10 μm, solid scale bars represent 50 μm.

upper Arroyo del Soldado Group was deposited in the Lower Cambrian (Gaucher, 2000; Gaucher et al., 2004c, 2007c). The Ediacaran fossil groups occurring in the unit are acritarchs and shelly fossils, which are briefly described below.

### 4.3.3.2.1. Acritarchs

Organic-walled microfossils (acritarchs, filamentous microfossils, colonial sphaeroids) were the first fossils reported from the Arroyo del Soldado Group (Gaucher and Schipilov, 1994; Gaucher et al., 1996, 1998). Two different assemblages have been recognised by Gaucher (2000), namely:

(a) A *Bavlinella–Soldadophycus* assemblage strongly dominated by *B. faveolata*, *S. bossii* and/or *S. major* occurs in the Yerbal and Cerro Espuelitas formations (Figure 4.3.2A–C, E). It represents a low-diversity, high-abundance assemblage including sphaeromorphic acritarchs, colonial sphaeroids and filamentous microfossils (Gaucher, 2000; Gaucher et al., 2004b). Sphaeromorphs are characteristically small ($<100\,\mu m$).
(b) A *Leiosphaeridia–Lophosphaeridium* assemblage is restricted to the Polanco Formation, showing somewhat higher diversity. Typical species are *L. minutissima*, *L. tenuissima* (Figure 4.3.2D), *Lophosphaeridium* spp., colonial sphaeroids (*Coniunctiophycus*, *Glenobotrydion*, *Myxococcoides*) and large, branched filaments (Gaucher, 2000). In contrast to the previous assemblage, microfossils of the *Leiosphaeridia–Lophosphaeridium* assemblage are sparse.

Gaucher (2000) interpreted the different assemblages as reflecting different palaeoenvironmental conditions. The *Bavlinella–Soldadophycus* assemblage represents an eutrophic community, similar to other *Bavlinella*-dominated assemblages worldwide (Mansuy and Vidal, 1983; Knoll and Swett, 1985; Palacios, 1989; Vidal and Nystuen, 1990; Gaucher et al., 2003, 2005c; Gaucher and Germs, 2006, 2007). On the other hand, the *Leiosphaeridia–Lophosphaeridium* assemblage probably represents a plankton community adapted to lower nutrient levels (oligotrophic).

Biostratigraphically, both communities are comparable to other Late Ediacaran acritarch assemblages, characterised by long-ranging species, comparatively small sphaeromorphs and low diversity (Vidal and Moczydlowska-Vidal, 1997; Knoll, 2000). The fact that four species are shared with the Las Ventanas Formation (*L. tenuissima*, *L. minutissima*, *S. bossi* and *S. major*: Figure 4.3.6) suggests that there is no large hiatus between both units (Gaucher et al., 2008a). However, important differences were noted by Gaucher et al. (2008a): (a) occurrence of *B faveolata* as a dominant species in the Arroyo del Soldado Group, unknown in the Las Ventanas Formation; (b) occurrence of large leiosphaerids up to $400\,\mu m$ in diameter in the latter unit; and (c) dominance of *Leiosphaeridia* in Las Ventanas Formation, unlike the Arroyo del Soldado Group.

Decreasing maximum leiosphaerid diameter characterises the transition from the Las Ventanas Formation into the overlying Arroyo del Soldado Group: from $400\,\mu m$ in the Las Ventanas Formation (Figure 4.3.2F) to $125\,\mu m$ in the Polanco Formation of the Arroyo del Soldado Group (Figure 4.3.2D). This trend has been reported in other shallow-water, marine Ediacaran successions elsewhere (Butterfield, 1997; Gaucher et al., 2005b). Leiospherids occurring in Early Ediacaran successions of Australia (Ediacaran Leiosphere Palynoflora) reach $240\,\mu m$ in diameter (Grey, 2005).

According to the biostratigraphic schemes of Vidal and Moczydlowska-Vidal (1997), acritarchs of the Arroyo del Soldado Group match the Kotlin–Rovno assemblage of Late Ediacaran age (Figure 4.3.6). This is in accordance with the above-mentioned assignment of the leiosphaerid assemblage of the Las Ventanas Formation to the older EELP (Grey, 2005; see Chapter 9.1).

### 4.3.3.2.2. Shelly fossils

A surprisingly diverse assemblage of skeletal fossils has been described by Gaucher and Sprechmann (1999) and Gaucher (2000) from the lower Arroyo del Soldado Group, which is the most diverse so far reported from the Ediacaran worldwide. It comprises calcified tubes, agglutinated shells, one species with a possibly phosphatic skeleton and other forms with yet undetermined shell mineralogy.

*Cloudina riemkeae* Germs (1972a) occurs mainly in the upper Yerbal Formation (Gaucher and Sprechmann, 1999; Gaucher, 2000). The fossils, which are haematised and rather poorly preserved (Figure 4.3.4D), occur in banded siltstones in life position (Gaucher and Sprechmann, 1999). This is shown by (1) the high angle between the shell axis and bedding plane, (2) lack of size sorting and preferred orientation and (3) absence of 'event beds' with higher shell concentration. The opposite is true for *C. riemkeae* occurring in the overlying Polanco Formation. We report here the occurrence of fragmented *Cloudina* shells in storm deposits of unit B of the Polanco Formation at the Calera de Recalde Syncline (Gaucher et al., 2004c; Figures 4.3.3I–J, 4.3.4A,C). Although a few specimens preserve their original calcitic shell, most of the specimens are replaced by quartz (Figure 4.3.3J). Fragments are small (diameter: 0.16–0.54 mm, mean = 0.30 mm, $N = 14$) and show a narrow

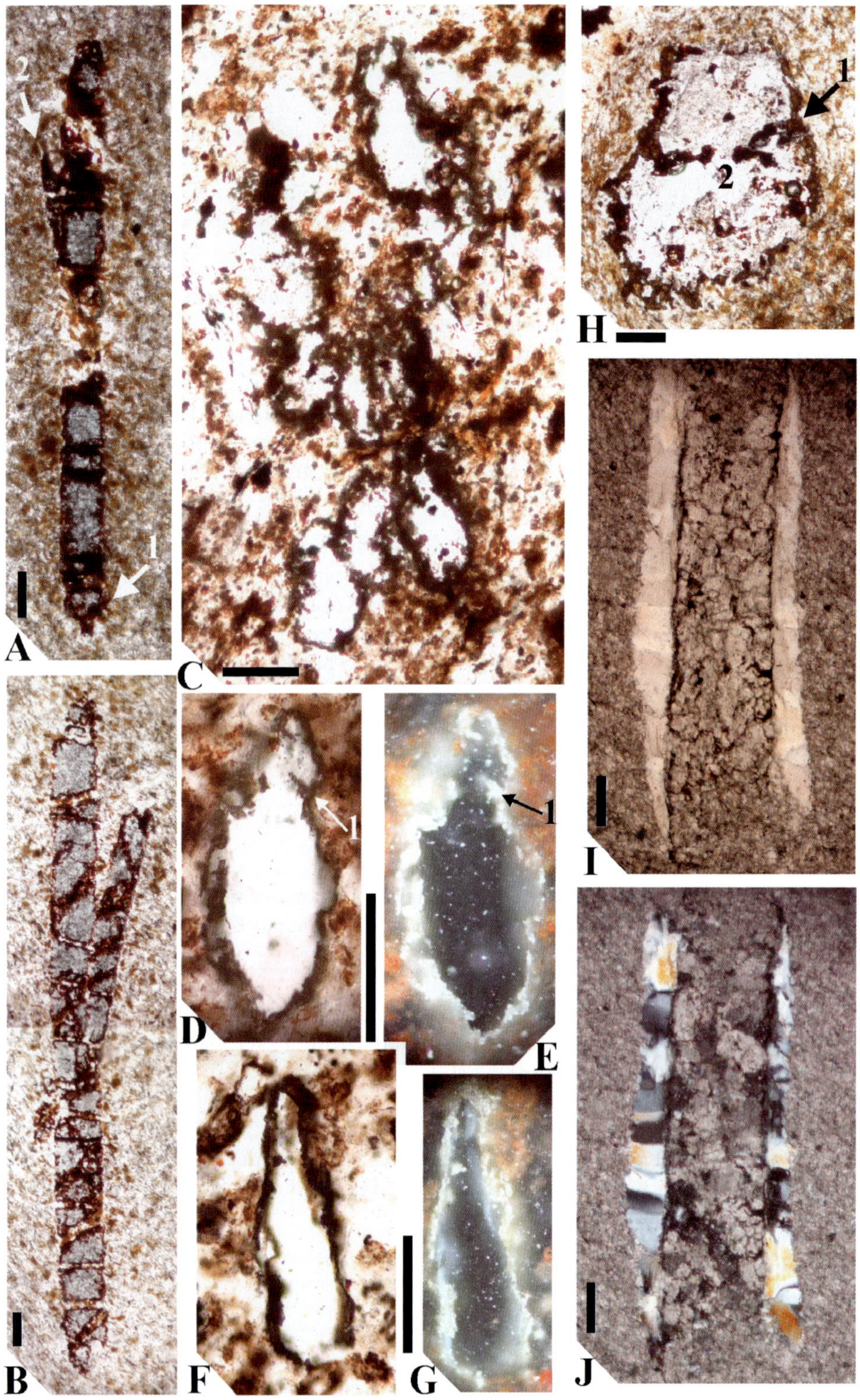


**Figure 4.3.3** *Waltheria*, *Titanotheca* and *Cloudina* in thin sections from the Arroyo del Soldado Group. (A) Holotype of *Waltheria marburgensis* Gaucher and Sprechmann (1999), Yerbal Formation. Note septate shell, sphaerical chamber (arrowed, 1) and small branch (arrowed, 2). (B) *W. marburgensis* Gaucher and Sprechmann (1999), same unit and locality as previous. (C) *Titanotheca coimbrae* Gaucher and Sprechmann (1999), group of attached, gregarious specimens, Yerbal Formation, 5 km to the north of Minas. (D, E) *T. coimbrae*, solitary specimen with curved neck and septum (arrowed, 1) in transmitted (D) and reflected (E) light; same locality as previous. Note highly reflective rutile grains that compose the shell. (F, G) *T. coimbrae*, flask-shaped specimen, in transmitted and reflected light, same locality as previous. (H) *T. coimbrae* figured by Gaucher and Sprechmann (1999), bilocular specimen with a septum (1) and a pore connecting both chambers (2), Yerbal Formation. (I, J) Fragment of *Cloudina riemkeae* Germs (1972a) from the Polanco Formation. Replacement of original shell minerals by quartz is clearly visible with cross nicols (J). All scale bars represent 0.1 mm.

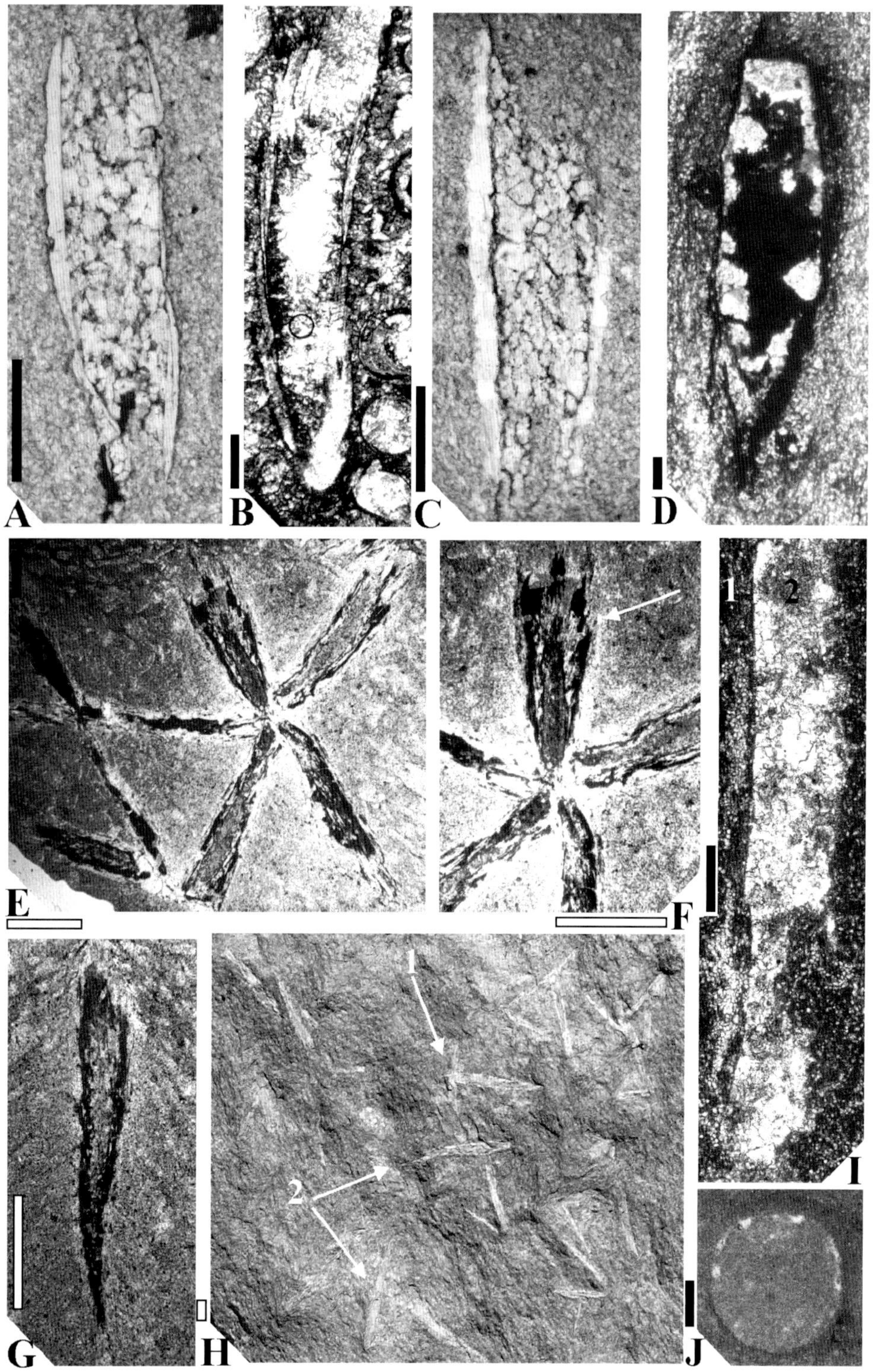


**Figure 4.3.4** *Cloudina* from the Arroyo del Soldado, Nama and Sierras Bayas groups and tubular-conical fossils of the Yerbal Formation (thin sections). (A, C) *Cloudina riemkeae* Germs (1972a) in limestones of the Polanco Formation (Unit B) with silicified shells and sparitic infill. (B) *C. riemkeae* Germs (1972a) from the type area in the Nama Group. Note cone-in-cone structure and same anatomy as specimen figured in A. (D) *C. riemkeae* in siltstones of the upper Yerbal Formation, showing haematised shell and infill composed of quartz and haematite (Gaucher et al., 2003). (E, F) Star-shaped aggregate of conical fossils in a bedding-parallel section of siltstone from the middle Yerbal Formation, Quebrada de los Cuervos Member (Gaucher, 2002). Note deeply nested cones (arrowed). (G) Solitary and sinuous cone from the same unit. Note sediment infill. (H) Conical-tubular fossils as observed on bedding planes of siltstones from the Quebrada de los Cuervos Member. Note (1) branching at 90° and (2) solitary, straight or sinuous tubes. (I) *C. riemkeae* Germs (1972a) in limestones of the Loma Negra Formation (longitudinal section). Note (1) recrystallised outer tube and (2) inner cavity filled with sparitic calcite (from Gaucher et al., 2005b). (J) *C. riemkeae* from the same locality as previous, cross section (from Gaucher et al., 2005b). Solid scale bars represent 0.25 mm, hollow scale bars represent 2.5 mm.

$\delta^{13}C$ of 2.4‰ PDB (Figure 4.4.2). Microfossils of the Cerro Espuelitas Formation allow to place it in the Ediacaran (Gaucher, 2000), thus showing that the negative excursion at its base is older than the basal Cambrian negative excursion (Derry et al., 1994; Amthor et al., 2003; Maloof et al., 2005). The latter is likely represented in the ASG by the negative excursion at the base of the Cerro Victoria Formation, as shown by trace fossils occurring in that unit (Sprechmann et al., 2004; Gaucher et al., 2007c; Figure 4.4.2). Therefore, negative excursions $N_2$ and $N_3$ are of different age, the positive, uppermost Ediacaran $\delta^{13}C$ excursion in between ($P_3$) being well represented in the Nama Group (Grotzinger et al., 1995), in the Corumbá Group (Misi et al., 2007) and elsewhere.

#### 4.4.2.2.6. Cerro Victoria positive ($P_4$) and negative ($N_4$) excursions

After ca. 130 m of carbonates yielding negative $\delta^{13}C$ values, slightly positive values around 0.3‰ PDB are recorded at the transition between units B and C of the Cerro Victoria Formation (Gaucher et al., 2007c). $\delta^{13}C$ in positive excursion $P_4$ reaches 0.64‰ PDB and shows fairly constant values for ca. 180 m of mainly stromatolitic carbonates (Figure 4.4.2). Up section (lower unit D), very low-amplitude, high-frequency oscillations around a mean value of 0‰ PDB are recorded in *Thalassinoides*-bearing dolostones (Sprechmann et al., 2004; Gaucher et al., 2007c). Finally, increasingly negative values of up to −1.1‰ PDB ($N_4$ excursion) characterise uppermost carbonates of the Cerro Victoria Formation (Figure 4.4.2). The described pattern is identical to the Lower Cambrian $\delta^{13}C$ global curve between 542 and 535 Ma (lower Nemakyt-Daldyn) presented by Maloof et al. (2005). Multiple, successive negative excursions in this interval (Maloof et al., 2005), however, prevent an unambiguous assignment of negative excursions $N_3$ and $N_4$.

### 4.4.2.3. Sierras Bayas Group

Gómez Peral et al. (2007) report $\delta^{13}C$ analyses for carbonates and organic matter of the Villa Mónica, Loma Negra and lower Cerro Negro formations. The Sierras Bayas Group is a shallow-water succession characterised by multiple erosional unconformities which, in practical terms, means that the $\delta^{13}C$ record is fragmentary and mostly reflects periods of high sea level.

Stromatolitic dolostones of the Villa Mónica Formation record a positive $\delta^{13}C$ excursion between −1.4‰ and +2.2‰ PDB (Gómez Peral et al., 2007), being truncated at the top by an erosional unconformity (Figures 4.2.1 and 4.4.2). Consistently positive $\delta^{13}C$ values up to 4.5‰ PDB characterise carbonates of the overlying Loma Negra Formation, which are truncated by the Barker palaeokarst (see Chapters 4.2 and 4.5; Figure 4.4.2). $\delta^{13}C$ of kerogens from the same unit vary between −28‰ and −27‰ PDB, $\Delta^{13}C$ ranging between 30.5‰ and 32‰ PDB (Gómez Peral et al., 2007). Marls of the basal Cerro Negro Formation resting atop the Barker Surface also show rising, positive $\delta^{13}C_{carbonate}$ values between 3.5‰ and 4.3‰ PDB (Figure 4.4.2) and corresponding $\Delta^{13}C$ values around 32‰ PDB. However, the fact that an important erosional surface separates the Loma Negra and Cerro Negro formations shows that the mentioned positive $\delta^{13}C_{carbonate}$ values in fact represent two different excursions (correlated here to $P_1$ and $P_3$; Figure 4.4.2) and not just one.

### 4.4.2.4. Piedras de Afilar Formation

Limestones occur at the top of the Piedras de Afilar Formation, and yielded $\delta^{13}C$ between 5.05‰ and 5.80‰ PDB, with little variation in 10 m of section (Pamoukaghlian et al., 2006). Rather negative $\delta^{18}O$ values between −18.1‰ and −17.5‰ are best explained by the thermal effects of nearby dolerites, which probably did not alter the carbon-isotopic composition.

## 4.4.3. Strontium-Isotope Chemostratigraphy

Strontium-isotope data are available for carbonates of the ASG (Kawashita et al., 1999a; Gaucher et al., 2004d, 2007c) and Sierras Bayas Group (Kawashita et al., 1999b). In this work, we report $^{87}Sr/^{86}Sr$ analyses of limestones from the Mina Verdún Group.

### 4.4.3.1. Mina Verdún Group

Ten analyses of $^{87}Sr/^{86}Sr$ ratios were carried out on limestones of the El Calabozo and lower Gibraltar formations at their stratotype. Six analyses from stromatolitic, pure limestones of the El Calabozo Formation yielded values between 0.7081 and 0.7189. Four analyses of limestones from the lower Gibraltar Formation at the same section yielded very radiogenic $^{87}Sr/^{86}Sr$ ratios mostly between 0.7153 and 0.7174. All values are too high, and obviously reflect either post-depositional resetting and/or input of radiogenic strontium from silicates. Only the least altered samples from the El Calabozo Formation, yielding ratios of 0.7081 and 0.7084, could be regarded as near-seawater values. However, according to Shields (2007a) and Halverson et al. (2007a), seawater $^{87}Sr/^{86}Sr$ ratios ranged between 0.7050 and 0.7065 in the upper Mesoproterozoic-Tonian (1,300–800 Ma). Therefore, the currently available $^{87}Sr/^{86}Sr$ data for the Mina Verdún Group cannot be used to constrain its depositional age.

### 4.4.3.2. Arroyo del Soldado Group

A total of 47 $^{87}Sr/^{86}Sr$ analyses are known for the ASG, 33 being from limestones of the Polanco Formation (Gaucher et al., 2004d, 2006, this work) and 14 from dolostones of the Cerro Victoria Formation (Gaucher et al., 2007c). The behaviour of the Sr-isotopic system in these two units differ markedly from each other. Whereas dolostones of the Cerro Victoria Formation yield too radiogenic values (>0.7106: Gaucher et al., 2007c), limestones of the Polanco Formation yield mostly near-primary $^{87}Sr/^{86}Sr$ ratios (Figure 4.4.1C). This is readily explained by the lower Sr concentrations that dolomite allows in its lattice, which renders it more susceptible to post-depositional alteration (Gaucher et al., 2007c). On the other hand, limestones of the Polanco Formation are characterised by high Sr concentrations of up to 2,250 ppm (mean = 1,025 ppm, standard deviation = 470 ppm and $N = 88$), and Rb concentrations typically less than 10 ppm, denoting a negligible proportion of radiogenic Sr from silicates. $^{87}Sr/^{86}Sr$ and $\delta^{13}C$ crossplots versus common alteration proxies (Figure 4.4.1C) show the unaltered nature of these isotopic signatures. Considering this, only the results from the Polanco Formation will be discussed, which are the ones useful for strontium chemostratigraphy.

All samples from the Polanco Formation show $^{87}Sr/^{86}Sr$ values within 0.7073 and 0.7086, the only exceptions being samples near the contact of the intrusive Minas Granite, with values up to 0.7156. Even there, one sample retained a near-primary value of 0.7077 (Gaucher et al., 2006). The variation of $^{87}Sr/^{86}Sr$ ratios in limestones of the Polanco Formation is not random, but shows secular variations of an amplitude of 0.0010 (Gaucher et al., 2004d; Figure 4.4.3). The lowermost Polanco Formation is characterised by declining $^{87}Sr/^{86}Sr$ ratios between 0.7085 and 0.7080 (Figure 4.4.3), dropping to a nadir of 0.7073 at peak $\delta^{13}C$ values of the uppermost unit A (Gaucher et al., 2004d). Throughout units B and C, $^{87}Sr/^{86}Sr$ values remain low (0.7073–0.7077), only increasing in unit D to a maximum of 0.7083. One possible exception could be the apparent surge of $^{87}Sr/^{86}Sr$ values from 0.7073 to 0.7086 at the base of unit B at Calera de Recalde Syncline (Figure 4.4.3), but this needs confirmation from other sections. A final drop to 0.7076 is observed at the top of the Polanco Formation (unit E; Figure 4.4.3).

The consistency of the observed pattern, along with the geochemical indications of a primary isotopic signature (Figure 4.4.1), suggest that the observed $^{87}Sr/^{86}Sr$ oscillations reflect secular variations of seawater composition. This has profound implications for both Neoproterozoic Sr-isotope chemostratigraphy and palaeoclimate models (see below).

### 4.4.3.3. Sierras Bayas Group

Gómez Peral et al. (2007) show, on the basis of different geochemical and isotopic proxies, that limestones of the Loma Negra Formation retain primary C- and Sr-isotopic signatures. Kawashita et al. (1999b) reports $^{87}Sr/^{86}Sr$ values between 0.7069 and 0.7087 for the Loma Negra Formation, with the least altered values between 0.7069 and 0.7075. As noted by Gaucher et al. (2005b), the $^{87}Sr/^{86}Sr$ ratios of the Loma Negra Formation are almost identical to the ones reported for the Polanco Formation of the ASG. Both units were previously correlated on the basis of litho- and biostratigraphy (Gaucher et al., 2005b; see Chapters 4.2 and 4.3; Figure 4.4.2).

No $^{87}Sr/^{86}Sr$ analyses are available for dolostones of the Villa Mónica Formation, which usually contain less than 45 ppm Sr (Gómez Peral et al., 2007).

## 4.4.4. Synthesis

The age of the Mina Verdún Group is still poorly constrained, but the $\delta^{13}C$ plateau around 2‰ PDB observed in the El Calabozo and Gibraltar formations is typical of late Mesoproterozoic to Tonian carbonates (1,300–800 Ma: Bartley et al., 2007). U-Pb dating of interbedded acid volcanics described by Gaucher et al. (2007a) will help improve the still poorly constrained pre-Cryogenian global $\delta^{13}C$ curve.

Poiré and Gaucher (2007) pointed out that dolostones of the Villa Mónica Formation show a $\delta^{13}C$ curve similar to carbonates of the Mina Verdún Group, both units being pre-Ediacaran in age and containing *Conophyton* stromatolites (Figure 4.4.2). However, similar, leiosphere-dominated acritarch assemblages in the Villa Mónica

Formation and overlying units of the Sierras Bayas Group militate against this interpretation (Gaucher et al., 2005b; see Chapter 4.3). Thus, the issue of the correlation of the Villa Mónica Formation remains controversial.

The ASG contains without doubt the most detailed late Ediacaran chemostratigraphic record in whole southwestern Gondwana between ca. 570 and 550 Ma. Positive excursion $P_1$ is recorded in the Yerbal and Polanco formations in the ASG, and in the Loma Negra Formation in the Sierras Bayas Group (Figure 4.4.2). In both units, *Cloudina riemkeae* and low-diversity acritarch assemblages occur (Gaucher et al., 2005b; see Chapter 4.3). The impressive negative excursion $N_1$ that follows may be correlated to the Shuram–Wonoka anomaly (Halverson et al., 2005), which ended at $551 \pm 1$ Ma and reached its nadir before $555 \pm 6$ Ma in south China (Condon et al., 2005; Zhang et al., 2005). $\delta^{13}C$ values in the Shuram–Wonoka anomaly are as low as −10‰ PDB or less (Halverson et al., 2005), but in the Polanco Formation $\delta^{13}C$ does not drop beyond −4.5‰ PDB. This, however, can be due to ocean stratification or other local effects (Calver, 2000; Frimmel, 2004; Shen et al., 2005). The sedimentary record of the Polanco Formation favors a Phanerozoic-type glaciation as the cause of the negative anomaly (Gaucher et al., 2004d; Gaucher, 2007; see Chapter 4.5), as also observed for other Neoproterozoic negative anomalies. Most interestingly, $^{87}Sr/^{86}Sr$ reaches minimum values before the onset of $N_1$, and stays low (0.7075) well into the following positive excursion ($P_2$; Figures 4.4.2 and 4.4.3). The decrease of $^{87}Sr/^{86}Sr$ values during a glacial event could be due to: (a) ice cover of continents, diminishing the input of $^{87}Sr$ to the oceans (Jacobsen and Kaufman, 1999); (b) pre-glacial, enhanced hydrothermal activity in mid-ocean ridges leading to plankton blooms, $CO_2$ sequestration and glaciation (Gaucher, 2007); and (c) a combination of both. Considering that $^{87}Sr/^{86}Sr$ ratios begin to fall well before the onset of negative anomaly $N_1$ (Figure 4.4.3), enhanced hydrothermal activity linked to rifting events seems a more attractive hypothesis.

The same sequence of events is observed during the terminal Polanco-Barriga Negra negative excursion $N_2$, which predates platform exposure in both the Arroyo del Soldado and the Sierras Bayas Group (Figure 4.4.2), leading to the regionally important Barker palaeokarst (Barrio et al., 1991; Poiré and Gaucher, 2007). A tentative correlation of this surface with the Vingerbreek Member in the Nama Group, constrained between $549 \pm 1$ and $545 \pm 1$ Ma (Grotzinger et al., 1995; Saylor et al., 1998), showing platform reworking, a negative $\delta^{13}C$ excursion to −2‰ PDB and evidences of a glacial event (Germs, 1995) is considered plausible (see Chapter 4.5).

Carbonates of the Cerro Victoria Formation are confidently assigned to the Lower Cambrian on the basis of trace fossils and carbon isotopes (Gaucher et al., 2007c, and references therein). The $\delta^{13}C$ curve reported for this unit is identical to the interval between 542 and 535 Ma (lower Nemakyt-Daldyn) in the high-resolution global $\delta^{13}C$ curve presented by Maloof et al. (2005). This is firm evidence that coeval Cambrian carbonates in the Cuyania-Precordillera Terrane in Argentina have potential counterparts in core areas of Gondwana (Finney, 2007), apart from Laurentia. The tectonic evolution of the area may explain the absence of carbonates – or other sedimentary rocks – younger than 535 Ma in the RPC (see Chapter 4.6).

## *ACKNOWLEDGEMENTS*

Research presented here has been funded by a PROSUL project of CNPq (Brazil) to A.N.S. and two CSIC (Uruguay) projects to C.G. We gratefully acknowledge valuable support by Leticia Chiglino, Gonzalo Blanco, Federico Cernuschi and Gabriela Martínez during field work. Léo A. Hartmann and Peter Sprechmann are thanked for constructive reviews. Finally, we are indebted to Cementos Avellaneda SA, Loma Negra SA, Cementos Artigas SA and Compañía Nacional de Cementòs for providing access to quarries and drillcore material in Argentina and Uruguay.

# PALAEOCLIMATIC EVENTS ☆

Claudio Gaucher[1] *and* Daniel Poiré[2]

## Contents



## 4.5.1. PALAEOCLIMATIC EVOLUTION OF THE RÍO DE LA PLATA PALAEOCONTINENT IN THE NEOPROTEROZOIC-CAMBRIAN

Neoproterozoic palaeoclimate in the Río de la Plata Craton (RPC) can be understood in terms of a warm, possibly tropical, 'background' climate punctuated by several cold or even glacial events. Evidences of a dominantly tropical palaeoclimate are:

(1) Thick carbonate deposits blanketed the marginal areas of the RPC in the Neoproterozoic, beginning in the late Mesoproterozoic-Tonian (Mina Verdún Group) and especially in the late Ediacaran-Early Cambrian (Sierras Bayas and Arroyo del Soldado groups).
(2) Clay minerals derived from the granitic-metamorphic basement were predominantly kaolinitic during certain periods (Poiré, 1987; Gaucher, 2000; Pamoukaghlian et al., 2004).
(3) Preliminary palaeomagnetic data suggest a tropical latitude for the late Ediacaran and Early Cambrian (Sanchez Bettucci and Rapalini, 2002).

On the other hand, clear evidences of ice-rafting and glacially influenced deposits were reported from the Playa Hermosa and Las Ventanas formations (Pazos et al., 2003, 2008; Gaucher et al., 2008a), which could be lateral equivalents (Blanco and Gaucher, 2005). These evidences will be discussed in detail below.

☆ Gaucher, C., Poiré, D., 2009. Palaeoclimatic events. Neoproterozoic-Cambrian evolution of the Río de la Plata Palaeocontinent. In: Gaucher, C., Sial, A.N., Halverson, G.P., Frimmel, H.E. (Eds.): Neoproterozoic-Cambrian tectonics, global change and evolution: a focus on southwestern Gondwana. Developments in Precambrian Geology, 16, Elsevier, pp 123–130.

[1] Departamento de Geología, Instituto de Ciencias Geológicas, Facultad de Ciencias Iguá 4225, 11400 Montevideo, Uruguay.
[2] Centro de Investigaciones Geológicas (CIG), Universidad Nacional de La Plata-CONICET, calle 1 n. 644, 1900 La Plata, Argentina.

*Developments in Precambrian Geology*, Volume 16
ISSN 0166-2635, DOI 10.1016/S0166-2635(09)01608-9

## 4.5.2. Diamictites of the Las Ventanas and Playa Hermosa Formations

### 4.5.2.1. Las Ventanas Formation

Conglomerates of the Las Ventanas Formation (Quebrada de Viera Member) are mostly orthoconglomerates without signs of glacial influence (Blanco and Gaucher, 2005).

A diamictite level occurs at the Mina Verdún Quarry, which is overlain with erosional unconformity by typical orthoconglomerates of the Quebrada de Viera Member (Gaucher et al., 2008a). Basalts and breccias of the La Rinconada Member are not present there, and the Quebrada de Viera Member overlies carbonates of the Mina Verdún Group (Poiré et al., 2003a, 2005b; Gaucher et al., 2008a; see Chapter 4.2) with erosional and angular unconformity. Similar stratigraphic relationships were also observed at Burgueño Quarry (Gaucher et al., 2007a, 2008a). A prominent palaeokarst is developed at both localities on top of carbonates of the Mina Verdún Group and filled with diamictites. These are made up of angular to subrounded clasts up to 25 cm in diameter (Figure 4.5.1A–C), composed of underlying carbonates, granitoids, rhyolite, metasiltstone, sandstone, quartzite and epidote-rich metavolcanics (Gaucher et al., 2008a). Clasts are embedded in a reddish or greenish, ferruginous mudstone matrix mainly composed of clay and silt. The matrix averages 35–40% by volume, but transitions to clast-free mudstones are common. Diamictites are massive and chaotic or slightly graded. Lonestones made up of limestone, metasiltstone and granite occur in laminated, red siltstones (Figure 4.5.1A–C). In a few cases, the long axis of the clasts is perpendicular to bedding plane (Figure 4.5.1C), allowing its interpretation as a 'bullet-clast' dropstone. Deformation of underlying laminae and onlap of overlying strata on the top of the dropstones confirm this interpretation. All petrographic features of the described diamictites, along with the occurrence of dropstones, suggest a glacial origin (Gaucher et al., 2008a). However, no cap carbonate, like those often seen on top of other Neoproterozoic glacial units (e.g. Kennedy, 1996; Hoffman and Schrag, 2002), occurs on top of the diamictites. This may be due to truncation during the events responsible for the overlying, clast-supported conglomerates or, alternatively, the absence of a cap carbonate may represent a primary feature of the glacial event in Uruguay, as observed in the type region for the 583 Ma Gaskiers glaciation (MacGabhann, 2005).

The main age constraints for the diamictite level of the lower Las Ventanas Formation are:

(1) Tholeiitic andesites tectonically intercalated in the Fuente del Puma Group and dated U-Pb SHRIMP at $590 \pm 2$ Ma by Mallmann et al. (2007) (see Chapter 4.6) probably correlate with basalts of the La Rinconada Member of the Las Ventanas Formation ($615 \pm 30$ Ma, K-Ar). If this is accepted, diamictites of the lower Las Ventanas Formation are younger than 590 Ma.
(2) Acritarchs of the middle and upper Las Ventanas Formation have been assigned by Gaucher et al. (2008a) to the Ediacaran Leiosphere Palynoflora (Grey et al., 2003; Grey, 2005), loosely constrained between the base of the Ediacaran Period and the Acraman impact (ca. 570 Ma; see Chapter 9.1).

Therefore, the mentioned data strongly suggest that the Las Ventanas diamictites were deposited during the Gaskiers glaciation ($582 \pm 0.4$ Ma: Bowring et al., 2003b; Bowring, in MacGabhann, 2005), as suggested by Blanco and Gaucher (2005).

### 4.5.2.2. Playa Hermosa Formation

The Playa Hermosa Formation (Masquelin and Sánchez Bettucci, 1993) crops out in a narrow strip along the coast near Piriápolis, a section described in detail by Pazos et al. (2003). The unit is made up of diamictites, breccias, conglomerates, sandstones and rhythmites, showing evidence of slope instability (i.e. slumping) and resedimentation processes (Pazos et al., 2003). Sandstone layers bear all the hallmarks of turbidites, and also indicate a relatively steep palaeoslope. Lonestones up to 1 m in length were interpreted as dropstones by Pazos et al. (2003, 2008), although sliding processes may also have played an important role in a tectonically unstable, steep-sloped basin.

Thus, the Playa Hermosa Formation differs markedly from the Las Ventanas Formation, because:

(1) it represents deep-water, slope sediments;
(2) resedimentation and turbidity currents were the most important depositional processes; and
(3) lithologically, conglomerates of the Playa Hermosa Formation contain up to 50% intraclasts and 30% quartzites (Blanco and Gaucher, 2005), in contrast to the granite-dominated conglomerates of the Las Ventanas Formation.

The contact between both units is tectonic, but it is reasonable to postulate that the Playa Hermosa Formation is the deep-water lateral equivalent of the shallower Las Ventanas Formation (Blanco and Gaucher, 2005).

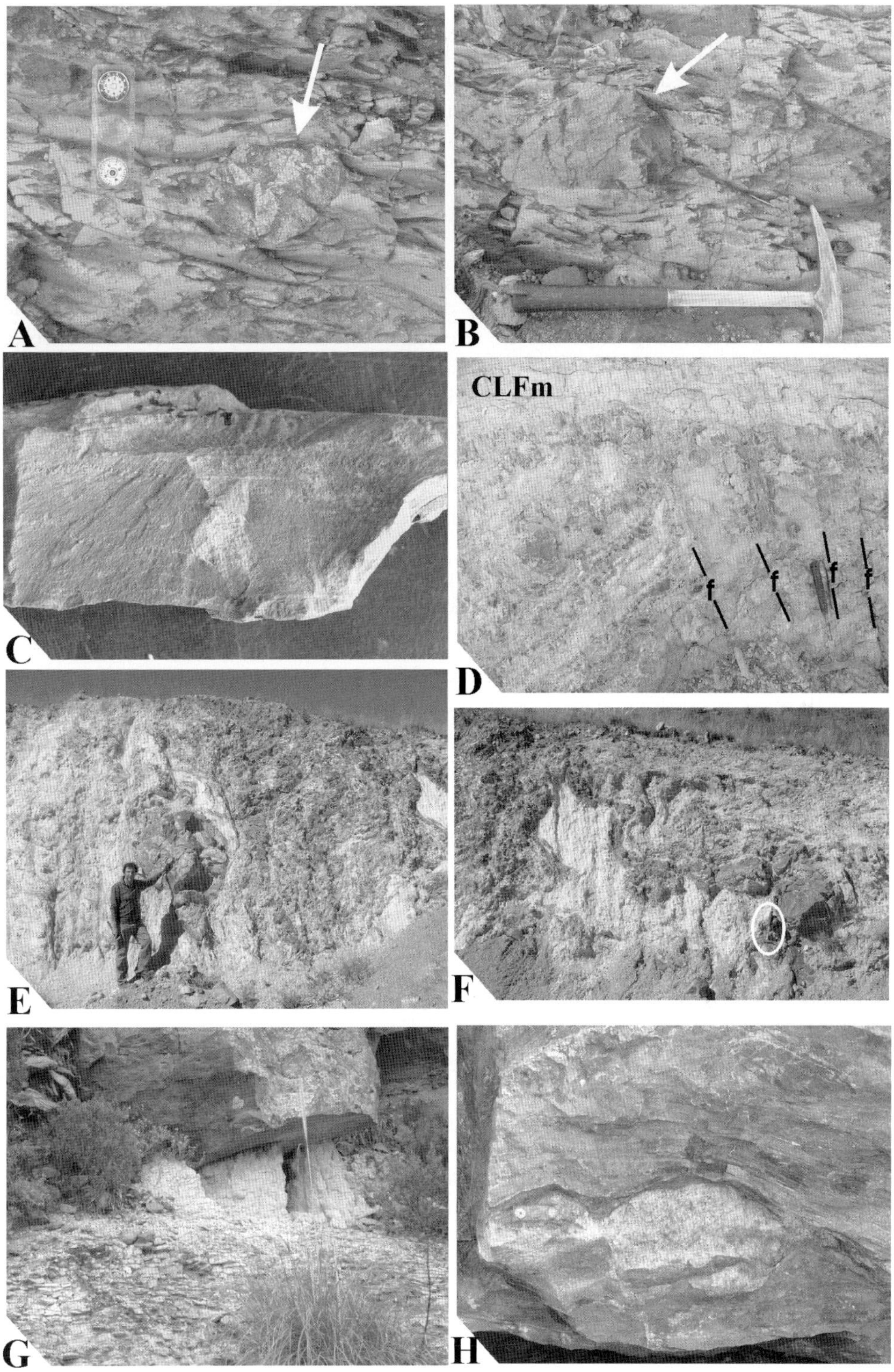


**Figure 4.5.1** Neoproterozoic diamictites in the RPC. (A and B) Outsized clasts in rain-out diamictite of the Las Ventanas Formation in the vicinity of Burgueño Quarry. Clasts (arrowed) are composed of granite (A) and metapelite (B). Scale in A is 8 cm long. (C) Granitic bullet clast (long axis is 2 cm long) in red siltstones, lower Las Ventanas Formation at Mina Verdún (from Gaucher et al., 2008a). (D) Normal faults (f) cross-cutting diamictites of the Colombo Member at Piedra Amarilla Quarry. Note truncation of faults by overlying sandstones of the Cerro Largo Formation (CLFm). (E) Large boulder (2 m in diameter) in the Colombo Member, Piedra Amarilla Quarry. (F) Convolute bedding (slump), Colombo Member at Piedra Amarilla Quarry. Hammer is indicated for scale. (G) Sierra del Volcán diamictites overlain by sandstones of the Balcarce Formation. Height of the outcrop is 1.5 m. (H) Boudinaged granite in limestones of the Carapé Group at Floridán Quarry; scale (centre left) is 8 cm long.

No direct age constraints are available for the Playa Hermosa Formation. However, both the correlation with the Las Ventanas Formation (Blanco and Gaucher, 2005) and the soft-sediment interaction with basic lavas of the La Rinconada Member suggest an age around 590–580 Ma (Pazos et al., 2008). Therefore, the unit may record the Gaskiers glacial event in a more distal, slope environment.

## 4.5.3. Diamictites in Tandilia

### 4.5.3.1. Sierra del Volcán diamictites

In the southeastern part of Tandilia, a thin (<8 m) diamictite unit known as Sierra del Volcán Formation occurs (Spalletti and del Valle, 1984). It rests with erosional unconformity on the Palaeoproterozoic Buenos Aires Complex, and according to Pazos et al. (2008) is overlain with angular unconformity by the Middle Ordovician to Lower Silurian Balcarce Formation (Seilacher et al., 2002; Rapela et al., 2007; Figure 4.5.1G). However, Poiré and Spalletti (2008) suggested a possible Ashgillian (Hirnantian) age for these glacial deposits on the basis of Lower Ordovician U-Pb ages of detrital zircon grains (van Staden et al., unpublished data) from the Sierra del Volcán Diamictite. The apparent angular unconformity between the Sierra del Volcán and the Balcarce formations (Figure 4.5.1G) is probably due to their different rheology. Therefore, this glacial unit should be referred to the Hirnantian, which has a very important record in southwestern Gondwana. The geochronological information provided by detrital zircons from the Sierra del Volcán Diamictite and the Balcarce Formation suggests that the kaolinite-rich Balcarce Formation could represent the onset of Silurian greenhouse conditions.

### 4.5.3.2. Colombo diamictites

In the Sierras Bayas Group, a diamictite horizon (Colombo Member) overlies a karstic surface on top of dolostones of the Villa Mónica Formation, and is in turn overlain with erosional unconformity by glauconitic sandstones of the lower Cerro Largo Formation. It is made up of sandstone, chert and breccia clasts up to 2.5 m in diameter, embedded in a greenish pelitic matrix (Figure 4.5.1E). Thickness of the Colombo Member never exceeds 10 m. So far, no evidences of a glacial origin have been reported for this unit. Furthermore, a number of features suggest that gravity flows in a tectonically unstable environment may be responsible for deposition of this diamictite level, namely:

(a) Large-scale slumps occur, as evidenced by large-scale convolute bedding (Figure 4.5.1F).
(b) Normal faults disturb bedding, but are truncated by the overlying sandstones of the Cerro Largo Formation (Figure 4.5.1D), showing that they represent syn-sedimentary extensional faulting during deposition of the Colombo Member.
(c) Clast-supported conglomerates and bedded diamictites occur, which are not of glacial origin.

The stratigraphic position of these mass-flow deposits is analogous to the Las Ventanas and Playa Hermosa formations in Uruguay. The geotectonic setting, like in the latter unit, was probably the border of a rift basin, characterised by a steep palaeorelief and tectonic instability (Masquelin and Sánchez Bettucci, 1993; Pazos et al., 2003; Blanco and Gaucher, 2005; Gaucher et al., 2008a). However, in the case of the Colombo Member, evidence of glacial processes is lacking. It is worth noting that the bulk of conglomerates of the Las Ventanas Formation is not of glacial origin, but were deposited in alluvial fans (Blanco and Gaucher, 2005; Gaucher et al., 2008a).

## 4.5.4. Other Units

Pazos et al. (2008) report the occurrence of lonestones in laminated limestones of the Zanja del Tigre Formation (Sánchez Bettucci and Ramos, 1999), now assigned to the Marco de los Reyes Formation of the Carapé Group (Bossi et al., 2007; see Chapter 4.1). Pazos et al. (2008) interpret these lonestones, composed of granite, quartzite and gabbro (Figure 4.5.1H), as dropstones. Apart from the great uncertainties regarding the age assignment of the Carapé Group (Bossi et al., 2007; Basei et al., 2008), there are serious objections to the glacial interpretation of Pazos et al. (2008). The occurrence of budinaged granites (Figure 4.5.1G) is a characteristic feature of limestones of the Carapé Group (Bossi and Navarro, 1991; Rossini, 2002; Bossi et al., 2007). Disrupted quartz-arenite layers also occur, resulting in boudinaged quartzite blocks within the less competent, sheared limestone. Deformation of the limestones is ductile and very significant (Rossini, 2002; Bossi et al., 2007), as shown by:

(a) vicinity (ca. 10 km) to the continental-scale Sierra Ballena Shear Zone (Figure 4.1.4);
(b) preservation of limestones as thin basement inliers between large granite intrusions (Figure 4.1.4);
(c) tight anisopachous folding with subvertical axial planes; and
(d) obliteration of all primary sedimentary structures.

That the lonestones are not of glacial origin is shown by granite 'clasts' often connected to each other (Figure 4.5.1H), typical of rheologically more resistant, boudinaged layers within a plastic matrix. Finally, $\delta^{13}C$ of the Marco de los Reyes limestones is predominantly positive (Chiglino et al., 2008), unlike glacially related units (e.g. Halverson et al., 2005).

## 4.5.5. Indirect Evidence of Late Ediacaran Glaciation

### 4.5.5.1. Middle Polanco Formation

Gaucher et al. (2004c,d) showed that an important $\delta^{13}C$ negative excursion encompassing 200 m of carbonates occurs in the middle Polanco Formation (Unit B; see Chapter 4.4). The following changes in measured parameters occur concomitantly to a drop in $\delta^{13}C$ (Figure 4.4.3):

(a) remarkable sea-level drop of $>50$ m, as shown by sedimentary structures;
(b) positive to negative $\delta^{13}C$ excursion, between +5.5‰ and −4.5‰ PDB;
(c) low $^{87}Sr/^{86}Sr$ values between 0.7073 and 0.7077 throughout Unit Polanco B, as compared to $^{87}Sr/^{86}Sr$ ratios for underlying carbonates of 0.7080−0.7085;
(d) lower organic matter contents; and
(e) lower diversity of organic-walled microfossils.

The most plausible explanation for these combined phenomena is a severe but non-global glacial event. Glaciation would have caused lowering of sea level, decrease in plankton bioproductivity, reduction of the amount of organic matter exported to the sediments, environmental stress leading to lower acritarch diversity, a negative $\delta^{13}C$ excursion and lowering of $^{87}Sr/^{86}Sr$ values due to diminished chemical weathering and continental run-off (see Chapter 4.4 for a full discussion). Absence of glacial deposits indicates that the glaciers did not reach the Polanco shelf, which probably was at tropical latitudes. According to palaeomagnetic data reported by Sanchez Bettucci and Rapalini (2002), the RPC was located at 17.5° latitude by 550 Ma.

The middle Polanco glacial event may have been a significant perturbation of the ocean-climate system. As shown in Chapter 4.4, the middle Polanco negative anomaly probably correlates with the Shuram−Wonoka anomaly (Halverson et al., 2005), which ended at $551 \pm 1$ Ma and reached its nadir before $555 \pm 6$ Ma in south China (Condon et al., 2005; Zhang et al., 2005). Both the occurrence of *Cloudina riemkeae* in Unit B of the Polanco Formation (see Chapter 4.3) and a U-Pb detrital zircon age of $566 \pm 8$ Ma in the overlying Barriga Negra Formation (Blanco et al., 2009) are consistent with this interpretation. A glacially related interpretation of the Shuram−Wonoka anomaly in Oman has been put forward by Cozzi et al. (2002), and is in agreement with our results.

### 4.5.5.2. Barker Surface

The uppermost Polanco Formation-basal Barriga Negra Formation is characterised by a negative $\delta^{13}C$ excursion, lowering of $^{87}Sr/^{86}Sr$ ratios (Figure 4.4.3), changes in acritarch diversity and – most significantly – drop in sea level leading to platform exposure and erosion (Gaucher, 2000; Gaucher et al., 2004c,d; Figure 4.5.2C and D). The same event is recorded in the Sierras Bayas Group in the uppermost Loma Negra Formation, which was karstified and filled with karstic breccias as a result of this regression (Barrio et al., 1991; Figure 4.5.2A and B). The karstic surface has been termed 'Barker Surface' by Poiré and Gaucher (2007). Age constraints for this event are provided by (a) *Cloudina riemkeae* occurring in the underlying Polanco and Loma Negra formations and (b) a U-Pb detrital zircon age of $566 \pm 8$ Ma for sandstones of the lower Barriga Negra Formation (Blanco et al., 2009). Therefore, the Barker Surface event is latest Ediacaran in age, allowing a preliminary correlation with the Vingerbreek Member in the Nama Group, constrained between $549 \pm 1$ and $545 \pm 1$ Ma (Grotzinger et al., 1995; Saylor et al., 1998). There, platform reworking, a negative $\delta^{13}C$ excursion to −2‰ PDB, striated pavements and other evidences of a glacial event have been described (Germs, 1995, and references therein; see Chapter 5.4).

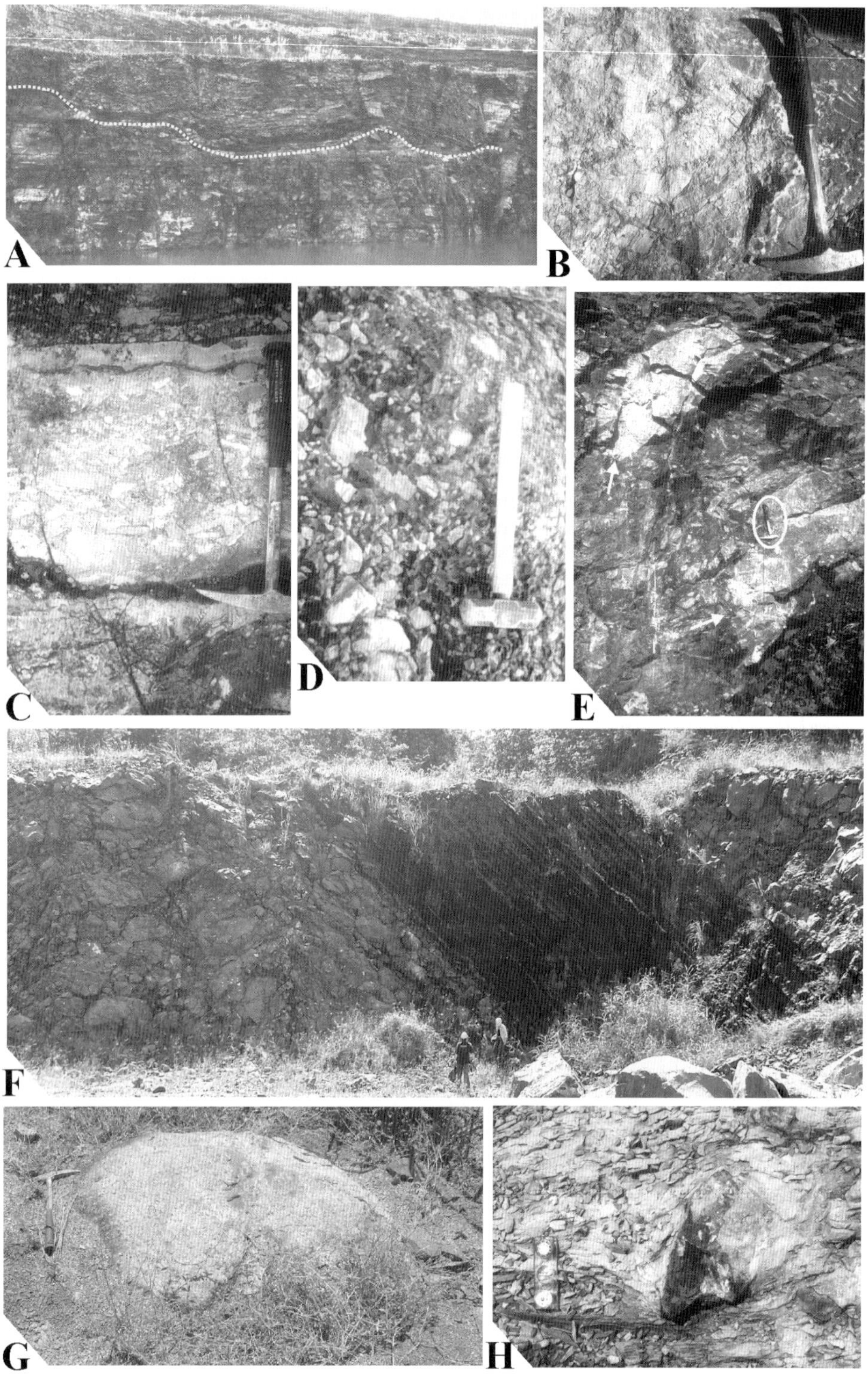


**Figure 4.5.2** Barker Surface and possible correlates in the Paraguay Belt. (A) Palaeokarst on top of the Loma Negra Formation (dotted) and covered by pelites and sandstones of the Cerro Negro Formation, Loma Negra Quarry near Olavarría. (B) Limestone breccias filling the palaeokarst at Polvorín Quarry (Olavarría area). (C) Graded carbonate breccias at the Polanco-Barriga Negra transition (from Gaucher, 2000). (D) Limestone breccia, lower Barriga Negra Formation. (E) Metric clasts (arrowed) in breccias of the lower Tamengo Formation, Laginha Quarry (see F). Hammer is shown as scale (circle). (F) Overview of breccias in the lower Tamengo Formation, separated by fine-grained, dark limestones and marls yielding negative $\delta^{13}C$ values. (G) Laginha diamictite: outsized clast 2 m in diameter, made up of oncolitic limestone, Laginha Quarry. (H) Laginha diamictite: outsized clast in a siltstone matrix, showing deformation features characteristic of dropstones (Laginha Quarry).

## 4.5.6. Record of Cambrian Global Warming

During deposition of the Cerro Victoria Formation (ASG), tropical climate prevailed once again in the RPC, as demonstrated by thick carbonate deposits (400 m), abundance of ooids and evaporitic minerals (Montaña and Sprechmann, 1993; Sprechmann et al., 2004; Gaucher et al., 2007c). Common occurrence of calcareous storm deposits confirms this interpretation (Gaucher, 2000). As noted by Gaucher et al. (2007c), climate stabilisation is indicated by the decreasing amplitude of $\delta^{13}C$ excursions in the ASG, being only 2‰ PDB in the Cerro Victoria Formation (see Chapter 4.4).

Lower Cambrian carbonate ramps are scarce in southwestern Gondwana. However, a number of carbonate successions broadly correlative with the Cerro Victoria Formation have been described, suggesting tropical conditions at least in part of Gondwana. These are:

(a) the thick carbonate deposits of the Precordillera terrane, and especially La Laja Formation, which according to Sial et al. (2003) is of Lower Cambrian age (Chapter 11.1); affinity of the Precordillera terrane is currently the subject of intense debate, but data presented by Finney (2007) seem to favour a Gondwanic affinity;
(b) the Nelson Limestone Formation, exposed in the Pensacola Mountains (Transantarctic Province) of Antarctica, comprises oolitic, pisolitic and archaeocyathan limestones of Lower to Middle Cambrian age (Laird, 1995); rhyolites occurring in the conformably overlying Gambacorta Formation yielded an Upper Cambrian age, confirming the biostratigraphic data (Schmidt, in Laird, 1995); and
(c) the Tucavaca Group of Bolivia comprises stromatolitic limestones conformably overlain by siliciclastics bearing *Treptichnus (Phycodes) pedum*, suggesting a Lower Cambrian age for the unit (Trompette, 1994).

Carbonate deposition was probably diminished due to high erosion rates in the mountain ranges generated during the Pan African–Brasiliano system of orogens (Trompette, 1994). Therefore, most Cambrian sedimentary successions in western Gondwana are siliciclastic and carbonate ramps were restricted to areas with low clastic input. Most interesting, Early Cambrian successions comprising carbonate ramps in western Gondwana occur in blocks located at tropical palaeolatitudes (Antarctica, RPC and Argentine Precordillera), according to the palaeogeographic reconstruction of Finney et al. (2003).

## 4.5.7. Discussion and Summary

It becomes clear that in the late Neoproterozoic the RPC was predominantly under tropical climate (Figure 4.5.3). If carbonate deposits of the Mina Verdún Group turn out to be Tonian in age (see Chapters 4.2–4.4), tropical conditions may have lasted for most of the Neoproterozoic. However, due to the gap in the sedimentary record, which encompasses at least the whole Cryogenian (Figure 4.5.3), no data exist on the effects of the most severe Neoproterozoic glacial events on the palaeocontinent. This might explain the 'paradox' put forward by Eerola (2002) regarding the absence of Neoproterozoic glacial deposits in the RPC and neighbouring areas. On the other hand, diamictites of the Las Ventanas and Playa Hermosa formations show that the RPC was briefly covered by ice during the Gaskiers glaciation. The limited extent of glacially related units suggests that this event was short-lived in the RPC, as also shown by U-Pb dating of ash beds at its type locality (Bowring et al., 2003b; MacGabhann, 2005).

One of the most interesting features of the Neoproterozoic palaeoclimatic record in the RPC is the strong evidence supporting post-Gaskiers glaciation (Figure 4.5.3). The glacial events represented by Unit B of the Polanco Formation and the Barker Surface event are significantly younger (>20 My) than the age currently accepted for the Gaskiers event, which ended at $582 \pm 0.4$ Ma (Bowring, in MacGabhann, 2005). Despite the fact that ice did not cover the RPC, compelling evidence supporting Phanerozoic-type glacial events includes: sea-level drop, shelf exposure and erosion, negative $\delta^{13}C$ excursions, diminished bioproductivity, impact on acritarch diversity and lowering of $^{87}Sr/^{86}Sr$ seawater ratios. Whereas the Polanco B glacial event is a likely correlative to the global Shuram–Wonoka anomaly (ca. 560–551 Ma), the Barker Surface is correlated to the Vingerbreek glacial event (Germs, 1995, and references therein), constrained between $549 \pm 1$ and $545 \pm 1$ Ma (Figure 4.5.3). These glacial events may have had a profound impact on the oxygenation of the atmosphere, evolution of eukaryotes and metazoans in particular (Fike et al., 2006).

On a regional scale, the glacial record of the RPC bears close resemblance to that encountered in the southern Paraguay Belt (SW-Brazil; see Chapter 2; Figure 4.5.3), which was also probably deposited on the Río de la Plata Palaeocontinent (Gaucher et al., 2003, 2008b; see Chapter 8). Glacial deposits of the Las Ventanas/Playa Hermosa Formation probably correlate to glacigenic diamictites of the Puga Formation in the southern Paraguay Belt (Boggiani and Alvarenga, 2004). The Polanco B glacial event may be represented by breccias and calcarenites of

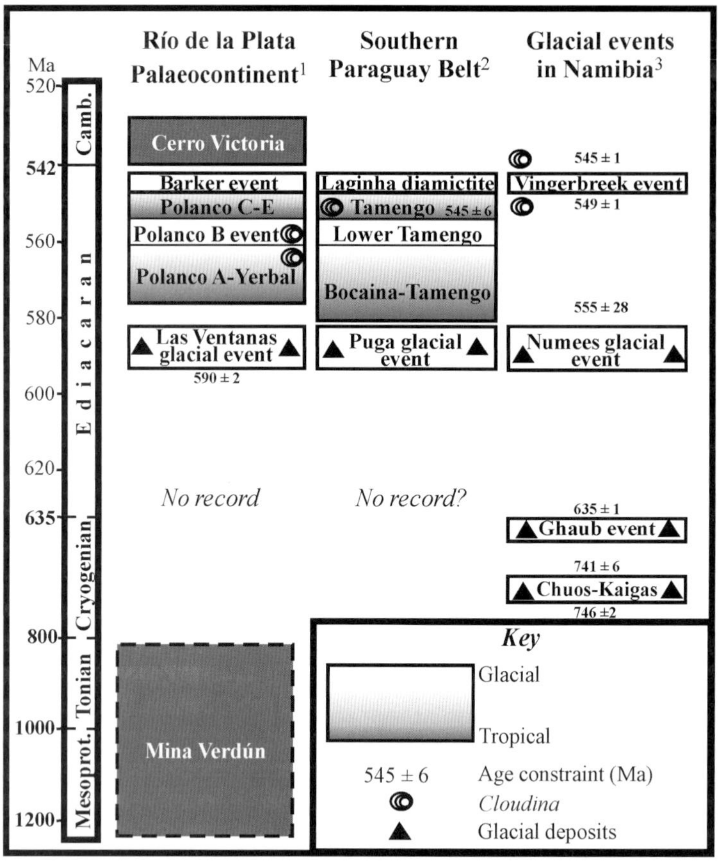


**Figure 4.5.3** Summary of Neoproterozoic-Cambrian palaeoclimatic events in the RPC, and comparison with events recorded in the southern Paraguay Belt and Namibia (Gariep and Damara belts). Note changes in the time scale (not linear). Absolute ages according to: (1) Mallmann et al. (2007), Barker event: Blanco et al. (2009; $<566 \pm 8$ Ma); (2) Babinski et al. (2007); and (3) from base to top: Hoffman et al. (1996), Frimmel et al. (1996c), Hoffmann et al. (2004), Fölling et al. (2000) and Grotzinger et al. (1995).

the lower Tamengo Formation (Corumbá Group; Boggiani, 1998; Gaucher et al., 2003; Figure 4.5.2E and F), which show a well-developed, negative $\delta^{13}C$ excursion to $-3‰$ PDB (Misi et al., 2007; Velásquez et al., 2008). A recent U-Pb SHRIMP zircon age of $543 \pm 2$ Ma for an ash bed at this level (Babinski et al., 2008) is considered too young for the following reasons (Figure 4.5.3): (a) 100 m of *Cloudina*-bearing limestones and marls occur above that level, which should be accommodated in the remaining 1 My of the Neoproterozoic (Amthor et al., 2003), implying unreasonably high sedimentation rates for the upper Tamengo Formation; (b) the dated ash bed occurs within a negative excursion in the Laginha Quarry, on top of which a $\delta^{13}C$ plateau around 3‰ occurs (Babinski et al., 2008). The opposite is observed in Namibia and Oman, where ash beds dated at 543 Ma occur within the top of this plateau (Grotzinger et al., 1995; Amthor et al., 2003). A correlation of the negative excursion at the base of the Tamengo Formation with the Shuram–Wonoka anomaly is thus preferred here.

A diamictite horizon (Laginha diamictite) occurs atop karstified limestones of the Tamengo Formation (Boggiani et al., 2004), at the same stratigraphic level of the Barker Surface (Figure 4.5.2G and H). *Cloudina* occurring in underlying limestones of the Tamengo Formation (Gaucher et al., 2003) confirms this correlation. A preliminary U-Pb SHRIMP zircon age of $545 \pm 6$ Ma (Babinski et al., 2006) for ash beds interbedded in the same unit is also consistent with this interpretation.


## ACKNOWLEDGEMENTS

This research has been partially funded by CSIC, Uruguay (project 'Estratigrafía de secuencias carbonáticas del Neoproterozoico del Terreno Nico Pérez'). Insightful reviews by Léo A. Hartmann and Peter Sprechmann are gratefully acknowledged. We are also indebted to Cementos Avellaneda SA, Cementos Artigas SA and Compañía Nacional de Cementos for providing access to quarries in Argentina and Uruguay. Paulo Boggiani is thanked for organising field trips to the Corumbá Group in 2004 and 2006.

CHAPTER 4.6

# PALAEOGEOGRAPHY☆

Claudio Gaucher[1], Jorge Bossi[2] *and* Gonzalo Blanco[3]

## Contents



## 4.6.1. Río de la Plata Craton: A Part of Rodinia

The apparent absence of Mesoproterozoic rocks in the Río de la Plata Craton (RPC) was interpreted in the past as evidence of this block being unrelated to Rodinia (e.g. Cordani et al., 2003). Other authors, however, pointed out that geological similarities between the Amazonia and the Río de la Plata palaeocontinents suggested that they were linked in the Mesoproterozoic and were part of Rodinia (e.g. Meert and Torsvik, 2003).

In Uruguay, the occurrence of Mesoproterozoic tectonomagmatic events in the RPC has been advocated by different authors in the last 15 years. While mapping in detail the Piedra Alta mafic dyke swarm, dated at 1,790 ± 5 Ma (U-Pb on beddeleyite: Halls et al., 2001; Ar-Ar: Teixeira et al., 1999), Bossi and Campal (1992) recognised and described a dextral megashear zone (Sarandí del Yí Shear Zone, SYSZ) with mylonites up to 8 km thick (Figure 4.6.1). The curvature of the eastern part of the dyke swarm is consistent with dextral shear, unlike all "Brasiliano – Pan African" structures in the area. The age of the SYSZ is probably late Mesoproterozoic as shown by post-emplacement thermal overprinting of the Piedra Alta mafic dike swarm (Bossi et al., 1993a) between 1,370 and 1,170 Ma, reflected in initial release portions of $^{40}$Ar-$^{39}$Ar spectra and Rb-Sr mineral isochrons (Teixeira et al., 1999). Bossi and Navarro (2001) report an $^{40}$Ar-$^{39}$Ar age of 1,240 ± 5 Ma for the same event (Figure 4.6.1). Cingolani, in Bossi et al. (1998), reported K-Ar ages of 1,253 ± 32 Ma for synkinematic muscovites that crystallised along southwest-vergent thrust planes in the Nico Pérez Terrane (Figure 4.6.1), thus consistent with the above-mentioned shear sense and timing of this tectonic event. The SYSZ is a continental-scale megashear (Unrug, 1996), which suggests significant reworking of the RPC during the Grenvillian Orogeny.

☆Gaucher, C., Bossi, J., Blanco, G., 2009. Palaeogeography. Neoproterozoic-Cambrian evolution of the Río de la Plata Palaeocontinent. In: Gaucher, C., Sial, A.N., Halverson, G.P., Frimmel, H.E. (Eds.): Neoproterozoic-Cambrian tectonics, global change and evolution: a focus on southwestern Gondwana. Developments in Precambrian Geology, 16, Elsevier, pp. 131 – 141.

[1] Departamento de Geología, Instituto de Ciencias Geológicas, Facultad de Ciencias Iguá 4225, 11400 Montevideo, Uruguay.
[2] Cátedra de Geología, Facultad de Agronomía, Garzón 780, Montevideo, Uruguay.
[3] Department of Geology, University of Johannesburg, PO Box 524, Auckland Park 2006, 13 Johannesburg, South Africa.

*Developments in Precambrian Geology*, Volume 16
ISSN 0166-2635, DOI 10.1016/S0166-2635(09)01609-0

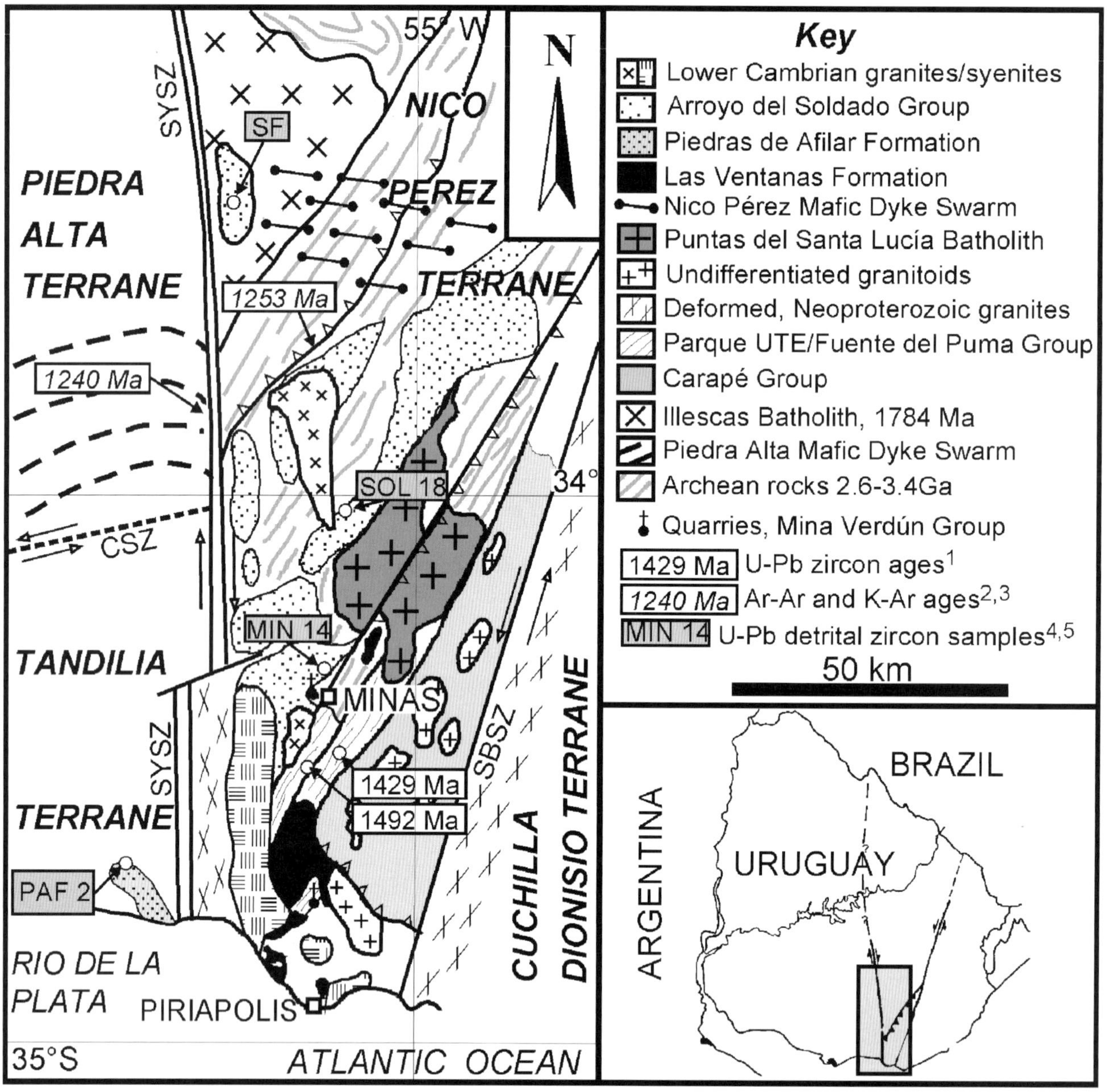


**Figure 4.6.1** Geological sketch map of the southern Nico Pérez Terrane and surrounding areas. Sources of Mesoproterozoic radiometric ages: (1) Oyhantçabal et al. (2005); (2) Teixeira et al. (1999) and Bossi and Navarro (2001); (3) Cingolani, in Bossi et al. (1998); (4) Gaucher et al. (2008b); and (5) Blanco et al. (2009), only sample SF. SYSZ: Sarandí del Yí Shear Zone (shear sense indicated for the main Mesoproterozoic event); SBSZ: Sierra Ballena Shear Zone; CSZ: Colonia Shear Zone. Modified from Gaucher et al. (2008b).

Bossi et al. (1998) assigned a number of volcanosedimentary successions in the southern Nico Pérez Terrane (Fuente del Puma and Carapé groups) to the Mesoproterozoic. This age assignment is confirmed by: (a) U-Pb zircon ages of 1,429 ± 21 Ma for metarhyolites of the upper Fuente del Puma Group and 1,492 ± 4 Ma for metagabbros intruding the succession elsewhere (Oyhantçabal et al., 2005; Figure 4.6.1); (b) a K-Ar age of 1,208 ± 10 Ma for metabasalts of the Fuente del Puma Group (Gómez Rifas, 1995), (c) Pb-Pb measurements of galenas from the same unit yield ages between 1,500 and 1,200 Ma (Garau, in Bossi and Navarro, 2001); (d) carbon isotope curves of carbonates previously assigned to the Fuente del Puma Group (now Parque UTE Group) and Carapé Group (*sensu* Bossi et al., 2007) are consistent with a Mesoproterozoic depositional age (Bossi et al., 2008; Chiglino et al., 2008); and (e) U-Pb detrital zircon ages of sandstones of the Carapé Group are consistent with a Mesoproterozoic age for part of this unit (Basei et al., 2008). It is worth noting that an extensional tectonic event is represented by the 1,588 ± 11 Ma mafic dike swarm of Tandilia (U-Pb on baddeleyite: Teixeira et al., 2002).

Independent evidence of one or more Mesoproterozoic tectonic events in the RPC is provided by U-Pb ages of detrital zircons from Neoproterozoic sandstone units deposited on the RPC in Uruguay and Argentina

(Figures 4.6.1 and 4.6.3). The Piedras de Afilar Formation in Uruguay shows a dominant Mesoproterozoic zircon population (62% of total) with four distinct peaks at 1.49, 1.35, 1.24 and 1.01 Ga (Gaucher et al., 2007b, 2008b; Figure 4.6.3). In the Cerro Largo Formation (Sierras Bayas Group), Mesoproterozoic zircons represent 15% of the population and define distinct peaks at 1.55, 1.23 and 1.05 Ga (Figure 4.6.3), roughly coincident with the Piedras de Afilar data. Rapela et al. (2007) report detrital zircon ages for a sample of the Villa Mónica Formation in the Barker area, with a subordinate Mesoproterozoic peak centred at 1,135 Ma. Finally, Blanco et al. (2007, 2009) report that 17% of detrital zircons from the Yerbal Formation (Arroyo del Soldado Group) yielded U-Pb ages between 1,063 and 1,009 Ma (Figure 4.6.3). This shows that extensive areas of Mesoproterozoic rocks were exposed in the RPC and acted as the source of detritus in the late Neoproterozoic. Gaucher et al. (2007b, 2008b) showed that the abundance of Mesoproterozoic detrital zircons in Neoproterozoic sandstones increases towards the west of the RPC. A proto-Andean, Mesoproterozoic belt fringing the western RPC has been suggested as the source of the Mesoproterozoic zircons (Gaucher et al., 2007b, 2008b), which may represent the southward extension of Mesoproterozoic units of the Amazonian Craton (Sunsás Orogenic Cycle; Figure 4.6.4).

It is significant that a number of Mesoproterozoic units and blocks fringe the Río de la Plata and Amazonian cratons to the west. Chew et al. (2007) suggest that the Arequipa-Antofalla Block accreted to Amazonia during the Grenvillian-Sunsás Orogeny. In the Sierra de Pie de Palo of the Cuyania Terrane (western Argentina), a Grenvillian (1.1 – 1.0 Ga) metamorphic basement known as Pie de Palo Complex is exposed. Recent data on the timing of tectonic events in the Pie de Palo range by Mulcahy et al. (2007) support the idea of this unit being a para-autochthonous block of Gondwanan affinity.

Thus, the available evidence strongly suggests that: (a) the RPC was part of the tectonic collage ultimately leading to the accretion of Rodinia and (b) the Amazonian Craton and the RPC were probably attached in the late Mesoproterozoic (Gaucher et al., 2003), representing the core of proto-South America. They were fringed to the west by a Grenvillian (Sunsás), proto-Andean orogenic belt. Remnants of Mesoproterozoic volcanosedimentary successions also occur at the eastern boundary of the RPC (Parque UTE and Carapé groups: Bossi et al., 2007, 2008).

## 4.6.2. Latest Cryogenian-Early Ediacaran Rifting

Evidence of Neoproterozoic rifting is compelling in the Nico Pérez Terrane in Uruguay. Mafic dyke swarms, anorogenic granitoids and rift-related sedimentary successions occur there, and are broadly constrained between 635 and 580 Ma. This is significantly younger than Neoproterozoic rifting in the Kalahari and Congo cratons (see Chapters 5.2 and 8), but older than the opening of the Iapetus Ocean between Laurentia and Gondwana.

### 4.6.2.1. Nico Pérez mafic dyke swarm

A mafic dyke swarm occurs in the central Nico Pérez Terrane (Figure 4.6.1). Dykes strike roughly EW and range in composition from basalts to tholeiitic andesites. They intrude an Archaean metamorphic complex, the late Palaeoproterozoic Illescas rapakivi granite and Mesoproterozoic peraluminous granites (Figure 4.6.1). A detailed geochemical study has been presented by Mazzucchelli et al. (1995) and Rivalenti et al. (1995), and a recent compilation of all available data is due to Bossi and Schipilov (2007). The age of the dykes is constrained by: (a) a K-Ar age of $581 \pm 13$ Ma on biotite at the contact between the dykes and the country rocks and (b) a poor Rb-Sr isochron of $665 \pm 203$ Ma (Rivalenti et al., 1995).

In either case, the dykes record rifting in the RPC around 600 Ma, thus immediately prior to deposition of shelf sediments of the Arroyo del Soldado and upper Sierras Bayas groups.

### 4.6.2.2. Early Ediacaran within-plate basalts

Porphyritic metabasalts occur in the Fuente del Puma Group *sensu* Bossi et al. (1998), for which Mallmann et al. (2007) report a U-Pb SHRIMP zircon crystallisation age of $590 \pm 2$ Ma. This age is within error of the above-mentioned K-Ar age of the Nico Pérez Dyke Swarm. The geochemical signature of the dated basalts falls between the within-plate and the E-MORB fields (Mallmann et al., 2007), thus consistent with a rifting event at 590 Ma. It is worth noting that the Fuente del Puma Group, or the roughly equivalent Lavalleja Metamorphic Complex, contains, according to Mallmann et al. (2007), at least two different groups of basaltic rocks, as deduced from their different geochemistry, Sm-Nd $T_{DM}$ model ages and Sr isotopic characteristics. As noted in Chapter 4.1, the unit undoubtedly includes a Mesoproterozoic volcanosedimentary succession (Parque UTE Group: Bossi et al., 2008), tectonically intercalated with the Neoproterozoic basalts.

### 4.6.2.3. A-Type granitic magmatism

The Puntas del Santa Lucía Batholith is a relatively large intrusion (Figure 4.6.1) including granodiorites, granites, syenites and rhyolites (A. Toselli, personal communication, 2007). Hartmann et al. (2002a) report a U-Pb SHRIMP zircon age of $633 \pm 10$ Ma for this batholith. Samples of the different facies of the intrusion plot in the within-plate (anorogenic) field in the diagrams of Pearce et al. (1984) (Figure 4.6.2; Table 4.6.1) probably representing the earliest record of extensional Neoproterozoic magmatism in the RPC (Blanco and Gaucher, 2005).

### 4.6.2.4. Las Ventanas Formation: early Ediacaran rift deposits

An extensional geotectonic setting was first suggested by Bossi and Navarro (1991) and Masquelin and Sánchez Bettucci (1993) for the possibly correlative Las Ventanas and Playa Hermosa formations in Uruguay (see Chapter 4.2). The latter authors mention evidences of syndepositional extensional tectonics. Pazos et al. (2003) also postulate the interaction of extensional faulting and glacial processes for the Playa Hermosa Formation.

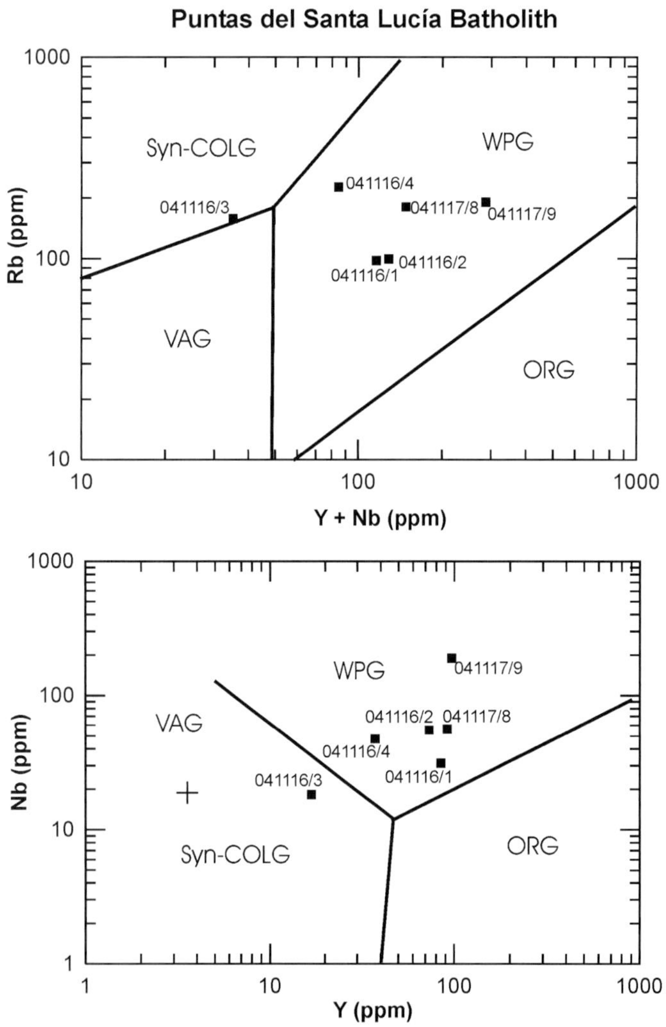


**Figure 4.6.2** Composition of six granitoid samples (Table 4.6.1) from the Puntas del Santa Lucía Batholith plotted in the diagrams of Pearce et al. (1984). WPG: within-plate granites; syn-COLG: syn-collisional granites; VAG: volcanic arc granites; ORG: ocean ridge granites. Note clustering of samples in the within-plate field.

**Table 4.6.1** X-ray fluorescence analyses of samples from the Puntas del Santa Lucía Batholith plotted in Figure 4.6.3.

| Sample | 041116/1 | 041116/2 | 041116/3 | 041116/4 | 041117/8 | 041117/9 |
|---|---|---|---|---|---|---|
| $SiO_2$ | 67.13 | 71.31 | 73.52 | 78.55 | 77.92 | 78.50 |
| $Al_2O_3$ | 12.61 | 12.09 | 12.08 | 10.80 | 12.64 | 11.54 |
| MgO | 0.10 | 0.06 | 0.82 | 0.20 | 0.02 | 0.04 |
| $TiO_2$ | 0.23 | 0.25 | 0.58 | 0.09 | 0.09 | 0.07 |
| $Fe_2O_3$ | 4.90 | 4.60 | 3.70 | 1.28 | 1.56 | 1.04 |
| CaO | 1.57 | 1.11 | 1.55 | 0.53 | 0.04 | 0.41 |
| $Na_2O$ | 3.79 | 3.57 | 2.99 | 2.96 | 4.12 | 3.53 |
| $K_2O$ | 5.18 | 5.12 | 4.08 | 4.28 | 4.73 | 4.29 |
| MnO | 0.10 | 0.08 | 0.04 | 0.01 | 0.06 | 0.00 |
| $P_2O_5$ | 0.06 | 0.02 | 0.21 | 0.03 | 0.01 | 0.01 |
| Sum % | 95.68 | 98.21 | 99.57 | 98.72 | 101.19 | 99.42 |
| Ba | 447.50 | 105.60 | 932.20 | 273.70 | 558.30 | 93.30 |
| Nb | 31.30 | 55.20 | 18.20 | 47.40 | 56.20 | 189.70 |
| Pb | 11.70 | 15.70 | 19.20 | 23.50 | 8.60 | 24.60 |
| Rb | 97.70 | 99.40 | 157.60 | 226.00 | 180.40 | 190.40 |
| Sr | 64.00 | 19.20 | 245.50 | 68.80 | 202.40 | 17.80 |
| Th | 1.70 | 28.00 | 13.80 | 36.90 | 19.00 | 26.50 |
| Y | 84.80 | 73.30 | 16.80 | 37.30 | 91.80 | 97.00 |
| Zr | 651.00 | 489.90 | 320.10 | 123.60 | 242.80 | 212.50 |
| Ce | 60.30 | 417.60 | 88.30 | 98.40 | 143.80 | 65.50 |
| Co | −0.90 | −0.60 | 3.20 | −0.10 | −2.40 | 0.50 |
| Cr | 17.50 | 17.50 | 18.60 | 15.80 | 15.70 | 10.70 |
| Cu | 3.20 | 2.70 | 3.60 | 2.10 | 1.60 | 1.40 |
| Ga | 25.20 | 31.30 | 21.10 | 25.60 | 19.40 | 31.90 |
| Hf | 17.90 | 16.30 | 9.90 | 4.50 | 11.50 | 10.40 |
| La | 158.10 | 599.60 | 39.20 | 48.40 | 104.20 | 20.20 |
| S | −352.90 | −1671.80 | 42.10 | 22.00 | −127.20 | 112.60 |
| Zn | 95.60 | 155.00 | 74.20 | 54.10 | 69.90 | 21.20 |
| Ni | 6.20 | 7.10 | 7.00 | 6.90 | 8.60 | 9.30 |
| V | 8.50 | 7.60 | 32.90 | 5.50 | 3.50 | 5.70 |

*Note*: Oxide concentration in wt.%, trace elements in ppm.

A rift setting was postulated by Blanco and Gaucher (2005) for Las Ventanas Formation. This hypothesis is favoured by:

(a) the bimodal, synsedimentary volcanism, with basalts at the base (La Rinconada Member) and rhyolites at the top (Quebrada de Viera Member) of the unit (Figure 4.2.4);
(b) strong palaeorelief, as shown by high-energy, immature conglomerate deposits;
(c) great thickness of alluvial fan conglomerates (Figure 4.2.4); and
(d) the progressive evolution from continental to open marine environments, suggesting a taphrogenic basin setting (Miall, 2000).

The onset of this rifting event is poorly constrained by a K-Ar age of 615±30 Ma for basalts of the La Rinconada Member (Sanchez Bettucci and Linares, 1996) and datings of a number of magmatic units probably associated with this extensional event. The basalts tectonically intercalated in the Fuente del Puma Group and dated U-Pb SHRIMP at 590±2 Ma by Mallmann et al. (2007) (see above) probably represent the same magmatic event. If this is accepted, the age of the basal Las Ventanas Formation is ca. 590 Ma, in accordance with biostratigraphic data (Gaucher et al., 2008a; see Chapter 4.2) and field relationships with other volcanic units (Pazos et al., 2008). On the other hand, mafic dykes crosscut conglomerates of the Las Ventanas Formation at the Mina Verdún Quarry, nearby Minas (Blanco and Gaucher, 2005) (Figure 4.6.1). These dykes yielded a minimum K-Ar age of 485±13 Ma (González et al., 2004), and could be part of the Nico Pérez Dyke Swarm (581±13 Ma; see above). Thus, a high sedimentation rate could be responsible for the accumulation of more than 3.5 km of conglomerates in ca. 10 Myr, a typical feature of rift basins (Miall, 2000).

The Las Ventanas rifting was clearly influenced by pre-existing lineaments, notably the SYSZ. This fact explains the roughly NS trend of the inferred rift axis (Figure 4.6.1), and the occurrence of a thicker basalt pile near the shear zone (Blanco and Gaucher, 2005).

In the Tandilia area only little evidences of rifting were preserved. A thin (< 10 m) diamictite level ("Colombo Diamictite") between the Villa Mónica and the Cerro Largo formations shows evidence of slumping, gravity flows and syndepositional, extensional tectonics (Figures 4.2.1 and 4.5.1D–F). The stratigraphic position of these mass-flow deposits suggest correlation with the Las Ventanas Formation in Uruguay.

Lower Ediacaran rifting in the RPC heralded the opening of the Brazilides Ocean and the development of the Sierras Bayas-Arroyo del Soldado-Corumbá shelf (Gaucher et al., 2003, 2005b), representing the drift phase of the platform (Figure 4.6.4).

### 4.6.2.5. Camaquã Basin and São Gabriel Arc: part of the Río de la Plata palaeocontinent?

The only area in Brazil currently accepted as part of the RPC is the Taquarembó Block, close to the Uruguayan border (Figure 4.1.2). The roughly NW-trending Ibaré Shear Zone in southernmost Brazil could represent the boundary of the RPC there. To the north of this shear zone, a juvenile magmatic arc (São Gabriel Orogen) occurs, represented by the Vacacaí and Cambaí groups. A juvenile Neoproterozoic source for the Yerbal Formation is suggested in NE-Uruguay (Isla Cristalina de Rivera) by $T_{DM}$ model ages of 1.2 Ga and $\varepsilon_{Nd}$ between −0.6 and −1.1, which are in contrast with $T_{DM}$ ages of around 2.1 Ga and $\varepsilon_{Nd}$ as low as −17 for outcrops of the same units in the south (Blanco et al., 2007, 2009). Blanco et al. (2009) suggested that the Vacacaí and Cambaí groups acted as main sources, both units showing positive $\varepsilon_{Nd}$ and $T_{DM}$ model ages between 0.9 and 1.2 Ga (Gastal et al., 2005). Palaeocurrent directions in the Yerbal Formation, from the northeast (Gaucher, 2000), are in agreement with this interpretation. Thus, according to these data, the São Gabriel Arc was accreted to the RPC before deposition of the Yerbal Formation, and according to a number of authors at ca. 700 Ma (Babinski et al., 1996; Saalmann et al., 2005).

The Camaquã Basin records episodical sedimentation between ca. 630 Ma and the Ordovician, and is developed mainly to the NE of the Ibaré Shear Zone (Paim et al., 2000; Borba et al., 2008). It represents a series of different, stacked basins that share the same depositional locus (Paim et al., 2000). Andesitic and rhyolitic volcanism has been dated U-Pb SHRIMP at 630 ± 3 Ma (pyroclastic clasts, Maricá Formation: Borba et al., 2008), 592 ± 5 Ma (andesites of the Hilário Formation: Remus et al., in Paim et al., 2000) and 573 ± 18 Ma or 549 ± 5 Ma for rhyolites of the Acampamento Velho Formation (Paim and Fonseca, 2004; Sommer et al., in Borba et al., 2008). Granitic magmatism in the area is constrained by U-Pb SHRIMP zircon ages between 592 and 541 Ma (Paim et al., 2000, and references therein). Whereas the 630 and 592 volcanic units match well with the extensional magmatism recorded in the Nico Pérez Terrane (see above), voluminous granitic magmatism and bimodal volcanics between 580 and 541 Ma in the Camaquã Basin contrast with tectonic quiescence in the RPC. This fundamental difference militates against the Camaquã Basin being associated to the RPC between 580 and 541 Ma. One possibility would be that the block containing the basin rifted from the RPC between 630 and 592 Ma, and evolved separately in the late Ediacaran. During the Lower Cambrian Búzios Orogeny, the basin was probably once again accreted against the RPC. However, this is at odds with the results of Blanco et al. (2007) showing that the juvenile São Gabriel Arc probably acted as a source of the Yerbal Formation in northern Uruguay. It is worth noting that the nature of the contact between the São Gabriel Orogen and its sedimentary cover may be tectonic, as suggested by the latter not being sourced in the São Gabriel juvenile rocks (Borba et al., 2008, and references therein). This might explain the conflicting results regarding the relationship between the Camaquã Basin and the RPC.

## 4.6.3. Ediacaran Drift Phase and Passive Margin

Extensive deposits of the Arroyo del Soldado and Sierras Bayas groups and correlative units (see Chapters 4.1 and 4.2) represent the drift phase of the platform. This passive margin probably reached as far north as Corumbá (Gaucher et al., 2003; Figure 4.6.4), where the Corumbá Group records essentially the same evolution.

An Atlantic-type continental shelf was interpreted as the geotectonic setting for deposition of the mentioned units (Poiré, 1987; Gaucher, 2000; Gaucher et al., 2003, 2005b; Bossi and Gaucher, 2004), on the bases of: (a) sandstone petrography, which are predominantly mature quartz arenites and subarkoses; (b) lateral persistence of lithofacies over hundreds of kilometres and markedly tabular geometry; (c) absence of synsedimentary volcanism; (d) occurrence of thick (up to 1,000 m) carbonates and chemical deposits (BIF, chert), denoting low palaeorelief and little siliciclastic input; and (e) palaeocurrent pattern.

More recently, detrital zircon ages and Sm-Nd model ages of sediments confirmed the geotectonic setting postulated for the Arroyo del Soldado and Sierras Bayas groups and Piedras de Afilar Formation. Sm-Nd analyses of sandstones and shales from the Arroyo del Soldado Group yield $T_{DM}$ model ages between 1.6 and 2.9 Ga and $\varepsilon_{Nd}$

values of approximately −10 (Blanco et al., 2007; Mallmann et al., 2007), showing provenance from old continental crust. U-Pb LA-ICP MS ages of detrital zircons from the mentioned units show (Figure 4.6.3) that the main source areas were Archaean, Palaeoproterozoic and Mesoproterozoic rocks (Gaucher et al., 2008b; Blanco et al., 2007, 2009). Neoproterozoic zircons are rare (Figure 4.6.3): one grain from a sandstone sample of the Yerbal Formation yielded an age of 664 ± 14 Ma (0.8% of concordant grains), another grain from the Cerros San Francisco Formation was dated at 605 ± 53 Ma (0.8% of concordant grains) and 28% of zircons from a sample of the Barriga Negra Formation define an age population of 631 ± 12 to 566 ± 8 Ma. These ages match the rift-related magmatic units mentioned above, such as the Puntas del Santa Lucía Batholith. The higher abundance of Neoproterozoic zircons in the Barriga Negra Formation can be explained by deeper erosion of basement units compared to the other studied formations, which allowed exhumation of deep-seated anorogenic intrusions. A source area dominated by Archaean and Palaeoproterozoic rocks is corroborated by $T_{DM}$ model ages of the Barriga Negra Formation between 1.6 and 2.7 Ga (Blanco et al., 2009). Due to the fact that conglomerates of the Barriga Negra Formation wedge out towards the basin and are absent in deep-water sections, Gaucher (2000) suggested that a large-amplitude sea-level drop was responsible for deposition of the unit. Shallow areas of the shelf were exposed and reworked, allowing rivers to cut deeply through the stratigraphy of the Arroyo del Soldado Group and its basement. This explanation is consistent with similar conglomerate levels occurring at the same stratigraphic position in the Sierras Bayas, Corumbá, Nama and Cango Caves groups (see Chapter 4.5).

Thus, available data are consistent with provenance of the Arroyo del Soldado and Sierras Bayas groups from the peneplainised RPC, and effectively rule out an input from a Neoproterozoic orogenic belt (Figure 4.6.4), as in the case of a foreland basin (e.g. Basei et al., 2000; Pecoits et al., 2008). The most important controls on deposition were climate and associated sea-level fluctuations.

## 4.6.4. Evolution of the Western Boundary of the Río de la Plata Palaeocontinent

The western boundary of the RPC has remained controversial due to the extensive Phanerozoic cover of the Paraná Basin. Almeida (2004) placed the western boundary at the Eastern Sierras Pampeanas, which has been confirmed by Rapela et al. (2007) by U-Pb SHRIMP dating of basement samples from boreholes drilled through the Paraná Basin in Argentina. All studies regarding the western boundary of the RPC have systematically considered Palaeoproterozoic (Transamazonian) ages between 2.2 and 2.0 Ga as typical of the craton. However, as demonstrated by detrital zircon dating of Neoproterozoic sedimentary covers of the RPC (Gaucher et al., 2008b), Mesoproterozoic belts are also an integral part of the craton (Figures 4.6.3 and 4.6.4). This means that the RPC may extend farther to the west than currently accepted, probably including a Grenvillian belt surrounding the Palaeoproterozoic-Archaean core (see above).

The main events linked to the evolution of the western margin of the RPC have been outlined by Rapela et al. (1998a) (Figure 4.6.5). The Puncoviscana Formation, a sedimentary succession of latest Ediacaran to Lower Cambrian age (Aceñolaza, 2003, 2004, and references therein; see Chapter 6), fringes the RPC to the west (Figure 4.6.4). This unit was deposited in a narrow ocean between the RPC and the Pampia Terrane (see Chapter 6) that probably opened as a result of late Ediacaran rifting between the RPC and Laurentia (Grunow et al., 1996). A prominent angular unconformity (Tilcaric unconformity) between the Puncoviscana Formation and the Middle Cambrian Mesón Group shows that closure of the Puncoviscana Basin took place in the Lower Cambrian (Aceñolaza, 2003). This short-lived orogenic event, known as the Pampean Orogeny, is also characterised by calc-alkaline granite intrusions between 530 ± 4 and 523 ± 2 Ma (U-Pb on zircon: Rapela et al., 1998a). Mulcahy et al. (2007) suggest that following the accretion of the Pampia Terrane to the RPC at 515 ± 2 Ma (Ar-Ar on mylonites from a thrust plane), eastward subduction stepped westward, establishing a new convergent margin along the Famatina arc, active until the Middle Ordovician. It is worth noting that while in the eastern RPC extensional tectonics led to the intrusion of a syenitic anorogenic complex, at the western boundary of the craton granulite-facies metamorphism was underway (Rapela et al., 1998a).

The paucity or absence of Palaeoproterozoic detrital zircons in late Ediacaran-Cambrian metasediments from the Eastern Sierras Pampeanas has been suggested as proof of these rocks being sourced in a pre-Pampean, Neoproterozoic arc and not in the RPC (Escayola et al., 2007). However, as has been recently shown by detrital zircon data from autochthonous Neoproterozoic covers in the eastern RPC (Blanco et al., 2007; Gaucher et al., 2008b), Palaeoproterozoic detritus becomes progressively less important towards the west of the RPC, and Mesoproterozoic zircons dominate at the western boundary (Schwartz and Gromet, 2004). Therefore, the logical explanation is that the metasedimentary rocks of the Eastern Sierras Pampeanas were sourced mainly in the proto-Andean, Grenvillian belt fringing the RPC to the west (Figure 4.6.4).

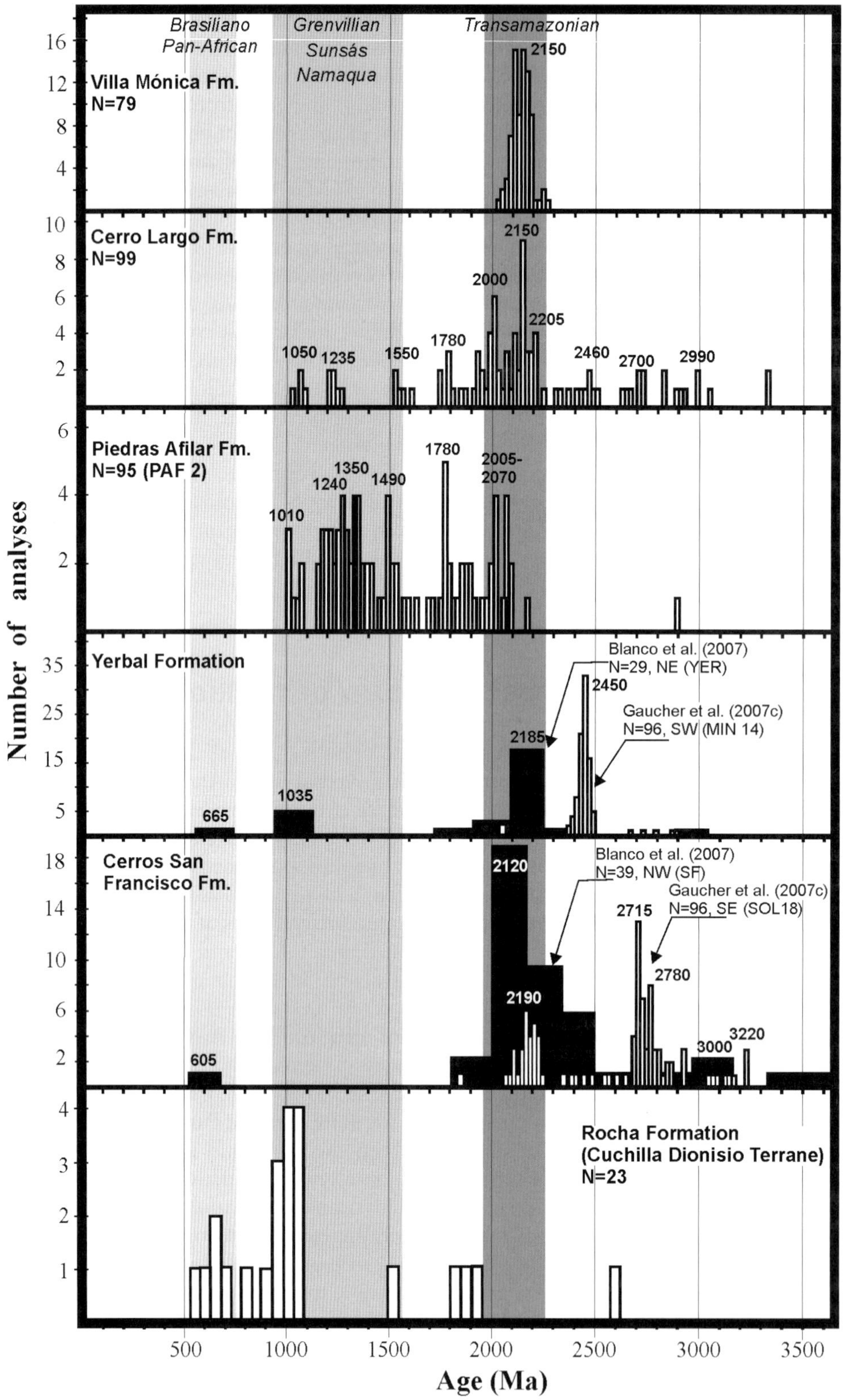


**Figure 4.6.3** Detrital zircon age spectra for Neoproterozoic-Lower Cambrian sandstones of the RPC and Cuchilla Dionsio Terrane. Location of samples in Uruguay: see Figure 4.6.1. Sample for the Rocha Formation from Basei et al. (2005), all other samples from Gaucher et al. (2008b) otherwise indicated (Blanco et al., 2009). Age of the main peaks (in Ma) and range of important orogenic events is indicated. Note (a) distinctively different spectra for the Rocha Formation compared to the samples from the RPC; (b) paucity of Neoproterozoic zircons in the RPC samples; and (c) significant Mesoproterozoic populations.

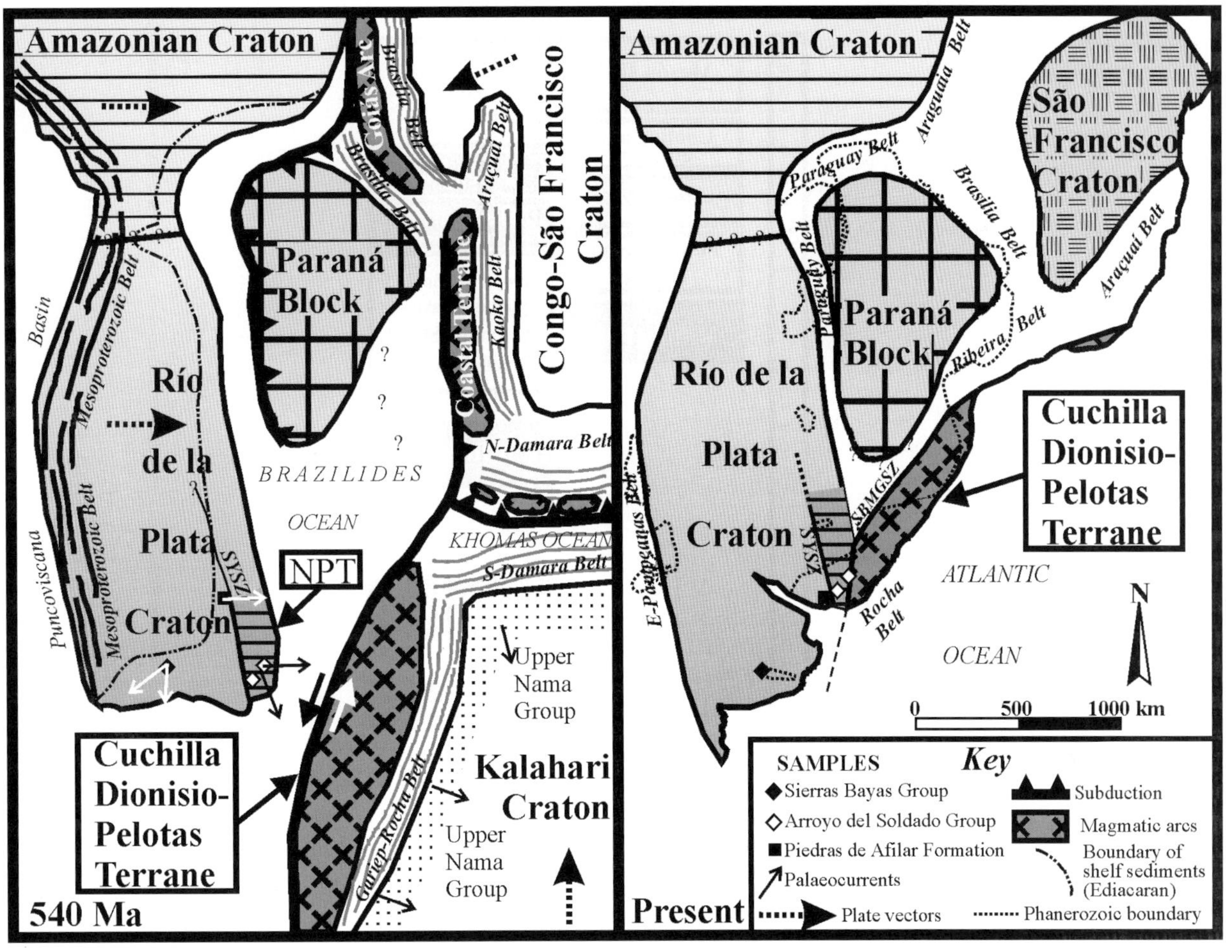


**Figure 4.6.4** Palaeogeographic reconstruction of the RPC in the earliest Cambrian and present configuration of cratons and belts. Note location of detrital zircon samples of Neoproterozoic-Lower Cambrian sedimentary successions discussed in the text, showing corresponding palaeocurrent pattern. Modified from Gaucher et al. (2008b) and sources therein.

Detrital zircon ages of Ordovician-Silurian sandstones from the Balcarce Formation show Cambrian to Neoproterozoic, Mesoproterozoic and late Palaeoproterozoic peaks (Rapela et al., 2007). In this case, the paucity of Transamazonian ages is easily explained by the docking in the Lower Cambrian of the Cuchilla Dionsio-Pelotas Terrane, which provided Cambrian-Neoproterozoic and late Mesoproterozoic detritus typical of that block (Basei et al., 2005). This is confirmed by west-directed palaeocurrents in the Balcarce Formation (Poiré, 1987; Rapela et al., 2007). The Cuchilla Dionsio Terrane was probably an important mountain range undergoing active erosion in Ordovician times, thus overshadowing the sediment input of the low-relief, Transamazonian units of the RPC.

## 4.6.5. Lower Cambrian Búzios Orogeny

Bossi and Gaucher (2004) proposed that the Cuchilla Dionisio-Pelotas Terrane is an allochthonous block juxtaposed to the RPC by means of tangential tectonics (see Chapter 8). Although it represents the roots of a magmatic arc active between 650 and 550 Ma, as already envisaged by Ramos (1988), its present position is out of context and does not reflect the origin of the magmatic arc. Detrital zircon ages of the Ediacaran Rocha Formation (Figure 4.6.3), sourced in the Cuchilla Dionisio Terrane, show prominent peaks at 1.0 Ga and between 764 and 596 Ma (31% of all zircons; Basei et al., 2005), typical ages of the basement and granitic intrusions of the terrane. This is in sharp contrast to the above-mentioned detrital zircon ages for coeval units deposited in the RPC, showing that the RPC and Cuchilla Dionisio Terrane were separated by a large (oceanic) basin in the Ediacaran. Given that the youngest units of the Arroyo del Soldado Group were deposited between 542 and 535 Ma (Sprechmann et al., 2004; Gaucher et al., 2007c; see Chapter 4.4), the tangential collision took place no earlier than the Lower Cambrian, possibly around 530 Ma (Figure 4.6.4). This is in accordance with Rb-Sr and U-Pb zircon ages of associated granites of 532 $\pm$ 11 Ma (Kawashita et al., 1999a) and 530 $\pm$ 14 Ma (Oyhantçabal

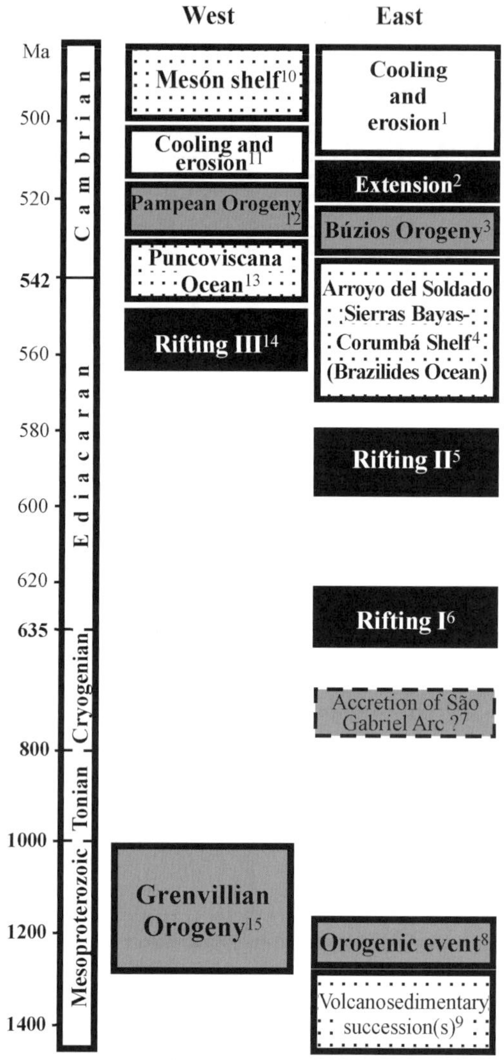


**Figure 4.6.5** Palaeogeographic evolution of the RPC in time between the Mesoproterozoic and the Cambrian. Sources of data: (1) Cingolani et al. (1990b), in Gaucher (2000); (2) Bossi et al. (1993b); (3) Bossi and Gaucher (2004) and references therein; (4) Gaucher (2000) and Gaucher et al. (2003, 2004c, 2008b); (5) Blanco and Gaucher (2005) and Mallmann et al. (2007); (6) Hartmann et al. (2002a); (7) Babinski et al. (1996) and Blanco et al. (2009); (8) Bossi et al. (1998), Teixeira et al. (1999) and Bossi and Navarro (2001); (9) Oyhantçabal et al. (2005) and Gaucher et al. (2006); (10) Aceñolaza (2003); (11) Mulcahy et al. (2007); (12) Rapela et al. (1998a) and Mulcahy et al. (2007); (13) Aceñolaza (2003, 2004) and references therein; (14) Grunow et al. (1996); and (15) Gaucher et al. (2008b) and references therein.

et al., 2005), K-Ar ages of recrystallised shales between $532 \pm 16$ and $492 \pm 14$ Ma (Cingolani et al., 1990b, in Gaucher, 2000) and remagnetisation of the Arroyo del Soldado Group around 525 Ma (Rapalini and Sanchez Bettucci, 2008). Typical Brasiliano compressional tectonics, such as in the Brasília and Araçuaí belts, are distinctively older than 530 Ma. Therefore, we assign the tangential collision of the RPC with the Cuchilla Dionisio Terrane to the Búzios Orogeny (Schmitt et al., 2004, 2008b), which represents the last event leading to Gondwana amalgamation. The Búzios Orogeny is the result of the collision of the block conformed by the Cuchilla Dionisio-Pelotas Terrane (*Arachania*; see Chapter 8) and the Kalahari Craton with the Río de la Plata, Amazonian, Paranapanema and Congo-Sao Francisco cratons. A reconstruction of this complex tectonic event was published by Gaucher et al. (2008b), and is shown in Figure 4.6.4.

## 4.6.6. Cambrian Anorogenic Magmatism

The onset of post-orogenic relaxation and extensional tectonics is marked by the intrusion of the Sierra de Animas Formation in Uruguay. This unit comprises syenites, trachytes, rhyolites and basalts, yielding an Rb-Sr isochronic age of $520 \pm 5$ Ma (Bossi et al., 1993a). The Sierra de Animas Formation shows a subalkaline to alkaline affinity (Bossi et al., 1993a) and is largely undeformed. It is worth noting that syenites were emplaced along the SYSZ. The undeformed nature of the Sierra de Animas Formation shows that compressive tectonics likely finished before 520 Ma. The reactivation of the Sarandí del Yí lineament is constrained by Rb-Sr ages of 530 ± 10 Ma on intensely deformed granites emplaced in the shear zone (Umpierre and Halpern, 1971; see Chapter 4.1), and by the age of the upper Arroyo del Soldado Group (542–535 Ma), which is affected by the fault (Gaucher et al., 2007c). A northward displacement of several hundreds of kilometres (Figure 4.6.4) is indicated for the Nico Pérez Terrane by detrital zircon age spectra of several Ediacaran units in the RPC (Gaucher et al., 2008b).

## 4.6.7. Palaeomagnetic Constraints

A recent compilation of Neoproterozoic palaeomagnetic data by Trinidade and Macouin (2007) lists only two palaeopoles for the RPC that fulfil the necessary quality criteria and represent primary magnetisation: one from the Sierras Bayas Group ("La Tinta Formation" pole: Valencio et al., 1980) and the other from the Campo Alegre Group. The latter unit is not associated to the RPC but to the Luis Alves Block, a microplate possibly accreted in the Cambrian (D'Agrella-Filho and Pacca, 1988; see Chapter 7.2). The age of the palaeopole from the Sierras Bayas Group is poorly constrained, given the uncertainties regarding the age of the Villa Mónica Formation. Furthermore, in a recent comprehensive work, Rapalini and Sanchez Bettucci (2008) demonstrate the non-primary nature of the "La Tinta" palaeopole. According to these authors, a number of Neoproterozoic units, including the Arroyo del Soldado Group, are remagnetised and cannot be used for palaeogeographic reconstruction (Rapalini et al., 2008). Other published palaeopoles lack the necessary tests to ascertain their primary nature or their age is poorly constrained (Sanchez Bettucci and Rapalini, 2002; Rapalini and Sanchez Bettucci, 2008). Therefore, the paucity of palaeomagnetic data for the RPC does not allow yet carrying out palaeogeographic reconstructions independently from the geological data.

## 4.6.8. Conclusions

Several lines of evidence suggest that the RPC was part of Rodinia, and that a proto-Andean Grenvillian belt is an integral part of the craton (Figures 4.6.4 and 4.6.5). The RPC rifted off Rodinia between 635 and 580 Ma. Evidences of rifting include mafic dyke swarms, anorogenic granites, within-plate basalts and rift sedimentary deposits. Coeval volcanism in the Camaquã Basin in southern Brazil suggest that this block may have rifted off the RPC, evolving separately until its accretion in the Lower Cambrian, although other data suggest that the São Gabriel juvenile arc was accreted to the RPC at ca. 700 Ma.

Following rifting, an extensive passive margin was established on the eastern margin of the RPC, including deposits of the Sierras Bayas, Arroyo del Soldado and Corumbá groups. Shelf deposits were deformed and weakly metamorphosed due to tangential collision of the Cuchilla Dionisio-Pelotas Terrane (Arachania) with the RPC. This collision, assigned here to the Búzios Orogeny, likely took place around 530 Ma, representing the last event leading to final accretion of Gondwana. Anorogenic, syenitic magmatism dated at 520 Ma marks the end of compressional tectonics in the eastern RPC and the onset of post-orogenic relaxation. Roughly at the same time (530–515 Ma) in the western RPC, the Pampean Orogeny led to deformation and metamorphism (up to granulite facies) of latest Ediacaran-Lower Cambrian siliciclastic successions (Puncoviscana Formation and equivalents; Figure 4.6.5). Once Pampia was accreted to the RPC, eastward subduction likely migrated westwards and continued until the Middle Ordovician.

## *ACKNOWLEDGEMENTS*

This research has been partially funded by CSIC, Uruguay (project "Estratigrafía de secuencias carbonáticas del Neoproterozoico del Terreno Nico Pérez"). Insightful reviews by Léo A. Hartmann and Peter Sprechmann are gratefully acknowledged.

CHAPTER 5.1

# Configuration of Pan-African Orogenic Belts in Southwestern Africa☆

Hartwig E. Frimmel

Effectively all Neoproterozoic rock successions in southwestern Africa are found, variably deformed and metamorphosed, in orogenic belts that emerged during the amalgamation of Gondwana towards the end of the Neoproterozoic Era. Various cratonic blocks became welded together along these tectonic belts in the course of what is commonly referred to as the Pan-African orogeny. The Archaean Congo Craton in central Africa is fringed to the west by the West Congo Belt, a fold-thrust-belt with top-to-(north)east transport direction and a corresponding foreland on the craton-facing side (Tack et al., 2001). Although an allochthonous thrust package of Palaeoproterozoic basement rocks follows to the west, the overall structural trend, timing of tectono-thermal events, metamorphic polarity and tectonic vergence point to a relationship between the West Congo Belt and the Araçuaí Belt in Brazil (Pedrosa-Soares et al., 2001, 2008; Alkmim et al., 2006). They are the eastern and western parts, respectively, of a Pan-African compressional zone ('confined orogen') between the São Francisco and Congo Cratons (Figure 5.1.1).

To the south of the Congo Craton is the Lufilian Belt (Kampunzu and Cailteux, 1999) in which four north-convex tectonic zones can be distinguished. These are, from south to north, an overriding basement plate (Katanga High), the Synclinorial Belt, the anticlinorial Domes Region and the External fold-thrust belt. Tectonic vergence is to the northwest, north and northeast (Figure 5.1.2), and corresponding foreland deposits occur to the northeast of the belt. A crustal-scale shear zone (Mwembeshi Shear Zone) separates the Lufilian Belt in the north from the Zambezi Belt in the south. Porada and Berhorst (2000) pointed out that contrary to older models, little vertical displacement took place along the Mwembeshi Shear Zone, but instead recognised it as a major sinistral transcurrent shear zone.

To the east of the Congo Craton lies the north-south trending Mozambique Belt. A commonly held view is that the Mozambique Belt represents a largely linear orogen that resulted from subduction and collision of East and West-Gondwana, thus extending along East Africa for more than 5,000 km (Stern, 1994). More recent structural, geochronological and petrological studies have shown, however, that the Mozambique Belt is probably a heterogeneous collage of microplates that moved in different directions and collided with each other at different times during the convergence of the East- and West-Gondwana continents (Kröner, 2001; Ring et al., 2002). A discussion of the tectonic and geodynamic evolution of the Mozambique Belt is, however, beyond the scope of this book in which we shall focus on southwestern Gondwana.

Of greater interest here is the continuation of the network of Pan-African tectonic belts from the West Congo and the Lufilian/Zambezi Belts towards southwestern Africa. Extrapolation of structural trends, geophysical evidence and sparse borehole information all point to a continuation of the Lufilian Belt, largely obscured by younger sediment cover, to the Damara Belt in central Namibia (Figure 5.1.2). This northeast-trending belt separates the Congo Craton in the north from the Kalahari Craton in the south. Some workers have argued for the existence of a separate Angola Craton between the Congo-São Francisco Craton and the Kalahari Craton because of highly diachronous stages of rifting as outlined in Chapter 5.2.

The Damara Belt has been subdivided into several tectonic zones (Miller, 1983a, 2008). These are, from south to north: the Southern Foreland (Zaris Subbasin of the larger Nama Basin, Figure 5.1.3), the Southern Margin Zone, the Southern Zone including the Okahandja Lineament Zone ('Khomas Trough'), the granite-rich Central

☆ Frimmel, H.E., 2009. Configuration of Pan-African orogenic belts in Southwestern Africa. In: Gaucher, C., Sial, A.N., Halverson, G.P., Frimmel, H.E. (Eds.): Neoproterozoic-Cambrian tectonics, global change and evolution: a focus on southwestern Gondwana. Developments in Precambrian Geology, 16, Elsevier, pp. 145–151.

Department of Geological Sciences, University of Cape Town, Rondebosch 7700, South Africa.

*Developments in Precambrian Geology*, Volume 16
ISSN 0166-2635, DOI 10.1016/S0166-2635(09)01610-7

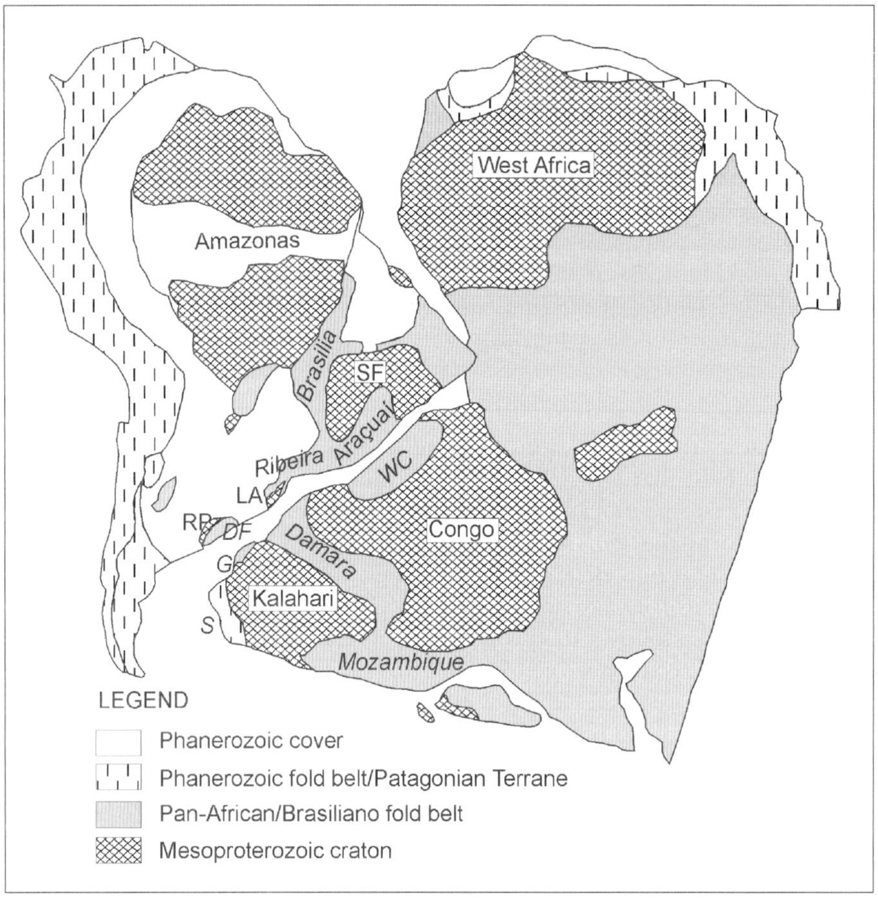


**Figure 5.1.1** Collage of Pan-African/Brasiliano tectonic belts around Archaean to Mesoproterozoic cratonic blocks in Africa and South America with focus on southern Africa; DF – Dom Feliciano Belt, G – Gariep Belt, LA – Luiz Alvez microplate, RP – Rio de la Plata Craton, S – Saldania Belt (basement of Permo-Triassic Cape Fold Belt); modified from Unrug (1996).

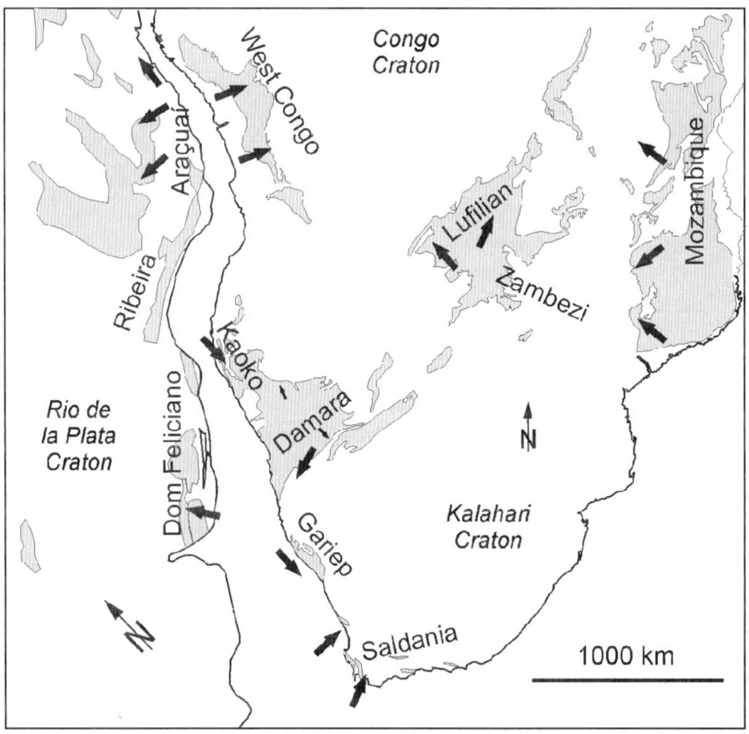


**Figure 5.1.2** Distribution of Neoproterozoic units in Pan-African/Brasiliano tectonic belts in southern Africa and southeastern South America (shown in an Upper Cretaceous Gondwana break-up position); main orogenic kinematic transport directions are indicated by arrows.

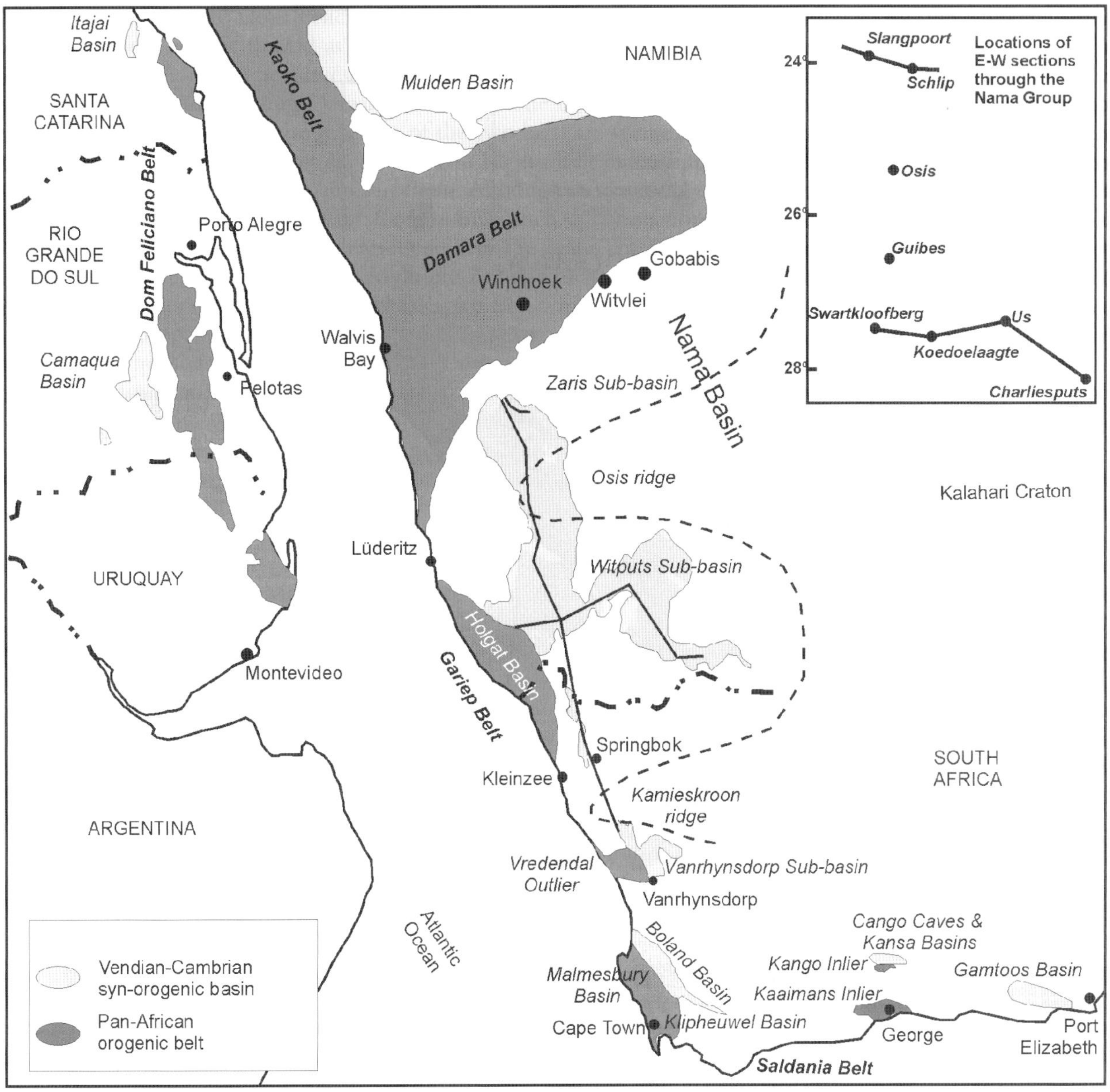


**Figure 5.1.3** Position of the Pan-African orogenic belts and corresponding Vendian to early Palaeozoic late- to post-orogenic basins in southwestern Africa and their counterparts in southeastern South America; also shown are the north-south and east-west (see insert) cross-sections through the Nama Basin, discussed in Section 5.4.1.

Zone that is divided into northern and central halves, the Northern Zone, the Northern Margin Zone and the Northern Platform (Mulden Basin, Figure 5.1.4). The Damara Belt has been interpreted as the result of the transpressive, northeast over southwest collision between the Congo and Kalahari Cratons during northwestward subduction of oceanic crust of unknown width beneath the continental Congo plate (Miller, 1983a). In the southern Central Zone and the Southern Margin Zone, the transport direction was top-to-southwest but in the intervening Southern Zone, the transport direction was top-to-southeast (Hälbich, 1977; Kukla, 1992). The Okahandja Lineament is a crustal-scale, 0.5–2 km wide feature that forms the leading edge of the Congo Plate between the Central and Southern zones. Although largely a zone of intense, near-vertical isoclinal folding, it also contains a sinistral shear component (Downing and Coward, 1981). Top-to-north transport is typical of the northern parts of the belt. A completely allochthonous stack of nappes, the Naukluft Nappe Complex, is rooted in the Southern Margin Zone but was emplaced post-tectonically under the influence of gravity southwards onto the Southern Foreland. Regional metamorphic field gradients follow two trends across the entire belt as will be discussed in greater detail in Section 5.5.2. Within the Central Zone, the metamorphic grade decreases from lower granulite-facies conditions at the Atlantic coast towards the east. In contrast, within the marginal zones, the metamorphic grade decreases away from the central zone towards the forelands in the north and south, with medium-pressure, Barrovian-type metamorphism recorded in the southern zones.

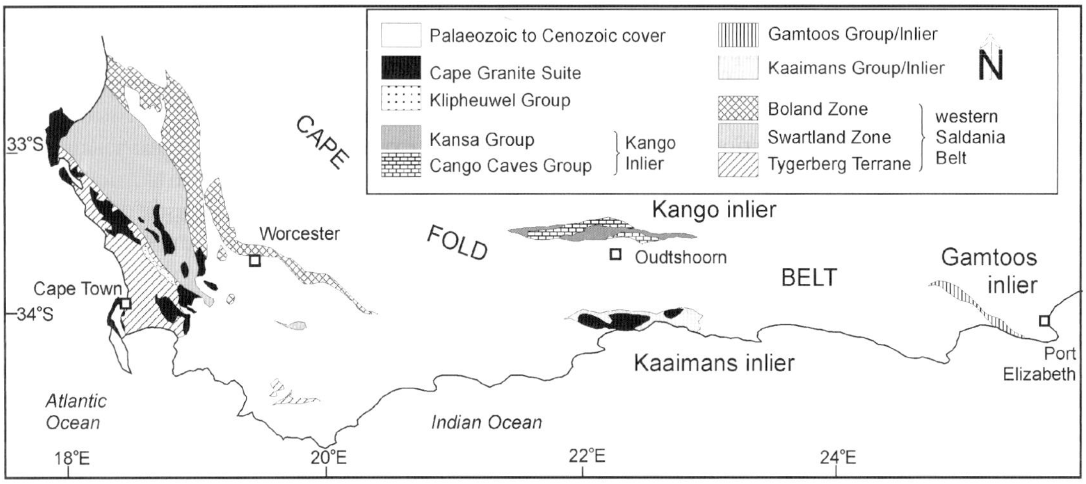


**Figure 5.1.5** Tectonic/stratigraphic subdivision of the Saldania Belt (modified from Frimmel et al., 2001a). Note that the stratigraphic subdivision of the western Saldania Belt remains uncertain.

deformation was in places focussed into major near-vertical faults, the most prominent of which is the Colenso Fault. The latter separates the Tygerberg Terrane, which consists of syn-orogenic foredeep deposits (Tygerberg Formation), from two tectono-stratigraphic zones that are regarded by some as para-autochthonous and by others as separate terranes (Swartland and Boland Terranes). The latter two are made up of possibly Cryogenian to Lower Cambrian volcano-sedimentary successions (Figure 5.1.5).

In spite of the diverse kinematic patterns in the various Pan-African belts of southwestern Africa, there are marked differences but also distinct commonalities in the timing of tectono-thermal events that shaped these belts, as recently summarised by Gray et al. (2006). The oldest metamorphism ($M_1$), reaching high-grade conditions, is dated between ca. 660 and 630 Ma and is recorded only in the exotic Coastal Terrane of the western Kaoko Belt. A second metamorphic event ($M_2$), again reaching high-grade conditions in the Orogen Core of the Kaoko Belt (eastern part of the Western Kaoko Zone) between 580 and 560 Ma, is also reflected by an earlier, syn-tectonic metamorphic event at a somewhat elevated pressure/temperature (P/T) ratio in the Marmora Terrane of the Gariep Belt. This event corresponds in age with syn-tectonic granite intrusions in the Coastal Terrane of the Kaoko Belt and with calc-alkaline magmatism in the central Damara Belt. Final continental collision in the Damara Belt at 542 Ma is post-dated by regional metamorphism ($M_3$) peaked at approximately 535 Ma (Miller, 1983a, 2008). In the Gariep Belt, continental collision led to the peak of regional, generally low-grade metamorphism at 545 Ma. Collision was accompanied and followed by syn- to post-orogenic granite emplacement, mainly in the central Damara Belt and the Saldania Belt, with syn-orogenic magmatism conspicuously lacking in the Gariep Belt. Exhumation beyond a cooling temperature of about 500°C was achieved earlier at about 530–520 Ma in the shallower crustal sections of the Gariep and Damara Belts, whereas it was delayed until about 510–500 Ma in the deeper crustal sections, now exposed in the Orogen Core of the Kaoko Belt. Finally, a series of K-Ar and Ar-Ar ages indicate cooling through 300°C between 500 and 480 Ma. A more detailed account of the tectono-thermal history of the various belts is given in Chapter 5.5.

While the Gariep, Damara and Lufilian, Kaoko and West Congo Belts can be explained by continental collision of the São Francisco, Congo, Kalahari and Rio de la Plata Cratons in various directions, the Saldania Belt (or at least its southern branch) takes a position that has been on the fringe of the pre-Gondwana plates and eventually of Gondwana itself. It thus represents a typical accretionary orogen, which can be considered the western continuation of the Ross Orogen in Antarctica.

In the past, a more or less synchronous basin evolution from continental rifting, via passive margin sedimentation, syn-orogenic flysch-type deposition to post-orogenic molasse sedimentation was assumed for the various Pan-African belts in the region. Previous geodynamic models considered the 'Damara-Ribeira Orogen' as an orogen/aulacogen system that started from a three-pronged continental rift system at about 1,000 Ma ago (Porada, 1979). While a Wilson Cycle model was applied to the Ribeira (including Kaoko) Belt and the Gariep Belt, the Damara Belt was regarded by some workers as having developed from a failed continental rift (Martin and Porada, 1977). This view has been challenged by numerous workers (e.g. Miller, 1983a) who preferred a Wilson Cycle model also for the inland branch, the Damara Belt *sensu stricto*. All of these older geodynamic models for the various Pan-African belts in southwestern Africa and their counterparts in eastern South America assumed, however, a more or less synchronous Neoproterozoic basin evolution. In the light of more recent chrono-, bio- and chemostratigraphic evidence, as well as detrital zircon provenance data, such a simplistic model has to be questioned. Today we know that the various Neoproterozoic basins differ with regard not only to the timing of sediment deposition, and in places volcanic activity, but also to the overall basin architecture and tectonic setting. Onset of deposition, which had been triggered by the break-up of an inferred Mesoproterozoic supercontinent (Rodinia or Palaeopangea), was diachronous, and the style of sedimentation was different, depending on the tectonic regime, palaeogeography and proximity to the respective basin margins.

Radiometric constraints on the maximum age of sedimentation, which is further outlined in the subsequent Chapter 5.2, indicate that continental rift sedimentation could not have started before ca. 770 Ma in the Gariep Belt. This is different to the situation further north and east. In the West Congo Belt, continental rifting is recorded in the volcano-sedimentary successions of the Zadinian and Mayumbian Groups, whose age is constrained between $999 \pm 7$ and $912 \pm 7$ Ma (Tack et al., 2001). In the Lufilian and Zambezi Belts, the entire Roan Group (Katanga Supergroup), which grades from more siliciclastic facies in the north to carbonate-dominated facies in the south, was deposited in a time window between $<880$ and 760 Ma (Wendorff, 2003).

In spite of the diachronous basin evolution recorded in the Neoproterozoic rock record of the various Pan-African belts, all of them, except for the Saldania Belt, contain ubiquitous sedimentary beds, specifically glaciogenic diamictite and corresponding post-glacial carbonate successions. In all of these belts, there are two distinct Neoproterozoic glaciogenic stratigraphic levels. The apparent similarities between these beds tempted many workers into lithostratigraphic correlation across the various basins (e.g. Hoffmann, 1989) and also on a global scale (Kennedy et al., 1998). The older diamictite units were typically correlated with the global Sturtian glaciation, whereas the younger ones with the Marinoan glaciation (Frimmel et al., 2002). Only in the last few years, with globally improving radiometric age control on these deposits, did it become apparent that the older and younger diamictite units in the different belts are not necessarily of the same age – an aspect that is further discussed in Chapter 5.3.

Correlation of the Pan-African belts of southwestern Africa with the Brasiliano belts of South America has been problematic for as long as the existence of a Gondwana supercontinent has been accepted. This is due to vast differences in the late Neoproterozoic to early Cambrian tectonic belts on either side of the South Atlantic. The connection of the West Congo Belt with the Araçuaí Belt is well established (Pedrosa-Soares et al., 2001, 2008). Towards the south, the relationships between the Pan-African/Brasiliano belts along the South Atlantic become less clear. As pointed out above, the Coastal Terrane of the Kaoko Belt could have an equivalent in the Ribeira Belt, or it could reflect a separate microplate – an issue that is still unresolved (e.g. Gray et al., 2008; Heilbron et al., 2008a). Further south, a geodynamic relationship between the western Damara-Gariep system and the Dom Feliciano Belt has been suggested for a long time (Porada, 1979). Those who advocated a Wilson Cycle model for the Damara and Gariep Belts assumed northwest-directed subduction of oceanic crust beneath the Rio de la Plata Craton and explained the large granite belt within the Dom Feliciano Belt in southeastern Brazil as the deeper parts of a volcanic arc system that resulted from the closure of an ocean between palaeo-southern Africa and palaeo-South America, the Adamastor Ocean (Hartnady et al., 1985). In that model, initial convergence between the Congo and northern Rio de la Plata Craton would have caused the orogeny in the Kaoko Belt (Dürr and Dingeldey, 1996), followed by the northeast-dipping subduction of the Khomas oceanic plate beneath the Congo Craton (Miller, 1983a) and eventually subduction of the southern Adamastor oceanic crust beneath the southern Rio de la Plata Craton, thus giving rise to the Gariep Belt (Frimmel et al., 1996b). Consequently, the modern South Atlantic would have opened up along a suture that traced the Adamastor Ocean, which would have taken the position of a Neoproterozoic palaeo-South Atlantic.

Over the past few years, a great number of new data has emerged from the various Pan-African/Brasiliano belts that flank the modern South Atlantic, which are inconsistent with the above sequence of postulated events. One of the particularly intriguing aspects in this regard is that the Dom Feliciano volcanic arc might not have emerged from the subduction of an Adamastor Ocean to the east but rather from the subduction of an oceanic basin to the west (Basei et al., 2000). The easternmost portion of the Dom Feliciano Belt, the Cuchilla Dionisio (Punta del Este) Terrane in eastern Uruguay (Bossi and Gaucher, 2004), has a tectono-metamorphic history and isotopic signature that are comparable to those of the Namaqua Belt. The overlying metasedimentary rocks (Rocha Group) have been recognised as foredeep deposits related to the closure of the Gariep Basin with correlatives in the Oranjemund Group of the Marmora Terrane (Basei et al., 2005, 2008). Consequently, the main suture between palaeo-South America and southern Africa would lie to the west of the Dom Feliciano Belt and not along the modern South Atlantic. This model explains the lack of a proper high-pressure metamorphic belt anywhere in the Dom Feliciano Belt or the Gariep Belt, the timing of sediment deposition in the upper parts of the Gariepian stratigraphy and the provenance of these sediments in a 640–590 Ma volcanic arc, i.e. the Dom Feliciano arc, and last but not least also the good preservation of oceanic crust in the Marmora Terrane of the Gariep Belt.

# Continental Rifting ☆

Hartwig E. Frimmel[1] *and* Roy McG. Miller[2]

## Contents



## 5.2.1. Introduction

It is generally assumed that a supercontinent formed towards the end of the Mesoproterozoic, welded together by tectonic belts of 'Grenvillian' age, taking the tectonothermal history of the Grenville Belt in Laurentia as reference. Different palaeogeographic configurations have been suggested for this supercontinent, ranging from various modifications of 'Rodinia' (Hoffman, 1991; Dalziel et al., 2000) to a Palaeopangea model (Piper, 2000). The term 'Grenvillian' in this regard may be misleading, because it refers to a very specific time window, but amalgamation of that supercontinent was not synchronous everywhere, with different stages of volcanic arc formation, accretion and continental collision in different areas, ranging in time from about 1.4 to 1.05 Ga. These include the 'Kibaran' (1.37 Ga), 'early Namaqua' (1.19–1.17 Ga) and 'late Namaqua' (1.06–1.03 Ga) orogenies in southern Africa. It remains doubtful whether all continental fragments were amalgamated into a single supercontinent at the end of the Mesoproterozoic (Cordani et al., 2003). Palaeomagnetic evidence suggests that the Congo Craton (linked with the São Francisco Craton) was not part of Rodinia (Tohver et al., 2006). Nevertheless, the history of the Neoproterozoic basin fills, now present in the various Pan-African belts of southwestern Africa, begins in almost all cases with continental break-up.

## 5.2.2. Pre-Rift Magmatism

Continental rifting was preceded and accompanied by largely bimodal magmatism in the thinning lithosphere, giving rise to a suite of alkaline melts that intruded the basement prior to, or reached the surface during early rift sedimentation. Pre-rift magmatism occurred in the pre-Gariep basement where a range of predominantly felsic intrusive bodies along a northeasterly trend from the Richtersveld in South Africa to southern Namibia are clearly post-tectonic with respect to the Namaqua structures (Ritter, 1980). Stratigraphically, these rocks are unified as the Richtersveld Suite. Based on a geochemical and geochronological study (Frimmel et al., 2001b), the genesis of this suite is best explained by crustal thinning in preparation of continental break-up between 831 and 771 Ma.

The various intrusive to hypabyssal rocks of the Richtersveld Suite are found as Richtersveld Igneous Complex in the type area, the Richtersveld in South Africa and its geomorphological continuation across the

☆ Frimmel, H.E., McG. Miller, R. 2009. Continental rifting. Neoproterozoic to Early Palaeozoic evolution of Southwestern Africa. In: Gaucher, C., Sial, A.N., Halverson, G.P., Frimmel, H.E. (Eds.): Neoproterozoic-Cambrian tectonics, global change and evolution: a focus on southwestern Gondwana. Developments in Precambrian Geology, 16, Elsevier, pp. 153–159.

[1] Department of Geological Sciences, University of Cape Town, Rondebosch 7700, South Africa.
[2] PO Box 11222, Windhoek, Namibia.

*Developments in Precambrian Geology*, Volume 16
ISSN 0166-2635, DOI 10.1016/S0166-2635(09)01611-9

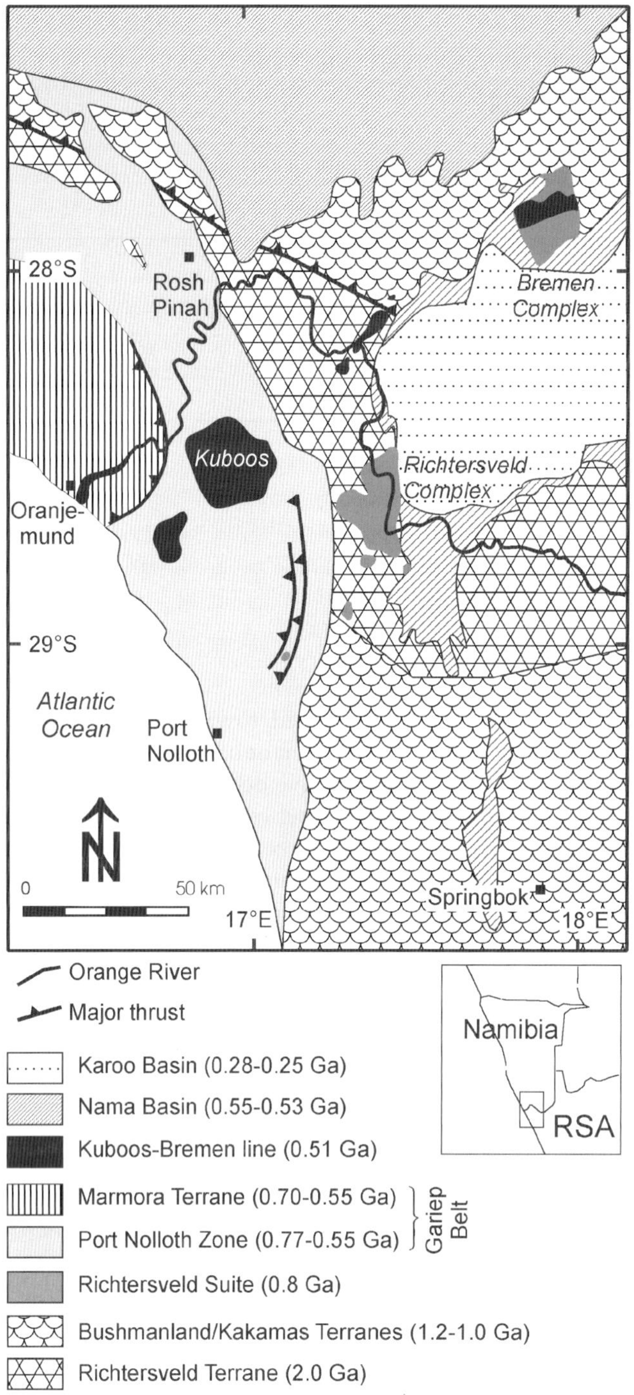


**Figure 5.2.1** Location of the 833–771 Ma Richtersveld Suite (Richtersveld Igneous Complex) and the post-orogenic ca. 507 Ma Kuboos-Bremen Suite within the main tectonic units of southernmost Namibia and adjoining South Africa (modified from Frimmel et al., 2004).

Orange River in southernmost Namibia, as well as in the outer Bremen Complex in southern Namibia (Figure 5.2.1). Undeformed, fine-grained, post-tectonic granite bodies (Uruchab Granite) in the pre-Gariep basement near the Aurus Mountains and along the margin of the Gariep Belt respectively northwest and north of Rosh Pinah, appear similar to the Richtersveld Suite granite both petrologically and in terms of relative timing of emplacement, but these granite bodies still await precise dating.

In the extreme south of Namibia, the rocks of the Richtersveld Suite intruded low-grade metamorphosed and tectonised granitoids of the 1.7–1.9 Ga Vioolsdrif Suite (Reid, 1982) and metavolcanic rocks of the 2.0 Ga Orange River Group (Reid, 1997). Further to the south in South Africa, the plutons, dykes and subvolcanic rocks of the Richtersveld Suite extend into the Bushmanland Terrane of the Namaqua-Natal Belt, which consists of high-grade gneisses and felsic intrusions, where the main period of metamorphism is dated at 1.03–1.06 Ga (Robb et al., 1999; Raith et al., 2003).

The igneous rocks making up the suite are best studied in the type area, where four major leucogranite and quartz syenite bodies are distinguished (Ritter, 1980). Geochemically, these felsic rocks bear all the hallmarks of anorogenic magmatism, with trace-element distributions conforming to those expected in a continental within-plate setting (Frimmel et al., 2001b). Mantle-derived basaltic melts are inferred as sources on the basis of incompatible trace-element ratios that are insensitive to magma differentiation by crystal fractionation, and crustal contamination was variable. Single zircon U-Pb age data indicate a prolonged phase of magmatism with individual pulses dated at 837±2, 801±8 and 771±6 Ma (Frimmel et al., 2001b). The youngest of these ages is of particular significance as it was obtained on granite that underlies the oldest sedimentary rocks of the Gariep Supergroup (Lekkersing granite), thus providing a maximum age for the onset of sedimentation in the Gariep Basin.

Pre-rift magmatism is known also from the other Pan-African belts on the southern African subcontinent. An older phase of rift magmatism is recorded in the West Congo Belt, where the Zadinian and Mayumbian Groups were formed by bimodal volcanism that is dated between 999±7 and 912±7 Ma (Tack et al., 2001). In the Lufilian Belt, the Nchanga Granite (877±11 Ma) and probably the poorly dated, ca. 865 Ma Lusaka Granite are overlain unconformably and with an erosional contact by the metasedimentary rocks of the Katanga Supergroup (Porada and Berhorst, 2000). The Kafue Rhyolite (879±19 Ma) and associated Nazingwe metabasalt have been interpreted as volcanic equivalents of this bimodal magmatism that took place in an extensional regime in preparation of continental break-up (Hanson et al., 1994). In the Zambezi Belt, the 804±10 Ma Basal Rushinga Igneous Complex compares well with the Richtersveld Suite. A reheating of the lower crust at that time has been documented for both the Zambezi Belt, where granulite-facies conditions were attained (Vinyu et al., 1997 cited in Porada and Berhorst, 2000), as well as the northern Namaqua Belt (Robb et al., 1999). Of similar age are various alkaline and carbonatitic complexes in the western branch of the African Rift (Rb-Sr whole rock ages between 830±51 and 803±22 Ma), which have been explained by crustal extension prior to break-up (Kampunzu et al., 1998).

### 5.2.3. Rift Sedimentation

The onset of sediment deposition in the various emerging continental rift basins is characterised by a basal conglomerate that is overlain by fluvial to fluvio-deltaic and alluvial plain deposits. With progressive rifting, a shallow marine influence is noticeable in places further up in the stratigraphy of these rift deposits. At the same time, syn-rift volcanic centres developed in many of the various rift basins (see the following section). In southwestern Africa, these rift deposits are stratigraphically unified as the Stinkfontein Subgroup of the Port Nolloth Group in the Gariep Belt and as Nosib Group in the Damara and Kaoko Belts. Their maximum age is constrained by the youngest ages obtained on pre-rift rocks. In the Gariep Belt, this age is 771±6 Ma (Frimmel et al., 2001b). The age of the onset of rifting in the Damara Belt is unknown but similarities between the broad stratigraphy of the Damara and Katanga Supergroups and several small Copperbelt-type chalcocite deposits in a 50-m thick grey shale at the top of the Nosib Group along the northern edge of the Kamanjab Inlier suggest that Damaran rifting may have been initiated at some time after 880 Ma, as is suggested for Katangan rifting. No record of rift sediments is known from the Saldania Belt.

Several former rift basins can be distinguished. In the Gariep Belt, these basins were orientated roughly north-south and separated by equally orientated basement highs. The westernmost of these basins evolved furthest into the Gariep Basin proper, whereas smaller grabens, east of the main basement high (Aurus Horst), including the economically important Rosh Pinah Graben, failed earlier (Frimmel et al., 2002). In the Damara Belt, a Southern Rift and a Northern Rift can be distinguished (Porada, 1979; Miller, 1983a, 2008). While the former developed into a Red Sea-like Khomas Sea, the latter failed. An early rift graben has not been satisfactorily recognised in the Kaoko Belt although Porada (1989) suggested the presence of a north-south orientated Sesfontein Rift and the north-northwest trending Kaoko Rift along the present coast line. The location of the latter is now occupied by the exotic Coastal Terrane and the obducted Orogen Core, in which a bimodal suite of high-grade metamorphic rocks yielded U-Pb zircon ages between 810 and 840 Ma (Konopásek et al., 2008), interpreted as dating rift volcanism.

In the Gariep Belt, the continental rift sediments comprise the Stinkfontein Subgroup, which is subdivided into the Lekkersing and the Vredefontein Formations. The subgroup is best developed in South Africa along the southern and southeastern front of the Gariep Belt. In a few places, best seen in the Richtersveld in South Africa, the entire subgroup is cross-cut by mafic dykes of the Gannakouriep Suite. South of the Orange River, the Lekkersing Formation, which reaches as much as 500 m in thickness, is made up mainly of medium-bedded quartz arenite with sharp, flat bedding planes. Continuous to discontinuous erosional channels are frequent. Thin, flat laminated beds alternate with planar cross-beds and massive beds. Palaeocurrent measurements indicate source areas to the east or southeast. The arenite is sparsely intercalated, in particular at the bottom of the formation, with conglomerate that contains sub-rounded to well-rounded pebbles of pre-Gariep basement provenance. Feldspathic arenite becomes widespread in the upper parts of the formation where it undergoes a conformable and continuous transition into the overlying, 300 m thick Vredefontein Formation. Internal sedimentary structures, accentuated by heavy mineral concentrations along laminations, are well preserved in the dominantly medium-bedded feldspathic arenite. Cross-bedding alternates with parallel laminated bedding. Together with small-scale asymmetrical ripple marks, it indicates palaeocurrent directions from the east or southeast. Soft sediment deformation features are abundant. Varied metavolcanic rocks within the upper Vredefontein Formation represent the first stage of syn-rift volcanism in the Gariep Belt (see the next section).

Sedimentary facies distribution suggests that the rift basin(s) became deeper towards the south. Marine ingression from that direction is indicated by an increasing carbonate content up-section and the formation of a shallow marine carbonate platform in the southern parts. The latter became completely reworked into a diamictite that is rich in dolomite clasts during a subsequent glaciation. This glaciation (Kaigas Formation, to be described in Chapter 5.3) interrupted the progressive flooding of the evolving rift shoulders.

Towards the south, rocks equivalent to the Stinkfontein Subgroup and Kaigas Formation occur along the South African west coast, where they have been mapped as Karoetjies Kop Formation of the Gifberg Group. This group is considered an equivalent to the Port Nolloth Group (Frimmel, 2008a) and is exposed mainly in the Vredendal Outlier (Figure 5.1.3). Towards the north of the Gariep Belt, comparable rocks form some isolated outcrops on the coast west of the main Namib sand dune fields. They are named after their locality Spencer Bay Formation and take an intervening position between the Gariep Belt and southern Damara Belt.

The *Nosib Group* at the base of the Damara Supergroup consists predominantly of coarse- to medium-grained, thinly to thickly bedded feldspathic quartzite, with subordinate metagreywacke, and a basal as well as several intraformational conglomerate beds. These rocks display very similar sedimentological features as described for the Stinkfontein Subgroup. They are exposed as the Nabis Formation in the eastern part of the Northern Platform in the Otavi Mountainland. In the Central Zone, they make up large parts of the Etusis Formation, whereas they are referred to as Kamtsas Formation in the Southern Margin Zone. Palaeocurrent directions have been reported both parallel to and at high angles to the graben edges (Miller, 1983a). One significant difference to the Stinkfontein Subgroup is the relatively abundant presence of former evaporite deposits. They are distinguished as Duruchaus Formation, which is a local facies variation of the Kamtsas Formation in the Southern Margin Zone. There, a succession of meta-arenite and -argillite, as well as conglomerate beds, contains metamorphosed playa and sabkha deposits near its top (Behr et al., 1983). Pseudomorphs after original evaporitic minerals are still preserved and evidence of large-scale Na- and Mg-metasomatism is evidence of circulating early-diagenetic brines. Dolomite-rich brines derived from these evaporitic deposits may have facilitated large-scale thrust emplacement, and played a major role in the final gravity emplacement of the Naukluft Nappe Complex (Behr et al., 1981, 1983).

In the eastern Kaoko Belt, the siliciclastic Nosib Group is overlain by a carbonate-dominated succession in which several sequences from deeper water rhythmitic argillite to shallow water stromatolite, grainstone and microbiolaminite have been distinguished (Hoffman et al., 1998). This succession, the Ombombo Subgroup, has been assigned to the Otavi Group (Hoffmann and Prave, 1996) and is described in more detail in Chapter 5.3. Thus a similar trend from purely continental siliciclastic sedimentation to marine ingression as in the southern Gariep Belt is recorded there as well. In both areas there is evidence of upward shallowing with siliciclastic deposits returning and eventually being overlain paraconformably or unconformably by a diamictite. Overall, rifting in southwestern Africa commenced some 100 and 200 myr later than evident in the West Congo Belt and the Zambezi/Lufilian Belts, respectively. In the West Congo Belt, a failed rift developed between 999 and 912 Ma (Zadinian and Mayumbian Groups), whereas in the Zambezi and Lufilian Belts, maximum age constraints for rift sedimentation in the Roan Basin are given as $879 \pm 19$ Ma and $877 \pm 11$ Ma (Porada and Berhorst, 2000). Three subgroups are distinguished within the Roan Group, though with different names in the Democratic Republic of Congo and Zambia. The oldest subgroup (Roches Argilo Talqueuses or Mindola Subgroup, respectively) is dominated by fluvial siliciclastic deposits, the middle subgroup (Mines or Kitwe Subgroup) reflects the first marine ingression with a basinal carbonaceous argillite facies in the south, an argillite-siltstone slope facies and a littoral carbonate facies with algal bioherms and sabkha deposits in the north. The probably youngest subgroup (Dipeta or Bancroft Subgroup) represents a mainly dolomitic carbonate platform, but the exact correlation of the various stratigraphic units between the Katangan and Zambian portions of the copper belt remains problematic (for review see Porada and Berhorst, 2000).

The overlying Mwashia Group rests with a tectonic contact on the above rocks. It is a mixed argillaceous to calcareous succession that contains conglomerate with fragments of the underlying Roan Group, abundant dolostone, allodapic carbonates and pseudomorphs after gypsum. From the spatial distribution of the various lithofacies, which reflect shallow water to partly exposed environments, it has been concluded that the Mwashia Group heralds the development of a new basin, the Kundelungu Basin, with a north-south orientated axis (Porada and Berhorst, 2000). This change in the crustal stress field was accompanied by widespread bimodal volcanism. As in the other Pan-African belts of the subcontinent, an older diamictite, the Grand Conglomerat (or Tillite Formation in Zambia), is recognised and this rests directly on the Mwashia Group at the base of the overlying Kundelungu Group. It is laterally continuous and reaches as much as 1,200 m in thickness. It contains interstratified feldspathic arenite and banded iron formation as well as mafic lava flows and hyaloclastites.

Continental flood volcanism of the Zadinian and Mayumbian Groups in the West Congo Belt was not accompanied by the development of a significant rift basin. Only after a considerable hiatus, spanning possibly as much as 140 myr, did a proper continental rift develop there. The basin fill comprises a basal conglomerate and overlying quartz arenite, feldspathic arenite and argillite, unified as Sansikwa Subgroup. Its deposition heralds a new stage of rifting, which eventually led to the deposition of the West Congolian Group. Analogues to the other Neoproterozoic successions further south, the siliciclastic rift fill of the Sansikwa Subgroup is overlain by an older glaciogenic diamictite, the Lower Mixtite Formation, which has been tentatively correlated with the ca. 750 Ma Kaigas glaciation (Frimmel et al., 2006).

### 5.2.4. Syn-Rift Magmatism

Almost all major continental rift zones that opened up at various stages in southern Africa during the Neoproterozoic Era experienced some form of syn-depositional magmatic activity. The earliest of these activities is recorded in the West Congo Belt. There up to 2,400 m thick tholeiitic metabasalt overlies siliciclastic rift sediments and a basal metarhyolite, all of which comprise the Zadinian Group (Tack et al., 2001). A maximum age for this group is given by the $999 \pm 7$ Ma Noqui Granit in the pre-Zadinian basement (Tack et al., 2001). It is followed by an up to 4,000 m thick succession of metarhyolite, volcanoclastic and volcano-sedimentary rocks that were intruded by hypabyssal granite (Mayumbian Group). The time of deposition is well constrained by U-Pb single zircon ages for the bottom and the top of the group of respectively $920 \pm 8$ and $912 \pm 7$ Ma (Tack et al., 2001). No further basin development followed until deposition of the West Congolian Group. The Lower Mixtite Formation of that group contains so far undated basaltic pillow lavas, dolerite sills and dykes of tholeiitic composition (Kampunzu et al., 1991). Similarly, basaltic pillow lavas, flows, hyaloclastites and dolerite dykes and sills are intercalated in the older diamictite unit in the Kundelungu Group, the Grand Conglomerat, in the Lufilian Belt. A series of metagabbro or amphibolite bodies are associated with the Bancroft Subgroup on the southern side of the Kafue anticline, but their contacts are tectonic and their age remains uncertain. They could well be younger than the Mwashia Group and related to the mafic volcanism in the Grand Conglomerat.

In southwestern Africa, Neoproterozoic syn-rift magmatism is known from the Port Nolloth Zone of the Gariep Belt and the Central and Northern zones of the Damara Belt. First evidence of syn-rift magmatism in the Port Nolloth Zone is found in the Richtersveld of South Africa in the upper part of the Vredefontein Formation in which minor metavolcanic rocks are intercalated (Middlemost, 1966). The composition of these originally porphyritic to aphyric lava flows and tuffs ranges from dacitic to basanitic, whereby intermediate varieties dominate. High alkali contents are probably related to syn-orogenic metasomatism and not typifying the original melt compositions (Frimmel and Macey, unpublished data, 2004). Bimodal, but predominantly felsic magmatism took place a few tens of kilometres further to the north, in the same basin, at the end of coarse-grained siliciclastic rift sediment deposition leading to the deposition of the *Rosh Pinah Formation*. The volcanic to volcaniclastic rocks interfinger or overlie the Kaigas Formation diamictite (Frimmel, 2008b). The age of this volcanism is given by U-Pb and Pb-Pb ages of $752 \pm 6$ Ma (Borg et al., 2003) and $741 \pm 6$ Ma (Frimmel et al., 1996c), respectively. A volcanic centre with proximal rhyolitic to rhyodacitic lava flows, agglomerates and subvolcanic granite and quartz as well as feldspar porphyry was located on the principal growth fault along the eastern margin of the Rosh Pinah Graben. Distal expressions of this volcanism are tuff, tuffite and ignimbrite deposits. Evidence of both subaerial and under-water extrusion exists in the form of fluvially reworked crystal tuff and lapillistone with inverse grading as well as hyaloclastite, respectively (Alchin et al., 2005). Mafic expressions of this magmatic episode are metabasalt and metagabbro bodies in the vicinity. Geochemically, both the acidic and basic rocks conform to compositions typical of continental within-plate settings (Frimmel et al., 1996b).

Further evidence of syn-rift magmatism in the Port Nolloth Zone and vicinity is the widespread occurrence of mafic dykes, unified as the *Gannakouriep Suite*. Most of them cross-cut the pre-Gariep basement, but a few dykes were emplaced also within the Stinkfontein Subgroup. In the Richtersveld, they are terminated by the Kaigas Formation diamictite, thus providing a relative minimum age constraint (>750 Ma). The only available radiometric age constraint is a Rb-Sr whole rock age of $717 \pm 11$ Ma (Reid et al., 1991), which is probably younger than the true emplacement age due to partial resetting during Pan-African metamorphism. The dyke

southern edge of the southern CZ was still an exposed rift shoulder when deposition of the Arandis Formation ceased.

The low-grade *Zerrissene Group* (Swart, 1992) in the Southern Kaoko Zone consists entirely of deep-water turbidites that are the lateral equivalents of the Usakos and Navachab subgroups, i.e. almost the whole of the post-Nosib Damara Supergroup. Thus, the Spes Bona Member is correlated with the Zebrapütz Formation (Miller, 2008), consisting of thinly bedded, distal, fine-grained, lobe fringe to basin-plain shales within which packages of distal fan greywackes are interbedded (Swart, 1992). Even the thinnest beds are laterally persistent. $T_c$ current ripples are common in many beds. Other common sedimentary structures include $T_{b-e}$ sequences, rip-up clasts, flute casts, load structures, concretions and dewatering structures. The Zebrapütz Formation is 350 m thick but its base is not exposed.

The lateral equivalent of the Okawayo Member, the Brandberg West Formation, reaches a maximum of only 20 m in thickness but is extremely persistent and uniform along strike. The lower half to two-thirds consists of graded, fine- to medium-grained, white, yellowish and brown limestone turbidites interbedded with dark, thin-bedded, partly graded shales. The remainder of the formation is made up of thin, light blue limestone beds which may have been interpreted as hemipelagic carbonate oozes (Swart, 1992). Individual layers are highly continuous.

The 300 m thick Brak River Formation correlates with the Oberwasser Member. This consists of fan fringe turbidites with greywacke:shale ratios ranging from 12 to $<1$. Interbedded in the formation are thick, poorly graded to ungraded and laterally continuous layers of medium- to coarse-grained, psammitic greywacke that have almost no pelitic tops and can be up to 11 m thick. In the greywacke-shale couplets, $T_{b-e}$ sequences, rip-up clasts, flute casts, load structures, concretions and dewatering structures are common (Swart, 1992).

These Zerrissene Group turbidites and the marked westward thinning of the lower Swakop Group succession in the CKZ (Miller, 1983a) are indicative of the development of a west-facing passive continental margin along the western edge of the CKZ and deep water west thereof (the Adamastor Ocean) after the Chuos glaciation.

Early stages of concomitant continental break up are also evident in the southern graben as it began to evolve into the deep Khomas Sea. Nevertheless, the graben still formed a restricted basin and thick *Blaukrans Formation* carbonaceous shales (now graphitic schists) accumulated above the basal Waldburg carbonates. As the graben, together with basement fragments along its northern edge, began to drift southwards, its southern margin, the SMZ, began to develop into a passive continental margin. Two huge fans of turbiditic sandstone that interfingered with the carbonaceous shales were deposited in the deeper water of this passive margin (*Hakosberg* and *Auas Formations*). The actual locus of continental separation was the Okahandja Lineament, which was the northern bounding fault to the southern graben.

The southern rift shoulder to the southern graben, the northeast-trending Dordabis-Morgen Ridge or Rift Shoulder was located on the SF (Miller, 2008). Within a shallow basin to the southeast of it, the *Court Formation* of the Witvlei Group was deposited on a southeast-dipping palaeoslope (Hegenberger, 1993). Isopachs of and facies within the Court Formation indicate that the rift shoulder was exposed throughout this depositional phase. After the Gobabis transgression, the basin shallowed progressively. The fine siliciclastic rocks of the Constance Member (shale, mudstone, siltstone) grade laterally and upwards into sandstone and siltstone beds with interbedded carbonates and oolitic carbonates. The member thickens to 200 m southeastwards away from the rift shoulder. An exposure surface and unconformity mark the top of the Constance Member rocks. The overlying Simmenau Member sandstones reach 160 m in thickness. They overstep the Constance Member rocks onto the Dordabis-Morgen Rift Shoulder where they become more proximal and conglomeratic. They pinch out against the rift shoulder.

Biostratigraphic data on the passive continental margin deposits in the Damara Belt are very sparse. Gaucher and Germs (2007) were the first to report organic-walled microfossils (acritarchs) from the Otavi and Swakop Groups. They discovered a *Bavlinella faveolata* assemblage in the Auros Formation of the Abenab Subgroup (Figure 5.4.1A, B) and a *Octoedryxium truncatum* assemblage in the Arandis Formation.

## 5.3.2.2. The Ghaub glaciation and final submergence of the rift shoulders

### 5.3.2.2.1. Evidence of a pre-Ghaub fall in sea level

Only the successions deposited in the marginal, shallow-water regions of the Damara Ocean provide such evidence, the upper Abenab Subgroup on the NP and NMZ and the upper Court Formation on the SF. There is an upward shoaling of the Gruis Formation accompanied by a south to north shift of the main grainstones upwards through the formation. The upper two cycles of the formation are remarkably well preserved below the erosive base of the Ghaub Formation. Nevertheless, lithology and carbon isotopes presage the impending glaciation and the accompanying sea-level fall. The $\delta^{13}C$ ratios range from +1 to +8‰ through most of the Abenab Subgroup but through the last three Ombaatjie cycles (7–9) fall to −6‰ (Figure 5.3.3). The marked negative $\delta^{13}C$ anomaly preceding the Ghaub glaciation is referred to as the Trezona anomaly (Halverson et al.,

2002). Above cycle nine on the highest point of the rift shoulder is a dolomite aeolianite derived entirely from cycle nine carbonate (Hoffman and Halverson, 2008). Similarly, the low-stand clastic wedge of the Franniaus Member at the top of the Ombaatjie Formation records a pre-Ghaub sea-level fall of 400 m in the foreslope rocks of the Fransfontein Ridge in the NMZ (Hoffman and Halverson, 2008).

In the Court Formation in the SF, the unconformity at the base of Simmenau Member cuts deep into the underlying units and has removed them completely in places (Hegenberger, 1993). It is assumed that the sandstones and conglomerates of the Simmenau Member correlate with the uppermost cycles of the Ombaatjie Formation on the NP but the finer grained top of the member remains enigmatic unless it is the first bit of evidence relating to subsidence and eventual drowning of the Dordabis-Morgen Rift Shoulder.

#### 5.3.2.2.2. The Ghaub Formation

Only small, thin patches of the glaciogenic Ghaub Formation are present in the western regions of the NP. The formation reaches 2,000 m in thickness in the western parts of the Otavi Mountainland (T1 of Söhnge, 1957) but tends to be patchily developed. Diamictite is the main rock type with minor interbeds of conglomerate, shale, sandstone and dolomite. The unsorted clast suite is dominated by carbonates from underlying Otavi Group formations and, in the Otavi Mountainland, quartzites from the Nosib Group. Basement rock types are minor. Facetted and striated clasts can be found. Spiny acritarchs assigned to the genus *Asteridium* (Figure 5.4.1D) occur in the matrix of Ghaub diamictite, differing from the assemblage occurring in underlying carbonates (Gaucher and Germs, 2007).

In the Fransfontein Ridge along the southwestern margin of the Kamanjab Inlier, the formation consists of a lower lonestone-bearing siltstone, a massive diamictite in the middle and an upper unit of allodapic dolomite choked with dropstones (Hoffman and Halverson, 2008). The transition to the deeper waters south of the NP becomes even more evident within the formation in the form of marine ice grounding-line diamictites, diamictite-derived gravity flows and ice-rafted debris. Thickness variations from 40 to 600 m are related to scour deepening. Specific depositional features suggest repeated advance and retreat of ice sheets and localised fast-flowing ice streams (Hoffman and Halverson, 2008). Evidence of glacial rain-out is greater in this region than on the platform.

In the NZ, the formation is well developed and follows conformably on the Okonguarri Formation. On the eastern edge of the Summas Mountains and south thereof it consists of several tens of metres of diamictite followed by a lesser thickness of thin, graded siltstones and shales that carry dropstones, which Hoffmann et al. (2004) interpreted as having been deposited in a proximal basinal to lower slope setting. In the Vrede Dome just south of the southwestern margin of the Fransfontein Inlier, the basal Ghaub scour cut out much of the Usakos Subgroup stratigraphy, and the Ghaub Formation rests directly on the Chuos glacials in places. In the dome, the Ghaub Formation is made up almost entirely of carbonate-clast diamictite (Maloof, 2000; Hoffman and Halverson, 2008).

In the northern CZ, the upper 200 m of Badenhorst's (1992) Oberwasser Member is actually schistose and calc-silicate-bearing Ghaub Formation, which is difficult to recognise because rain-out clasts are not abundant and are almost exclusively of carbonate that have thick calc-silicate rims resulting from post-depositional metamorphism. Approximately the same thickness of turbiditic greywackes of the upper part of Swart's (1992) Brak River Formation (Zerrissene Group) is now placed in the Ghaub Formation (Miller, 2008). Clasts are rare and tend to be concentrated in streaks at the base of layers. Isolated granite dropstones reach a diameter of 1.75 m. Swart (1992) suggested that the shale at the top of the formation may have been deposited during glacial retreat and during the post-glacial rise in sea level when coarse debris became trapped on the shelf. This shale section may, therefore, be equivalent to the Keilberg Member cap carbonate on the NP.

Hoffmann et al. (2004) were the first to recognise and date a likely correlate of the Ghaub Formation in the Karibib area of the southern CZ. They obtained a single-zircon age of 635.5 ± 1.2 Ma from an ash near the top of the formation. The main component of the formation in this area is the *Kachab Member* dropstone interval consisting of fine-grained, very thinly bedded, siliciclastic rocks with scattered carbonate and rare granitic dropstones (phyllitic schists of Badenhorst, 1992). Miller (2008) reclassified the diamictites along the southern margin of the southern CZ and placed them in the Ghaub Formation rather than the Chuos Formation as was formerly the case (Martin, 1965b; Jacob, 1974; Miller, 1983a, and references therein). Thus, the Ghaub Formation oversteps the underlying formations in a southerly direction and is the first unit in the Damara succession to be deposited on the subsiding and finally submerged Rooikuiseb Rift Shoulder.

Of considerable significance are three subaqueous and locally pillowed, mafic volcanic units of limited lateral extent interbedded in the Ghaub Formation, the Daheim, Omusema and Lievental Members. The *Daheim Member* metabasalts occur near the northern edge of the southern CZ, are the furthest of these units from the southern graben and have the composition of alkaline continental basalts (Miller, 1983b). On the farm Navachab

58 near Karibib, occasional dropstones of granite and carbonate occur in the Daheim metabasites (Miller, 2008). The *Omusema Member* metabasalts were erupted almost exactly on the bounding, graben-margin fault, the Okahandja Lineament (Miller, 1979), separating the southern CZ from the southern graben. The Okahandja Lineament was the locus of continental break up and the evolving southern passive margin of the Congo Craton as the initial rift-phase fill of the southern graben drifted southwards and the northern edge of the graben became the SZ ocean or Khomas Sea (Miller, 2008). The Omusema metabasalts also have the composition of alkaline continental basalts (de Kock and Botha, 1988). The *Lievental Member* metabasalts are interbedded with basal Kuiseb Formation schists just south of the Okahandja Lineament, i.e. near the northern margin of what was the deepening Khomas Sea. These metabasalts are low-K tholeiites with MORB compositions and have the same stratigraphic relationship to the overlying Tinkas turbidites as the Ghaub diamictites have to these turbidites (Miller, 2008). Thus, the metabasalts in the Ghaub Formation specifically reflect the crustal setting within which they were erupted, within-plate continental in the southern CZ and oceanic in the SZ.

The age of the *Naos Formation* in the SMZ is uncertain and although initially considered to be part of the Chuos Formation (Martin, 1965b; Miller, 1983a), its preferred correlation at present is with the Ghaub Formation (Hoffmann, 1983; Hoffmann, 1989). It is a thick and extensive succession of light grey, well foliated, variably pebbly to highly conglomeratic, semipelitic mica schist with a wide variety of extrabasinal clasts. These rocks range from clast-supported conglomerates to schist with only rare, widely separated clasts. Minor but locally important rock types are layers of quartzite and metavolcanic amphibolite and amphibole-chlorite schists (Hoffmann, 1983), the mafic rocks having the composition of continental, within-plate tholeiites (Miller, 1983b). Local layers of laminated, siliceous, exhalative iron formation are associated with the amphibolites (Breitkopf, 1988).

In the Witvlei Group on the SF, no Ghaub equivalent diamictite occurs. However, a carbonate succession that bears all the hallmarks of a cap carbonate is well developed in the Buschmannsklippe Formation (Hoffmann, 1989; Hegenberger, 1993).

### 5.3.2.2.3. Post-Ghaub cap carbonates

The cap carbonate to the syn-Marinoan Ghaub glaciogenic rocks was identified by Hoffmann and Prave (1996) for the first time and named the *Keilberg Member* of the Maieberg Formation on the NP. Detailed descriptions of this cap are provided by Hoffman and Halverson (2008) for the western parts of the NP and Hegenberger (1993) for the SF. Hoffmann (1989) suggested that the Bildah Member of the Buschmannsklippe Formation in the latter region is this cap carbonate.

The basal tan to pinkish or pale grey Keilberg Member lies with a sharp contact on the diamictites (Hoffmann and Prave, 1996; Hoffman and Schrag, 2002; Hoffman and Halverson, 2008). The Maieberg Formation forms a single, thick and extensive depositional sequence, initially transgressive and then gradually upward shallowing (T2 and T3 of Söhnge, 1957). This accumulated during the global flooding event that followed the Ghaub glaciation. The formation reaches 1,800 m in thickness in the Otavi Mountainland and 400 m east of Sesfontein (Söhnge, 1957; Hoffman and Halverson, 2008).

The Keilberg cap and the Maieberg section immediately above the cap contain features that are common and virtually identical to the equivalent Marinoan-age cap carbonates on five continents (Hoffman and Halverson, 2008). Generally 10–15 m thick, but swelling to 75 m in the western parts of the NP, up to 40 m thick in the Otavi Mountainland and between 2 and 8 m thick in the Fransfontein Ridge, the Keilberg Member is a good, weather-resistant marker bed. In the most instructive exposures, the basal metre is composed of recrystallised and cemented dolomite siltite or grainstone laminated by low-angle, metre-scale cross-beds. Above this, narrow, vertical, convex-up, stromatolitic columns about 2–5 cm across and with a similar spacing are characteristically developed in many places and produce a honeycomb-like texture. The microbial lamination of the columns is usually poorly visible. A laminated, concave-up dolomicrite infill between the columns is generally far more apparent and appears as evenly spaced 'tubes' in outcrop (Figure 5.3.4). Commonly, the infill has a marginal rind of diagenetically introduced quartz. Replacement sparry calcite forms a core to the tubes in many places. Rarely, as on the farm Abenab 707 in the Otavi Mountainland, galena, sphalerite and sparry calcite form the core filling.

The stromatolites, termed plumb stromatolites by Hoffman and Halverson (2008), are overlain by a zone of giant wave ripples in peloidal dolomite with individual ripple sets reaching 150 cm in thickness. Crestal spacing is 1–1.5 m. According to Allen and Hoffman (2005) and Hoffman and Halverson (2008), such ripples were produced at water depths of up to 400 m by waves with periods of 21–30 s. This sequence illustrates the initial transgressive nature of the Keilberg Member with the stromatolites having formed in the shallower photic zone and the giant wave ripples in deeper water. In places just above the Keilberg Member, vertically standing crystal fans of calcite pseudomorph original seafloor aragonite (Figure 5.3.4e). These fans are often only 5 mm in height but can develop into stacked bouquets of much larger fans up to 2 m high.

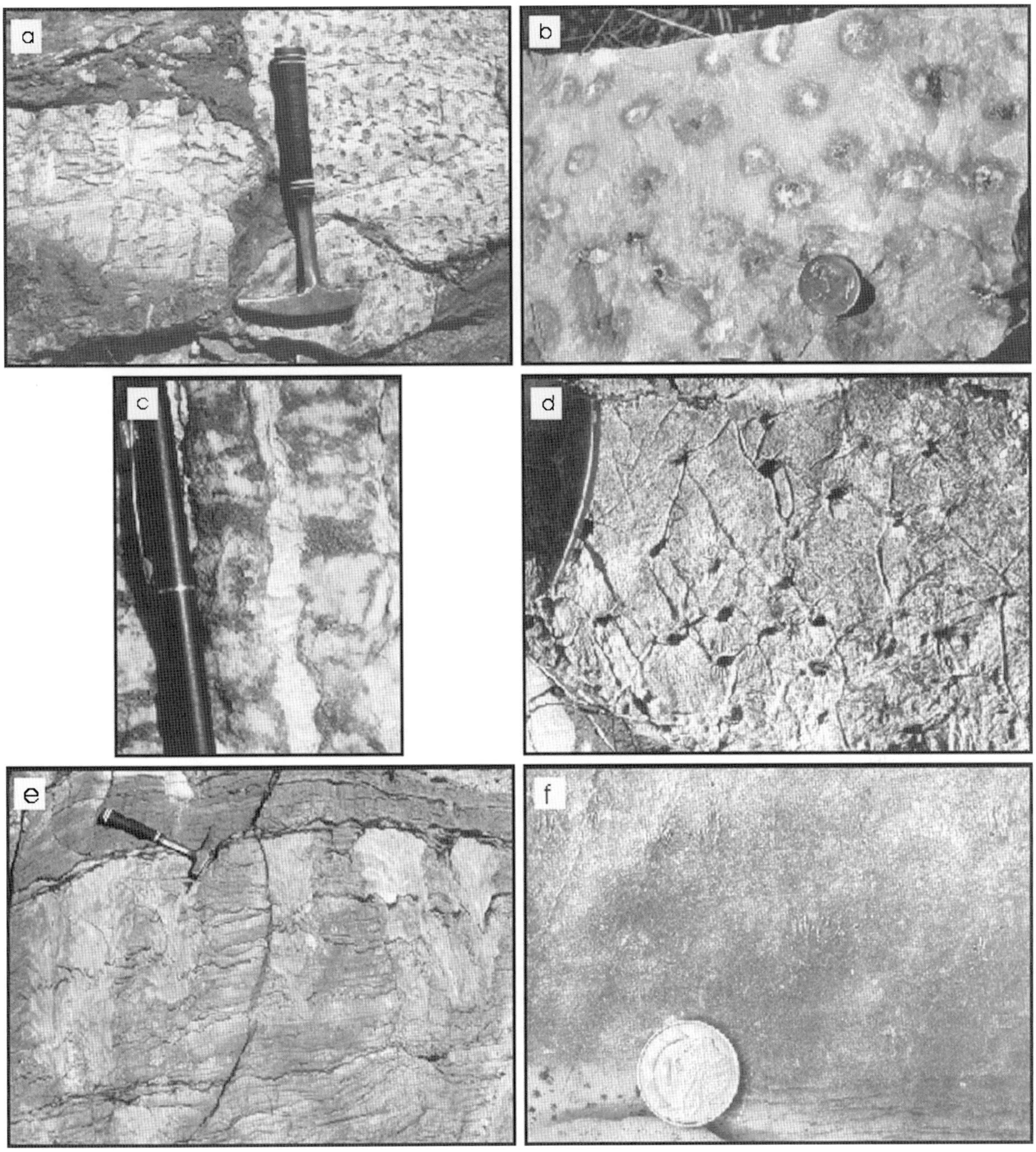


**Figure 5.3.4** Features of, and associated with, the Keilberg and Bildah cap carbonates to the glaciogenic, 635-Ma Ghaub Formation: (a) basal Mulden conglomerate west of Khorixas containing large angular blocks eroded out of the Keilberg Member. The blocks display the plumb stromatolites and the prominent intervening tube structures, vertical section in left block, plan view in right block with negatively weathering tube structures; (b) fresh exposure surface of the Keilberg Member showing the dolomicrite in the centres of the tube structures replaced by sparry calcite and black metallic sulphides, farm Abenab 707, Otavi Mountainland; diameter of coin 22 mm; (c) close-up photograph showing the slight convex-up curvature of the plumb stromatolite laminae and the concave-up laminae of the intervening, lighter coloured tube structure, Keilberg Member, western Northern Platform; (d) bedding-plane surface showing positively weathering tube structure in the Bildah Member of the Witvlei Group, farm Eilenriede 53, southwest of Gobabis; (e) unusually large, stacked calcite fans after seafloor aragonite stratigraphically above the plumb stromatolites; Maieberg Formation, farm Tweelingskop 676, northwest of Khorixas; (f) the small, more usual occurrence of the pseudomorphic calcite fans above the plumb stromatolites; Maieberg Formation, farm Auros 595, Otavi Mountainland, diameter of coin 22 mm.

In the Fransfontein Ridge, the characteristic features of the Keilberg cap carbonate are present – allodapic dolostone, laminated dolostone, plumb stromatolites with the intervening 'tube structures', giant wave ripples with a crestal spacing of 4.5 m, and pseudomorphs after seafloor aragonite fans. In the NZ and CZ, a thin, tan-coloured dolostone that is widespread and often only 2 m thick and with locally preserved cross-bedding at the base of the Karibib Formation is the equivalent of the Keilberg Member (Swart, 1992; Hoffmann et al., 2004; Miller, 2008).

The pale grey to whitish or pinkish dolomite of the *Bildah Member* at the base of the Buschmannsklippe Formation on the SF is a cap carbonate that might correspond to the Keilberg Member. This reaches 60 m in thickness. Bedding is thick to massive but the unit has a fine, internal cryptalgal lamination. The tube structures and generally less obvious plumb stromatolites occur extensively in the member. The Bildah Member has all the characteristics of the Keilberg dolostone (Hoffmann, 1989; Hoffmann et al., 2004). North of Gobabis, a basin-margin facies consists of layers of sandy to pebbly dolomite, oolitic carbonate and dolomitic to pebbly and cross-bedded sandstone that interfinger with the laminated Bildah dolomite. Desiccation polygons and small-scale karst features in the sandy and pebbly dolomite indicate local, temporary exposure. The upper 5–10 m of the member are thinly bedded, become more calcareous, are slightly argillaceous and form a transition zone into the overlying La Fraque Member where the calcite fans also occur (Hegenberger, 1993). While the Bildah Member could be a post-Ghaub cap carbonate, a younger, Ediacaran age for it is possible because of high $^{87}Sr/^{86}Sr$ ratios (Saylor et al., 1998; see Section 5.4.2.2).

### 5.3.2.3. Advanced spreading

#### 5.3.2.3.1. Tsumeb Subgroup, Karibib and Tinkas formations, Vaalgras Subgroup, Court Formation, rocks of the Deep-Level Southern Zone

The top of the Keilberg Member grades over approximately 2 m through flaggy dolomite into thinly bedded to laminated, argillaceous microbialaminite, limestone and dolomite rhythmites. They are interbedded with allodapic dolomitic turbidites, beds with large-scale, low-angle, swalley cross-stratification and clast-bearing, intraformational mass-flow beds and constitute the remainder of the *Maieberg Formation*. The top of the Keilberg Member is a flooding surface with the maximum flooding in the lower part of the middle of the Maieberg Formation. Cross-bedded grainstones are developed in the upper part of the formation. The top of the formation is a tepeed exposure surface that is fractured and intensely silicified (Hoffman et al., 1998b; Hoffman and Halverson, 2008). In the Otavi Mountainland, the lower and thicker part of the formation is calcareous and the upper part dolomitic and dark to medium grey in colour. First studies of the acritarch assemblages in the Otavi and Swakop Groups (Gaucher and Germs, 2007) revealed that the middle Maieberg Formation carbonates, in contrast to the pre-Ghaub carbonates, contain an *Octoedryxium truncatum* assemblage (Figure 5.4.1C).

Following on the Maieberg Formation, the *Elandshoek Formation* (T4 and T5 of (Söhnge, 1957) has a relatively uniform thickness of 500–550 m in the western NP where it is made up of cherty, grainstone-dominated dolomite depositional cycles several metres thick. Cycles of pinkish dolomite in the lower 60 m are topped by light to dark grey microbialaminite and tepeed exposure surfaces. Flooding surfaces terminate cycles above this. Columnar stromatolites are common in the upper half of the formation, which is better bedded. In this half, some cycles reach 70 m in thickness. Bedding-parallel silicification is a characteristic feature. In the Otavi Mountainland, the formation reaches 1,500 m in thickness, forms much of the relief in the region, is extensively karsted and is made up of massive to bedded dolomite with a rather uniform light grey colour.

The overlying *Hüttenberg Formation* starts with almost 300 m of partly marly and rather recessive ribbonite that is followed by about 200 m of dolomite grainstone in which oolitic layers are fairly common. In the Otavi Mountainland, Söhnge (1957) divided the Hüttenberg Formation into a lower T6 dolomite with chert layers, a middle T7 dolomite and an upper T8 dolomite with chert. The T6 is 310 m thick at Tsumeb but thickens southwards to over 1,000 m towards the southern edge of the NMZ near the Kombat Mine. It consists of light to medium grey, bedded dolomite. Numerous chert layers serve to distinguish it from the T5 dolomite. Stromatolitic marker layers and lateral colour changes are characteristics. The T7 dolomite is 290 m thick and consists of light and dark grey grainstone and mudstone layers with black carbonaceous material and black chert bands. A black, 200 m thick limestone interfingers west of Tsumeb. The T8 dolomite reaches 300 m in thickness and consists of massive to bedded and cyclically graded grainstone and mudstone layers with interbedded silicified oolite beds and columnar stromatolites. The typical T7 and T8 lithotypes become difficult to recognise and seem to disappear southwards across the NMZ.

The carbonate succession overlying the Ghaub Formation diamictite in the Fransfontein Ridge and the NZ as well as the CZ is the *Karibib Formation*. In the Fransfontein Ridge, the Keilberg Member is overlain by 25–314 m of marly limestone rhythmite equivalent to the middle Maieberg Formation. This is succeeded by about 400 m of platy allodapic dolomite and debris flows similar to lower slope and basin-margin fan carbonates. The Karibib Formation lacks the sequence boundaries that characterise the Elandshoek and Hüttenberg formations but the carbon isotope ratios demonstrate that the Karibib Formation is the equivalent of the complete carbonate succession of the Tsumeb Subgroup, i.e. the Maieberg, Elandshoek and Hüttenberg Formations together (Hoffman and Halverson, 2008).

The Karibib Formation in the NZ consists of grey, bedded limestone overlain by dolomite. It is almost 1,000 m thick in the Okonguarri Antiform (Swart, 1986) but thins westwards and pinches out completely on the western flank of the Mitten Fold (Miller, 1980). In the area of the Vrede Dome, the formation shoals upwards and northwards from deep-water rhythmite through rhythmite slump breccia to cross-bedded grainstones. In this region, the formation thins markedly southwards off the shelf into the basin.

The Karibib Formation is thick and extensively developed in the nCZ. The thin, brown Keilberg-equivalent dolostone is directly overlain by dark grey, parallel-laminated metapelites (Hoffmann et al., 2004) above which the main carbonate units of the formation occur. These are the Harmonie Member at the base followed by the Otjongeama and Arises River Members. The Harmonie Member is normally about 200 m thick and consists of light brown calcitic marble with interbedded calc-silicate layers that make up approximately 60% of the succession. Thickness is variable and this member can be represented by a single calc-silicate layer or it may be absent altogether. The Otjongeama Member is similar but contains some dolomite and only 25% interbedded calc-silicate layers. Its average thickness is 500 m but it reaches 1,600 m in places. Both members contain intraformational breccia beds of limestone rhythmites. Hoffmann et al. (2004) described the Harmonie and Otjongeama Members as originally having been limestone-marl-pelite rhythmites. The Arises River Member is a light grey to white, very coarse-grained, medium- to thick-bedded or massive calcitic marble with sparsely disseminated graphite flakes and calc-silicate minerals but no calc-silicate layers. It averages 120 m in thickness but is absent in places and up to 650 m thick in others.

The above stratigraphy can be followed southwards into the southern CZ but lower units are overstepped southwards by upper units and the typical northern CZ stratigraphy pinches out rapidly southwards. The southern CZ appears to have been a relatively shallow shelf, at least towards the end of Karibib Formation times. Oolitic to pisolitic lenses occur near the top of the formation in places (Badenhorst, 1992). The succession of light and dark grey marble, white dolomite, ribbon marble, brown marble and various mass-flow carbonate breccias is entirely dolomitic. Thickness ranges from a few metres to 700 m. Hoffmann et al. (2004) pointed out that, like the Maieberg Formation on the NP, the lithology of the Karibib Formation in the northern and southern CZ records the gradual upward shallowing of the initial post-glacial highstand system tract.

Towards the southern edge of the southern CZ and the developing southern passive margin of the Congo Craton, the *in situ* carbonates of the Karibib shelf grade into the Karibib-derived debris and turbidite apron of the *Tinkas Formation* that straddles the Okahandja Lineament. The Tinkas Formation is divided into the Kuduleck and Quelle Members. The Kuduleck Member is closest to the Karibib shelf, is up to 500 m thick and consists of purer carbonate turbidite and mass flow layers with some interbedded and graded schists. Further from the shelf and into the deeper water on the northern edge of the deepening Khomas Sea of the SZ, distal siliciclastic and marly turbidites have been metamorphosed to the thin-bedded succession of calc-silicate and schist layers of the Quelle Member. Although almost 500 m thick in places, the Quelle Member was deposited on the continental slope and rise forming the northern margin of the Khomas Sea and it thins and pinches out southwards into this deep-water environment now represented by the SZ.

The *Vaalgras Subgroup* in the SMZ is the corresponding passive margin succession on the southern side of the Khomas Sea. Although the Samara Formation (formerly Corona Formation) forms the structural base of the Hakos Group southwest of the Gamsberg Mountain in the western part of the SMZ, it is considered to post-date the Naos diamictites and to have many similarities to the Buschmannsklippe Formation with which it is correlated (Hoffmann, 1989). It consists of fine-grained, massive to thickly bedded, white, yellowish, pale pink, pale purplish to maroon or pale grey dolomite and limestone. These carbonates alternate with greenish, chlorite-mica phyllite or garnet-bearing schist. Also present are white, thinly bedded, ripple cross-bedded and mud-cracked quartzites. Distinctive cross-beds in the pale-coloured dolomites are defined by disseminated grains of detrital quartz. The interbedded greenish phyllites and schists suggest that the Samara Formation is a lateral equivalent of the chlorite-garnet schists of the Melrose Formation. The upper three formations of the Vaalgras Subgroup consist of varying proportions of schist and very thinly bedded quartzite together with interbedded amphibolites and minor lenses of marble and chlorite-tremolite schist (picritic metabasalts). Individual quartzite layers have a very uniform thickness along strike and were probably deposited in outer fan regions on the lower slope or continental rise. At different levels within the uppermost of these three formations, the Gomab River Formation, are between one and five layers of a reddish, ash-fall feldspar porphyry, the Hartelust Rhyolite Member, that have the composition of alkali rhyolite (Miller, 1983a). The lower intercept of a zircon discordia gave an age of 609 +5/−13 Ma for this rhyolite (Nagel, 1999).

Evidence for the final subsidence and drowning of most of the Dordabis-Morgen Rift Shoulder at the southern margin of the Khomas Sea (Miller, 2008) is provided by the *Buschmannsklippe Formation* which was deposited across the rift shoulder on a northwesterly dipping palaeoslope (Hegenberger, 1993). The final subsidence and drowning of each of the Damaran rift shoulders is illustrated in Figure 5.3.5. The formation was probably continuous with the (now tectonically disrupted) formations in the SMZ. Thus, the upper Witvlei Group formed part of the upper passive margin succession on the northern edge of the Kalahari Craton. The formation reaches 500 m in thickness. The Bildah cap carbonate at the base is overlain by the La Fraque and Okambara Members.

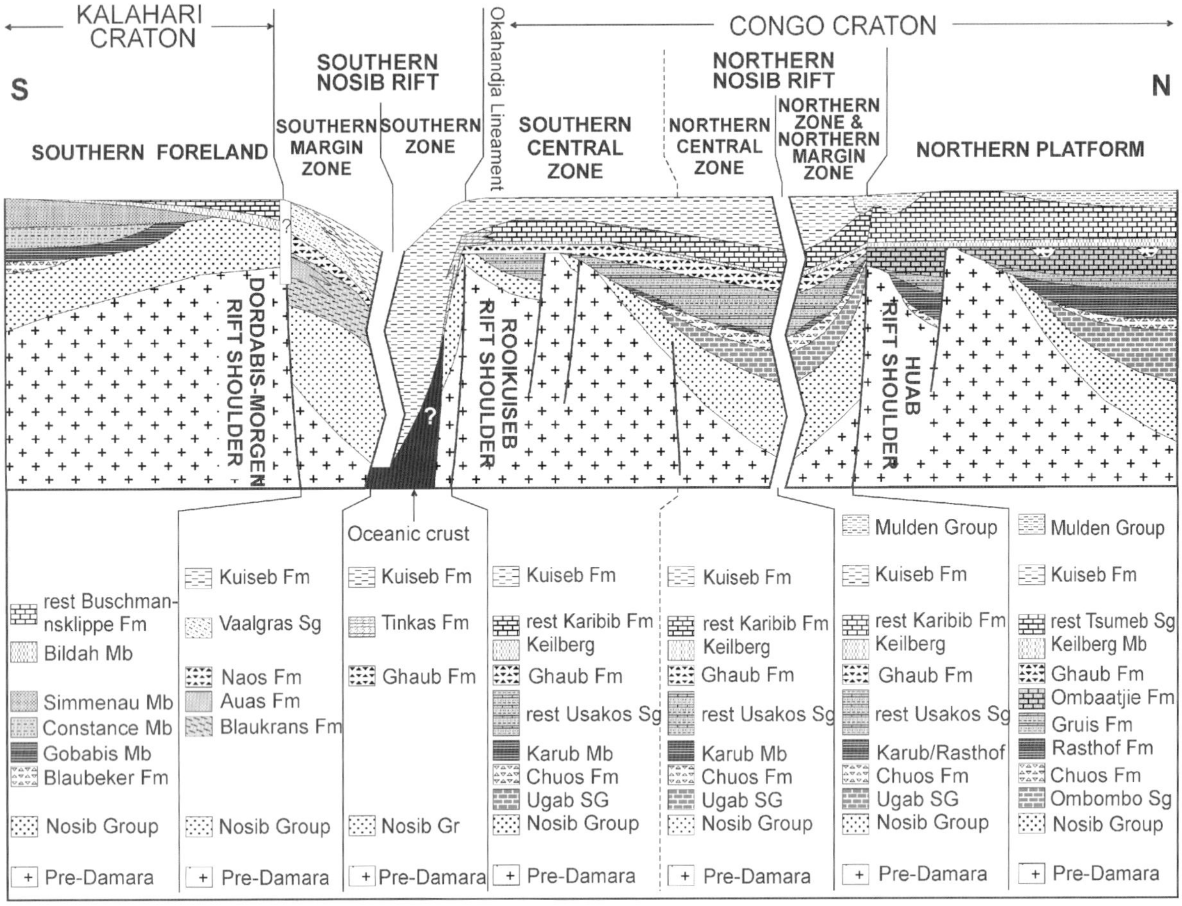


**Figure 5.3.5** Schematic north-south section across the Damara Belt from the Northern Platform to the Southern Foreland showing: (i) the initial post-rift exposure of three rift shoulders; (ii) deposition in sub-basins with palaeoslopes dipping away from the rift shoulders; (iii) onlap of successive stratigraphic units onto the rift shoulders; and (iv) the almost simultaneous subsidence and drowning of all three rift shoulders during or immediately after the Ghaub glaciation. The northern Nosib-age rift, encompassing the northern Central Zone and the Northern Zone, deepened continuously during deposition but never ruptured. The southern rift ruptured along its northern bounding fault, the Okahandja Lineament, which led to the development of oceanic crust below the Southern Zone (the Nosib-age rift fill now occurs in the Southern Margin Zone). The amount of post-Ghaub spreading in the Southern Zone may have only been about 875 km at a spreading rate of 25 cm/year in the 35 Ma between the Ghaub glaciation (635 Ma) and the postulated termination of spreading at about 600 Ma; Fm — formation, Gr — group, Sg — subgroup.

The La Fraque Member reaches 170 m in thickness and consists of calcareous, thinly bedded to laminated marly shale, siltstone and sandstone with several interbedded, massive, light grey to reddish grey carbonate layers. The reddish brown colours near the base change to greenish and greenish brown upwards. Locally, thin lenses of flat-pebble conglomerate are associated with the carbonate layers. Towards the northeast, as the proportion of coarser siliciclastic material progressively increases in a basin margin setting, the member becomes difficult to distinguish from the overlying Okambara Member.

The Okambara Member is made up of a vertical succession of three carbonate-bearing facies, respectively up to 45, 50 and 60 m thick, and a marginal siliciclastic-dominated facies. Reddish to grey limestone is the carbonate in the lower carbonate facies and dolomite the carbonate in the other two. Evidence for a general upward shoaling is provided by an upward increase in the abundance of edge-wise and flat-pebble conglomerates, rill marks, ripple marks, hummocky cross-stratification, low-angle planar and trough cross-bedding, sandy carbonate, layers of microbialaminite, well developed stromatolites, and oolites and pisolites. Several of the uppermost dolomite layers contain scattered to abundant chert nodules, up to 1 cm in diameter, that are reminiscent of cauliflower quartz. These suggest final deposition under intertidal conditions with a degree of desiccation. Hegenberger (1993) suggested that the features of the Okambara Member point to it having been deposited in a shallow subtidal to intertidal environment subjected to frequent exposure and periodic storms.

The SZ has such a thick fill of intensely deformed schists that little can be gleaned about the base of the Damara Supergroup in this region. However, the Deep-Level SZ does provide some indication of the basal Damaran lithology. The poorly exposed Deep-Level SZ has an aeromagnetic signature similar to basement to the north of its bounding lineament, the Okahandja Lineament (Corner, 2008). The exposed lithology consists of a series of stacked thrust sheets of marble, quartzite, conglomerate, various kyanite-bearing schists and basement gneisses (Kasch, 1986; Miller, 2008). The metasedimentary rocks are broadly equated with the basal Swakop Group rocks but no specific correlation has been made.

#### 5.3.2.3.2. Kuiseb Formation

The final spreading-phase unit in the Damara Supergroup is the schistose Kuiseb Formation which is by far the thickest unit of the Swakop Group. This was deposited in the deepest central parts of the Damara Belt from the NZ to the SZ as well as in the western part of the CKZ. In the adjoining WKZ, the equivalent of the Kuiseb Formation is the informal Amspoort Formation (Miller, 2008). The Kuiseb Formation reaches a maximum thickness of 10 km in the NZ (Miller, 1980). It consists of several thick interfingering sequences of graded, semipelitic to pelitic metagreywackes (Kukla 1992; Kukla and Stanistreet, 1993). Locally, thin quartzites or dolomitic marbles are interbedded near the base of the formation. With the westward increase in the grade of metamorphism, the schist becomes more gneissic and the abundance of cordierite and garnet give parts of the formation a kinzigite-like appearance in places. Slight migmatisation sets in the region of the confluence of the Khan and Swakop Rivers and increases to such an extent towards the southwest that in the area between Conception Bay and Meob Bay, almost the whole formation is a gigantic, deformed and disrupted migmatite (Miller, 1983a, 2008).

The SZ has a thick succession of schists consisting of spreading-phase Kuiseb Formation metagreywackes overlain by syn-tectonic, arc-trench metagreywackes of the Hureb Formation. The base of the Kuiseb Formation is only exposed at the northern and southern margins of the SZ. Along what was the southern passive margin to the Khomas Sea, the intensely tectonised mica schists of the Kuiseb Formation overlie the Gomab River Formation of the Hakos Group tectonically. However, there is so much mica schist in the Vaalgras Subgroup that it is likely that much of this schist is a southern and basal, passive-margin facies of the Kuiseb schists that interfingers with the more readily recognisable passive margin lithologies. Above the Gomab River Formation in the Windhoek area, the fan-like quartzites of the Kleine Kuppe Formation interfinger with the basal Kuiseb schists.

In the opposing passive margin succession along the Okahandja Lineament, the calc-silicate and marble layers of the Tinkas passive-margin fan interfinger intimately with thin schist layers. The fan itself is underlain by schist. The Nosib arkoses of the Lievenberg and Westfalenhof domes form the deepest stratigraphic level exposed just south of the Okahandja Lineament. An intermittent layer of metabasalt up to 7 m thick of the Lievenberg Member lies directly on the Nosib arkoses. The metabasalt is overlain by only 2 m of the Tinkas Formation. Schist follows this and contains lenses of Lievental metabasalt up to 250 m above the base of the schist. The metabasalts have the composition of MORB. Using the top of the Nosib arkoses and the post-Ghaub Tinkas Formation as marker units, it is clear that equivalents of the Ugab and Usakos subgroups were not deposited above the Nosib rocks of the Lievenberg and Westfalenhof domes in the region just south of the present Okahandja Lineament. These domes must have been part of fault blocks on the southern edge of the Rooikuiseb Rift Shoulder, i.e. part of the Okahandja Lineament fault system. The Lievenberg/Westfalenhof fault block only subsided below sea level with the rest of the rift shoulder during or immediately subsequent to the Marinoan-age Ghaub glaciation. The schists interbedded with the Lievental metabasalts are, therefore, considered to be part of the Kuiseb Formation (Miller, 2008).

The interfingering relationships described above indicate that distal greywacke deposition in the Khomas Sea began at least during Karibib times and possibly even earlier. The MORB-type Lievental metabasalts at the base of the Kuiseb Formation in this region heralded the final continental break up in the SZ and the formation of the earliest ocean floor.

The intensity of deformation of the Kuiseb Formation in the SZ precludes determination of its thickness. However, interbedded with the Kuiseb schists are metabasalt, metagabbro and metabasite dykes of the highly linear, 350 km long and 1 km wide Matchless amphibolite belt (Matchless Member – see Section 5.3.4).

It remains unknown when deposition of the spreading-phase Kuiseb Formation ceased and exactly when reversal of plate motion took place and subduction began. These events can be bracketed, however, between the 609 Ma Hartelust Rhyolite Member at the top of the Gomab River Formation (Nagel, 1999) and the 595 Ma oldest syn-tectonic garnet generated during the initial stages of transpressive continental collision in the Kaoko Belt (Goscombe et al., 2005a).

## 5.3.3. Post-Rift Evolution in the Gariep Belt

A different sedimentation history is recorded in the Port Nolloth Zone of the Gariep Belt. Although the Neoproterozoic basin fill there has been metamorphosed under greenschist facies conditions and, in places, intensely deformed, the nature of the sedimentary protoliths is in most cases still very well recognisable. Lithostratigraphically, the sediments that follow the siliciclastic rift deposits of the Stinkfontein Subgroup have been grouped together as the Hilda Subgroup, with the intervening, stand-alone Kaigas Formation, in the Port Nolloth Group (Figure 5.3.6).

The *Kaigas Formation* is only locally developed along the eastern and northeastern margin of the Gariepian rift basin (Rosh Pinah Graben). It reaches a maximum thickness of 115 m but pinches out laterally over only a few hundred metres. It comprises laterally discontinuous, medium to thick bedded diamictite and subordinate massive, locally cross-bedded or graded feldspathic arenite and argillite. Ripple cross-laminations, indicating palaeocurrent directions from the east, mudstone rip-up clasts, load and flute casts and sinuous ripple marks are locally present. The sedimentary features of the diamictite, such as sharp upper and lower bedding contacts, lack of internal structures, except for crudely graded bedding, correlation between clast size and bedding thickness, rapid decrease in clast size towards the basin centre and the limited geographical distribution suggest rapid deposition as debris flows. Transport by turbidity currents in a subaqueous fan environment is indicated by common upwards fining beds in the intercalated arenite (arkose and greywacke). The presence of westward-directed gravity flow deposits stepping back onto the basement foreland suggests a phase of marine transgression along an active fault scarp. This is also indicated by local dolomite clasts in the diamictite. It should be emphasised, however, that in contrast to a younger diamictite in the Port Nolloth Group, the clast lithology of the Kaigas Formation diamictite is, in general, dominated by basement-derived rock types, such as granite, gneisses, amphibolite, schists, and by reworked arenite from the underlying Stinkfontein Subgroup. Carbonate clasts are, in general, very rare and localised. A component of melt-out glacial debris in the diamictite, such as dropstones, bimodal clast distribution, sandwiched and facetted clasts and outsized erratic blocks in argillaceous beds, point to a glacio-marine or fluvio-glacial origin of parts of the Kaigas Formation.

Locally, the Kaigas Formation diamictite grades into an agglomerate or is overlain by rocks of the volcano-sedimentary Rosh Pinah Formation. Elsewhere it is overlain by a carbonate-dominated succession with intercalated carboniferous argillite and arenite beds that constitute the *Pickelhaube Formation*. The latter, which attains a thickness of not more than 280 m, starts with laminated, variably dolomitised, medium to dark grey limestone, followed by a varied sequence of predominantly argillite and marl with minor feldspathic quartz arenite. Intercalated are a number of limestone and dolostone beds with the latter becoming thicker towards the

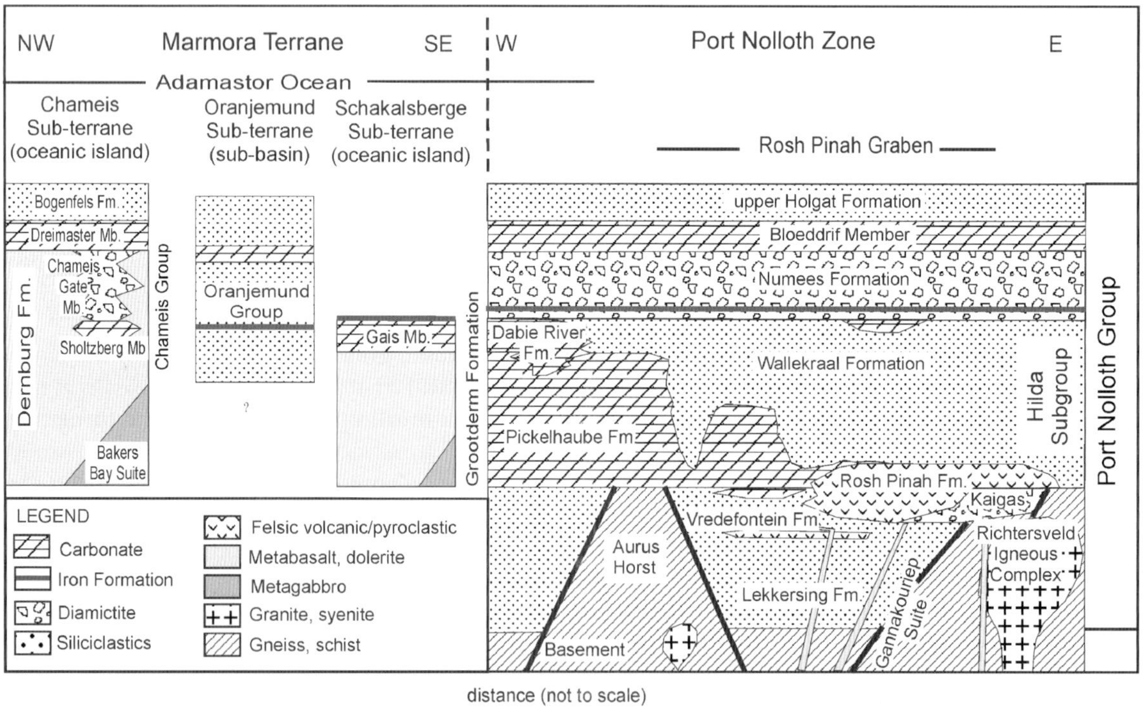


**Figure 5.3.6** Lithostratigraphy of the Gariep Supergroup in the Gariep Belt (from Frimmel, 2008b).

middle of the formation. A further succession of intercalated shale and siltstone follows, which, in turn, is paraconformably overlain by massive dolostone of the Dabie River Formation.

The feldspathic arenite in the Pickelhaube Formation contains scour structures, load casts, rip-up shale clasts, sand dykes, ripple cross-laminations and small-scale channels indicating a palaeocurrent direction from the southeast – analogous to the palaeocurrent directions in the underlying Stinkfontein Subgroup. The absence of conglomerate, scarcity of cross-bedding, presence of thin bedding and the increasing carbonate content of the Pickelhaube Formation arenites indicate quieter, submerged conditions compared to the underlying Stinkfontein Subgroup and Kaigas Formation, as can be expected in the distal parts of a fan complex. A Pb-Pb carbonate age of 728 ± 32 Ma obtained on limestone of the lower Pickelhaube Formation (Fölling et al., 2000) agrees well with the position of cap carbonates above the ca. 750 Ma Kaigas glaciogenic deposits.

Some of the clastic beds may represent turbidites and minor debris flow deposits. The original limestone deposits are largely dolomitised in the more proximal positions. Laminated allodapic limestone (calcarenite) is a ubiquitous feature. Basement-derived clasts occur where the formation onlaps directly onto basement. The carbonate beds in the middle of the formation show extensive syn-sedimentary brecciation that is reminiscent of tempestite. This is in contrast to slump breccias and carbonate debris flow deposits some of which contain extrabasinal clasts. The latter are common in the thicker carbonate beds.

A regional unconformity within the Hilda Subgroup marks a major sequence boundary and is overlain by the clastic succession of the *Wallekraal Formation*. Although the highly tectonised nature of the Port Nolloth Zone makes it difficult to map out this unconformity, evidence of it can be found in almost all parts of the Port Nolloth Zone whereby the clastic rocks of the Wallekraal Formation cut down to various levels of the older stratigraphy and, in places, even to the pre-Gariep basement. The formation reaches a thickness of as much as 180 m but it is laterally discontinuous and, in places, not developed at all. It is dominated by coarse-grained siliciclastic sedimentary rocks consisting of mature, well sorted, quartz-pebble conglomerate, gritstone and arkose that grade into mudstone in locally upward-fining cycles. Lateral and vertical facies changes reflect vastly different energy regimes, with boulder beds with individual, well-rounded basement-derived boulders reaching almost a metre in diameter occurring in some places and laminated argillite displaying ripple marks occurring in others. Locally, decimetre-thick, cream-coloured dolomite beds are intercalated with the arenaceous fraction. Isolated, angular dolomite blocks (exceeding 20 m in diameter) and clasts (olistostromes) are conspicuous features, in many places set in a carbonate breccia that displays crudely defined, upward-fining graded bedding. These are interpreted as seismogenic deposits from the basin-bounding growth fault. Overall, lithology and sedimentary features are consistent with deposition in a submarine fan environment dominated by high-energy gravity-flow processes. Lateral discontinuity of the conglomerate and arenite units indicates deposition in channels in the upper fan or in supra-fan lobes in the mid-fan region. Low-energy turbidity current deposition on channel levees or abandoned lobes is inferred for the fine-grained beds from parallel, even laminations and normal grading. The good rounding of pebbles, the isolated dolomite blocks and the dominance of gravity flow mechanisms indicate re-sedimentation from a shallow-water platform or shelf environment, probably represented by the Pickelhaube Formation.

Massive dolomitic limestone and/or, in places brecciated, dolostone of the *Dabie River Formation* makes up the top of the Hilda Subgroup. The up to 160 m thick formation is lithologically distinguished from the other carbonate-bearing successions of the Hilda Subgroup by the presence of stromatolites displaying *Conophyton*-like forms, several centimetres to decimetres in height. Pisolites, oolites and oncolites are also present. The formation is almost exclusively calcareous, with original limestone variably dolomitised. The carbonate rocks are typically massive, light to medium grey and, in places, intensely brecciated. Some of the carbonate breccias are interpreted as debris flow deposits, whereas others are ascribed to gravitational slumping. Cyclical emergence and submergence is indicated by desiccation cracks and by the interbedding of limestone and dolostone. A shallow-water, rimmed shelf environment, such as a barrier bar or shelf lagoon, passing seaward (westward) into a shelf margin, comprising reef build-ups and oolitic to pisolitic shoals is envisaged for the depositional environment. Reef rocks formed, or are preserved, particularly in those areas that escaped the pre-Wallekraal erosion. Consequently, the Dabie River Formation carbonates rest in many places paraconformably above dolostone of the Pickelhaube Formation and not necessarily on top of the clastic Wallekraal Formation rocks.

Above the Hilda Subgroup follows the *Numees Formation*, a laterally continuous, relatively uniform unit that is dominated by diamictite, reaching 500–600 m in thickness. The formations thins towards the east, where it onlaps directly onto pre-Gariep basement. Its lower contact to the Hilda Subgroup is in most places tectonic, but locally, where the Dabie River Formation carbonates are not developed, the siliciclastic Wallekraal Formation grades conformably into the Numees Formation. The lithology of the clasts, which reach several cubic metres in size, comprises pegmatite, leucogranite, granite-gneiss, quartzite, schist, dolomite, minor limestone of the immediately underlying Dabie River Formation and reworked diamictite. Dolomite clasts are particularly abundant in the lower parts of the formation but become rare in the upper parts where basement-derived clasts dominate. Clast and matrix grain sizes decrease upwards.

CHAPTER 5.4

# Syn- to Late-Orogenic Sedimentary Basins of Southwestern Africa ☆

Gerard J.B. Germs[1], Roy McG. Miller[2], Hartwig E. Frimmel[3] *and* Claudio Gaucher[4]

## Contents



## 5.4.1. Introduction

In response to the inversion from spreading to basin closure and subsequent orogeny, a number of syn- to late-orogenic basins developed in southwestern Africa. These are from north to south the Mulden, Khomas, Holgat, Nama, Malmesbury, Klipheuwel *s.l.* and Boland Basins and further to the southeast, the Cango Caves, Kansa and Gamtoos Basins (Figure 5.1.3).

These basins are not true basins as such but represent remnants of basins. They occur on the margin of the Kalahari palaeocontinent with the exception of the Mulden Basin which occurs largely landward of the western edge of the Congo palaeocontinent. Various basins related to the Kalahari palaeocontinent were originally interconnected, but the exact relationships between some of them, especially in the Saldania Belt, remain enigmatic. This chapter is a synthesis of what is known about the litho-, chemo-, biostratigraphic and radiometric data of the above-mentioned basins. Thus far, only incomplete sequence stratigraphic studies have been undertaken of the Nama Basin, for example of the Nama Group (Saylor et al., 1995, 1998, 2005) and of the Vanrhynsdorp Group (Gresse and Germs, 1993). An attempt has also been made to unravel the evolution of each basin and determine their inter-relationships. At present the Nama Basin provides most data, whereas relatively little is known of the Boland Basin.

☆ Germs, G.J.B., Miller, R.McG., Frimmel, H.E., Gaucher, C., 2009. Syn- to late-orogenic sedimentary basins of southwestern Africa. Neoproterozoic to Early Palaeozoic evolution of Southwestern Africa. In: Gaucher, C., Sial, A.N., Halverson, G.P., Frimmel, H.E. (Eds.): Neoproterozoic-Cambrian tectonics, global change and evolution: a focus on southwestern Gondwana. Developments in Precambrian Geology, 16, Elsevier, pp. 183–203.


[1] University of Johannesburg, Private Bag X607, 6620 Oudtshoorn, South Africa.
[2] PO Box 11222, Windhoek, Namibia.
[3] Department of Geological Sciences, University of Cape Town, Rondebosch 7700, South Africa.
[4] Departamento de Geología, Instituto de Ciencias Geológicas, Facultad de Ciencias, Iguá 4225, 11400 Montevideo, Uruguay.


*Developments in Precambrian Geology*, Volume 16
ISSN 0166-2635, DOI 10.1016/S0166-2635(09)01613-2

## 5.4.2. Stratigraphy and Depositional Environment

### 5.4.2.1. The Mulden Basin

Detailed descriptions of the Mulden Group, which accumulated in the Mulden Basin, have been given by Hedberg (1979) and Miller (1997, 2008). The Mulden Group is a northern, post-$D_1$ to pre-$D_2$ molasse succession deposited primarily on the Northern Platform and eastern Northern Margin Zone (Etoshafontein Subgroup; Miller, 2008) above the Otavi Group rocks with a paraconformable to slightly disconformable contact. This contact becomes markedly discordant in the west and northwest, where intramontane units also occur in the western Northern Margin Zone of the Damara Belt (Welkom Subgroup; Miller, 2008), the Eastern and Central Kaoko Zones (Sesfontein Formation and unnamed units) and the Western Kaoko Zone (Toscanini Formation). The best evidence for post-Kuiseb, pre-Mulden $D_1$ folding, uplift and erosion is provided by outcrops west of Khorixas. In this region, Mulden rocks fill deeply incised, steep-sided valleys in Otavi rocks as much as 150 m deep; they also rest unconformably on Kuiseb Formation schists (Frets, 1969; Miller, 1983a, 1997, 2008). The age of the Mulden Group is constrained approximately between 580 and 541 Ma by $^{40}Ar$-$^{39}Ar$ thermochronological data (Gray et al., 2006) and palaeontological data from the Owambo Formation of this group (see below).

The Etoshafontein Subgroup is made up from bottom to top of the Tschudi Formation consisting largely of lithic to arkosic sandstone, the pelitic Kombat Formation and the Owambo Formation. The former two formations crop out only locally in the Otavi Mountainland and on the northern edge of the Kamanjab Inlier, whereas the latter is known only from boreholes in the Owambo Basin. Vibroseis surveys suggest a maximum thickness of about 4,200 m in the basin (Hedberg, 1979). The subgroup is proximal in the west and distal in the east, and the source region was primarily the rising mountain belt in the Central and Western Kaoko Zones as well as $D_1$ fold structures in the Otavi rocks forming the southern margin of the Mulden Basin. A thin shale-dominated unit at the base of the Tschudi Formation thickens to 440 m into the basin. In the Tschudi Formation sandstones, Otavi-derived lithic fragments give way to feldspar higher up reflecting a more local source initially (Otavi Group carbonates and cherts) and more distal arkosic Nosib and/or basement granitic sources subsequently either as a result of progressive unroofing and/or emplacement of basement-cored nappes in the Central Kaoko Zone.

In places, the upper Tschudi Formation feldspathic sandstones rest directly on Otavi Group carbonates. During development of the pre-Mulden unconformity, significant karsting of the Otavi Group carbonates took place. Where the upper Tschudi arkosic sands were deposited directly onto this unconformity, they were also able to pour down and fill palaeokarst features forming the so-called "pseudo-aplites" of the Otavi Mountainland, which later became the locus for Tsumeb-type polymetallic mineralisation (Lombaard et al., 1986).

The Kombat Formation in the Otavi Valley consists of dark grey phyllite and silty and sandy phyllites that are often pyrrhotitic or pyritic. In the Mulden Basin, it can be divided into three parts, a lower Kombat Formation, a middle Black Shale Member and an upper Kombat Formation (Miller, 1997).

In the Mulden Basin, the lower Kombat Formation reaches 330 m in thickness, fines upwards and is transitional from the underlying Tschudi Formation. The Black Shale Member is 93 m thick, fine-grained, highly carbonaceous and is an excellent stratigraphic, seismic and conductivity marker unit throughout the basin. Black to very dark grey, almost silt-free shales form the lower 49 m. The next 44 m consist of two cycles that fine upwards from light grey siltstones with moderate resistivities to medium to dark grey shales with a low resistivity. The upper Kombat Formation reaches 400 m in thickness, fines upwards and consists of several upward-fining cycles of pyritic, grey, grey-green and green sandstone, siltstone and shale.

The first extensively red sandstone forms the base of the Owambo Formation. Three units make up the 1,092 m of the formation in the ST-1 well (Miller, 1997), the lower and upper being red and grey in colour, the middle unit being grey and black. The formation consists of over 100 depositional cycles that fine upwards from sandstone or siltstone bases to shale tops. The number of cycles that are capped by grey carbonate increases upwards through the formation. Limestone is the dominant carbonate in the middle unit and dolomite in the upper unit. Siliciclastic rock types are pyritic in the lower two units. Rhombohedral vugs after gypsum (?) become more abundant upwards through the formation.

The Welkom Subgroup occurs in the Khorixas area of the Northern Margin Zone. Initially mapped by Clifford (1967), Frets (1969) and Guj (1974), the Welkom Subgroup bears broad similarities to the Etoshafontein Subgroup and is made up of the Braklaagte Formation at the base followed by the Renosterberg and Gaseneirob Formations (Miller, 2008). The Braklaagte Formation consists of a local, basal, polymictic, proximal conglomerate followed by more extensively developed, thinly bedded greywackes and phyllites that thin southwards. The Renosterberg Formation is a monotonous succession of well-exposed, coarse-grained, immature, light brown to grey, cross-bedded, feldspathic quartzite that reaches 1,600 m in thickness. Sourced in the west, the quartzite grades eastwards into greywacke, arkose and shale. The Gaseneirob Formation follows concordantly on the

Renosterberg Formation and is the main unit of the Mulden Group east of Khorixas. It consists mainly of thinly bedded to laminated, silvery grey to pale greenish grey chlorite–muscovite phyllitic schists.

The Mulden successions in the Eastern and Central Kaoko Zones all occur in local intramontane basins formed by synforms of Otavi Group rocks. Rock types include a variety of contrasting polymictic conglomerates, pebbly to gritty sandstones, calcareous sandstone, micaceous sandstone, feldspathic sandstone, arkose, siltstone, lithic greywacke, subgreywacke, chlorite–muscovite phyllite and dolomite, all in colours of grey, green and maroon. Each synform has a unique stratigraphy. Some, such as that occupied by the Sesfontein Formation, are dominated by green phyllite, whereas others in the Baines Mountains near the Kunene River are dominated by conglomerates and sandstones.

The Toscanini Formation near the southern end of the Coastal Terrane of the Western Kaoko Zone is different. This is made up entirely of green, chlorite–muscovite semipelitic to pelitic schist with a single, well developed, vertical, north-south cleavage (Miller, 1983a). The formation overlies upper amphibolite facies rocks of the Coastal Terrane (Möwe Bay Formation) but was itself only subjected to greenschist facies metamorphism.

The Mulden Group is largely a Kaoko Belt-derived molasse. No datable rock types have been found in the succession but the tectono-metamorphic evolution of the Kaoko Belt provides a reasonable indication of the age of the Mulden Group. The earliest syn-tectonic garnet in the Coastal Terrane has an age of 595 Ma and the earliest syn-tectonic granite an age of 580 Ma (Seth et al., 1998). The peak of high-grade metamorphism in the pre-Mulden basement was during the early stages of syn-$D_1$ sinistral, transpressional, wrench tectonics in the Kaoko Belt and is dated at $575 \pm 10$ Ma (Goscombe et al., 2005a). The ensuing erosion and rapid decompression of the kyanite-bearing Central Kaoko Zone provided the Mulden sediments. Continuous deformation lasted some 30 myr with $D_1$ structures and the Mulden succession being deformed during $D_2$. Late- to post-$D_2$ granites in the Western Kaoko Zone range from 554 to 541 Ma in age (Goscombe et al., 2005b). The younger of these granites intruded the Toscanini Formation post-tectonically (Miller, 2008). The age of the Mulden Group is thus constrained between 580 and 541 Ma. This is supported by recent micropalaeontological evidence. Organic-walled microfossils indicate an Ediacaran age for the Owambo Formation (Gaucher and Germs, 2007).

No systematic chemostratigraphic studies have yet been undertaken of the Mulden Group. The samples collected thus far are all depleted in $^{13}C$ with some $\delta^{13}C$ values ranging as low as −8‰ (Kaufman et al., 1991; Gaucher and Germs, 2007).

The decreasing trend from very high $\delta^{13}C$ values in the Hüttenberg Formation (Tsumeb Subgroup, Otavi Group) to the negative $\delta^{13}C$ values of the overlying Owambo Formation may be indicative of a glaciation (Kaufman et al., 1991; Germs, 1995). For these negative values and organic-walled microfossils (Vidal, written communication, 1979), the basal Tschudi conglomerate of the Mulden Group (occurring in deeply incised palaeovalleys) has been correlated with the Numees Formation diamictite in southern Namibia (Kaufman et al., 1991; Germs, 1995), which probably is of Gaskiers (580 Ma) or Moelv age (ca. 560 Ma) – a possibility that has found support from micropalaeontological studies (Gaucher et al., 2005a; Gaucher and Germs, 2007).

Neither body nor trace fossils have yet been found in the Mulden Group. Up to now only organic-walled microfossils have been found in the Owambo Formation in borehole ST-1 (Vidal, written communication, 1979; Gaucher and Germs, 2007, Figure 5.4.1E,F). Vidal analysed 12 core samples and reported the occurrence of *Bavlinella faveolata* at level 4582′, *Leiosphaeridia* sp. (5963′), *Vendotaenia* sp. (5963′, 5993′ and 6094′) and poorly preserved sphaeromorphic acritarchs at 4582′ and 4592′. A recent study of organic-walled microfossils of the Owambo Formation in borehole ST-1 (Gaucher and Germs, 2007) shows an assemblage of filamentous microfossils and sphaeromorphic acritarchs occurring in samples 4264′, 5569′, 5679′ and 5847′. The assemblage consists of the following taxa:

(a) Abundant, non-septate, non-branched filaments assigned to the genus *Siphonophycus* (Schopf) emend. (Knoll et al., 1991) and comprising the species *S. solidum* (width 16–32 μm), *S. kestron* (width 8–16 μm) and *S. typicum* (width 4–8 μm), the last mentioned species being the most common and widespread (Figure 5.4.1F). These microfossils are usually interpreted as the empty sheaths of filamentous cyanobacteria. A few specimens show that *Siphonophycus* was a mat-forming cyanobacterium, and might represent the main stromatolite-building organism in the Owambo Formation.
(b) *Tortunema wernadskii* (Schepeleva) emend. (Butterfield et al., 1994), representing pseudoseptate filamentous microfossils. This species is subordinate with respect to *Siphonophycus*.
(c) *Leiosphaeridia minutissima* (Naumova) emend. (Jankauskas et al., 1989), which occurs as strongly corroded vesicles (Figure 5.4.1E).
(d) *Leiosphaeridia tenuissima* Eisenack, up to 130 μm in diameter and showing strongly corroded, folded vesicle walls.

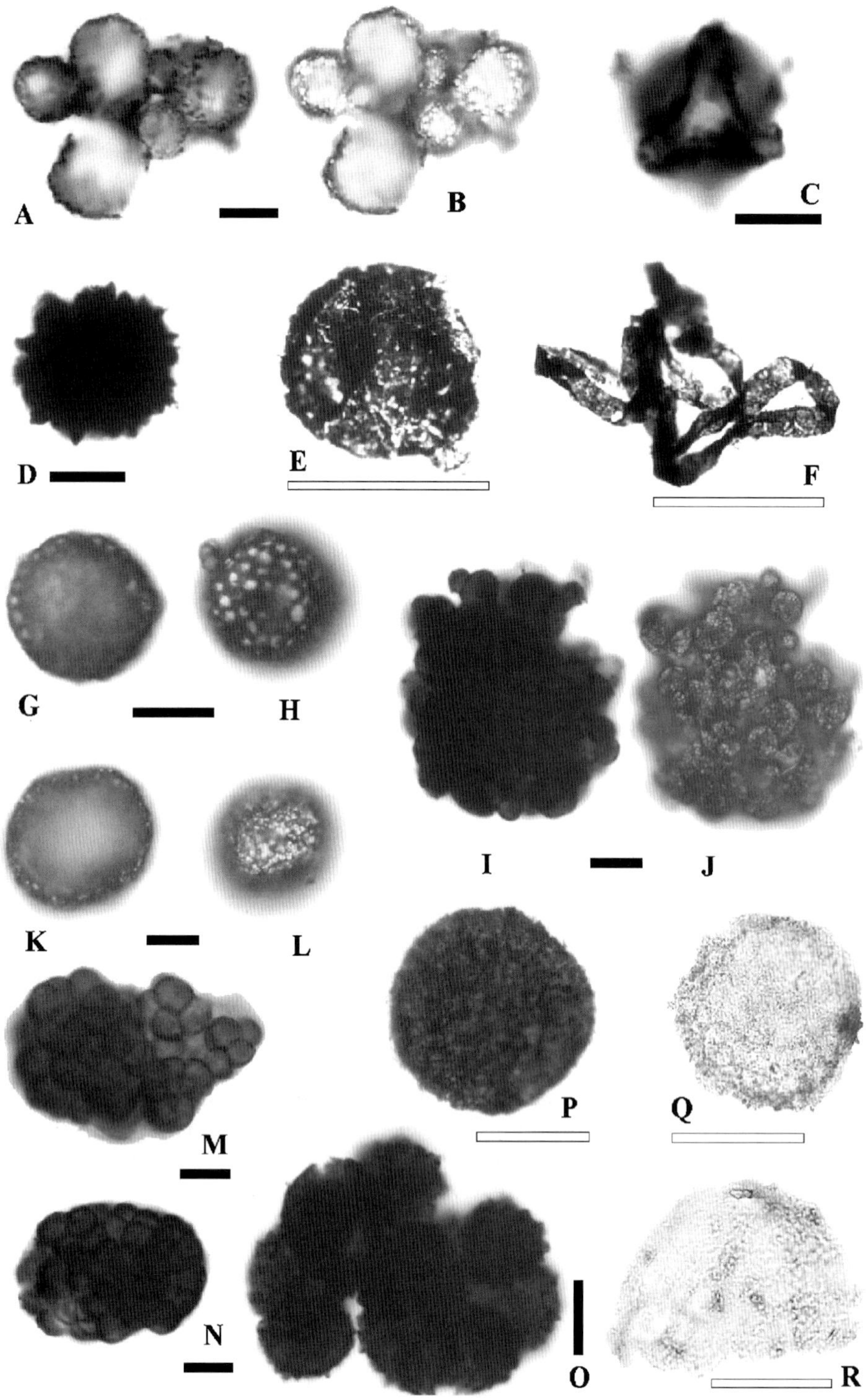


**Figure 5.4.1** Organic-walled microfossils (acritarchs) isolated by means of acid maceration from the Otavi (A–D), Mulden (E, F), Port Nolloth (K–M, R), Cango Caves (I, J, O, Q) and Gamtoos (G, H, P) groups. (A, B) *Bavlinella faveolata* (Schepeleva) Vidal (1976), Auros Formation, in transmitted (A) and reflected light (B). (C) *Octoedryxium truncatum* (Rudavskaja) Vidal (1976), Maieberg Formation. (D) *Asteridium* cf. *tornatum* (Volkova) Moczydlowska (1991), Ghaub Formation (reworked?). (E) *Leiosphaeridia minutissima* (Naumova) Jankauskas (1989), Owambo Formation, ST-1 borehole. (F) *Siphonophycus typicum* (Hermann) Butterfield et al. (1994), folded filament, Owambo Formation, ST-1 borehole. (G, H) *B. faveolata*, Kaan Formation, in reflected light and at different focus levels. (I, J) *B. faveolata*, cluster of colonies, Nooitgedagt Member, in transmitted (I) and reflected (J) light. (K, L) *B. faveolata*, Wallekraal Formation, in reflected light and at different focus levels (Gaucher et al., 2005a). (M) *Soldadophycus major*, Holgat Formation. (N) *Soldadophycus bossii*, Kombuis Member. (O) *Asteridium tornatum*, Kombuis Member. Note spiny processes near the edge of the vesicles. (P) *Leiosphaeridia tenuissima* Eisenack (1958), Kleinrivier Formation (Gaucher and Germs, 2006). Note more intense carbonisation compared to Q and R. (Q) *L. tenuissima*, Huis Rivier Formation. (R) *L. tenuissima*, incomplete, Holgat Formation (Gaucher et al., 2005a). Hollow scale bars represent 50 μm, solid scale bars represent 10 μm.

This assemblage is characterised by the following features (Gaucher and Germs, 2007): (i) low diversity; (ii) abundance of filamentous algae of the genus *Siphonophycus*, suggesting proximity to shore (Butterfield and Chandler, 1992) and (iii) similar composition of the assemblage throughout the section studied. Filaments represent pieces of benthic cyanobacterial mats carried offshore by currents. A marked trend is observed, starting with rare, small filaments (mainly *Siphonophycus typicum*) at the base, and passing into more abundant and larger filaments up section. This is a good indication of a shallowing-upward trend, as also evidenced by the common occurrence of stromatolites in the upper Owambo Formation, but not in the lower part of the unit in the ST-1 borehole. Concomitantly, leiosphaerids become smaller and relatively less common towards the top, also suggesting a shallowing-upward trend.

### 5.4.2.2. The Khomas Basin (Southern Zone of the Damara Belt)

Although it has not been possible to date $D_1$ structures in the Damara Belt, Miller (2008) has argued that the initiation of transpressive subduction of the ocean floor of the Khomas Sea (Southern Zone) beneath the Central Zone was almost contemporaneous with the early stages of transpressive deformation in the Kaoko Belt. Be this as it may, once the Central Zone had become an elevated and active continental margin, it will have shed sediment into the closing Khomas Sea. Thus, a forearc-trench succession of deep-water greywackes (Kukla, 1992; Kukla and Stanistreet, 1993), the *Hureb Formation*, was deposited on top of the deep-water, spreading-phase succession of greywackes of the Kuiseb Formation in the Khomas Sea. In contrast to earlier interpretations, Miller (2008) suggested that a first cleavage developed in the Kuiseb metagreywackes in the Southern Zone before any syn-tectonic sediments were shed onto them. In an area east of Okahandja near the northern edge of the Southern Zone, Kasch (1988) recorded Kuiseb schists with three syn-tectonic cleavages separated by an unconformity from overlying metagreywackes with only two cleavages. This younger metagreywacke succession is the lower Hureb Formation. This succession is, in turn, separated by an unconformity from, and overlain by, metagreywackes with only one cleavage, the upper Hureb Formation (Miller, 2008). Kukla (1992) and Kukla and Stanistreet (1993) described the development of a single cleavage in semi-consolidated sediments of this forearc-trench succession. Post-$D_1$ granites yielded an age of 565 Ma (de Kock et al., 2000). Continental collision and final closure of the Khomas Sea took place at 542 Ma (Miller, 1983a, 2008). Consequently, the timing of deposition of the syn-orogenic Hureb Formation is bracketed between 565 and 542 Ma.

In the Gobabis area, the upper depositional sequence, a transgressive–regressive succession, is formed by the *Buschmannsklippe Formation* (Figure 5.3.5; Hegenberger, 1993). The formation thickens northwestward to a maximum of 200 m and consists of a basal buff to pink dolostone (Bildah Member) with metre-high domes, slump structures and vertical tube-like structures (Hegenberger, 1987). Similar structures have been described from many other Neoproterozoic post-glacial cap carbonates, including the post-Ghaub Keilberg Member in the Otavi platform (see Section 5.3.2.2). The dolostone grades upwards into pink, thin-bedded limestone, purple shale, siltstone and sandstone (La Fraque Member). It contains conspicuous upward-divergent calcite fans that have been interpreted as pseudomorphs after sea-floor aragonite (Grotzinger and Knoll, 1995). Sedimentary features indicative of storm reworking are abundant. The overlying Okambara Member reflects shoaling upwards through upper shoreface into a restricted tidal flat facies. The contact with sandstone of the overlying basal Nama Group has been described as gradational (Saylor et al., 1998). A comparable succession occurs in the Naukluft Nappe Complex (Hoffmann, 1989). Based on lithological similarities and distinct negative $\delta^{13}C$ ratios between $-3$ and $-6‰$ (Saylor et al., 1998), a correlation of the Buschmannsklippe Formation with either post-Ghaub, Cryogenian cap carbonates or with Ediacaran cap carbonates could be feasible. Due to a high $^{87}Sr/^{86}Sr$ of 0.7081 (Saylor et al., 1998) and the local gradational contact with the Nama Group, a correlation with the post-$D_1$ Owambo Formation in the Mulden Basin as well as with the Holgat Formation in the Gariep Belt (see below) is a possibility. In that case, the negative $\delta^{13}C$ excursion might represent the Shuram event as defined by Brasier et al. (2000).

### 5.4.2.3. The Holgat Basin

Predominantly massive, glaciogenic diamictite of the Numees Formation (Port Nolloth Group, Figure 5.3.6), as described in Section 5.3.3, represent the oldest sediments that accumulated in the Holgat Basin. The Holgat Formation generally overlies the Numees Formation with a sharp conformable contact where it is not tectonised. From the base upwards, the Holgat Formation consists of a basal cap carbonate, the Bloeddrif Member, overlain by siliciclastic rocks (upper Holgat Formation). The Bloeddrif Member, which attains a maximum thickness of 100 m, is light grey, cream or pale pink in colour and poor in organic matter. It displays the characteristics of a typical cap carbonate, that is vertical tube-like structures of infilled micritic sediment and cement. They are usually a few centimetres across and several decimetres high and appear analogous to those in the Bildah Member of the

Buschmannsklippe Formation mentioned above. The upper siliciclastic metasedimentary rocks comprise upwards-fining cycles of medium-bedded sandstone, greywacke and arkose with minor siltstone, mudstone and intraformational conglomerate. These rocks are interpreted to represent turbidites.

Sedimentary successions analogous to those of the Holgat Formation in the Port Nolloth Zone are also found in the stratigraphic highest position within the Marmora Terrane. They do not occur in the Schakalsberge Subterrane, but are prominent in the Oranjemund and Chameis Subterranes. Particularly in the latter, these rocks are well exposed and form the upper part of the Chameis Group. They have been reclassified as Bogenfels Formation (Frimmel, 2000c), only taking that part of the original Bogenfels Formation, as defined by Martin (1965b), that is located in the Marmora Terrane. The Bogenfels Formation rests upon mafic rocks of the Dernburg Formation. It starts with a carbonate-dominated succession, the Dreimaster Member, consisting of a 10 m thick, thinly laminated, medium grey limestone that is variably dolomitised and rich in $H_2S$ – analogous to the lower Bloeddrif Member. The upper unit of the Dreimaster Member is made up of a few tens of metres of thickly bedded, micritic, light creamy white to medium grey dolomite, local ferruginous black chert and dolomite breccia in which the underlying rock types appear re-worked. The Dreimaster Member is overlain by a varied succession of quartzite, meta-arkose, phyllite, chlorite schist and calcareous metapelite. The thickness of this siliciclastic succession is not constrained because of intense deformation but is estimated to be a few hundred metres – similar to the upper Holgat Formation. Graded bedding is common in medium to thickly bedded arenitic units that grade into argillaceous rocks towards the top of each cycle. Analogues siliciclastic rocks are also found in the upper part of the Oranjemund Group in the Oranjemund Subterrane, but they have not yet been subdivided further. Carbonate rocks are very sparse in the latter.

The similarity in lithology between the Bogenfels and Holgat formations points to sediment deposition of both formations in the same basin. Conspicuously, no such sedimentary deposits are known from the lowest tectonic unit of the Marmora Terrane, the Schakalsberge Subterrane. This field observation and the lack of an older foliation and early metamorphic overprint, as found in the older units of the Marmora Terrane, has led Frimmel and Fölling (2004) to suggest that the Holgat/Bogenfels deposits were laid down during the closure of the larger Gariep Basin in a foredeep position (Holgat Basin) in front and on top of the advancing oceanic thrust sheets of the Marmora Terrane. Such an early syn-orogenic timing of sedimentation is supported by a Pb-Pb carbonate age of $555 \pm 28$ Ma of the Bloeddrif Member limestone (Fölling et al., 2000), which is younger than Ar-Ar ages (575 Ma) obtained on an older syn-tectonic hornblende generation associated with first accretion of oceanic crust in the Marmora Terrane (Frimmel and Frank, 1998).

Further indirect support for such a Late Ediacaran age comes from Sr isotopes: $^{87}Sr/^{86}Sr$ ratios of Bloeddrif Member limestones are essentially primary and vary between 0.7080 and 0.7085 (Fölling and Frimmel, 2002). This range fits well into the youngest zone of the Ediacaran, that is the Kotlin-Rovno Zone (ca. 560–543 Ma), which yielded ratios between 0.7075 and 0.7085 (Walter et al., 2000). Such an age is also indicated by micropalaeontological evidence. Organic-walled microfossils have been found in unmetamorphosed Holgat Formation siltstone near the eastern margin of the Holgat Basin (Gaucher et al., 2005a). The microfossil assemblage is characterised by low diversity (six species) and dominated by *Soldadophycus bossii*, *Soldadophycus major* (Figure 5.4.1M and *Myxococcoides siderophila* with minor *L. tenuissima* (Figure 5.4.1R), *Synsphaeridium* sp. and *Coniunctiophycus conglobatum*. The agglutinated foraminifer *Titanotheca* also occurs. This microfossil assemblage is consistent with a Late Ediacaran age (see Chapter 9.1).

Comparison of three chemostratigraphic profiles through the post-glacial Bloeddrif Member cap carbonate shows that enrichment in carbonate $^{13}C$ is in the first instance a function of proximity to the basin margin and thus to the extent of ocean water stratification (Frimmel and Fölling, 2004). However, in all cases, the profiles start with a negative $\delta^{13}C$ excursion at the contact with the underlying diamictite of the Numees Formation (Figure 5.3.7). An elevated input of fresh water from melting ice may be indicated by the distribution of O isotopes in the immediate cap carbonates (Fölling and Frimmel, 2002).

Further south, in the Vredendal Outlier, equivalents of the Holgat Formation are found in the uppermost parts of the Bloupoort Formation of the Gifberg Group (Figure 5.3.7). There the diamictite unit within this formation is overlain by cap carbonates that are lithologically identical to those of the Bloeddrif Member, and they are also followed by siliciclastic deposits very similar to those of the upper Holgat Formation.

### 5.4.2.4. The Nama Basin

Sediments of the classic Nama Group, which were deposited in the Nama Basin, occur in the area extending from Schlip and the Naukluft Mountains (southern Namibia) southwards to Springbok (northwestern South Africa). Stratigraphic correlations show that the Nama sedimentary rocks also occur in the Witvlei and Gobabis regions (Hegenberger, 1993), which are located northeast of Schlip and east of Windhoek, and also in the Vanrhynsdorp region in South Africa where the Nama Group is named the Vanrhynsdorp Group (Germs and Gresse, 1991). Thus, the basin extended over a distance of at least 1,000 km (Figure 5.1.3).

### 5.4.2.4.1. Classic Nama Group

The classic Nama Group comprises a predominantly carbonatic foreland succession overlain by red-bed siliciclastic molasse and is subdivided, from old to young, into the Kuibis, Schwarzrand and Fish River Subgroups (Figures 5.4.2–5.4.5). Clastic sedimentary rocks in the Kuibis Subgroup are mainly white quartzite, those in the Schwarzrand Subgroup are predominantly greenish and less quartzitic, whereas the Fish River Subgroup lacks

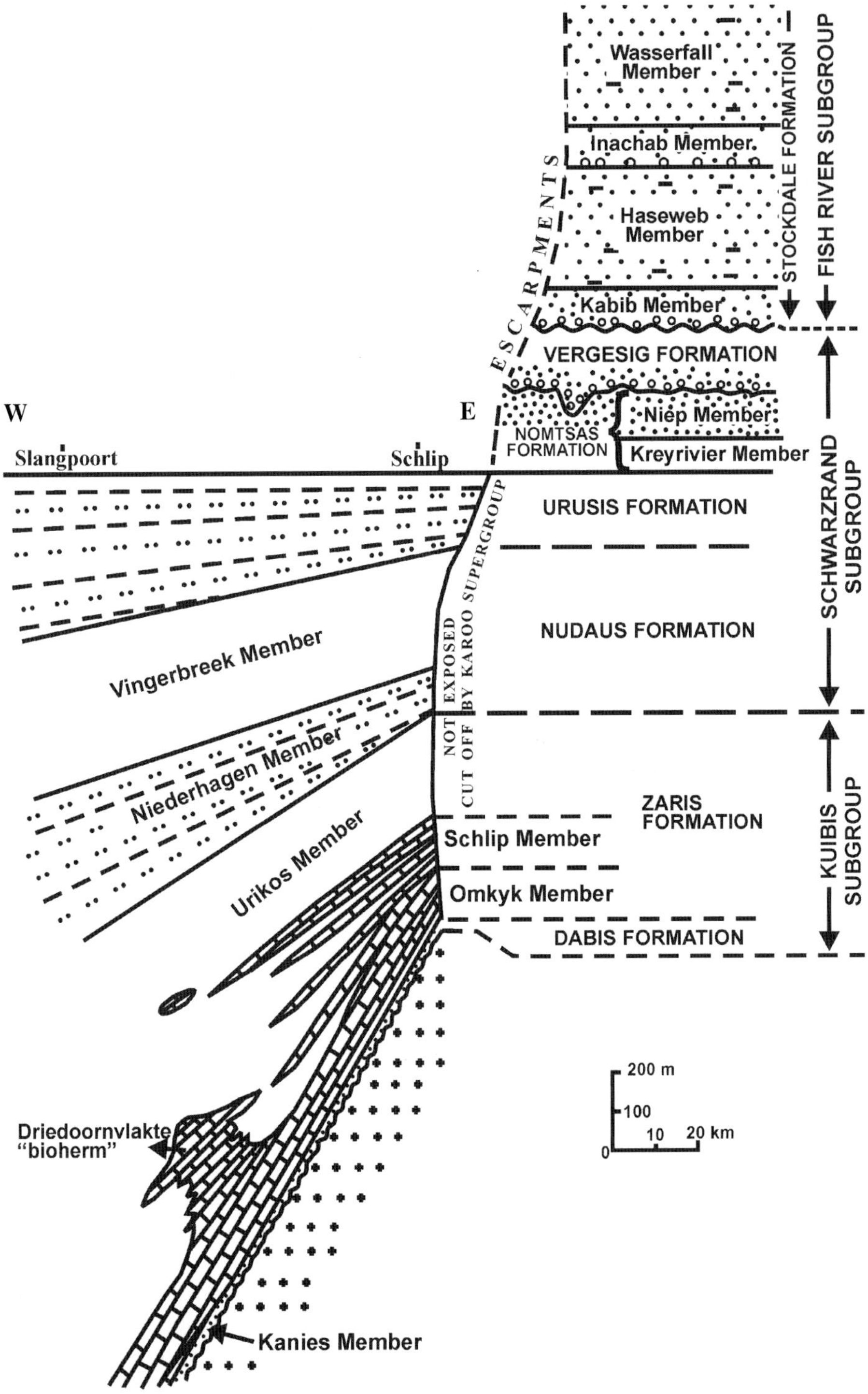


**Figure 5.4.2** East-west section through the Zaris sub-basin of the Nama Basin (Germs, 1983; Germs and Gresse, 1991). For location of section see Figure 5.1.3.

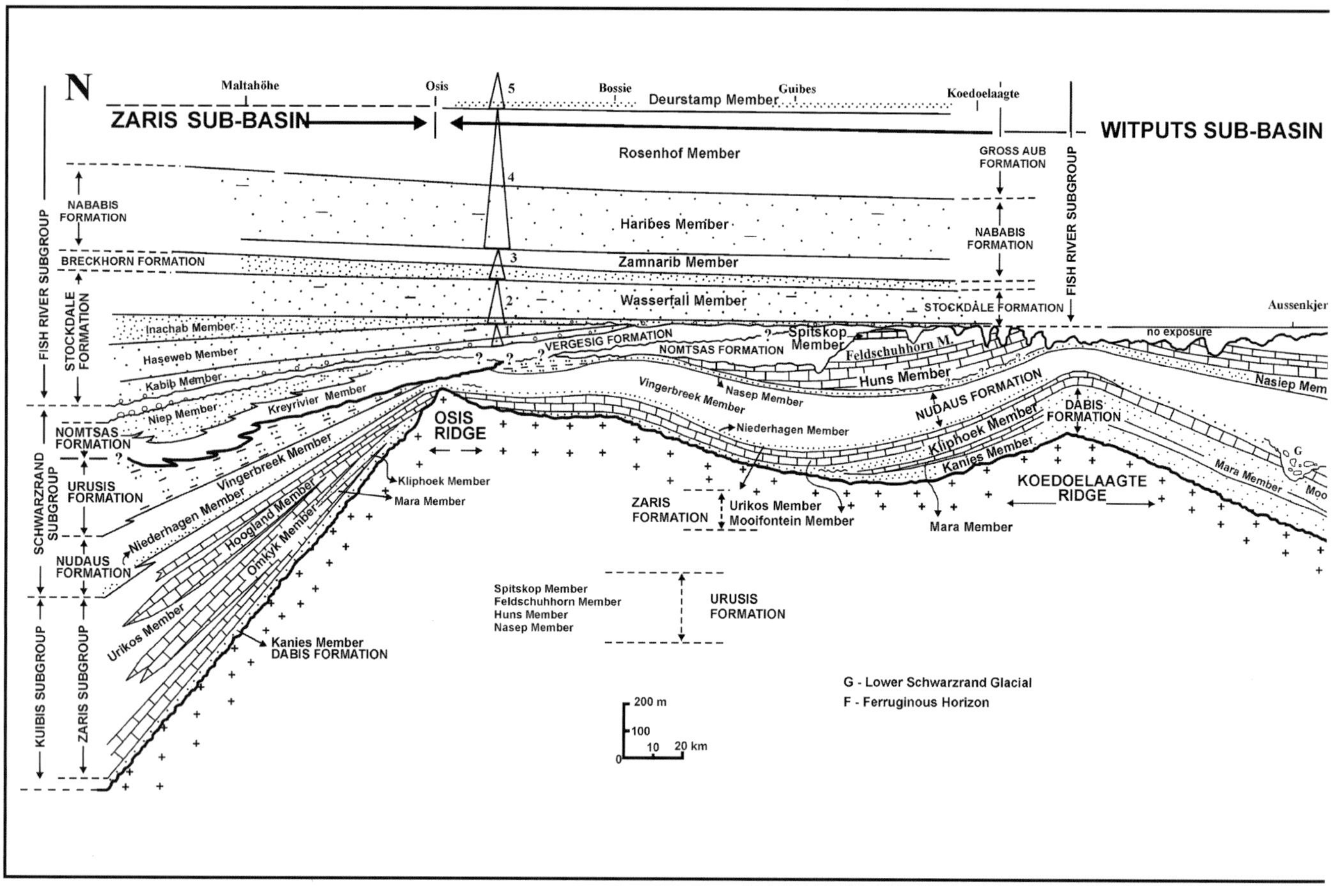


**Figure 5.4.3** North-south section through the Nama Basin (Germs, 1983; Germs and Gresse, 1991). For location of section see Figure 5.1.3.

major carbonate beds and is dominated by red, mainly coarse-grained, feldspar-bearing siliciclastic rocks. Thus, the group displays an overall upward decrease in maturity. Possible evaporites occur in the southern lower Nama Group and in the northern upper Nama Group (Germs, 1972b, 1983). A few thin layers of chert-like silicified volcaniclastic rocks are present in the Kuibis and Schwarzrand Subgroups (Germs, 1972b; Grotzinger et al., 1995; Saylor et al., 2005). Major unconformities occur at the base of the Nama Group and at the bases of the Nomtsas and Vergesig Formations of the upper Schwarzrand and of the Fish River Subgroups (Figure 5.4.3). The generalised depositional environments of the various stratigraphic units are shown in Figure 5.4.5. For more detail, the reader is referred to Germs (1983), Saylor et al. (1995, 1998), Grotzinger et al. (2000, 2005), Saylor (2003), Adams et al. (2004) and DiBenedetto and Grotzinger (2005). A short description of the lithostratigraphy and depositional environments of the classic Nama Group follows.

*The Kuibis Subgroup* represents the oldest subgroup (Late Ediacaran) of the Nama Group. A major unconformity occurs at the base of the subgroup and can be traced over a distance of some 1,000 km. The Kuibis sediments were generally transported from the Kalahari Craton in the east. During Kuibis and much of Schwarzrand times, the Osis Ridge subdivided the classic Nama basin into a northern Zaris sub-basin and a southern Witputs sub-basin (Figure 5.1.3). For some distance north and south of the Osis Ridge, the Kuibis Subgroup consists of two cycles. Each cycle has a conglomerate or pebbly quartzite at its base passing stratigraphically upwards into shale and limestone, respectively (Figure 5.4.3). Towards the Osis Ridge, the base of the second cycle truncates the first cycle (Germs, 1972b, 1983). South of the Orange River, the carbonates of the two cycles, the Mara and Mooifontein Members, wedge out and are replaced by two quartz sandstone-shale units derived from the southern Kamieskroon Ridge (Figure 5.4.3). Towards the northwest and southwest, that is towards the deeper parts of the Zaris and Witputs sub-basins, the predominantly fluvial Kuibis Subgroup sandstones pass into marine shale and limestone. In the Zaris sub-basin, the Kuibis Subgroup represents a largely northwest dipping microbial (thrombolitic and stromatolitic) dominated and storm-and-wave dominated carbonate ramp (Saylor et al., 1998; Grotzinger et al., 2000, 2005; Adams et al., 2004). In a northwesterly direction, the carbonate ramp interfingers with shales of the Urikos Member, which were predominantly deposited offshore in a relatively deep basin. No detailed studies of the Kuibis Subgroup carbonates of the

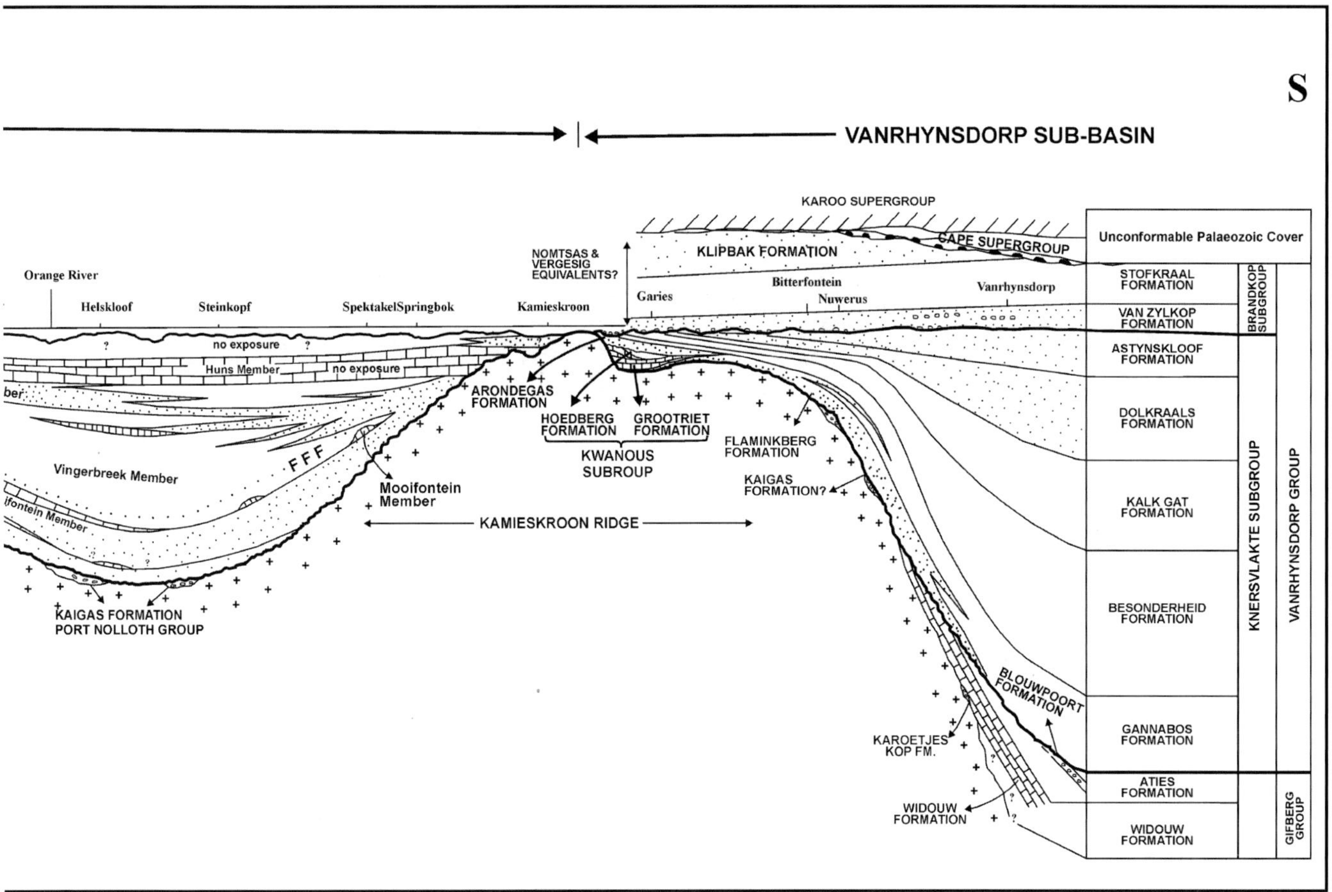


southern Witputs sub-basin have yet been undertaken. South of the Osis Ridge and northeast of Witputs, northwesterly trending patch reefs and slump features in shelf lagoonal micrites of the Mooifontein Member possibly formed on fault controlled topographic highs in the Witputs sub-basin.

*The Schwarzrand Subgroup* generally overlies the Kuibis Subgroup conformably but in some localities unconformably (Figures 5.4.3 and 5.4.4). The subgroup is composed of clastic sedimentary rocks and thick limestone units (e.g. the Huns and Spitskop Members), the latter only occurring in the southern Witputs sub-basin. The clastic rocks are greenish in the lower Schwarzrand Subgroup. They decrease in maturity and become reddish towards the top of the subgroup. The first red beds occur in the Nomtsas Formation.

In the southern classic Nama Group at and near the base of the Schwarzrand Subgroup, glacial phenomena, such as erosional palaeovalleys cutting into Kuibis quartzite with glacial pavements and glacial grooves at their bottoms, occur over a fairly extensive area (Schwellnus, 1941; Martin, 1965b; Germs, 1972b, 1983, 1995). The glacial features are overlain by sedimentary rocks of the Vingerbreek Member of the Nudaus Formation. In the palaeovalleys these are conglomerates. The unconformity associated with this lower, Late Ediacaran, ca. 547 Ma Schwarzrand glaciation is widespread in the southern Nama and is characterised by a negative $\delta^{13}C$ excursion to −2.0‰ (Kaufman et al., 1991; Saylor et al., 1998; Figure 5.4.6). The unconformity can be correlated with the unconformity at the base of the Groenefontein Formation of the Cango Caves Group and with similar unconformities occurring in South America (Gaucher et al., 2008). The shale-rich Vingerbreek Member can be traced over a large area and represents a major deepening of the Nama basin. South of the Orange River, storm deposits and turbidites occur in the Vingerbreek Member indicating sedimentation in deeper water than in the north.

In the Witputs sub-basin thick limestone units, named the Huns and Spitskop Members, occur in the overlying Urusis Formation. They pinch out towards the east and also northwards towards the Osis Ridge (Figure 5.4.3); the Huns Limestone Member extending further eastwards than the older Mooifontein Member of the Kuibis Subgroup (Figure 5.4.4). Detailed facies and sequence stratigraphic studies of the siliciclastic-carbonate units comprising Urusis Formation were carried out by Saylor (2003). Similar to the Kuibis Subgroup, thrombolites (clotted internal structures) and stromatolites (laminated internal structures) occur in these Urusis

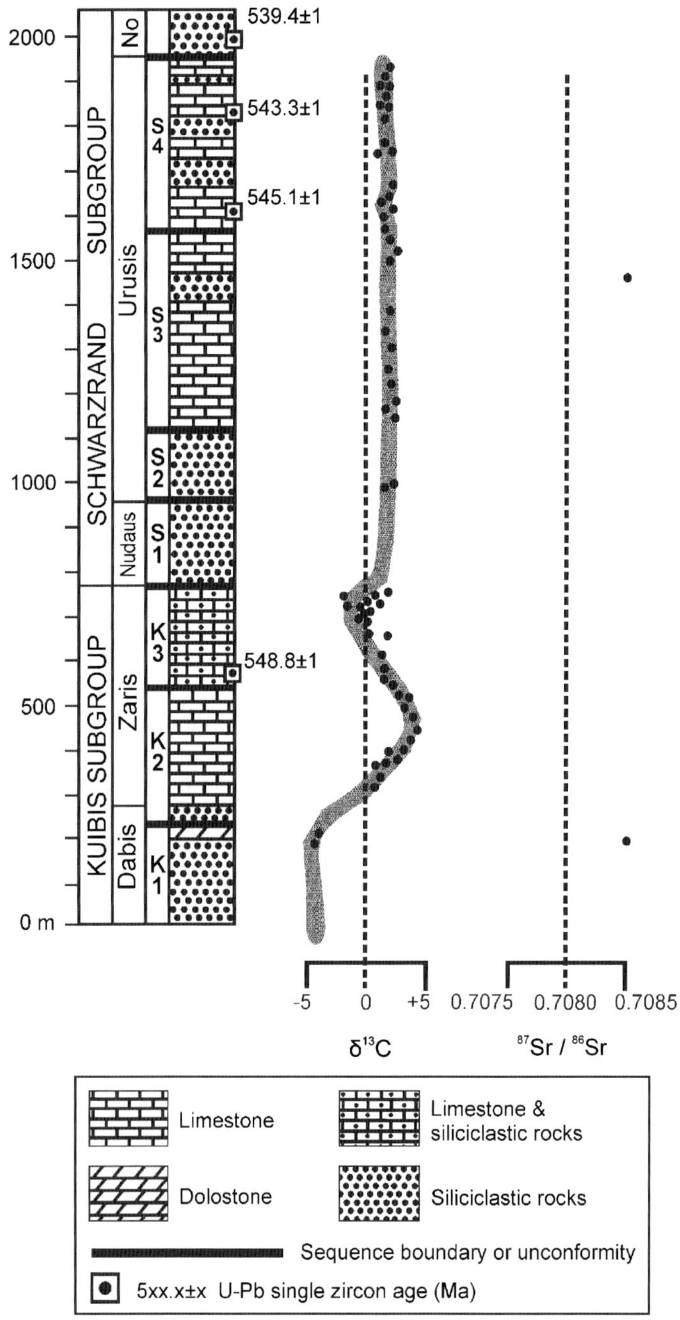


**Figure 5.4.6** Stratigraphic column of the Nama Group in Namibia showing carbon isotope excursions, strontium isotope ratios, U/Pb zircon age of silicified ash beds and sequence boundaries (modified after Saylor et al., 1998); No – Nomtsas Formation.

$^{87}Sr/^{86}Sr$ ratio of the upper Kuibis and Schwarzrand Subgroup carbonates is 0.7085 (Saylor et al., 1998), which is in good agreement with comparable Late Ediacaran values elsewhere (Halverson et al., 2007a).

The classic Nama Group is well known for its important fossils, especially the lower Nama Group, which contains, among others, representatives of the Ediacaran fauna (Figure 5.4.7H,I; Germs, 1995; Grotzinger et al., 1995; Narbonne et al., 1997). The stratigraphic distribution of fossils occurring in the Nama Group north and south of Osis is listed in Table 5.4.1. Some of these fossils, especially trace fossils, warrant to be re-examined. Of particular interest are *Swartpuntia germsi* and *Pteridinium carolinaense* (Figure 5.4.7H) in the Spitskop Member of the Schwarzrand Subgroup (Narbonne et al., 1997). The occurrence of these vendiobionta so high in the Nama succession indicates that the Ediacaran fossils had a substantial age range until approximately 6 Ma prior to the Precambrian-Cambrian boundary, that is near the beginning of the Cambrian explosion (Grotzinger et al., 1995). Apart from the skeletal fossil *Cloudina* (Figure 5.4.7C–F; Germs, 1972b; Grant, 1990), other skeletal fossils, such as *Namacalathus* (Figure 5.4.7A, B; Grotzinger et al., 2000; Watters and Grotzinger, 2001) and *Namapoikia*

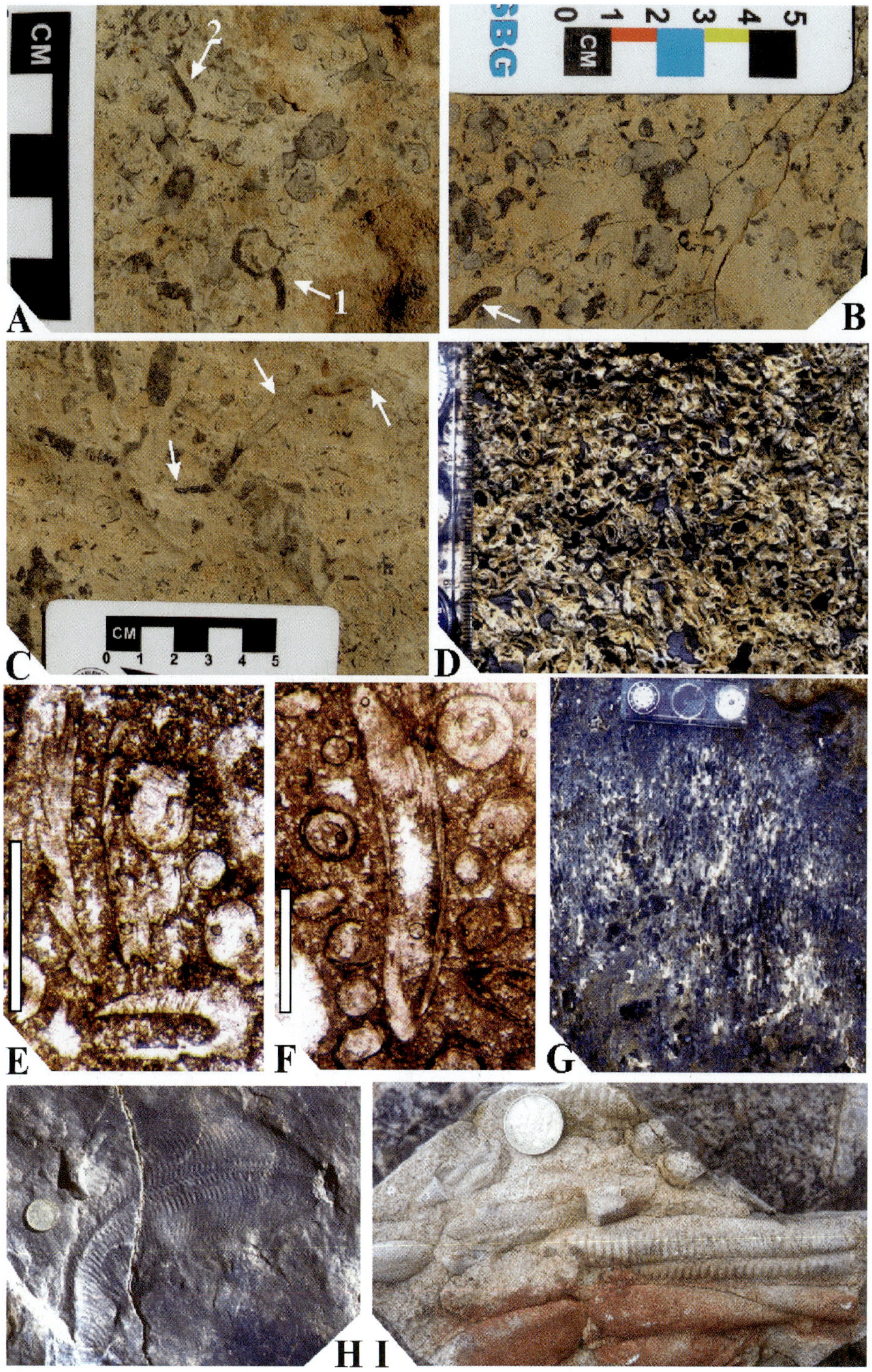


**Figure 5.4.7** *Cloudina*, *Namacalathus*, *Namapoikia* and *Pteridinium* from the Nama Group. (A) *Namacalathus hermanastes* Grotzinger et al. (2000) from the Omkyk Member at farm Omkyk/Zwartmodder, Namibia. Note (1) longitudinal section with stem and (2) co-occurring *Cloudina riemkeae* Germs (1972a). (B) Cross sections of *Namacalathus hermanastes*, same unit and locality as previous. Note radial symmetry of specimen in the centre and *Cloudina* tube (arrowed). (C) *Cloudina hartmannae* Germs (1972a), large sinuous tube (arrowed), same unit and locality as previous. (D) *Cloudina hartmannae* event accumulation, Omkyk Member at Driedoornvlakte buildup, Namibia. Scale in mm. (E, F) *Cloudina riemkeae* in thin section, Kuibis Subgroup. Note thick-walled specimens in E, thin-walled specimen in F and numerous cross sections. Scale bars equal 1 mm.
(G) *Namapoikia rietoogensis* Wood et al. (2002), a possible tabulate coral encrusting Omkyk carbonates at Driedoornvlakte buildup. (H) *Pteridinium carolinaense* St Jean (1973), Spitzkopf Member at farm Swartpunt (see Narbonne et al., 1997).
(I) *Pteridinium* event (storm) accumulation showing three-dimensional preservation, Kliphoek Member at Farm Plateau.

**Table 5.4.1** Stratigraphic distribution of fossils, stromatolites and thrombolites of the Nama Group: (a) northern Nama Basin, north of Osis, (b) southern Nama Basin, south of Osis, mainly after Germs (1983, 1995; Grotzinger et al., 2000; Wood et al., 2002; Geyer, 2005).

| Sub-group | Stratigraphy | | Ediacaran fossils | Skeletal fossils calcified meta-phytes | Trace fossils, biolaminites* | Organic-walled microfossils | Stromatolites thrombolites |
|---|---|---|---|---|---|---|---|
| | Formation | Member | | | | | |
| Fish River | Gross Aub | Deurstamp | | | *Trichophycus (Treptichnus) pedum* | | |
| | | Rosenhof | | | *Enigmatichnus africani, Gordia Planolites, Paleophycus, Treptichnus pollardi, Trichophycus pedum* | | |
| | Nababis | Haribes | | | *Skolithos* | | |
| | | Zamnarib | | | *Skolithos, Trichophycus tripleurum* | | |
| | Breckhorn | | | | | | |
| | Stockdale | Wasserfall | | | | | |
| | | Inachab | | | | | |
| Schwarzrand | Nomtsas | Niep | | | | | |
| | | Kreyrivier | | ? | *Curvolithos, ?Diplichnites, Neonereites biserialis, N. uniserialis, Trichophycus pedum, T. coronathum* | *Vendotaenia* | |
| | Urusis | Spitskop | *Pteridinium car. Swartpuntia germsi* | *Cloudina Namacalathus* | *Neonereites uniserialis, Strepthichnus narbonnei* | | *Acaciella or Kulparia* thrombolites |
| | | Feldschuh-horn | | | | | |
| | | Huns | *Cyclomedusa Nasepia* | *Cloudina* cal. *metaphytes* | *Brooksella, Curvolithos, ?Didymaulichnus* | *Chuaria circul. Leiosphaerids Vendotaenia* | *Acaciella* or *Kulparia Boxonia, Gymnosolen* or *Katavia,* thrombolites |
| | | Nasep | *Paramedusium ?Pteridinium* | | *Trichophycus (Treptichnus), Archaeichnium* | | |
| | Nudaus | Vingerbreek | *Pteridinium* | | | | |
| | | Niederhagen | *Pteridinium Rangea* | | *Arumberia** | | Flat concentric-type stromatolites |
| Kuibis | Zaris | Urikos | | | | | |
| | | Mooifontein | | *Cloudina* cal. *metaphytes* | | *Bavlinella fav ?Comasphaeri-dium* | *Conophyton*-like stromatolites thrombolites |
| | Dabis | Kliphoek | *Ernietta, Namalia Orthogonium, Beltanelliformis, Pteridinium, Rangea* | | *Bergaueria (Intrites), Buchholzbrunichnus kröneri, ?Skolithos Arumberia** | | |
| | | Mara | | *Cloudina* | | | |
| | | Kanies | | | | | |

*Notes: Trichophycus* is synonymous to *Treptichnus*. Some of the fossils, especially trace fossils, warrant re-examination.

(Figure 5.4.7G; Wood et al., 2002) as well as stromatolites, have also been found in Nama carbonates (see also Chapter 9.2).

Generally, simple trace fossils occur below the Nomtsas Formation, whereas more complex trace fossils are found in this and overlying formations (Germs, 1972c, 1983; Crimes and Germs, 1982; Geyer, 2005; see Chapter 9.3). However, some complex trace fossils have been found stratigraphically below the Nomtsas Formation in the Nasep Member (Germs, 1972c; Geyer and Uchman, 1995) and in the Spitskop Member in southern Namibia (Jensen and Runnegar, 2005). The occurrence of *Streptichnus narbonnei* in the Spitskop Member may indicate that the Ediacaran-Cambrian boundary may be in this member and not at the base of the Nomtsas Formation (Jensen and Runnegar, 2005).

The Rosenhof Member of the Gross Aub Formation is the major source of Fish River trace fossils. The trace fossils found in this member are *Trichophycus*, *Treptichnus* (*Trichophycus*) *pedum*, *Treptichnus* (*Trichophycus*) *pollardi*, *Paleophycus*, *Gordia*, *Skolithos* and *Enigmatichnus africani*. The diverse ichno-assemblages consist of traces of deposit feeders as well as others that were probably created by suspension feeders and the agrichnial farmers (Geyer, 2005).

A biologically impoverished assemblage of organic-walled microfossils occurs in the lower Nama Group (Germs et al., 1986; Table 5.4.1). Of special significance is the occurrence of *Vendotaenids* also in the Cambrian Nomtsas and Vergesig Formations of the uppermost Schwarzrand Subgroup (Geyer, 2005).

Thrombolites (clotted internal structures) and stromatolites (laminated internal structures) occur in carbonate deposits of the Kuibis Subgroup and in the southern Urusis Formation of the Schwarzrand Subgroup. These structures indicate that during the terminal Proterozoic (Ediacaran) colonisation of cyanobacterial mats by higher algae and metazoans was an important process in the formation of thrombolites (Grotzinger et al., 2000). Siliciclastic biolaminites are quite common in the Nama Group. These laminites are thinly laminated sediments resulting in the interplay of epibenthic biomass production in microbial mats, metabolic processes therein and sedimentation. Biolaminites are particularly widespread in the Precambrian due to the absence of predators. Elephant skin *Arumberia*-type biolaminites occur especially in the Kliphoek Member (Kuibis Subgroup) and Niederhagen Member (Schwarzrand Subgroup). Other types of biolaminites have also been reported from the Vingerbreek Member in the Nudaus Formation (Bouougri and Porada, 2007).

### 5.4.2.4.2. Gobabis–Witvlei Nama Group

The Nama Group in the Gobabis–Witvlei area (Figure 5.1.3) generally overlies unconformably the Witvlei Group, although it may be locally conformable (Hegenberger, 1993). There only the *Dabis and Zaris Formations* have been preserved. Both these formations accumulated in the Zaris sub-basin. The Dabis Formation is represented by the grey quartzite-containing Weissberg Member and the Zaris Formation by the Zenana Member (consisting of a basal carbonate and an upper quartzite) and the overlying Grünental Member. The shale-rich Grünental Member comprises clastic sedimentary rocks and lenses of black limestone containing the fossils *Cloudina* and *Namacalathus* (Germs, personal observations, 1973).

The carbonate-rich lower Zenana Member probably pinches out eastwards and the Grünental Member becomes more sandy in the same direction (Hegenberger, 1993). The Gobabis–Witvlei Nama Group was deposited in a fluvial to shallow marine environment. The Weissberg Member was deposited in a proximal fan to shallow clastic subtidal environment. The member thickens northwestwards and therefore was predominantly supplied from the southeast. The Zenana Member mainly accumulated in a sandy tidal to subtidal carbonate environment and the Grünental Member in a clastic fluvial to shallow subtidal clastic and carbonate environment.

The only fossils thus far found in the Nama Group of the Witvlei-Gobabis area are the skeletal fossils *Cloudina* and *Namacalathus* in the Grünental Member of the Buschmannsklippe Formation.

### 5.4.2.4.3. Vanrhynsdorp Group

The Vanrhynsdorp Group as defined by Gresse (1992) is exposed in the northwestern part of South Africa (Figure 5.1.3). It correlates with the classic Nama Group and was deposited in the Vanrhynsdorp sub-basin (Figure 5.4.3), which was separated by the Kamieskroon Ridge from the Witputs sub-basin. Sediments of the Vanrhynsdorp sub-basin differ from those in the northern Witputs and Zaris sub-basins, and they are therefore grouped into the Vanrhynsdorp Group and not in the Nama Group (Germs and Gresse, 1991; Gresse and Germs, 1993). The stratigraphic units of the Vanrhynsdorp Group and their generalised depositional environments are shown in Figure 5.4.8.

Similarly to the Nama Group, the sedimentary rocks of the Vanrhynsdorp Group display a change in colour stratigraphically upwards from generally greyish (Flaminkberg Formation) to greenish (Kwanous and Knersvlakte Subgroups) and finally reddish (Brandkop Subgroup) and in this order show an increase in immaturity. The *Flaminkberg Formation* is the oldest unit of the Vanrhynsdorp Group (Figure 5.4.8). It overlies basement gneiss of

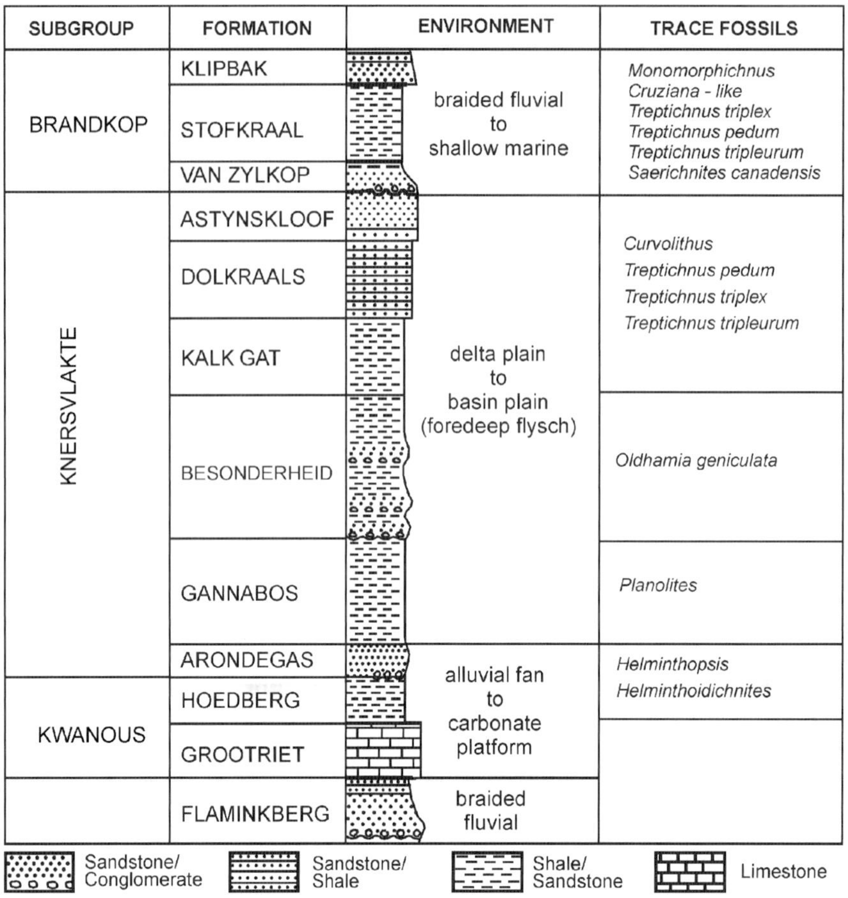


**Figure 5.4.8** Generalised lithostratigraphy, depositional environments and trace fossils of the Vanrhynsdorp Group (Gresse, 1992; Gresse and Germs, 1993; Buatois et al., 2007). Note that *Treptichnus* is synonymous to *Trichophycus*.

the Namaqua Belt and locally the Gifberg Group (Gariep Supergroup). It comprises coarse-grained, immature quartz sandstone and conglomerate. Shallow marine greyish quartz sandstone and greenish shale in the south (ca. 100 m thick) thin rapidly to braided fluvial reddish quartz sandstone, less than 5 m thick in the north, where they are overlapped by the Kwanous Subgroup and Arondegas Formation (Knersvlakte Subgroup). The *Kwanous Subgroup* consists of black limestones of the Grootriet Formation, overlain by wavy laminated, dark (pyritic) shales intercalated with limestone of the Hoedberg Formation. The *Knersvlakte Subgroup* most probably overlies unconformably the Kwanous Subgroup. Its base, the Arondegas Formation, grades from a residual (alluvial fan) arkose (overlying granite) in the northeast to turbiditic sandstone beds interfingering with shelf and basinal shales (Gannabos Formation) in the south and southwest. The Gannabos Formation is unconformably overlain by the Besonderheid Formation which contains turbiditic conglomerate, sandstone and shale. Silicified ash beds occur in both the Gannabos and Besonderheid Formations. The Besonderheid Formation fines upwards into the shalier Kalk Gat, Dolkraals and Astynskloof formations. The *Brandkop Subgroup* unconformably overlies the Knersvlakte Subgroup. Its base is partly defined by a large slump, coinciding with the base of the conglomeratic Van Zylkop Formation. The polymictic composition of the Van Zylkop channelised conglomerates, ranging from well-rounded quartz, chert and volcanic pebbles to flat shale shingles and the occurrence of faceted pebbles suggest a possible glacial origin in the source area (Von Backström, 1960). The basal Van Zylkop conglomerate is probably a correlate of the basal Nomtsas conglomerate of the Nama Group. The remainder of the Brandkop Subgroup comprises reddish shale, siltstone and sandstone (Gresse, 1992; Gresse and Germs, 1993; Gresse et al., 1996).

As with the Nama Group, during deposition of the Vanrhynsdorp Group, a reversal of palaeocurrents took place. The basal Flaminkberg sands were transported from the Kalahari palaeocontinent in the east, whereas the Gannabos and Besonderheid and younger sediments were supplied from the Gariep Belt in the west. Evidence of a rising northwestern tectono-metamorphic source is provided by extra- and intraformational conglomerates in the Besonderheid Formation of the Knersvlakte Subgroup.

The Flaminkberg Formation was deposited on a coastal alluvial braid-plain sloping towards the southwest and the carbonate-containing Kwanous Subgroup in an oolite barrier/lagoonal environment. The sedimentary rocks

of the unconformably overlying Arondegas Formation at the base of the Knersvlakte Subgroup represent alluvial fan deposits. A change in depositional slope to the southeast preceded deposition of the turbidites of the Besonderheid Formation (Knersvlakte Subgroup), which were derived from a northwestern source (Gresse, 1992; Gresse and Germs, 1993). The Besonderheid Formation accumulated on a continental slope and/or basin plain. After deposition of the turbiditic Besonderheid Formation, the Vanrhynsdorp sub-basin shallowed, resulting in the accumulation of the overlying Kalk Gat, Dolkraals and Astynskloof Formations of the upper Knersvlakte Subgroup and of the Brandkop Subgroup in southeastward prograding delta fan systems. Abundant mud cracks and ripple marks in the reddish upper Brandkop Subgroup indicate shallow water and subaerial exposure on the delta plain.

The Flaminkberg Formation and the overlying Grootriet and Hoedberg Formations of the Kwanous Subgroup are Ediacaran in age as they are correlatives of the Late Ediacaran Kuibis and lower Schwarzrand Subgroups (Figure 5.4.3). The trace fossils occurring in the Vanrhynsdorp Group (Figure 5.4.8) will be discussed in Chapter 9.3. They indicate that from the base of the Besonderheid Formation (Knersvlakte Subgroup) stratigraphically upwards the Vanrhynsdorp Group is Cambrian in age (Germs, 1983; Gresse, 1992; Gresse and Germs, 1993; Buatois et al., 2007). The age of the Gannabos and Arondegas Formations below the Besonderheid Formation is uncertain. No micro- or body fossils have been found in the Vanrhynsdorp Group.

### 5.4.2.5. The Malmesbury/Boland Basin

The main exposure of the western branch of the Saldania Belt north and northeast of Cape Town has been known as the Malmesbury Group in spite of the division of that area into three tectonic domains (Hartnady et al., 1974) or terranes (Von Veh, 1983). These are from the northeast to the southwest the Boland, Swartland and Tygerberg Terranes. It remains, however, uncertain whether these fault-bounded blocks represent true allochthonous terranes. At least the latter appears allochthonous and is separated from the others by a major shear zone, the Colenso Fault. The Boland "Terrane" is most likely autochthonous on Mesoproterozoic metamorphic basement of the Namaqua–Natal Belt and is therefore referred to as the Boland Zone in the following. The tectonic position of the intervening Swartland "Terrane" is uncertain and this will be referred to as the Swartland Zone (Figure 5.1.5). It is also uncertain whether the low-grade metasedimentary rocks of the Boland Zone (Boland Subgroup according to SACS, 1980) are of the same age or younger than those of the more complex, polydeformed Swartland Zone (Swartland Subgroup). In a revised stratigraphic scheme for the western Saldania Belt, Belcher and Kisters (2003) challenged the above terrane model and distinguished across all three tectonic units between an older, redefined Swartland Group and the younger Malmesbury Group. According to these authors, both groups are separated by an inferred unconformity and an older deformation phase is recognised only in the Swartland Group. This interpretation is, however, questionable because a critical outcrop that displays this alleged older deformation phase in the Swartland Group is most likely part of the Mesoproterozoic basement of the Namaqua–Natal Belt (H.E. Frimmel, unpublished data, 2008). Thus, the lithostratigraphy of the Saldania Belt remains controversial.

The Malmesbury Group, as defined by the South African Commission for Stratigraphy and presented by Gresse et al. (2006), includes the *Tygerberg Formation*, which is the only stratigraphic unit in the westernmost tectonic unit, the Tygerberg Terrane. The formation consists of rhythmic turbiditic alternations of greywacke, phyllitic shale and siltstone and immature quartzite with some thin impure carbonate and conglomerate beds. A local volcanic succession, the Bloubergstrand Member, occurs near Cape Town. The metavolcanic rocks of this member consist of fine red tuffs and dark red-brown to green amygdaloidal andesitic lavas (Von Veh, 1983). Detrital zircon ages from the Tygerberg Formation are mainly within the 1,050–900 Ma bracket, indicating a Namaqua–Natal Belt provenance, and within the 700 and 575 Ma range (Armstrong et al., 1998). The latter range is similar to detrital zircon ages obtained for the upper Oranjemund Group in the Marmora Terrane (Basei et al., 2005). No source rocks of this age range are known from the potential hinterland in South Africa, and by analogy with the interpretation of the Oranjemund data, the source is inferred to be in the Dom Feliciano Belt of South America. The youngest detrital zircon age of approximately 560 Ma provides a maximum constraint for the time of sedimentation, whereas 552–540 Ma granites of the Cape Granite Suite (see Section 5.6.4), which are intrusive into the Tygerberg Formation metasedimentary rocks, set a minimum constraint. Thus, sedimentation of this formation postdates the first phase of orogenic deformation but predates the main phase of continental collision and associated magmatism at ca. 540 Ma. The extent and shape of the depositional basin is unknown, but it may well be possible that the basin linked up with the Holgat Basin to the north. The Tygerberg Formation experienced, however, a different tectonic history to that of the Vanrhynsdorp Group and its correlate, the Boland Subgroup both of which were deformed along NNW-SSE trending fold axes sometime between 500 and 480 Ma.

The central Swartland Zone or Terrane is composed of the Berg River, Klipplaat and Moorreesburg Formations, which together constitute the Swartland Subgroup (Gresse et al., 2006). Due to a lack of primary contacts, bedding and way-up criteria, the stratigraphic relationships between these units remain unresolved. The most important rock types in these formations are chlorite schist, quartz schist, phyllite and impure limestone lenses. The metavolcanic Bridgetown Formation occurs at the boundary between the Swartland and Boland zones. It is a highly sheared succession of greenschist with minor serpentinite, dolomite, chert and graphitic schist. A tholeiitic composition of within-plate affinity characterises the mafic rocks (Slabber, 1995).

Due to lithological resemblance between the Flaminkberg Formation and overlying Kwanous Subgroup of the Vanrhynsdorp Group, Belcher and Kisters (2003) suggested that the siliciclastic rocks in the Boland Zone are correlates of the approximately 550–543 Ma lower Vanrhynsdorp Group (Figure 5.4.9). In the Boland Subgroup, rocks are described as the Piketberg (predominantly highly foliated conglomerate, feldspathic grit, feldspathic

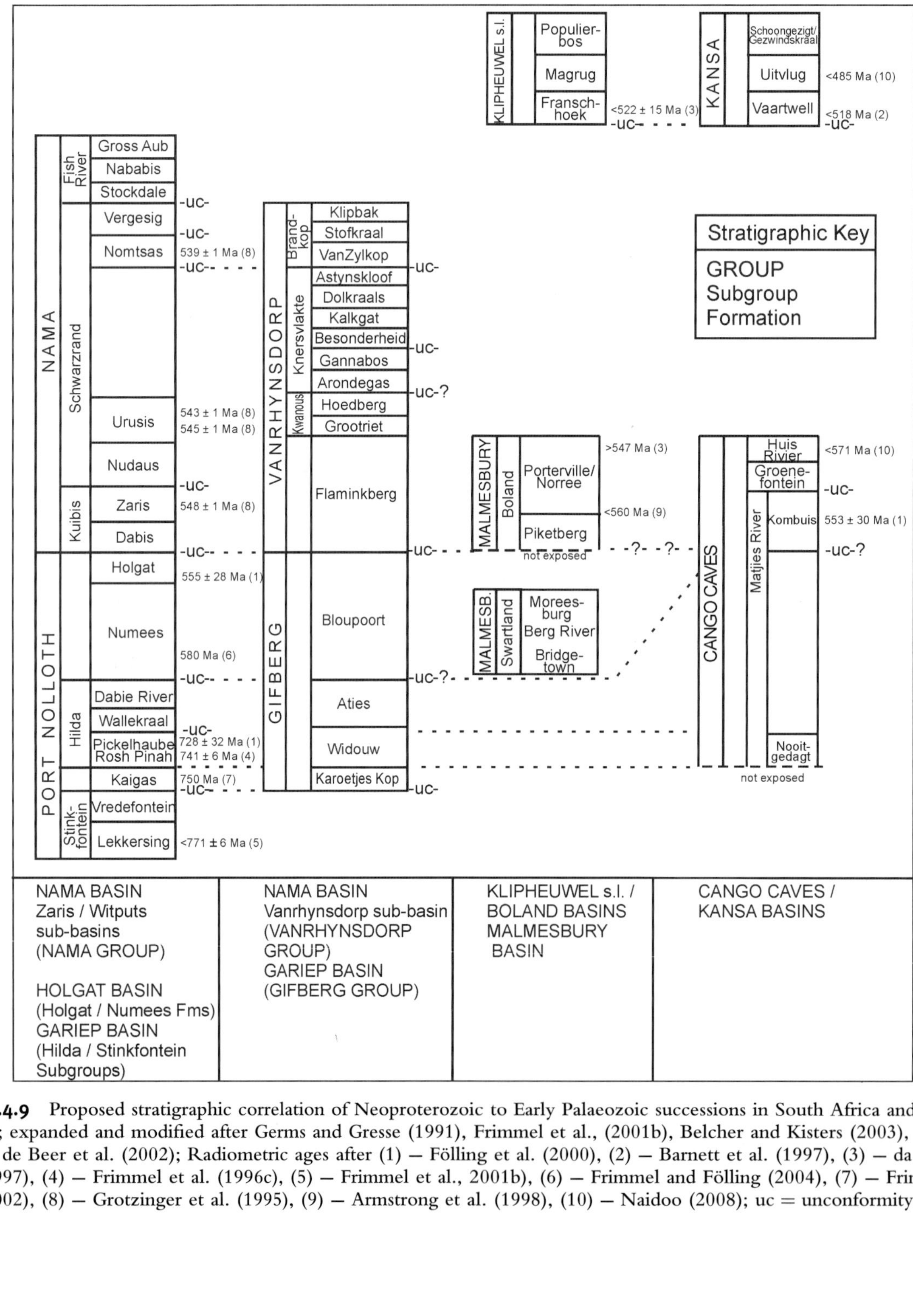


**Figure 5.4.9** Proposed stratigraphic correlation of Neoproterozoic to Early Palaeozoic successions in South Africa and southern Namibia; expanded and modified after Germs and Gresse (1991), Frimmel et al., (2001b), Belcher and Kisters (2003), Frimmel (2008a), de Beer et al. (2002); Radiometric ages after (1) – Fölling et al. (2000), (2) – Barnett et al. (1997), (3) – da Silva et al. (1997), (4) – Frimmel et al. (1996c), (5) – Frimmel et al., 2001b), (6) – Frimmel and Fölling (2004), (7) – Frimmel et al. (2002), (8) – Grotzinger et al. (1995), (9) – Armstrong et al. (1998), (10) – Naidoo (2008); uc = unconformity.

quartzite, greywacke and sericite schist) and Porterville (mainly of phyllitic shale and greywacke with quartzite, limestone and chert in the lower part) Formations. A greenstone unit, the Voëlvlei unit, occurs locally in the Porterville Formation. In the area around Worcester, slightly different facies have been distinguished as Norree (older) and Brandwacht (younger) Formations. The latter formation contains the Brewelskloof Andesite Member.

## 5.4.2.6. The Klipheuwel *s.l.* Basin

In the southern part of the Swartland Zone or Terrane, rocks of the Swartland Group are overlain unconformably by feldspathic conglomerate and sandstone, slate and phyllite all of which are unified as *Franschhoek Formation* (Theron et al., 1992). Cross bedding and grading are preserved in sandstones, whereas polymictic conglomerate contains orientated and elongated pebble- to boulder-size clasts of vein quartz, quartzite, chert, phyllite, greywacke, hornfels, granite, quartz porphyry, jasper and volcanic rocks, some apparently derived from the underlying Malmesbury Group and Cape Granite Suite. The siliciclastic rocks of this formation, which are ascribed to fluvial sedimentation, unconformably overlie and thus postdate the Cape Granite Suite (Dunlevey, 1983). They show intense deformation that is comparable to that displayed by the Malmesbury Group rocks (Hartnady, 1969).

Elsewhere, the Boland Zone successions and the Cape Granite Suite are overlain unconformably by typical red bed successions of up to 2,000 m thickness, the Magrug Formation (Figure 5.4.9). The formation grades up from white to reddish conglomerate, grit and coarse-grained sandstone near the base to fine- and medium-grained sandstone that is intercalated with purple siltstone and shale. The conglomerates contain clasts derived from the Malmesbury Group and Cape Granite Suite and also many volcanic and volcaniclastic clasts. The overlying Populierbos Formation consists of at least 50 m of red and purple shale with dewatering structures and scattered lone-stones. An alluvial fan-lacustrine palaeoenvironment appears feasible for these sedimentary rocks. In places the siliciclastic rocks are folded and contain a northwest-striking subvertical cleavage.

The Magrug and Populierbos Formations constituted the original Klipheuwel Group, but Belcher and Kisters (2003) included the Franschhoek Formation as the lowest unit of their extended Klipheuwel Group (Figure 5.4.9). The age relationship between the Franschhoek Formation and the two formations of the original Klipheuwel Group remain uncertain, however.

The overall upward-fining succession of the original Klipheuwel Group in the type area (Klipheuwel–Klapmuts) contrasts with an overall upward-coarsening succession of green and red mudstone, siltstone, sandstone and conglomerate in an area further south (Elands Bay). This latter sequence represents progradation of distal braided alluvial plain sediments onto deltaic and marine shelf sediments (Vos and Tankard, 1981), seemingly in-facies with the unconformable basal Piekenier Formation of the overlying Table Mountain Group.

## 5.4.2.7. The Cango Caves Basin

The Neoproterozoic *Cango Caves Group*, previously named Goegamma Subgroup of the Kansa Group (Le Roux and Gresse, 1983), occurs in the southern branch of the Saldania Belt and accumulated in the Cango Caves Basin. The group is predominantly a carbonate-clastic turbidite succession. Chemostratigraphy (Frimmel et al., 2001a; Fölling and Frimmel, 2002) and organic-walled microfossils (Gaucher and Germs, 2006) indicate that the Cango Caves Group is Neoproterozoic in age. From base to top, the Cango Caves Group includes the Matjies River, Groenefontein and Huis Rivier Formations (Le Roux and Gresse, 1983; Figure 5.4.9).

The *Matjies River Formation* comprises shale, greywacke and limestone of the Nooitgedagt Member and the overlying predominantly carbonate-containing Kombuis Member. The Nooitgedagt Member was deposited as coarsening upward deltaic sediments and shallow marine deposits. The carbonate-rich Kombuis Member predominantly accumulated on a shelf. Both members are sourced in the east (Le Roux, 1997). The latter author considers the formation to represent a continuous sedimentary succession without a major unconformity. However, C and Sr isotope chemostratigraphy (Fölling and Frimmel, 2002; Figure 5.3.7) and a difference of 100°C in thermal overprint between the Nooitgedagt and Kombuis Members (Frimmel et al., 2001a) point to a pre-Ediacaran (>630 Ma) age for the Nooitgedagt Member and an Ediacaran age (<630 Ma) for the rest of the Cango Caves Group, and the existence of an unconformity between the two members. The only available radiometric age for this formation is a Pb-Pb carbonate age of $553 \pm 30$ Ma for the Kombuis Member limestone (Fölling et al., 2000), which is characterised by a negative $\delta^{13}C$ excursion. This is in good agreement with the carbonates of the Holgat Formation, but differs drastically from those of the underlying Nooitgedagt Member. The latter yielded a marked positive $\delta^{13}C$ excursion to ratios as high as +9.5‰ (Figure 5.3.7).

An unconformity follows at the base of the overlying turbiditic *Groenefontein Formation*, which consists mainly of sandstone and shale with some limestone lenses (Rozendaal et al., 1999). This change in depositional environment from the Kombuis carbonate platform deposits to the clastic turbiditic Groenefontein sediments reflects a deepening of the Cango Caves Basin. Consequently, it is understandable that the sharp contact between the two units previously has been interpreted as a subaqueously formed drowned unconformity (Rozendaal et al., 1999). However, the finding of up to 150-m-deep and 400-m-wide karstic features associated with this unconformity between the Kombuis Member and the Groenefontein Formation suggests that it was subaerially formed and therefore does not represent a drowned unconformity (Praekelt, 1996). Thus, it may be possible that this unconformity formed during the same drop in sea level that caused the Vingerbreek unconformity in the Nama Basin as a result of a minor terminal Neoproterozoic, ca. 547 Ma glaciation.

A proposed marine condensed section in the Groenefontein Formation signifies a change from transgression to regression and deposition of the upward-coarsening turbiditic Huis Rivier Formation (Gresse et al., 1996). The *Huis Rivier Formation* only occurs in the western part of the Kango Inlier. It most probably was eroded prior to deposition of the Kansa Group in the eastern part of the inlier. This formation is younger than 571 Ma as indicated by the youngest detrital zircon age (Naidoo, 2008). The succession displays mega-flute casts and other prominent sole marks as well as $T_a$-type Bouma sequences (Le Roux and Gresse, 1983). The Huis Rivier Formation was deposited in a continental slope or basal plain environment with sediment supply from southwestern and perhaps northwestern sources (Le Roux and Gresse, 1983).

Up to now only one possible body fossil has recently been found in the basal Groenefontein Formation of the *Cango Caves Group* on farm Groenefontein (Praeckelt, personal communication, 2006) but no trace fossils have been found yet. Three different, low-diversity microfossil assemblages can be distinguished in the Cango Caves Group (Gaucher and Germs, 2006):

(i) A *Bavlinella*-dominated assemblage in the Nooitgedagt Member (Figure 5.4.1I,J). This interval is lithologically characterised by organic-rich limestones, marls and shales and by great abundance of microfossils.
(ii) A *Soldadophycus*-dominated assemblage mainly occurring in the Kombuis Member (Figure 5.4.1N). This interval is characterised by the transition from carbonates into shales.
(iii) A *Leiosphaeridia*-dominated assemblage in the Groenefontein and Huis Rivier Formations (Figure 5.4.1Q). These units are characterised by shale and sandstone interbeds mainly of turbiditic origin.

The assemblages that most resemble the microflora of the Cango Caves Group are those of SW Gondwana, that is those of the Gamtoos Group (Gaucher and Germs, 2006), Nama Group and Holgat Formation of Namibia (Germs et al., 1986; Gaucher et al., 2005a); Arroyo del Soldado Group of Uruguay (Gaucher, 2000); Corumbá Group of Brazil (Gaucher et al., 2003) and the Eleutério, Pico de Itapeva and Pouso Alegre basins of Brazil (Teixeira and Gaucher, 2004). All of these successions are Late Ediacaran, post-Gaskiers in age ($582 \pm 0.4$ Ma; Bowring et al., 2003b; Gaucher and Germs, 2006; see Chapter 9.1).

Another Neoproterozoic succession, the *Kaaimans Group*, is exposed in a separate unroofed mega-anticline of the Cape Fold Belt, the Kaaimans Inlier, further south, along the south coast in the George area (Figure 5.1.3). It consists of a highly tectonised, south-dipping, tectono-metamorphic succession of quartzite, quartz schist, phyllite and minor impure carbonate rocks. Although Gresse (1983) distinguished different formations, their stratigraphic position remains unknown and, based on an inverted metamorphic gradient, it has been suggested that the whole succession might be overturned (Frimmel and van Achterbergh, 1995). The group was intruded by late- to post-orogenic granite of the Cape Granite Suite (George and Woodville Plutons), and both the Kaaimans Group rocks and the granites experienced northwards-directed compression and thrusting prior to the Cape Orogeny. The entire Kaaimans Group may well represent a separate terrane that was accreted onto the southern margin of Gondwana towards the end of the Pan-African orogenic episode.

### 5.4.2.8. The Kansa Basin

After the Cango Caves Group was folded along NNW-SSE trending axes, uplift took place and the Kansa Basin was formed in which the Kansa Group sediments accumulated. The *Kansa Group* (Figure 5.4.9) commences with the Vaartwell Formation, which is younger than 518 Ma as indicated by the youngest detrital zircon age (Barnett et al., 1997). The *Vaartwell Formation* contains conglomerates of the Rietkloof and overlying Andriesberg members. Both of these members were derived from sources in the southwest. The unstratified polymictic Rietkloof conglomerates are composed of a large percentage of granite, gneiss and chert pebbles and cobbles (ca. 50–60%), set in an argillaceous matrix. They also include some shale, greywacke and sandstone/quartzite clasts, possibly from the Groenefontein Formation, but almost no vein quartz. In contrast, the Andriesberg Member is composed almost exclusively of white vein quartz. Locally, the two members grade into each other. The Vaartwell Formation conglomerates fine up into sandstone and shale of the Uitvlug and Schoongezicht/Gezwinds Kraal Formations. The latter two formations were deposited by renewed delta incursions from the east.

The *Uitvlug Formation* consists of cross-bedded fine to coarse-grained greywacke, sandstone and subarkose that are greenish when fresh. Tabular cross bedding dominates and ripple marks are rare. Thin intercalated shales and grits occur. The youngest detrital zircon grains have an age of ca. 485 Ma and thus indicate an Ordovician age for this formation (Naidoo, 2008). The *Gezwinds Kraal Formation* forms a gradational lower contact with the Uitvlug Formation. The Uitvlug Formation rocks weather cream-white, whereas those of the Gezwinds Kraal Formation show a reddish-brown colour. The Gezwinds Kraal Formation comprises quartz wacke with shale. Shale is more prominent and sandstone is finer grained up section. The Gezwinds Kraal is the eastern facies of the Schoongezigt Formation and distinguished by the presence of quartz pebbles. The *Schoongezigt Formation* is more massively bedded than the Gezwinds Kraal Formation and contains conglomeratic lenses with predominantly quartz wacke. Conglomerate and sandstone with thin shale intercalations of the *Schoemans Poort Formation* overlie the Kansa Group with an angular unconformity and are, in turn, overlain unconformably by the Table Mountain Group (Le Roux and Gresse, 1983).

### 5.4.2.9. The Gamtoos Basin

The lithostratigraphy of the *Gamtoos Group* was originally established by Amm (1935) and later reinterpreted and redescribed by Frankel (1937), Haughton et al. (1937), Bell (1980) and Hill and Nolte (1989). The latter subdivided the group into four formations. These are from bottom to top the Lime Bank, Kleinrivier, Kaan and Van Stadens Formations. The lower boundary of the Gamtoos Group is not exposed, whereas its upper boundary with the Sardinia Bay Formation is faulted or unconformable (Le Roux, 2000). The structural complexity of this group, including many thrust faults and ubiquitous recumbent folds, and poor exposures make it difficult to estimate the thickness of the formations reliably. For this reason no stratigraphic columns of the Gamtoos Group, indicating thicknesses, have yet been published.

The group is characterised by predominantly calcareous units (Lime Bank and Kaan formations) and predominantly arkosic and pelitic units (Kleinrivier and Van Stadens formations; Le Roux, 2000). Phyllite is the rock type common to all the individual formations. The carbonates are interpreted as shallow marine deposits over which alluvial fans and turbidites accumulated as the basin margin was rapidly uplifted. The arkosic rocks are tentatively interpreted as shallow marine deposits.

It has been suggested that the commonly used lithostratigraphic units of the Gamtoos Group represent thrust-bounded tectono-stratigraphic and not true lithostratigraphic units (Shone et al., 1990). This has been confirmed by recent organic-walled microfossil studies of the group. Thus far, no body and trace fossils have been found in the *Gamtoos Group* but various organic-walled microfossils have been discovered (Gaucher and Germs, 2006). Similar to its stratigraphic correlate, the Cango Caves Group, three different low diversity assemblages of microfossils have been found in the Gamtoos Group:

(i) A *Bavlinella*-dominated assemblage in the Lime Bank Formation and most of the Kaan Formation (Figure 5.4.1G,H). These intervals are lithologically characterised by organic-rich limestone, marl and shale, and by many microfossils.
(ii) A *Soldadophycus*-dominated assemblage that occurs mainly in the upper Kaan Formation and lower Van Stadens Formation. This interval is characterised by the transition from carbonates into shales.
(iii) A *Leiosphaeridia*-dominated assemblage in the Kleinrivier Formation (Figure 5.4.1P). This formation is characterised by shale and sandstone, mainly of turbiditic origin.

## *ACKNOWLEDGEMENTS*

This work summarises results obtained in the course of many research projects that received support from a variety of institutions and funding agencies, especially the Geological Survey of Namibia and the South African National Research Foundation (NRF). G. Germs acknowledges financial support from the Sasol Fund of the Faculty of Science of the University of Johannesburg. C. Gaucher is indebted to Anton and Susan Horn for their generous hospitality and for providing access to fossiliferous outcrops at Farm Zwartmodder. G.J.B. Germs and C. Gaucher thank Rogert Swart at NAMCOR (Windhoek) for providing access to core material of ST-1 borehole. A. Tankard and M. Okrusch are thanked for providing helpful reviews of the original manuscript.

CHAPTER 5.5

# Orogenic Tectono-Thermal Evolution ☆

Thomas M. Will[1], Roy McG. Miller[2] *and* Hartwig E. Frimmel[1,3]

## Contents



## 5.5.1. The Kaoko Belt

Early reconnaissance studies in the Kaoko Belt (Guj, 1970; Miller, 1979, 1983a; Porada, 1979) have recently been augmented by much more detailed work (Dingeldey et al., 1995; Dürr and Dingeldey, 1996; Prave, 1996; Franz et al., 1999; Seth et al., 1998, 2000, 2002; Stanistreet and Charlesworth, 2001; Passchier et al., 2002; Goscombe et al., 2003a,b, 2005a,b; Will et al., 2004; Kröner et al., 2004; Masberg et al., 2005; Konopásek et al., 2005; Gray et al., 2006; Goscombe and Gray, 2007; Jung et al., 2007).

Several authors (e.g. Dürr and Dingeldey, 1996; Trompette, 1997; Alkmim et al., 2001; Goscombe et al., 2003a; Will et al., 2004; Konopásek et al., 2005; Gray et al., 2006; Goscombe and Gray, 2007) have proposed that the Kaoko Belt is the result of oblique convergence between the Rio de la Plata Craton in the west and the Congo Craton in the east. The associated strain partitioning caused the observed, probably, crustal-scale sinistral shear zones and the east-southeast verging nappes and thrusts in the Kaoko Belt. According to Goscombe et al. (2005a), transpression occurred some 580–550 Ma ago. Downing and Coward (1981) and Coward (1983) recorded structural evidence pointing to low-angle, transpressive convergence in the southern Central Zone (sCZ) and Southern Margin Zone (SMZ) of the Damara Belt. $D_1$ structures in the sCZ are cut by 565 Ma granites and collision occurred at 542 Ma (Miller, 2008). The timing of the initiation of convergence in the Damara Belt is uncertain but it followed that in the Kaoko Belt by only a few million years (Miller, 2008).

The tectono-stratigraphic zones of the Kaoko Belt and their lithological make-up are shown in Figure 5.5.1. The *Eastern Kaoko Zone* (EKZ) is the carbonate-dominated platform and molasse foreland and consists of very low-grade, upright folded Neoproterozoic sedimentary rocks that overlie the western margin of the Congo

☆ Will, T.M., Miller, R. McG., Frimmel, H.E., 2009. Orogenic tectono-thermal evolution. Neoproterozoic to Early Palaeozoic evolution of Southwestern Africa. In: Gaucher, C., Sial, A.N., Halverson, G.P., Frimmel, H.E. (Eds.): Neoproterozoic-Cambrian tectonics, global change and evolution: a focus on southwestern Gondwana. Developments in Precambrian Geology, 16, Elsevier, pp. 205–218.

[1] Geodynamics and Geomaterials Research Division, University of Würzburg, Am Hubland, D-97074 Würzburg, Germany.
[2] PO Box 11222, Windhoek, Namibia.
[3] Department of Geological Sciences, University of Cape Town, Rondebosch 7700, South Africa.

*Developments in Precambrian Geology*, Volume 16
ISSN 0166-2635, DOI 10.1016/S0166-2635(09)01614-4

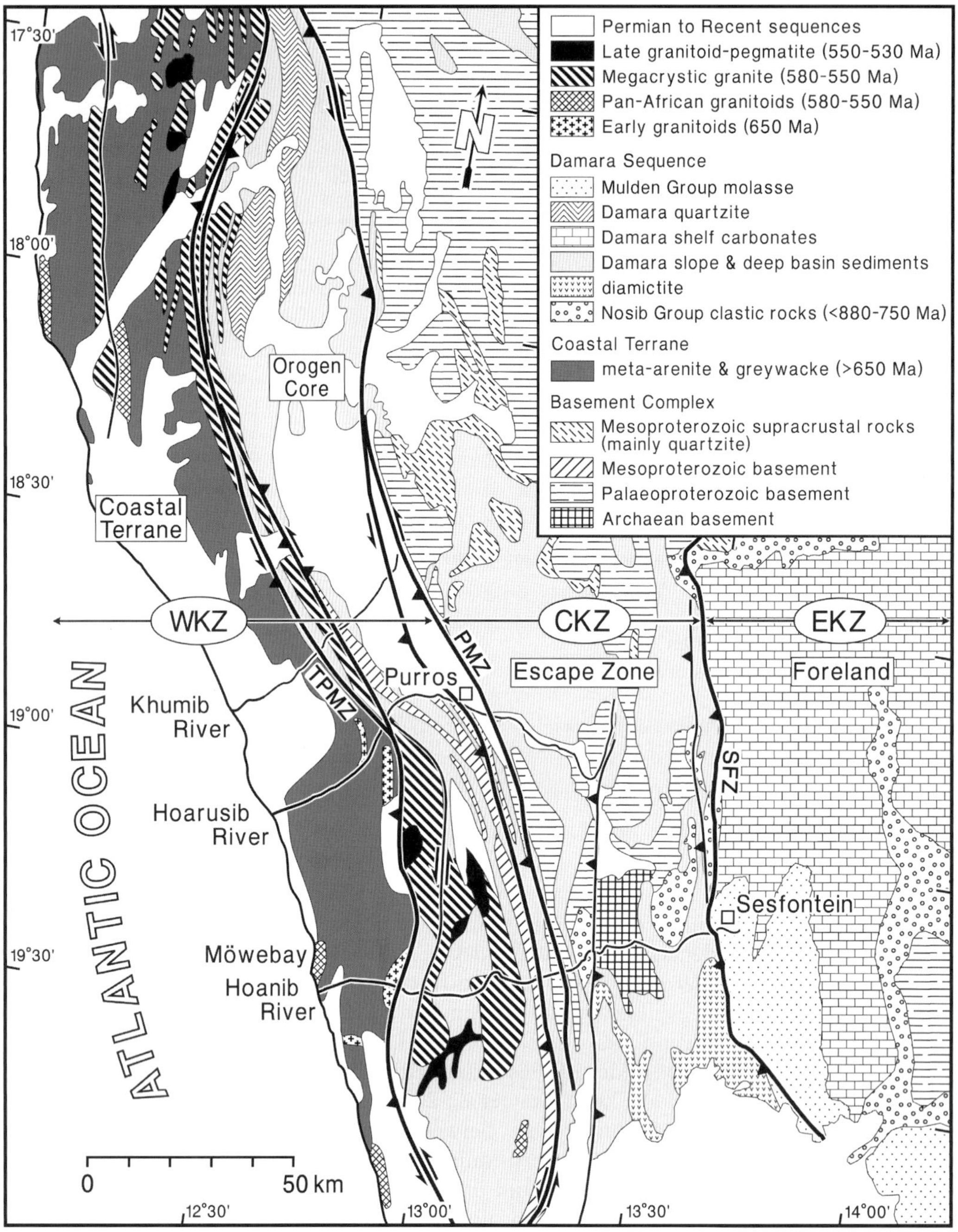


**Figure 5.5.1** Lithological make-up of the various tectono-stratigraphic zones of the Kaoko Belt; CKZ – Central Kaoko Zone, EKZ – Eastern Kaoko Zone, WKZ – Western Kaoko Zone, PMZ – Purros Mylonite Zone, SFZ – Sesfontein Shear Zone (or Sesfontein Thrust), TPMZ – Three Palms Mylonite Zone.

Craton. The western margin of the EKZ is defined by the Sesfontein Thrust (ST), a shallowly west-dipping fault zone, which formed under brittle-ductile conditions at the end of the Pan-African orogeny (Goscombe et al., 2003a).

The *Central Kaoko Zone* (CKZ), termed 'Escape Zone' by Goscombe et al. (2005b), contains alternating metapelitic and metapsammitic Neoproterozoic slope and deep basin deposits overlying Archaean to Palaeoproterozoic granitic gneisses that are locally intruded by gneissose Palaeo- to Mesoproterozoic granites (Seth et al., 1998). Pan-African granitoids have not been found in the CKZ. From east to west there is a continuous increase in metamorphic grade ranging from lower greenschist- to upper amphibolite-facies conditions, typical of Barrovian-type metamorphism (Gruner, 2000; Goscombe et al., 2003b, 2005b; Will et al., 2004). Both the basement and the sedimentary cover experienced an intense fabric development during transpressional, sinistral, northwest-over-southeast shearing and they are tightly folded into a series of E-vergent thrusts and nappes (Guj, 1970; Dingeldey et al., 1995; Dürr and Dingeldey, 1996; Goscombe et al., 2003a;

Konopásek et al., 2005). The CKZ is bounded by the ST in the east and the Purros Mylonite Zone (PMZ) in the west.

The *Western Kaoko Zone* (WKZ) is made up of the Orogen Core in the east and the Coastal Terrane in the west and consists of upper amphibolite- to granulite-facies metasedimentary rocks, Neoproterozoic granitoids and slivers of Mesoproterozoic basement rocks. Locally it is migmatised. Early syn-tectonic to late syn-tectonic granitoids range from 580 to 550 Ma in age (Seth et al., 1998; Goscombe et al., 2005b). Sinistral, crustal-scale shearing and isoclinal folding caused an intense wrench-style deformation in all rocks (Goscombe et al., 2003a).

The *Orogen Core* between the PMZ and the Three Palms Mylonite Zone (TPMZ) contains Mesoproterozoic basement slivers and Damaran cover rocks. It is a composite of three domains with distinct lithostratigraphy and tectono-metamorphic evolution (Goscombe et al., 2005b). The latter authors distinguish between the Hartmann and Hoarusib Domains in the northern and the southern parts of the Orogen Core. Both consist of strongly deformed upper amphibolite- to granulite-facies metamorphic rocks. The Khumib Domain lies between these two high-grade areas and comprises chevron-folded, lower amphibolite-facies turbidites.

The *Coastal Terrane*, the westernmost part of the Kaoko Belt, is an intimate mixture of upper amphibolite- to granulite-facies supracrustal paragneisses and orthogneisses (Seth et al., 1998; Franz et al., 1999; Goscombe and Gray, 2007). As pointed out in Chapter 5.3, the supracrustal rocks are distinct from the Damara Supergroup elsewhere in the Kaoko Belt and are dominated by migmatised meta-greywacke and feldspathic metapsammitic rocks. Carbonates and quartzites are not present. The Pan-African igneous rocks include metamorphosed and deformed calc-alkaline I-type granitoids of intermediate composition (Seth et al., 1998; Masberg et al., 2005) and unmetamorphosed but pervasively sheared, S-type granites (Goscombe and Gray, 2007). Granitoid intrusion occurred between 650 and 630 Ma ago and between 580 and 540 Ma ago (Seth et al., 1998; Franz et al., 1999; Goscombe et al., 2005a), the last age being that of post-tectonic intrusions (Miller, 2008). The older granitoids occur in the Coastal Terrane only, whereas the younger granitoids occur in both the Orogen Core and the Coastal Terrane. The early Neoproterozoic high-grade thermal event at 650 Ma is a unique feature of the Coastal Terrane and is conspicuously absent in the rest of the Kaoko Belt.

The Kaoko Belt contains evidence of three distinct orogenic phases during a protracted Neoproterozoic to Cambrian history (Dingeldey et al., 1995; Dürr and Dingeldey, 1996; Passchier et al., 2002; Goscombe et al., 2003b, 2005b; Konopásek et al., 2005). The latter authors recognise three metamorphic events and distinguish between a 'thermal phase', $M_1$, at 660–640 Ma, a 'transpressional phase', $M_2$, at 580–550 Ma and a 'shortening phase' $M_3$ between 540 and 510 Ma ago. $M_1$ calc-alkaline magmatism and metamorphism are restricted to the Coastal Terrane, whereas $M_2$ and $M_3$ affected the entire Kaoko Belt.

### 5.5.1.1. Structure of the Kaoko Belt

The gross geometry of the Kaoko Belt is dominated by the two major crustal-scale sinistral oblique shear zones that form the borders to the Orogen Core, the TPMZ in the west and the PMZ in the east (Figure 5.5.1). The significance of the PMZ was recognised by Guj (1970) and was documented further by a series of structural and petrological studies (e.g. Dürr and Dingeldey, 1996; Stanistreet and Charlesworth, 2001; Goscombe et al., 2003a,b; Will et al., 2004; Konopásek et al., 2005), whereas the importance of the TPMZ has only been recognised recently (Goscombe et al., 2003a, 2005b; Goscombe and Gray, 2007).

The PMZ is a north- to NNW-trending high-strain ($\gamma > 10$), 4–5 km wide, ductile mylonite and ultramylonite zone with a lateral displacement well exceeding 50 km (Goscombe et al., 2003a). The shear zone is dominated by a subvertical high-temperature, upper amphibolite-facies metamorphic foliation associated with NNE trending subhorizonal to shallow N-plunging mineral and stretching lineations. Miller (1983a) suggested that the Ogden Rocks mylonites, bounding the western edge of the Southern Kaoko Zone, represent the southern continuation of the PMZ. Goscombe et al. (2003a) noted that the orientation of the PMZ and its mylonitic L-S fabric are parallel to, and continuous into, the adjacent regions of the CKZ and WKZ. Sense-of-shear indicators reveal a compressional obliquely reverse, sinistral shearing along shallow N-plunging lineations (Dürr and Dingeldey, 1996; Goscombe et al., 2005b; Konopásek et al., 2005).

The TPMZ is a north- to NNW-trending anastomosing array of high-strain mylonite and ultramylonite zones within the WKZ, separating the Orogen Core from the Coastal Terrane. Dip and dip directions change along strike. In the south the TPMZ dips at shallow to moderate angles (10–65°) to the west, in the central NNW-trending segment it dips more steeply (60–90°) to the west, whereas further north it dips moderately (40–55°) to the east (Goscombe and Gray, 2007). Subhorizontal to SSE-plunging mineral stretching lineations along the entire length of the shear zone are subparallel to lineations in the adjacent Coastal Terrane. In contrast to the high-grade PMZ, the TPMZ contains evidence for a protracted kinematic history as indicated by overprinting shear fabrics of successively lower grades, which points to continued progressive shearing at decreasing crustal levels into the final stages of orogenesis (Goscombe and Gray, 2007). These authors claim that the TPMZ is an obliquely extensional shear zone with a consistent sinistral sense-of-shear along south-plunging lineations.

The pervasive regional foliation in the Coastal Terrane parallels the strike of the orogen. In the Hoarusib and Khumib Domains of the Orogen Core, the foliation dips moderately to the SW. Along the coast, further north, it dips steeply to the west. Mineral stretching lineations are subhorizontal or plunge at shallow angles to the south and SE. This foliation is axial-planar to early tight to isoclinal folds; a later east-vergent generation of folds have axes subparallel to the stretching lineations (Goscombe and Gray, 2007).

The dominant planar element in the Orogen Core between the TPMZ and the PMZ is a regionally extensive NNW-SSE striking and steeply west dipping (60−90°) bedding-parallel foliation with a stretching lineation that is subhorizontal or plunges at shallow angles (0−30°) to the NNW (in places to the SSE; Goscombe et al., 2003a). Sense-of-shear indicators are consistently sinistral. According to these authors the L-S fabric formed early during $M_2$ and was the first deformation episode that affected the Orogen Core. This view was questioned by Konopásek et al. (2005) who argued that the subvertical foliation was caused by the transposition of an earlier metamorphic foliation. Their suggestion that the fabric is the result of two deformation episodes is based on the observation that several shallowly SSW- and NNE-dipping metamorphic foliations that are only moderately folded are preserved in the area. According to Goscombe et al. (2003a), the first generation of folds in the Orogen Core are close to isoclinal early $M_2$ folds with a strong subvertical axial-planar foliation and axes parallel to the NNW-trending stretching lineation. Subsequent deformation refolded the dominant L-S fabric and produced a spaced crenulation cleavage or biotite foliation.

The CKZ is characterised by large-scale antiforms and nappes (Guj, 1970; Dingeldey et al., 1995; Goscombe et al., 2003a; Konopásek et al., 2005). The dominant foliation is orogen parallel and strikes NNW-SSE. The western part of the CKZ, close to the PMZ, is characterised by tight to isoclinal folds (Konopásek et al., 2005) and a steep westerly dipping foliation with shallow north- to NW-plunging lineations, whereas the area further east towards the west-dipping ST is typified by a shallow west-dipping metamorphic foliation with WNW-ESE trending mineral and stretching lineations (Dürr and Dingeldey, 1996). Thus, there is a progressive change in the lineation orientation from the PMZ in the west towards the ST and the foreland in the east. Open and recumbent folds that refolded a steep foliation occur in the eastern part of the CKZ (Konopásek et al., 2005). Kinematic indicators suggest non-coaxial shear with a top-to-ESE displacement at a high angle to the orogen in this area (Dürr and Dingeldey, 1996). According to these authors, the ST cuts a sinistral shear zone, whereas in other places north-south striking sinistral shear zones occur that predate and postdate the ESE-directed thrust.

From these cross-cutting relationships the above workers concluded that both shearing events occurred coevally in a transpressional deformation regime. Furthermore, they interpreted the entire Kaoko Belt as a half flower or pop-up structure that was formed during transpression, with contemporaneous strain partitioning into a NNW-SSE sinistral transcurrent shear component in the west and an ESE-directed overthrusting component in the east. Goscombe et al. (2003a) interpreted the overall structure of the orogen in the same way. However, they concluded that the sinistral strike-slip wrench tectonics predated both the folding and the east-directed thrusting during the continuous, progressive $D_2$ deformation event. In their opinion an early transpressional phase dominated by wrench tectonics changed progressively into a more convergent style, collisional phase. The first stage caused the strike-slip deformation, the associated subvertical foliations and shallow to subhorizontal mineral and stretching lineations in the orogen. The second stage was responsible for the development of the multiple fold generations, the progressively steepening stretching lineations, the east-verging, large-scale nappes and folds and, finally, for the east-directed thrusting and shortening at a high angle to the belt.

A post-transpression brittle/ductile $D_3$ phase produced large-scale upright, open folds and kink bands across the belt during a minor, north-south directed shortening event (Goscombe et al., 2003a; Konopásek et al., 2005), which is correlated with high-angle convergence in the central Damara Orogen to the south.

### 5.5.1.2. Metamorphic pressure−temperature−time evolution

Both, the basement gneisses and the supracrustal cover sequence were deformed and metamorphosed during the Pan-African orogeny. The EKZ, east of the ST experienced low-grade metamorphism, whereas medium- to high-grade conditions were achieved in the CKZ and WKZ (Guj, 1970; Dingeldey et al., 1995; Gruner, 2000; Stanistreet and Charlesworth, 2001; Goscombe et al., 2003b, 2005b; Will et al., 2004).

Early investigations (Guj, 1970; Ahrendt et al., 1983; Dingeldey et al., 1995) described a gradual increase in metamorphic grade from greenschist-facies conditions in the eastern CKZ to upper amphibolite-facies conditions in the vicinity of the PMZ and, finally, to granulite-facies metamorphism in the WKZ. Along the Gomatum-Hoarusib valleys, a traverse across the entire width of the CKZ, the PMZ and most of the Orogen Core were studied in detail and distinct metamorphic zones could be constrained along the transect (Gruner, 2000;

Goscombe et al., 2003b; Will et al., 2004). East of the PMZ the regional metamorphic zones define a medium- to high-temperature, medium-pressure Barrovian-type metamorphism, whereas a high-temperature, low-pressure Buchan-type metamorphism was recognised west of the PMZ. Based on mineral assemblages in metapelitic rocks, Will et al. (2004) distinguished seven subparallel NNW to NW trending metamorphic zones from east to west, which are characterised by the following mineral assemblages:

| | |
|---|---|
| Garnet zone: | garnet + chlorite + biotite + muscovite (Mu) ± paragonite |
| Staurolite zone: | staurolite + garnet + biotite + muscovite ± chlorite |
| Kyanite zone: | kyanite (Ky) + staurolite + garnet + biotite + muscovite |
| Ky-Sill-Mu zone: | kyanite + (fibrous) sillimanite (Sill) + biotite + muscovite ± garnet |
| Sill-Mu zone: | garnet + sillimanite + muscovite + biotite |
| Sill-Ksp zone: | garnet + sillimanite + K-feldspar (Ksp) + biotite |
| Gt-Crd-Sill-Ksp zone: | garnet + cordierite (Crd) + sillimanite + K-feldspar |

Plagioclase and quartz are present in all zones. The extent of the Sill-Mu zone corresponds broadly to the width of the PMZ along the traverse.

In the CKZ a nearly continuous prograde Barrovian-type succession from the garnet (500 ± 30°C/9 ± 1 kbar) through the staurolite (580 ± 30°C/7–8 kbar) and kyanite zones (590 ± 30°C/6.5–8 kbar) to the kyanite-sillimanite-muscovite zone (650 ± 20°C/9 ± 1.5 kbar) has been documented from east to west (Gruner, 2000; Will et al., 2004). These pressure–temperature estimations correspond to a peak *T*/*P* ratio of ~20°C/km in the CKZ (Figure 5.5.2). West of the PMZ, the high-*T*/low-*P* Buchan-type metamorphism in the Orogen Core is evidenced by sillimanite-K-feldspar (690 ± 40°C/4.5 ± 1 kbar) and garnet-cordierite-sillimanite-K-feldspar zones (750 ± 30°C/4.5 ± 1 kbar). The inferred peak *T*/*P* ratio is ~50°C/km (Will et al., 2004). This and the high temperatures are consistent with the occurrence of migmatites. The pressure difference of ca. 4.5 kbar across the PMZ implies an exhumation of 12–15 km of the CKZ with respect to the Orogen Core in the WKZ. In addition to conventional geothermobarometric methods, which were used to infer the peak metamorphic conditions given above, Will et al. (2004) used phase petrological constraints to infer *P*–*T* paths for each individual zone and demonstrated that each *P*–*T* trajectory follows a clockwise loop (Figure 5.5.2). Goscombe et al. (2003b) determined similar *P*–*T* conditions for the Barrovian-style metamorphism in the CKZ and values of 843 ± 64°C/8.1 ± 1.6 kbar and 811 ± 58°C/6.2 ± 0.7 kbar for rocks from the Orogen Core to the west of the PMZ. In addition, the latter authors showed that the peak pressure $M_2$ assemblages in these high-grade rocks from the Orogen Core are surrounded by symplectic spinel-corundum-K-feldspar-biotite coronas that point to decompression by up to 3 kbar.

The metamorphic profile along the Gomatum-Hoarusib traverse is representative for most of the belt (Guj, 1970; Miller, 1979; Dingeldey, 1997; Goscombe et al., 2003b) except for the Orogen Core. There, a tight clustering of successive metamorphic zones encloses each of the Hartmann, Khumib and Hoarusib Domains (Goscombe et al., 2005b).

According to Goscombe et al. (2003b), the peak metamorphic conditions described above were reached early during the transpression phase, $M_2$, between 580 and 550 Ma. Sm-Nd garnet-whole rock geochronology carried out by these authors indicates that the metamorphic matrix mineral assemblages in the CKZ and the Orogen Core

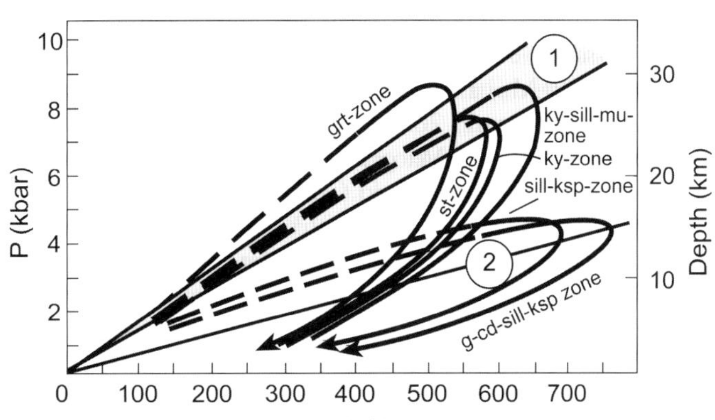


**Figure 5.5.2** Inferred *PT* paths for the metamorphic zones mapped along the Gomatum-Hoarusib traverse in the Kaoko Belt. Two different metamorphic types can be distinguished: a Barrow-type metamorphic evolution, with a peak *PT* ratio of some 20°C/km east of the Purros Mylonite Zone (1) and a Buchan-type metamorphism with a peak *PT* ratio of some 50°C/km west of that mylonite zone (2); modified after Will et al. (2004).

peak at 4.2 kbar and 600°C (Kasch, 1987). Similarly, $P-T$ conditions of 4.5 kbar and 630°C were recorded by the same author for the northeastern edge of this region close to the Okahandja Lineament.

In the area between Conception Bay and Meob Bay at the Atlantic coast, the whole schist succession of the SZ, probably made up of both Kuiseb and Hureb Formations, has undergone extensive partial melting over an exposed area on the western edge of the Namib erg of at least 430 km$^2$. This whole region is a single, huge garnet-sillimanite-biotite migmatite mass that became variously, and locally extensively, re-homogenised due to en bloc flow and remixing of the liquid–solid mush. Disrupted fragments of folded and unmolten schist and amphibolite float as autoliths in the migmatite.

The red Goanikontes anatectic granite dates the peak of post-tectonic $M_2$ regional metamorphism at 534 ± 7 Ma. Assuming an age of about 555 Ma for $D_2$ in both the southern CZ and the SZ, a period of 20 myr separated initial syn-$D_2$ garnet growth from the post-tectonic peak of $M_2$. An estimated decompression of 3 kbar in this period corresponds to a realistic erosion rate for the southern Damara mountain belt of approximately 0.5 mm/year.

In addition to the above metamorphic features, extensive contact metamorphism affected the country rocks around voluminous post-tectonic, largely granitic intrusions, e.g. the Omangambo Pluton at the boundary between the NZ and SKZ or the Donkerhuk Granite batholith in the SZ.

## 5.5.3. The Gariep Belt

### 5.5.3.1. Structural evolution

Within the arcuate-shaped arrangement of stacked thrust sheets that constitute the Gariep Belt, two principal tectono-stratigraphic units are distinguished. These are the para-autochthonous Port Nolloth Zone (PNZ) in the east and the allochthonous Marmora Terrane in the west, whose main deformational features (Frimmel, 2008b) are briefly summarised in the following.

A first penetrative schistosity ($s_1$) developed during a first phase of contractional deformation ($D_1$). It is almost horizontal and subparallel to bedding ($s_0$) in the western PNZ, but steepens towards the eastern margin of the belt. Open to isoclinal $F_1$ folds, thrusts, stretching lineations and sheath folds related to $D_1$ all follow the arcuate shape of the belt (Gresse, 1994). In the northern and northeastern part of the PNZ, $F_1$ folds are northeast-vergent (Davies and Coward, 1982), whereas in the eastern and southeastern parts, they are southeast-vergent (Von Veh, 1993; Hälbich and Alchin, 1995).

The second major phase of deformation ($D_2$) led to tight, southwest- to west-verging, non-cylindrical $F_2$ folds and associated axial-planar crenulation cleavage ($s_2$) as well as a NNW-plunging mineral lineation ($L_2$) in the respectively northeastern and eastern marginal domains of the PNZ. There they have been interpreted as both backfolds, associated with back-thrusts (Hälbich and Alchin, 1995) and as shear folds related to plane sinistral wrenching (Von Veh, 1993). In the inner parts of the PNZ, the $F_2$ folds verge towards the foreland (Gresse, 1994).

In the Marmora Terrane, the penetrative foliation $s_1$ dips to the west and is axial-planar relative to tight to isoclinal, east-vergent $F_1$ folds. Locally, $s_1$ was refoliated around open to tight, northeast-trending $F_2$ kink bands. The contact with the underlying PNZ is a major top-to-southeast thrust fault (Schakalsberge Thrust). At the northern boundary of the Marmora Terrane, $F_1$ folds within the Chameis Subterrane are truncated by the Schakalsberge Thrust, highlighting its syn-$D_2$ timing. The contact between the Oranjemund and the Schakalsberge Subterranes is a west- to northwest-dipping fault plane along which the former was thrust on top of the latter. The contact between the Chameis and Oranjemund Subterranes is a zone of intense $F_2$ folding with top-to-southeast transport and subsequent back-folding that resulted in locally steeply southeast-dipping crenulation.

The preferred kinematic model for the evolution of the main contractional structures in the Gariep Belt involves two stages (Figure 5.5.4): (i) eastward transport of the oceanic rocks of the Marmora Terrane, internal stacking of these and eventually obduction onto the continental to passive margin deposits of the PNZ. The variable angle between principal direction of compression and the pre-existing arcuate north- to northwest trending continental margin led to strain partitioning with northeast-, east- and southeast-directed transport along the northeastern, eastern and southeastern basin margin, respectively (Frimmel, 2000b); (ii) subsequent sinistral transpression with overall top-to-southeast transport. The resulting folds and thrusts are east-verging along the northeastern lateral ramps and southeast-verging along the southern frontal ramps in the outer PNZ (Gresse, 1994).

Late-orogenic sinistral transtension ($D_3$) led to small-scale, NNW-trending $F_3$ open gravity folds, mainly recorded in the PNZ, and E-W striking normal faults, N-S striking reverse faults, veins and fractures, and a set of

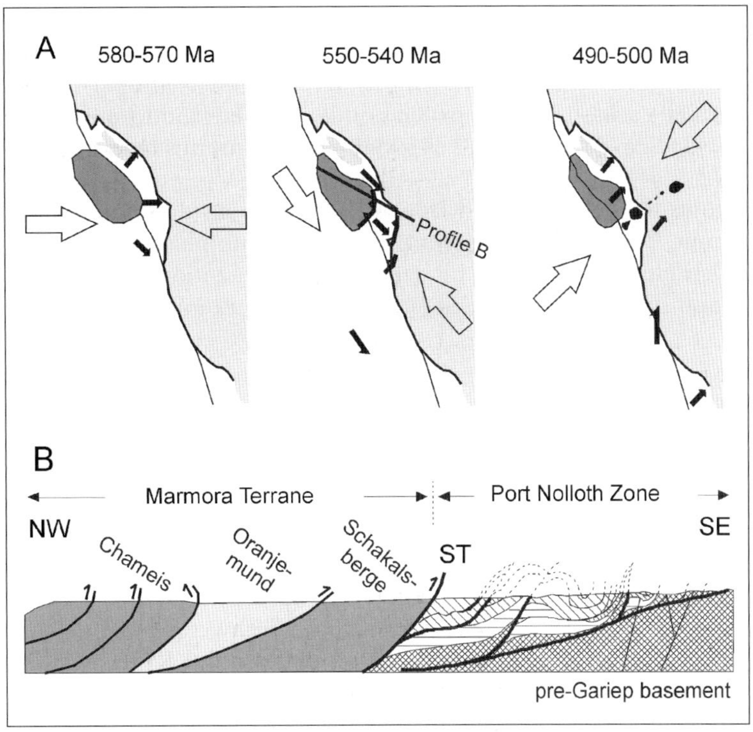


**Figure 5.5.4** A: Principal syn-orogenic stress fields and transport directions in the Gariep Belt (from Frimmel, 2000b); B: schematic NW-SE profile through the Gariep Belt (modified after Von Veh, 1993); ST, Schakalsberge Thrust.

Riedel shears and conjugates that strike ESE and NE, respectively (Von Veh, 1993). This transtensional stage was accompanied by the emplacement of the alkaline igneous bodies along the Kuboos-Bremen Line. Finally, a further phase of compressional deformation affected the region after the deposition of the Nama/Vanrhynsdorp Group molasse sediments ($D_4$). It caused northeast-verging open to tight folds and related faults that become the dominant structures in the southern extension of the Gariep Belt along the South African west coast and in Nama Group rocks to the east and southeast of the Gariep Belt (Gresse, 1994).

## 5.5.3.2. Metamorphic history

The Marmora Terrane differs from the PNZ not only with regard to stratigraphy and deformation, but also in its metamorphic history. In the PNZ, only one stage of regional metamorphism is discerned, the peak of which was attained after $D_2$ but prior to $D_3$ and it is ascribed to crustal thickening following the loading of the Marmora Terrane on top of the PNZ. Most of the PNZ is within the biotite zone (middle greenschist facies). Only locally, such as west of Rosh Pinah, does garnet occur, whereas the presence of syn-tectonic andalusite indicates a low-pressure regime. For that area, a peak temperature of about 520°C at 3.0–3.5 kbar has been obtained (Frimmel, 1995). In contrast, upper greenschist-facies kyanite occurs in highly aluminous metapelite in the southern extension of the PNZ in the westernmost Vredendal Outlier. The mafic Gannakouriep Suite dykes, which cross-cut the lower parts of the Port Nolloth Group, show albite-epidote-amphibolite-facies overprint within the PNZ. Away from the orogenic front towards the east, this metamorphic overprint decreases gradually.

The timing of syn-tectonic metamorphism is constrained by Ar-Ar muscovite ages of around 545 Ma (Frimmel and Frank, 1998), which corresponds to $M_2$ and the $D_3$ deformational stage in the southern parts of the Damara Belt. A younger age group between 505 and 480 Ma, obtained on muscovite and biotite in shear zones of the pre-Gariep basement along the South African west coast, indicate that metamorphic conditions prevailed until, or where attained again, in post-Nama times. This also explains a gradual increase in metamorphic grade within the siliciclastic rocks of the Vanrhynsdorp Group from virtually unmetamorphosed in the northeast of the Vanrhynsdorp Basin towards middle greenschist facies in the southwest (Gresse, 1992).

A different and more complicated, metamorphic history is recorded by the mafic rocks of the Marmora Terrane. There three amphibole generations could be distinguished in terms of composition (Frimmel and Hartnady, 1992) as well as Ar-Ar age dates (Frimmel and Frank, 1998). The oldest generation is silicic edenite and attributed to seafloor metamorphism ($M_1$). Imprecise Ar-Ar data obtained on it place some very loose constraints on its age sometime around 610 Ma. The second amphibole generation differs from the first by having a more barroisitic composition (higher glaucophane component) that indicates formation at lower amphibolite-facies temperatures and higher pressure (4–6 kbar). It yielded Ar-Ar plateau ages of 574±9 Ma, which corresponds to the likely age for $M_1$ in the southern Damara Belt. This $M_2$ event in the Marmora Terrane has been explained by accretion of the oceanic crustal rocks. The third generation is actinolite and formed during regional, greenschist-facies overprint ($M_3$). It yielded Ar-Ar age data of 545 Ma, which conforms both in grade and age to the regional metamorphism in the PNZ.

The syn-orogenic deposits of the Holgat and Bogenfels Formations differ by displaying only one very low-grade metamorphic overprint. They lack the higher pressure metamorphism ($M_2$) that is recorded in the mafic rocks of the underlying Dernburg Formation. This supports the contention of them having been deposited in a foredeep to foreland position subsequent to the onset of oceanic crust accretion (Frimmel and Fölling, 2004).

## 5.5.4. The Saldania Belt

The extent of metamorphism in the Saldania Belt is generally of low-grade. Poor outcrop and a similarly low-grade subsequent metamorphic overprint during the 290–220 Ma Cape orogeny (Frimmel et al., 2001a) has hindered a precise reconstruction of the Neoproterozoic to Cambrian metamorphic evolution in the belt. In the western part of the belt, regional metamorphism attained only the chlorite zone (lower greenschist facies). Chloritoid in the uppermost contact to sandstone of the overlying Table Mountain Group has been attributed to the Cape Orogeny (Frimmel et al., 2001a). Only at one locality, northwest of Malmesbury, are biotite-rich higher grade rocks exposed. While they have been interpreted by Belcher and Kisters (2003) as an older part of the Neoproterozoic succession that experienced an earlier Pan-African metamorphic event and was used by these authors to distinguish between different stratigraphic units in the Saldania Belt, these rocks are more likely part of the Mesoproterozoic basement (Namaqua-Natal Belt) because of their upper amphibolite-facies metamorphic mineral assemblage (muscovite-absent biotite-paragneiss) and an additional, high-temperature ductile deformation that is absent elsewhere in the Saldania Belt.

In the eastern sector of the belt, an increase in metamorphic grade from north to south from the chlorite to the biotite zone has been observed in the Kaaimans Inlier (Frimmel and van Achterbergh, 1995). Middle greenschist-facies conditions were obtained also for the Nooitgedagt Member, the lowermost unit of the Cango Caves Group, based on the fractionation of $^{13}C$ between coexisting carbonate and organic carbon in limestone samples (Frimmel et al., 2001a). In contrast, the carbonates of the overlying Kombuis Member yielded temperatures that are about 100°C lower (sub-greenschist facies, ca. 290°C), which provides strong support for a hiatus between the two members. This is probably the same hiatus that has already been noted between the Dernburg and Bogenfels/Holgat Formations in the Gariep Belt. Such a correlation is supported by chemo- and biostratigraphic data (see Chapter 5.4) and it is therefore concluded that the post-Nooitgedagt deposits in the Cango Caves Group were laid down after the first orogenic phase and regional metamorphism, which, by analogy with the Gariep Belt, may have taken place sometime between 570 and 580 Ma.

Locally, in the immediate vicinity of some of the Cape Granite Suite rocks, distinct contact metamorphic aureoles exist, prime examples of which can be found at the beach in Seapoint (Cape Town), first described by Charles Darwin, and around the George Pluton in the Kaaimans Inlier (Frimmel and van Achterbergh, 1995). A younger orogenic phase, which caused an omnipresent northeast-directed compressional deformation in all rocks of the Saldania Belt, including the syn-orogenic deposits of the Kansa Group, has been dated at 483–506 Ma (Gresse et al., 1988; Frimmel and Frank, 1998) and preceded the deposition of the overlying Schoemanspoort Formation and Table Mountain Group.

## *ACKNOWLEDGEMENTS*

This work summarises results obtained in the course of many research projects that received support from a variety of institutions and funding agencies, especially the Geological Survey of Namibia, the South African National Research Foundation (NRF) and the German Research Foundation (DFG). K.-P. Kelber kindly assisted with the drafting of some of the diagrams. Others, though slightly modified, are taken from Miller (2008); copyright approval by the Director of the Geological Survey of Namibia for their reproduction is gratefully acknowledged. A. Tankard and M. Okrusch are thanked for providing helpful reviews of the original manuscript.

CHAPTER 5.6

# Syn- to Post-Orogenic Magmatism ☆

Roy McG. Miller[1] *and* Hartwig E. Frimmel[2]

## Contents



## 5.6.1. Plutonic Rocks in the Kaoko Belt

Only granites and very minor diorite and syenite are recorded in the Kaoko Belt. Two age groups are present. The older group of granites, with ages of $657 \pm 4$ and $630 \pm 8$ Ma, intruded the Coastal Terrane while spreading in the rest of the Kaoko Belt and the Damara Belt was still active. They are porphyritic and biotite rich, and are markedly elongate, parallel the regional structural trend and were transformed into augen gneisses by intense $D_1$ and $D_2$ deformation. Associated aplite veins are isoclinally folded parallel to the foliation. The granites are calc-alkaline in composition in agreement with arc-like I-type granites, and elemental and isotopic data indicate derivation from a primary, mantle-derived magma followed by extensive crustal contamination (Seth et al., 1998; Goscombe et al., 2005a).

Syn- to late-tectonic intrusions form the bulk of the younger group of granites in the Western Kaoko Zone. They were emplaced during $M_2$ metamorphism and subsequent to $M_3$ (Goscombe et al., 2003a). This group includes older, strongly foliated granodiorite and monzogranite, younger, weakly foliated monzogranite and diorite, and late-tectonic granite and dykes of two-mica pegmatite and aplite. U-Pb single zircon ages of these granites range from $580 \pm 8$ to $552 \pm 2$ Ma (Seth et al., 1998; Franz et al., 1999). Zircon from a late-tectonic pegmatite yielded an age of $549 \pm 2$ Ma (Goscombe et al., 2005a). Most of the granites in this group are calc-alkaline but the late granites are alkaline in composition. Their trace element distribution corresponds to those of volcanic arc, syn-collisional and orogenic granites. Isotopic compositions suggest derivation largely from pre-Pan-African granitic crust (Seth et al., 1998).

The West Wind Diorite occurs at the southern end of the Western Kaoko Zone. It postdates the late-$D_2$ foliation in the Toscanini Formation schists (Miller, 2008) and yielded a conventional zircon-titanite U-Pb concordia intercept age of $541^{+19}_{-17}$ Ma (E.A. Retief, unpublished data, 1988). This is the same age as $D_3$ doming in the southern Central Zone (CZ) of the Damara Belt.

In the Southern Kaoko Zone, a few stocks of foliated, late syn-tectonic, porphyritic, biotite-rich, Salem-type granite (Salem granite being the original name given to this type of granite – Martin, 1965b) as well as tourmaline-bearing granite occur in the higher grade parts of this region south of Brandberg. The contacts of the porphyritic granite stocks are largely concordant with the enclosing schists, but a thermal overprint on the regional metamorphic assemblages is present. These post-tectonic granites include minor flow-foliated hornblende-biotite diorite (the oldest of these post-tectonic rocks), non-porphyritic foliated granite, coarsely porphyritic Salem-type

☆Miller, R. McG., Frimmel, H.E., 2009. Syn- to post-orogenic magmatism. Neoproterozoic to Early Palaeozoic evolution of Southwestern Africa. In: Gaucher, C., Sial, A.N., Halverson, G.P., Frimmel, H.E. (Eds.): Neoproterozoic-Cambrian tectonics, global change and evolution: a focus on southwestern Gondwana. Developments in Precambrian Geology, 16, Elsevier, pp. 219–226.

[1] PO Box 11222, Windhoek, Namibia.
[2] Department of Geological Sciences, University of Cape Town, Rondebosch 7700, South Africa.

*Developments in Precambrian Geology*, Volume 16
ISSN 0166-2635, DOI 10.1016/S0166-2635(09)01615-6

granite and granodiorite with a strong autoclastic marginal foliation, weakly foliated to unfoliated Salem-type granite, muscovite leucogranite, two-mica syenogranite and porphyritic hornblende syenite (the youngest of these rocks). A U-Pb zircon age of 540±3 Ma (Van de Flierdt et al., 2003) for quartz diorite of this suite is similar to the age of the post-tectonic West Wind Diorite further north.

One of the two stocks of porphyritic hornblende syenite, the Voetspoor Syenite, is slightly alkaline, metaluminous and strongly enriched in LREE with a minimum zircon evaporation age of 530±3 Ma (Seth et al., 2000). Elemental and isotopic chemistry and an Nd model age of 1.2–1.3 Ga suggest that the syenite was sourced from phlogopite-bearing mantle that was enriched by fluids from a subduction zone.

## 5.6.2. Plutonic Rocks in the Damara Belt

The plutonic suite in the Damara Belt covers approximately 74,000 km$^2$ and consists of over 200 plutons, plugs and stocks of syn- to post-tectonic granitic rocks (Figure 5.6.1) as well as thousands of granite and pegmatite dykes. Distinctive features of the granitic suite are the large-scale limitation of specific granites to specific stratigraphic levels within the Damara succession (Smith, 1965; Jacob, 1974; Haack et al., 1980; Miller, 1983a, 2008), the sequence of intrusion, the lopolithic shape of many of the early intrusions and the balloon-like shape of the late intrusions. The bulk are syn- and post-collision granites that occur in the southern leading edge of the high-temperature active continental margin of the Congo Craton, the CZ, which was located above the shallow portion of the north-dipping Damaran subduction zone. Shortly after the post-tectonic peak of regional $M_2$ metamorphism, the 350 km long, linear batholith of the Donkerhuk Granite intruded along the northern edge of the SZ directly in front of the leading edge of Congo Craton. The granites in the low-grade Northern Margin Zone (NMZ) and Southern Marginal Zone (SMZ) are post-collision in age. Two age groups are present: the older ranging from 565 to about 540 Ma and the younger from 480 to 460 Ma. A smattering of intrusions, including the huge Donkerhuk Granite batholith, has ages between 534 and 495 Ma.

The earliest Damaran intrusions are differentiated, layered metagabbro bodies (Neikhoes Metagabbro) that form much less than 1% of the plutonic suite. These only occur as disrupted fragments, xenoliths and megaxenoliths in the oldest diorites. A few small plugs of hornblendite are associated with the gabbro occurrences. The remainder of the plutonic suite is overwhelmingly dominated by granite. Area-wise, the proportion of gabbro-diorite:tonalite-granodiorite:granite in the Damara Belt is 2:2:96 with monzogranite being by far the most abundant rock type present (Miller, 1983a).

In the southern CZ, the sequence of intrusions is gabbro, diorite (565 Ma), porphyritic Salem-type biotite granite (555 Ma), leucogranite and finally pegmatite. The leucogranite intrusions range in age from late syn-tectonic, that is syn-$D_3$ (542 Ma), to significantly post-tectonic (508 Ma). In the northern CZ, no gabbro is present but there were two cycles of diorite – porphyritic Salem-type biotite granite-leucogranite – pegmatite intrusion, the older post-$D_1$ to syn-$D_2$ (550–514 Ma), the younger syn-$D_3$ to post-tectonic (480–460 Ma). $D_2$ and $D_3$ events were younger in the northern CZ than in the southern CZ. A few relatively large bodies of late- to post-tectonic porphyritic biotite granite are present. Distinctly I-type basic and granitic rocks that postdate the $M_2$ peak of regional metamorphism form four isolated and widely spaced stocks (Meob Bay Norite, Gawib Granodiorite, Achas Granite, Waldau Dome Gabbro) that have intruded along or near the leading edge of the Congo Craton, the Okahandja Lineament.

Largely autochthonous, basement- and Nosib-derived red gneissic granite, red homogeneous granite (Ozombanda Granite) and some white leucogranites in the southern CZ are confined to basement and Nosib stratigraphic levels. The Ozombanda Granite is commonly dammed up directly beneath the Swakop Group marbles and can almost completely fill antiforms, but small plugs and numerous veins of it also intruded to higher stratigraphic levels. Uraniferous alaskites are concentrated in the Khan Formation and to a significantly lesser extent in the overlying Chuos Formation southwest of Usakos.

Dioritic rocks, the voluminous porphyritic Salem-type biotite granite and the larger leucogranites occur in synforms in the stratigraphic position of the Kuiseb Formation schists. Large bodies of early diorite and tonalite of the Goas Suite occur in synforms just south of Karibib. The remaining diorites and tonalites are rare, occur in the northern CZ, are of limited volume and always occur as thin sheets of limited lateral extent directly below large, thick, lopolithic plutons of Salem-type porphyritic granite. A thin sliver of Kuiseb schist usually separates the diorite from the underlying Karibib marbles. Several large synforms east of Swakopmund are completely filled by Salem-type granite. In the northern CZ, isoclinally folded, syn-$D_2$ sheets of leucogranite occur in the Kuiseb schists stratigraphically well above the lopoliths of Salem-type granite. In the same region, stanniferous pegmatites and balloon-shaped syn- to post-$D_3$ granites are the stratigraphically highest intrusions. The 350-km-long batholith of the two-mica, garnetiferous Donkerhuk Granite is a post-$D_3$ intrusion that is located within the Kuiseb and Hureb Formation schists along the northern edge of the SZ.

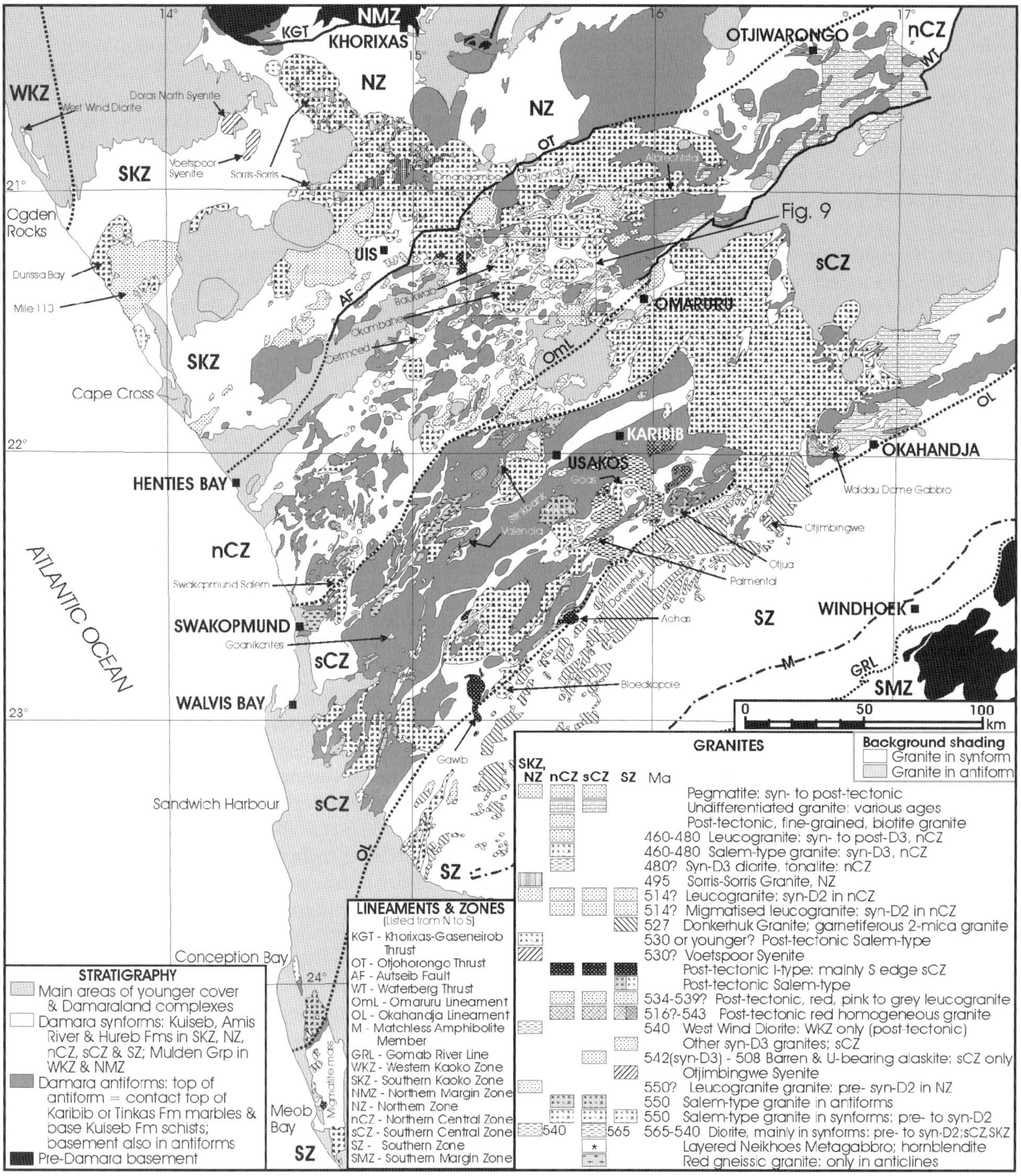


**Figure 5.6.1** Distribution, composition and stratigraphic position of the syn- to post-orogenic plutonic suites in the Damara Belt (modified after Miller, 2008).

The oldest reliable dates for the plutonic rocks in the Damara Belt are for the early post-$D_1$ to pre- or syn-$D_2$ diorites. These are $564 \pm 5$ Ma (Jacob et al., 2000) and $558 \pm 5$ Ma (de Kock et al., 2000), which agree well with those of the syn-transpressional granites in the Western Kaoko Zone. Slightly younger pre- to syn-$D_2$ granites have not been dated precisely yet but are estimated to be 555–550 Ma in age (Miller, 2008). Boudinaged, sheet-like Salem-type granite in the southern CZ (Stinkbank Granite) has been dated at $549 \pm 11$ Ma (Johnson et al., 2006). Syn-$D_3$ leucogranite in the southern CZ has an age of $542 \pm 6$ Ma (Tack et al., 2002). This magmatic pulse coincided with the final stage of continental collision (Miller, 1983a, 2008). The Rote Kuppe Granite near Karibib is the oldest post-$D_3$ granite and has an age of $539 \pm 6$ Ma (Jacob et al., 2000). The peak of post-tectonic $M_2$ regional metamorphism is dated by the anatectic Goanikontes red granite with an age of $534 \pm 7$ Ma (Briqueu et al., 1980). The Donkerhuk Granite batholith that intruded the northern edge of the SZ and postdates $M_2$ yielded whole-rock Rb-Sr ages of $523 \pm 8$ and $527 \pm 3$ Ma (Blaxland et al., 1979; Haack and Gohn, 1988). Other post-tectonic granites in the southern CZ range from $526 \pm 17$ Ma (Tack et al., 2002) to $508 \pm 2$ Ma (Briqueu et al., 1980). Single titanite grains from post-tectonic lamprophyre dykes at the Navachab gold mine gave a U-Pb age of $496 \pm 6$ Ma and the auriferous veins themselves an age of $494 \pm 8$ Ma (Jacob et al., 2000). In the NZ, the alkaline Sorris-Sorris Granite and the huge, post-tectonic pluton of Salem-type granite that it intrudes both have ages of 495 Ma (Hawkesworth et al., 1983; P.F. Hoffman and S.A. Bowring, unpublished data, 2005).

$D_1$ in the northern CZ is considered to be the same age as $D_1$ further south. The oldest post-$D_1$ and pre-$D_2$ Salem-type granite has an Rb-Sr age of $553 \pm 22$ Ma (Haack et al., 1980). However, $D_2$ and $D_3$ and the granites associated therewith are significantly younger than in the southern CZ. Syn-$D_2$ granite has a whole-rock Rb-Sr age of $514 \pm 22$ Ma and numerous syn- to post-$D_3$ granites range from $479 \pm 16$ to $459 \pm 15$ Ma in age (Haack et al., 1980). The youngest of these granites postdates continental collision by 80 myr.

The layered, pre-diorite metagabbro (Neikhoes Metagabbro) has a composition similar to subalkalic tholeiite (de Kock, 1989). In contrast, the dioritic rocks, the Salem-type granites and many of the leucogranites are calc-alkaline (McDermott, 1986). All early diorite and most Salem-type granite bodies are extensively contaminated along their margins by numerous xenoliths of the enclosing schists. In places almost complete assimilation makes it difficult to recognise the extent of marginal contamination. Despite this contamination, McDermott (1986) divided the Damaran granites on chemical grounds into crustal melt-derived, distinctly calc-alkaline rocks, and within-plate granites based on their Shand index ($Al_2O_3/(CaO+Na_2O+K_2O)$-Al/CNK) and content of the high-field-strength (HFS) elements Nb+Y (Figure 5.6.2, Table 5.6.1). The leucogranites and most of the Salem-type granites plot in the field of crustal melts at the high-Al end of the calc-alkaline trend. The diorites are distinctly calc-alkaline, as are the Otjimbingwe and Voetspoor syenite intrusions. The within-plate granites together with some Salem-type granites (Oetmoed, Baukwab and Albrechtstal Plutons) have the highest contents of HFS elements and are classed as A-type granites (Jung et al., 1998).

Most of the Damaran granites are mildly peraluminous, have Al/CNK ratios of between 1 and 1.1, and, with the exception of the Donkerhuk Granite, are not typical S-type granites. Some of the small-volume, so-called leucogranites contain abundant cordierite and garnet and variable amounts of biotite and are, however, clearly the product of partial melting of Damaran metasedimentary rocks. Similarly, the basement- and Nosib-derived alaskites and red granites (included with the leucogranites) of the southern CZ have metasedimentary sources. As shown on Figure 5.6.3, the Nb-Y-Rb relationships of the Damaran granitoids indicate a wide range of geotectonic settings from within-plate, syn-collision, post-collision to volcanic arc granites.

Trace element modelling of the Damaran granites shows that fractional crystallisation of hornblende and, to a lesser extent, plagioclase was dominant during crystallisation of the dioritic melts and that the rare earth element (REE) distribution was controlled by hornblende. In contrast, accessory phases controlled the distribution of REE in the within-plate granites. The high HFS element abundances in the within-plate granites suggest that they were derived from a calc-alkaline source that has not undergone extensive intracrustal reworking, that is not metasedimentary (McDermott, 1986; McDermott et al., 2000), which is consistent with a lower crustal source of Late Mesoproterozoic age (Figure 5.6.4, Table 5.6.1).

The distinct tholeiitic composition of the earliest Damaran syn-orogenic plutonic rocks, the mantle-derived Neikhoes Metagabbro of the Goas Suite, indicates emplacement before any significant subduction had taken place. The diorite bodies, in contrast, are clearly calc-alkaline and represent the first intrusions with a significant subduction component in their composition. The southern diorite intrusions of the Karibib area occur close to the leading edge of the overriding Congo Craton, the Okahandja Lineament, and above the shallow part of the

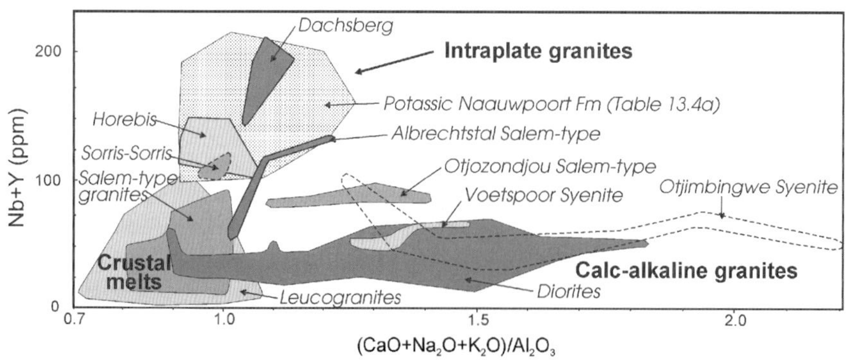


**Figure 5.6.2** Plot of Shand index (CNK/Al) versus Nb+Y for the Damaran and Southern Margin Zone granitic rocks (modified after McDermott, 1986).

**Table 5.6.1** Geochemical distinguishing features of the three main granitoid types in the Damara Belt (after McDermott, 1986).

| | Crustal-melt granitoids | Calc-alkaline diorites | Within-plate granitoids |
|---|---|---|---|
| Major elements | Peraluminous | Metaluminous | Peraluminous to metaluminous; extremely low CaO |
| Trace elements | High Rb; low Nb, Y, Zr (high Th, U in some) | Low Rb/Sr; low HFS | High Nb, Zr, Hf, Y; high REE |
| Sr isotopes | High initial $^{87}Sr/^{86}Sr$ (>0.710) | Low initial $^{87}Sr/^{86}Sr$ (<0.710) | Variable initial $^{87}Sr/^{86}Sr$ |
| Model Nd ages | ~2.0 Ga | ~1.0–1.4 Ga | ~1.2 Ga |
| O isotopes | High $\delta^{18}O$ | Variable $\delta^{18}O$ | Low $\delta^{18}O$ |

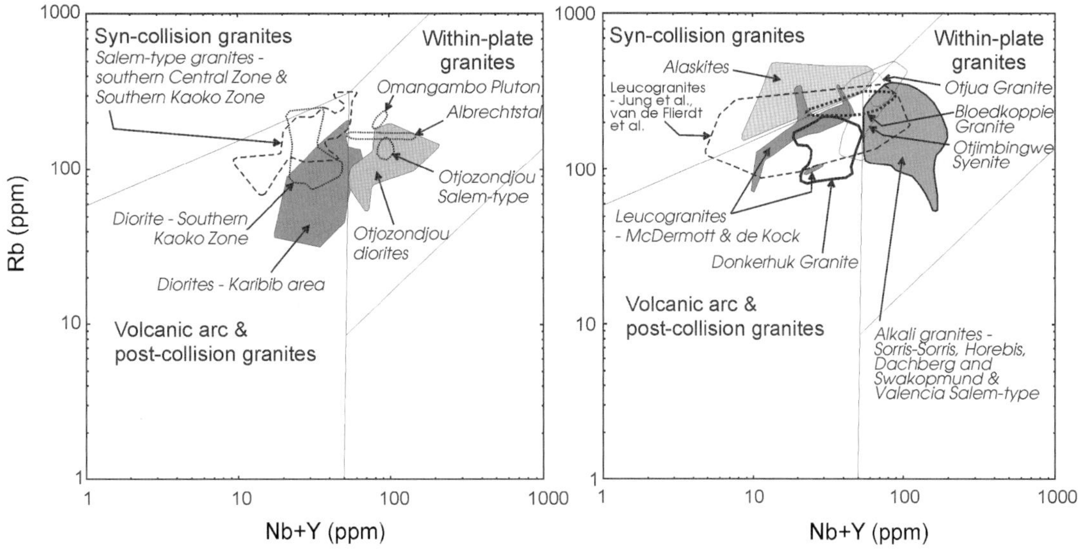


**Figure 5.6.3** Rb versus (Nb+Y) discriminant diagrams (Pearce et al., 1984) for various Damaran syn- to post-tectonic granitic rocks (modified after Miller, 2008).

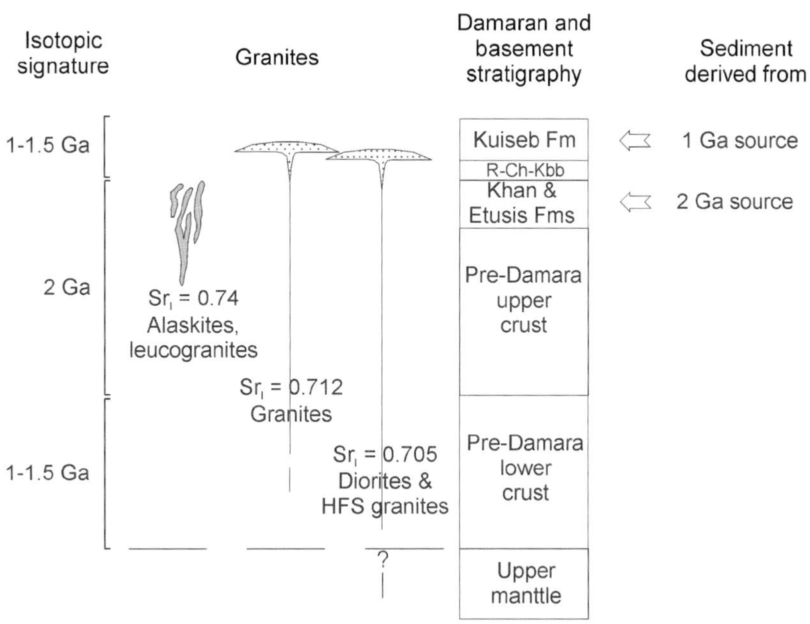


**Figure 5.6.4** Schematic crustal section from the central Damara Belt illustrating the source regions suggested for the Damaran granitic rocks. Average initial $^{87}Sr/^{86}Sr$ ratios are also given for the main groups of granitic rocks (modified after Hawkesworth et al., 1983).

subduction zone. They plot in the volcanic arc field in Figure 5.6.3 but the compositions of the northernmost diorite bodies from Otjozondjou, which occur above the deeper parts of the subduction zone, plot in the field for within-plate granites in line with their high contents of HFS elements. Similarly, most of the syn-tectonic Salem-type granite compositions from the southern CZ plot in the field for volcanic-arc granites although there is a slight spread into the field of within-plate granites. However, the Salem-type granites from the northern CZ (Otjozondjou and Albrechtstal) and the Northern Zone (Omangambo) conform to within-plate granites. Two relatively small plutons of Salem-type granite from the southern CZ, the so-called Swakopmund and Valencia bodies, have contents of HFS elements that place them in the within-plate field.

It is thus apparent that although diorites and Salem-type granites above the shallow and deep parts of the subduction zone are indistinguishable from each other in the field, the former have a volcanic-arc character whereas the latter have a more within-plate character. Many syn- to post-$D_3$ A-type granites in the northern CZ (Haack et al., 1980; Jung et al., 1998) postdate continental collision in the Damara Belt by some 60–80 myr. This is the youngest regional magmatic event to have taken place in the Damara Orogen and these ages are the key to the suggestion that the northern CZ has a late-stage, back-arc evolutionary overprint (Miller, 2008).

Most of the leucogranites of the southern CZ, which are largely late to post-tectonic, correspond to post-collision granites. Although the alaskites are post-tectonic relative to $D_3$, the final regional collisional deformation event, they correspond compositionally to syn-collisional granites (Figure 5.6.3).

It has tentatively been suggested by several authors that the peraluminous Donkerhuk Granite could be the product of partial melting of the Kuiseb schists. However, the Mesoproterozoic model age of the Kuiseb schists (Hawkesworth et al., 1983) and the Palaeoproterozoic model age of 2 Ga of the Donkerhuk Granite (McDermott, 1986) preclude this possibility. The latter model age indicates that the granite has been derived from Palaeoproterozoic crust as is the case with most of the other Damaran leucogranites. The Donkerhuk Granite has a distinctly post-collision character. In contrast, the Otjimbingwe Syenite which occurs in the same region but is older and syn-tectonic has a model age of 1.4 Ga (McDermott, 1986).

## 5.6.3. Post-Orogenic Plutonism in the Gariep Belt

No evidence exists for any syn-orogenic magmatism in the entire Gariep Belt. Post-kinematic with respect to the main phase of orogenic deformation are a series of alkaline intrusive bodies that are located along a northeast-trending line over at least 270 km, the Kuboos-Bremen Line (Figure 5.2.1; Söhnge and de Villiers, 1948). They cut across the penetrative foliation and main structures of the deformed Neoproterozoic metasedimentary successions and include, from southwest to northeast, the Swartbank Pluton (largely covered), the Kuboos Pluton, the Tatasberg Pluton (all in South Africa), the Grootpenseiland ring complex at the Orange River (Smithies and Marsh, 1996), the Marinkas Kwela complex and associated carbonatite, the Kanabeam ring complex (Reid, 1991), the Mount Ai-Ais breccia pipe, the younger Bremen complex, the Haruchas stock and the Garub pipe in southern Namibia. Alkali-granite and alkali-syenite dominate the plutons in the southwestern sector, whereas syenite, nepheline syenite, monzodiorite, phonolite and locally also carbonatite are more important further northeast.

An emplacement depth of about 7 km has been estimated from contact metamorphic phase relationships around the Kuboos Pluton (Frimmel, 1995), whereas further to the northeast, in the foreland, the melts reached the surface. Thus, the Gariepian crust was probably still thicker in the orogen than in the foreland at the time of magmatism. The best available geochronological datum for any of these bodies is a U-Pb single zircon age of $507 \pm 6$ Ma for the Kuboos Pluton (Frimmel, 2000a). Comparable post-orogenic, alkali granitic to syenitic bodies occur in pre-Gariep basement or rocks of the Port Nolloth Zone close to the northwestern margin of the belt near the Atlantic coast north of Bogenfels, but no further geochemical, petrological or geochronological data are available.

## 5.6.4. Plutonism in the Saldania Belt

A series of syn- to post-tectonic granitoid bodies intruded into Neoproterozoic metasedimentary rocks of the various tectono-stratigraphic zones of the Saldania Belt (Figure 5.6.5). Based on petrography, geochemical characteristics and available age data, three intrusive phases that correspond largely to S-, I- and A-type granites, respectively, can be distinguished (da Silva et al., 2000; Scheepers and Schoch, 2006). Systematic differences across the boundary between the Tygerberg Terrane and the Swartland Zone agree with the assumption of the Colenso Fault representing indeed a terrane boundary. Mafic and intermediate plutons are less common. Olivine gabbro, gabbronorite, gabbro, diorite and granodiorite represent the oldest intrusives in the Swartland Zone. Their age is not properly constrained but estimated at around 630 Ma (Scheepers and Schoch, 2006). Gabbro, monzogabbro, monzodiorite, syenodiorite and monzonite in the Tygerberg Terrane are younger and represent the onset of post-tectonic A-type magmatism.

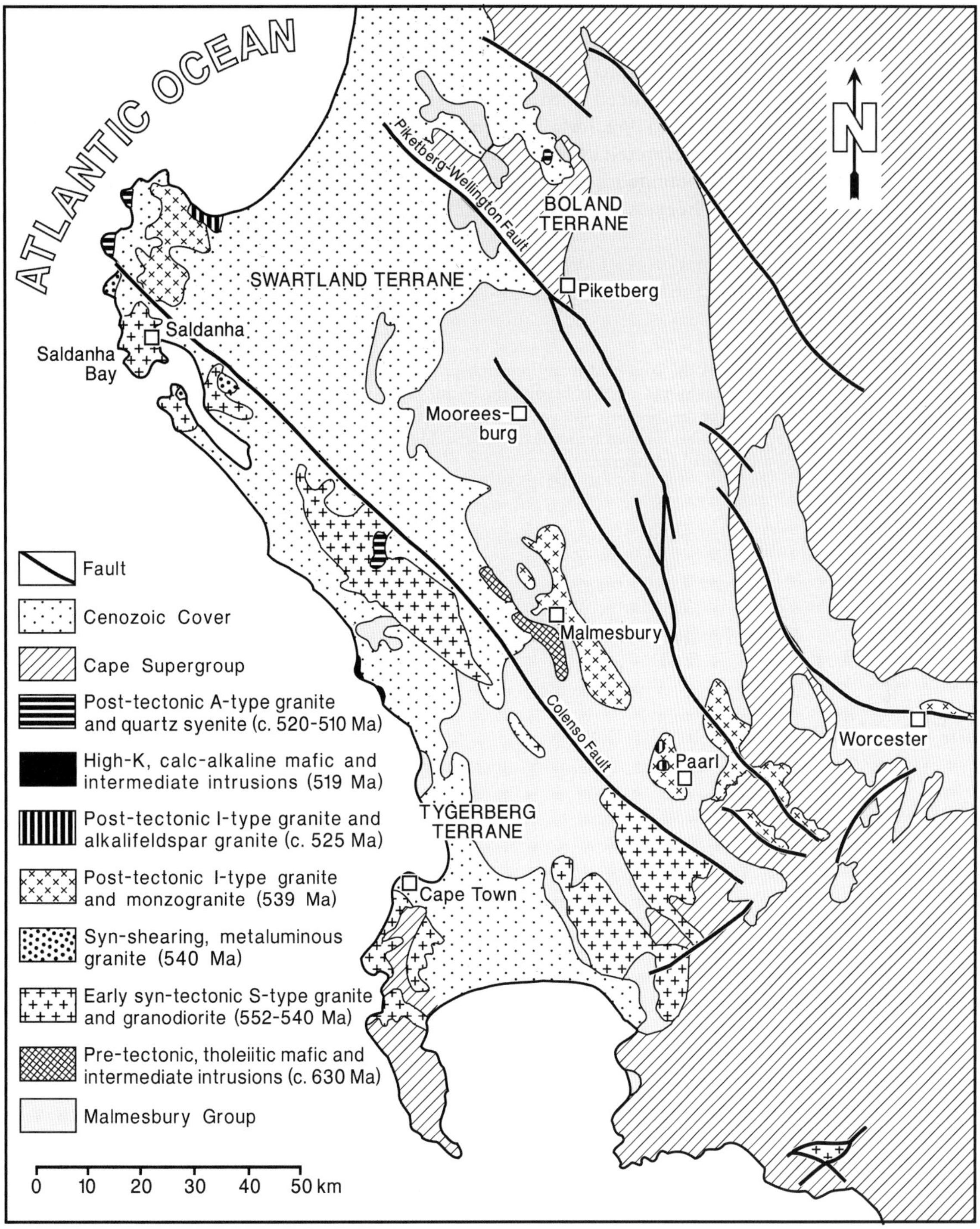


**Figure 5.6.5** Distribution of various generations of intrusive bodies in the western branch of the Saldania Belt (modified after Scheepers, 1995).

Peraluminous to metaluminous S-type granites (including granodiorite) dominate volumetrically, are found in the Tygerberg Terrane and represent the oldest intrusives of the Cape Granite Suite (Cape Peninsula, Saldanha and Darling Batholiths). The oldest group of these granites, with U-Pb single zircon ages ranging from $552 \pm 3$ Ma (Scheepers and Armstrong, 2002) to $540 \pm 4$ Ma (Armstrong et al., 1998), were emplaced along sinistral strike-slip shear zones and derived from garnet-rich metasedimentary material (Scheepers, 1995). At ca. 540 Ma, the tectonic regime changed to dextral strike slip and coeval granites, though of S-type, were derived from metaluminous, garnet-poor sources.

Post-tectonic granites with I-type characteristics were emplaced in a transtensional regime only to the northeast of the Colenso Fault (Vredenburg Batholith, Paarl Pluton). Their ages range from $539 \pm 5$ to ca. 525 Ma (da Silva et al., 1997; Armstrong et al., 1998). Their composition has been compared with shoshonitic or high-K calc-alkaline volcanism typical of island arcs or active continental margins. From $\varepsilon_{Nd}$ values between −3.89 and −1.44, a juvenile contribution to the magma has been inferred (da Silva et al., 1997).

A third phase of magmatism commenced with the intrusion of mafic to intermediate high-K calc-alkaline magmas (Yzerfontein pluton), dated at 519 ± 7 Ma (Jordaan et al., 1995), followed by anorogenic, A-type granite and quartz syenite. The latter are interpreted as products of dry, $CO_2$-rich melts that reached a high crustal level and they occur across the terrane boundaries. Ignimbrite ("Saldanha quartz porphyry"), which yielded a U-Pb single zircon age of 516 ± 3 Ma (Scheepers and Poujol, 2002), marks the final expression of Saldanian magmatism on the surface.

## ACKNOWLEDGEMENTS

This work summarises results obtained in the course of many research projects that received support from a variety of institutions and funding agencies, especially the Geological Survey of Namibia and the South African National Research Foundation (NRF). K.-P. Kelber kindly assisted with the drafting of some of the diagrams. Others, though modified, are taken from Miller (2008); copyright approval by the Director of the Geological Survey of Namibia for their reproduction is gratefully acknowledged. A. Tankard and M. Okrusch are thanked for providing helpful reviews of the original manuscript.

CHAPTER 5.7

# Mineral Deposits☆

Hartwig E. Frimmel[1] *and* Roy McG. Miller[2]

## Contents



## 5.7.1. Introduction

Comprehensive records of the early days of mining in Namibia are given by Wagner (1916) and Bürg (1942). More recent data on mines and mineral occurrences are given in a series of published and open-file reports that were collated by the Geological Survey of Namibia as a handbook in 1992. Different styles of mineralisation affected the Neoproterozoic rocks in southwestern Africa at the various stages of their tectonic evolution. These are briefly summarised in the following.

## 5.7.2. Rifting- and Spreading-Phase Deposits

Copper mineralisation occurs in the end-Nosib mafic volcanic rocks on the northeastern margin of the Northern Graben at the Neuwerk prospect. In the same graben succession, graphite in upper Nosib arkoses at the Okanjande and Black Range prospects must initially have been crude oil accumulations sourced from sapropelic shales at deeper levels in the graben (Miller, 2008). Significant volumes of low-grade, siliceous, slightly manganiferous magnetite–haematite ore occur in the Chuos Formation in the Owihende, Ongaba, Sebraskop and Okatjize deposits (Roesener and Schreuder, 1992). A comparable deposit with higher Mn content that has been mined is at Otjosondu (Schneider, 1992; Bühn et al., 1993). Copper-enrichment in marbles at Hohewarte and associated with amphibolite at Kainkagchas near Usakos may be of volcano-exhalative origin (Schneider and Seeger, 1992).

Economically important are a series of sedimentary-exhalative Pb-Zn-Ba ore bodies that are located within evolved rift-graben deposits. Of foremost significance is the Rosh Pinah Pb-Zn ore district in the Port Nolloth Zone of the Gariep Belt. Since 1969, the Rosh Pinah Mine has produced some 25 Mt of ore at an average grade of 7 wt% Zn, 2 wt% Pb, 0.1 wt% Cu and 11 ppm Ag, with remaining combined ore reserves and resources of 21.4 Mt at 7.5% Zn and 1.9% Pb. Largely stratiform ore bodies, consisting mainly of sphalerite, galena, tennantite, tetrahedrite, chalcopyrite, bornite and stromeyerite, are hosted by hydrothermally silicified carbonaceous argillite, with subordinate mineralisation in dolomitised limestone and arenite all of which belong to the Rosh Pinah Formation (Page and Watson, 1976; Alchin et al., 2005). Nearby felsic volcanic rocks probably did not serve as a major metal source. The contemporaneous volcanism provided, however, the heat engine for a productive hydrothermal convection system within the restricted Rosh Pinah Graben in which Pb was leached from the basement-derived, arkose-rich rift-graben fill (Frimmel et al., 2004). As the timing of mineralisation and

☆Frimmel, H.E., Miller, R. McG., 2009. Mineral deposits. Neoproterozoic to Early Palaeozoic evolution of Southwestern Africa. In: Gaucher, C., Sial, A.N., Halverson, G.P., Frimmel, H.E. (Eds.): Neoproterozoic-Cambrian tectonics, global change and evolution: a focus on southwestern Gondwana. Developments in Precambrian Geology, 16, Elsevier, pp. 227–229.

[1] Department of Geological Sciences, University of Cape Town, Rondebosch 7700, South Africa.
[2] PO Box 11222, Windhoek, Namibia.

*Developments in Precambrian Geology*, Volume 16
ISSN 0166-2635, DOI 10.1016/S0166-2635(09)01616-8

volcanism overlaps with the Kaigas glaciation, it has been postulated that an anoxic bottom water interface, required for the formation of massive sulphide ore bodies, developed in consequence of eustatic sea level drop and the cutting off of the Rosh Pinah Graben from the main Gariep Basin (Frimmel and Jonasson, 2003). Basin-bounding growth faults along the eastern margin of the graben acted as channels for the mineralising fluids.

Although of secondary origin, the supergene Skorpion Zn deposit (Borg et al., 2003) in the same district was derived from a Rosh Pinah-type protore (Frimmel et al., 2004). From 2003 until 2008, a total of 5.78 Mt of ore (at 12.0 wt% Zn) were mined at the Skorpion mine and combined ore reserves and resources amount to 24 Mt at 11.3 wt% Zn. The supergene, main-stage mineralisation was induced by acidic, low-temperature meteoric waters that dissolved detrital feldspar and carbonate cement within the host rocks with subsequent filling of the thus created secondary porosity with a range of Zn-hydrosilicates and Zn-carbonates (Borg et al., 2003).

In the Damara Belt, several comparable sedimentary-exhalative deposits occur in the vicinity of significant lineaments that evolved from syn-rift growth faults. These include a Pb-Zn occurrence at Klein Kachas near Usakos, and a series of deposits within or close to the Karibib Formation along the Omaruru Lineament, which formed the southern bounding fault to the northern graben, namely, Pb-Zn at the Namib Lead Mine, Pb-Zn and Cu in separate deposits at Joumbira, and pyrite north of the Rössing Mountain and at Odussa (Wartha and Genis, 1992; Schneider and Seeger, 1992). Stratiform, sediment-hosted base metal mineralisation is known from three prospects, Tsongoari-West, Tsongoari-East and Omupokko, near the eastern margin of the Central Kaoko Zone. Reported reserves of 5.8 Mt of ore grade at 6.4% Pb, 0.8% Zn, 0.4% Cu and 47 g/t Au and are located in metapelitic to metapsammitic rocks of the Otjorongwari Formation (Ugab Subgroup) within thin, laterally continuous marble beds (Gauert, 2005). The first stage of Pb and Zn mineralisation in the Otavi Mountainland, giving rise to Berg Aukas-type deposits, most likely was synsedimentary at the final stages of continental rifting (Frimmel et al., 1996a).

Twelve volcano-exhalative, Besshi-type, cupreous pyrite deposits are located along the Matchless Amphibolite Belt with its MORB-type metabasalts and metagabbros. Of these, the Otjihase (Goldberg et al., 2006) and Matchless (Adamson and Teichmann, 1986; Klemd et al., 1987) deposits are being mined. Although a syngenetic formation has been questioned (Maiden, 1993) for the latter deposit, a pre-metamorphic hydrothermal alteration related to the mineralisation has been recognised there (Häussinger and Okrusch, 1993) and provides strong support for synsedimentary ore formation on the former ocean floor.

## 5.7.3. Syn- and Post-Tectonic Deposits

There are over 600 mineral occurrences in the Otavi Mountainland (Söhnge, 1964). Two types of deposits have been distinguished there, Tsumeb and Berg Aukas type, based on contrasting Pb isotopic compositions (Hughes, 1987), metal association and timing of mineralisation (Söhnge, 1957). The polymetallic (Cu, Pb, Zn, Ga, Ge, Ag) Tsumeb-type ores of the Tsumeb, Kombat, Tsumeb West, Khusib and other smaller deposits were emplaced late during $D_3$ deformation and at about the peak of $M_2$ metamorphism into Mulden-derived feldspathic arenite (pseudo-aplite) and solution-collapse breccias filling pre-Mulden karst cavities (Innes and Chaplin, 1986; Lombaard et al., 1986; Frimmel et al., 1996a). The Pb-Zn-Ag deposits of the Berg Aukas type have both stratiform, massive sulphide components interpreted to be of sedimentary-exhalative origin (Frimmel et al., 1996a) as well as karst-generated breccia pipes containing primary sulphides (unpublished mining company reports and Emslie, 1979; Misiewicz, 1988). These deposits bear many characteristics of Mississippi Valley-type deposits, with the mineralising fluids having been orogenic brines that were expelled into the foreland (Frimmel et al., 1996a). Many of the mineralised solution-collapse breccias are not deformed but the early mineral phases at Tsumeb were deformed during a tightening (late $D_3$?) of the Tsumeb Syncline. Verwoerd (1957) reported deformed early galena in the Abenab West deposit. Intense syn-Kalahari karsting and alteration of the Tsumeb- and Berg Aukas-type ores (unpublished mining company reports and van der Westhuizen, 1984) added significant amounts of vanadium to the supergene products of this period of weathering for which an Oligocene age has been obtained (Boni et al., 2007).

Orogenic gold deposits occur in several different host rocks and stratigraphic levels from the pre-Damara basement to the Karibib Formation in the southern Central Zone. Of these deposits only the one at Navachab, which is hosted largely by the Okawayo Member, is currently being mined (Downing, 1983; Kisters, 2005). The Otjikoto deposit towards the eastern end of the northern Central Zone is currently being drilled and appears to be similar.

The locations of igneous U and Sn deposits appear to bear a relationship to the depth of the underlying former Benioff Zone. The uraniferous alaskites, such as those of the Rössing Mine (Berning, 1986; Mouillac et al., 1986), are located in the high-grade metamorphic western parts of the southern Central Zone above the shallow regions of the former Benioff Zone. The alaskites are the final differentiation products of limited-volume partial melting of basement granitic gneisses and the Nosib Group arkoses derived from this basement, all of which have slightly anomalous U contents. The rising alaskitic melts were dammed up against the overlying Swakop Group marbles and are concentrated in the Khan Formation at the top of the Nosib Group. The Rössing Mine has been operating since 1976 with 491.7 Mt combined resources and reserves at 0.03% $U_3O_8$ remaining. The nearby U deposit at Langer Heinrich is secondary, with carnotite being the main ore mineral, and associated with extensive Tertiary calcretisation. The Langer Heinrich mine opened in 2007 and estimated ore resources and reserves amount to 127 Mt at 0.06% $U_3O_8$.

Three belts within the highest stratigraphic levels of the Kuiseb Formation schists in the northern Central Zone and Southern Kaoko Zone host cassiterite-bearing pegmatites which also carry small amount of tantalite (Thamm, 1943; Richards, 1986). These belts are located, therefore, over deeper levels of the former Benioff Zone. The large, post-tectonic, unzoned, low-grade, cassiterite-bearing pegmatites at Uis were emplaced into sigmoidal tension fractures during a local, late-stage shearing event (Richards, 1986).

Still further north in the northern parts of the Southern Kaoko Zone and over even deeper parts of the Benioff Zone are the tin-tungsten stockworks of the Brandberg West and Goantagab deposits (Pirajno et al., 1987) and the gold stockwork at Ondundu in the Northern Zone (Reuning, 1937). Pegmatites in the northern and southern Central Zones and the Southern Kaoko Zone have been mined on various scales for beryl, various lithium minerals, and gemstones (mainly tourmaline) with waste mica, feldspar and quartz being the main by-products (minor pollucite in the Rubicon Pegmatite) and tin minerals, columbite-tantalite and bismuth minerals being rare curiosities. The Khan Mine pegmatite is an unusual occurrence of a Cu-bearing pegmatite that graded up to 6% Cu and contains giant crystals of hornblende and clinopyroxene in a pegmatitic quartzofeldspathic matrix (Ramdohr, 1938).

Apart from local mobilisation of base metal sulphides into thrust and fault planes, no significant syn-orogenic mineralisation is known from the Gariep Belt. Local post-orogenic skarn development around the Kuboos Pluton is of no economic significance. Similarly poor is the mineral endowment of the Saldania Belt. There the only exception is subeconomic W-Mo-REE mineralisation associated with granitoids of the Cape Granite Suite in the northern Boland Zone at the Riviera skarn deposit near Piketberg with reported 46 Mt of ore grading 0.22% $WO_3$ and 0.02% Mo (Walker, 1994; Rozendaal and Scheepers, 1995).

## *ACKNOWLEDGEMENTS*

This work summarises results obtained in the course of many research projects that received support from a variety of institutions and funding agencies, especially the Geological Survey of Namibia and the South African National Research Foundation (NRF). A. Tankard and M. Okrusch are thanked for providing helpful reviews of the original manuscript.

CHAPTER 5.8

# Geodynamic Synthesis of the Damara Orogen *Sensu Lato*☆

Roy McG. Miller[1], Hartwig E. Frimmel[2,3] *and* Thomas M. Will[3]

The three-pronged system of Pan-African orogenic belts in southwestern Africa, the coastal Kaoko and Gariep Belts and the intracontinental Damara Belt evolved through successive phases of intracontinental rifting, continental break-up, spreading, reversal of plate motion, subduction and continental collision. This evolution is recognised in contrasting tectono-stratigraphic zones bounded by major faults or thrusts or lineaments coinciding with large-scale facies changes. In the Kaoko Belt, these zones are, from east to west, the Northern Platform, Eastern Kaoko Zone, Central Kaoko Zone and Western Kaoko Zone, the latter including the Coastal Terrane. From north to south, the zones in the Damara Belt are the Northern Platform, Northern Margin Zone, Northern Zone, Central Zone, which is divided into northern and southern parts, Southern Zone, Southern Margin Zone and Southern Foreland, whereas in the Gariep Belt, the allochthonous, oceanic Marmora Terrane is distinguished from the eastern, para-autochthonous Port Nolloth Zone. The latter extends southwards via the Vredendal Inlier into the Swartland and Boland zones of the Saldania Belt. The Tygerberg Terrane in the latter belt may be a continuation of the Marmora Terrane.

The initiation of intracontinental rifting has been dated between 771 and 750 Ma in the Gariep Belt but was earlier in the Damara and Kaoko Belts (Kröner and Correia, 1980; Konopásek et al., 2008). In the Kaoko Belt, no early rift graben has been recognised but two parallel, northeast-trending rift grabens developed in the Damara Belt, one encompassing the present Northern Zone and the northern Central Zone, the other initially located where the present Southern Zone is. A series of half-grabens developed to the north and south and between the main grabens. The fill of both grabens (Nosib Group) exceeds 6,000 m in thickness and includes feldspathic sandstones, playa lake and evaporitic sabkha deposits, and, in the northern graben, alkaline volcanic rocks and originally oil-prone shales (now graphitic schists) in depth. Further south, in the Gariep Belt, a failed rift graben (Rosh Pinah graben) with alkaline volcanism and sedimentary-exhalative mineralisation was separated by a basement horst from a half-graben further west. No remnants of early rift-graben fills are known from the Saldania Belt.

The waning phases of continental rift volcanism at approximately 750 Ma coincided with deposition of the first shelf carbonates (Ombombo Subgroup, Rössing Formation) and directly thereafter with a regional, possibly global, Sturtian-age glacial event (Chuos glaciation in northern Namibia, Kaigas glaciation in the Gariep Belt). Only in northern Namibia is siliceous banded iron formation associated with the glaciogenic diamictite. This was followed by the accumulation of more shelf carbonates that were deposited in shelf basins on the rift shoulders that deepened away from the main rifts leaving exposed rift shoulders separating the main rifts and the shelf basins. They include at their base typical post-glacial cap carbonates in northern (Rasthof and Berg Aukas Formations) and central Namibia (Gobabis Member).

Following the Chuos/Kaigas glaciation, deposition of the remaining units of the Abenab, Usakos and Kudis Subgroups and the Court Formation in the Damara Belt took place over a period of approximately 115 myr, whereas post-glacial carbonate deposition in the Gariep Belt was possibly followed by a longer hiatus in sedimentation. After the initial post-glacial flooding, the Abenab Subgroup records a progressive shallowing with a final sea-level fall of possibly as much as 400 m ahead of the next major glaciation that is represented by the 635 Ma Ghaub Formation diamictite in the Damara and Kaoko Belts. Direct evidence of this glaciation, which can be correlated with the global Marinoan 'Snowball-Earth' event, is abundant in northern Namibia but

☆Miller, R. McG., Frimmel, H.E., Will, T.M., 2009. Geodynamic synthesis of the Damara Orogen sensu lato. Neoproterozoic to Early Palaeozoic evolution of Southwestern Africa. In: Gaucher, C., Sial, A.N., Halverson, G.P., Frimmel, H.E. (Eds.): Neoproterozoic-Cambrian tectonics, global change and evolution: a focus on southwestern Gondwana. Developments in Precambrian Geology, 16, Elsevier, pp. 231–235.

[1] PO Box 11222, Windhoek, Namibia.
[2] Department of Geological Sciences, University of Cape Town, Rondebosch 7700, South Africa.
[3] Geodynamics and Geomaterials Research Division, University of Würzburg, Am Hubland, D-97074 Würzburg, Germany.

*Developments in Precambrian Geology*, Volume 16
ISSN 0166-2635, DOI 10.1016/S0166-2635(09)01617-X

probably lacking further south in the Gariep Belt where the major unconformity beneath the Wallekraal Formation may be related to this glaciation. The Usakos and Kudis Subgroups record a rapid deepening of both Damaran grabens. The northern graben developed into the Outjo Sea in which 4,000 m of sediment was deposited. The southern graben developed into the Khomas Sea with the deep-water fans being deposited on, and interfingering with, anoxic carbonaceous pelites of the Blaukrans Formation on the developing southern passive margin of this graben. Concomitantly, the distal and relatively thin turbidites of the Zebra River, Brandberg West and Brak River Formations were deposited in the Southern Kaoko Zone at the western end of the northern graben where it opened out into the evolving north-south trending oceans of the Kaoko/Gariep Basins. The successive stratigraphic units in the Abenab and Usakos Subgroups overstepped each other onto the marginal rift shoulders in their respective shelf basins and gradually reduced the remaining areas of exposure of these rift shoulders. Similarly, the Court Formation was deposited in a shelf basin on the Southern Foreland that dipped away from the southern rift and the units in the formation overstep each other onto the exposed rift shoulder marginal to this shelf basin.

Prior to the completion of this phase of deposition in the Damara Orogen, granulite-facies $M_1$ metamorphism of, and granite intrusion into, forearc or trench greywackes was active above a westward or eastward dipping (various interpretations have been published) subduction zone west of the western Kaoko Basin in what may have been part of the Ribeira Belt in South America. These rocks would become the Coastal Terrane of the Western Kaoko Zone in time.

The Ghaub glaciation, the final submergence of the rift shoulders and the MORB composition of the basalts at the base of the Kuiseb Formation schists at the northern edge of the Southern Zone, mark: (i) the location and time at which continental rupture took place along the northern bounding fault of the southern graben, the Okahandja Lineament; (ii) the first development of oceanic crust below the Khomas Sea; (iii) the migration of the fill of the southern graben from its initial position in the Southern Zone to its present position in the Southern Margin Zone; (iv) the thermal sag of the opposing passive continental margins on either side of the widening Khomas Sea and (v) the formation of separate Congo and Kalahari Cratons on either side of the Khomas Sea.

The major marine transgression that followed the Ghaub glaciation resulted in a sea-level rise of about 400 m and deposition of distinct cap carbonates (Keilberg Member and Bildah Member). The submergence of the rift shoulders at the northern and southern edges of the Damara Ocean (extending from the Northern Platform to the Southern Margin Zone) produced a change in the palaeoslope of these edges such that the post-Ghaub successions at these edges thickened towards the Damara Ocean and not away from it as in the underlying units. Deep-water turbidites continued to be deposited on the opposing passive margins of the Khomas Sea (Southern Zone) – at least until 609 Ma as indicated by U-Pb ages of zircon grains from a rhyolite at the top of the Southern Margin Zone succession. The southern boundary of the northern graben, the Omaruru Lineament, was a growth fault throughout the spreading phase and the locus of sedimentary-exhalative activity.

The metagreywackes of the Kuiseb Formation make up the final spreading-phase stratigraphic unit in the Damara Supergroup. These extend from the Northern Margin Zone to the Southern Zone and reach a thickness of almost 10 km in the Northern Zone. Although the northern graben continued to deepen and contains possibly over 17 km of sedimentary and volcanic rocks, it never ruptured. In contrast, continental rupture in the southern graben led to the formation of a mid-oceanic ridge, now the Matchless Amphibolite Belt, and oceanic crust below the Southern Zone. Although the Kuiseb Formation greywackes swamped the active ridge most of the time, there were two periods during which volcanic rocks of MORB affinity reached the seabed and associated black smokers produced Besshi-type Cu-Fe sulphide deposits.

At about the same time, the failed rift graben of the older Gariep Basin became re-activated with renewed sedimentation in the younger Gariep Basin. The termination of spreading in the Damara Orogen has not been dated but this is presumed to have been at about 600 Ma on the basis of the earliest syn-tectonic granite ages. Assuming a post-Ghaub spreading rate of 25 cm $yr^{-1}$ in the Southern Zone, the total amount of ocean-floor spreading in the Khomas Sea would have been some 875 km in the 35 myr between the Ghaub glaciation and the end of spreading. Adding to this the gradual but indeterminate widening of the grabens in the 115 myr between the end of rifting at 750 Ma and the Ghaub glaciation at 635 Ma, the Khomas Sea may have had a final width of only about 1,200–1,500 km. Inversion of the inferred Gariep back-arc basin took place sometime between 600 and 580 Ma, followed by first accretion of oceanic crustal rocks by approximately 575 Ma (Figure 5.8.1). At about the same time, accretion of the exotic Coastal Terrane onto the rest of the Kaoko Belt occurred at an early stage of transpressive northwest-over-southeast orogenesis and continental collision along the crustal-scale Three Palms Mylonite Zone. This resulted in a first regional high-grade metamorphic event between 595 and 575 Ma ($M_2$ in the Kaoko Belt, $M_1$ in the Damara Belt) and first syn-tectonic granite emplacement (580 Ma) in the Kaoko Belt (Figure 5.8.1). Although not obvious, $D_1$ uplift occurred on the Northern Platform and exposure and karsting of the Otavi Group carbonates took place along the broad margins of the Northern Platform as far east as the Otavi Mountainland. In contrast, sedimentation probably continued in the

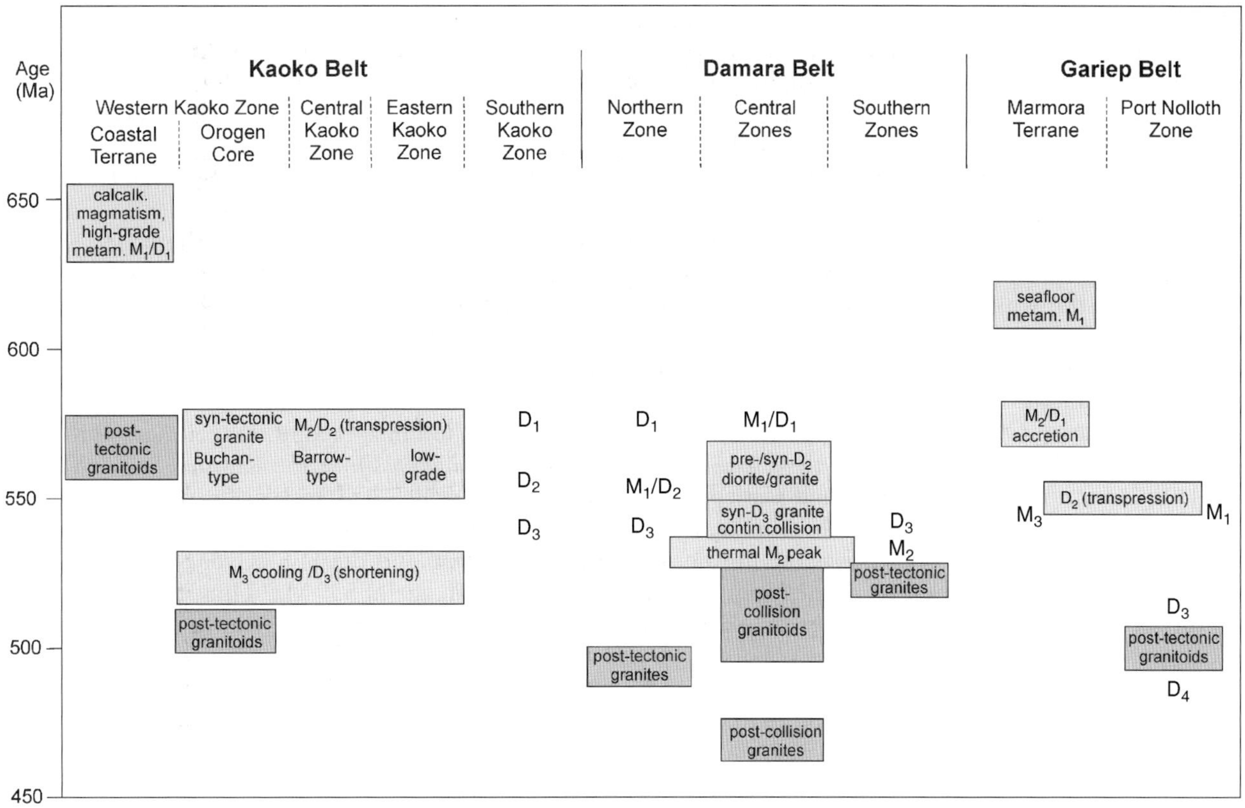


**Figure 5.8.1** Summary of the main tectonic, metamorphic and magmatic pulses in the various zones of the Kaoko, Damara and Gariep Belts.

south where younger glaciogenic deposits (Numees Formation and equivalents), possibly related to the global ca. 580 Ma Gaskiers glaciation, were laid down both in the continental as well as the oceanic realms of the Gariep Belt.

Lower pressure and eventually lower temperature assemblages that form coronas on peak-$M_2$ minerals indicate progressive decompression in the Kaoko Belt during the formation of huge, eastward-vergent, basement-cored, $D_2$ nappes in the Central Kaoko Zone as the compression direction gradually rotated to west over east. The erosional debris resulting from decompression was deposited as a northern molasse, the Mulden Group, on the Northern Platform and in intramontane basins in the Eastern Kaoko Zone of the Kaoko Belt and the Northern Margin Zone of the Damara Belt. Brittle, late-$D_2$, west-over-east deformation thrust the rocks of the Central Kaoko Zone over the Otavi and Mulden rocks of the Eastern Kaoko Zone. In the Northern Margin Zone and the Northern Zone of the Damara Belt, only a few open $D_1$ folds with kilometre-scale wavelengths have been identified. In the Northern Margin Zone, late $D_2$ folds in Mulden rocks can be traced downwards across the basal unconformity into tighter, early $D_2$ folds in the underlying Damara succession. Late-tectonic granites and pegmatites in the Kaoko Belt date the end of $D_2$ and thus the end of continental collision in this region at 550 Ma. A gentle phase of north-south compression in the Kaoko Belt coincided with final continental collision in the Damara Belt. Post-tectonic granites have ages of 540 Ma.

At about the same time that subduction was initiated in the Kaoko Belt, or shortly thereafter, northeastward subduction of the Kalahari Craton and the ocean floor of the Southern Zone beneath the Congo Craton began. The leading edge of the overriding Congo Craton was the Okahandja Lineament. Uplift and erosion of the active continental margin of the overriding plate, represented by the southern Central Zone and the zones to the north of it, shed an arc-trench succession of greywackes into the closing Khomas Sea of the Southern Zone, the Hureb Formation.

The age of $F_1$ structures and the accompanying sillimanite-bearing, $s_1$-parallel assemblages in the southern Central Zone is unknown but the oldest post-$D_1$ granitic intrusions in this region have an age of 565 Ma. Transpressive deformation in the Southern Margin Zone produced huge, southwest-vergent $D_1$ nappes (transport direction at a low angle to the strike of the belt) with roof and floor thrusts and complex internal duplex structures. Serpentinites intruded up the thrust planes. $D_2$ produced recumbent, southwest-verging sheath folds in the southern Central Zone and renewed southwest-directed thrusting and nappe formation followed by further serpentinite intrusion in the Southern Margin Zone. In stark contrast to the tectonic transport direction in the southern Central Zone and the Southern Margin Zone, $D_2$ thrusts and sheath folds in the stratigraphically high-level schists of the intervening Southern Zone indicate a tectonic transport direction perpendicular to strike, that is to the southeast. $M_1$ in the Southern Zone and Southern Margin Zone was syn-$D_2$ and medium-pressure mineral assemblages (9–10 kbar at 575–590°C) are recorded from near the junction of these two zones. An age of 555 Ma is assumed for $D_2$ in the Damara Belt. Syn-$D_3$ granites date continental collision, final closure of the Khomas Sea and $D_3$ doming in the southern Central Zone and the Southern Margin Zone at 542 Ma. Collision in the Gariep Belt appears to have had the same age.

Tectonic loading by the approaching Damara and Gariep Orogens depressed the northern and western margins of the Kalahari Craton below sea level initiating deposition of the foreland siliciclastic and carbonate successions of the Kuibis Subgroup of the Nama Group. The base of the group has an age of approximately 555 Ma. Fossil remains of an Ediacaran fauna are preserved in sandstones and the shelly fossils *Cloudina* and *Namacalathus* as well as a sponge-like metazoan in stromatolitic and thrombolitic carbonate reefs. Ages of detrital white micas indicate that the Nama siliciclastic sediments formed a distal flysch derived from the approaching Damara and Gariep Orogens. Zircon grains from three ash beds yielded terminal Neoproterozoic ages of 549, 545 and 543 Ma, the youngest just below the Precambrian/Cambrian boundary. Once the Khomas Sea had closed, the detritus from the active Damara and Gariep continental margins, which had initially been deposited as the Hureb Formation, was deposited on the Southern Foreland as the distal reddish molasse sediments that form the upper part of the Nama Group, that is the northern Nomtsas Formation and the Fish River Subgroup. Zircon from an ash bed almost at the base of the Nomtsas Formation yielded an age of 539 Ma. Thus, the Precambrian/Cambrian boundary is encompassed by the 4 myr unconformity at the base of the Nomtsas Formation. Pre-trilobite trace fossils occur in the Nomtsas Formation and Fish River Subgroup. While the unconformity at the base of the Kuibis Subgroup appears to be related to the $D_2$ phase of deformation in the Damara Belt, the unconformity at the base of the Nomtsas Formation is related to $D_3$. The various ridges that characterise the architecture of the Nama Basin formed during Nama sediment deposition and are thus most likely forebulges.

Voluminous 565–555 Ma pre- to syn-$D_2$ granitoids intruded the Central Zone of the Damara Belt, the active continental margin of the Congo Craton. The peak of Damaran $M_2$ metamorphism, following decompression by 2–3 kbar between $M_1$ and $M_2$, was post-tectonic, caused extensive partial melting of basement and cover rocks in the west and is dated by an anatectic red granite in the southern Central Zone and by the growth of very low-grade, $<2\,\mu m$ white micas in the Mulden and Nama rocks at 535 Ma. Smaller volumes of post-tectonic granitoids intruded the Central Zone and Southern Kaoko Zone between 540 and 508 Ma.

The northern Central Zone appears to have been a back-arc region with much younger, post-collision deformation and plutonic episodes. Syn-$D_2$ leucogranites in this region have been dated at 514 Ma. Widespread intrusion of syn- to post-$D_3$ granites accompanied the diapiric rise of hot marble domes (Karibib Formation) in the period 460–480 Ma, that is some 60–80 myr after continental collision.

The final, significant tectonic event in the Damara Belt was the gravity-induced emplacement of the thin-skinned nappe package of the parautochthonous and allochthonous Naukluft Nappe Complex onto the Schwarzrand Subgroup rocks of the northwestern Southern Foreland at 495 Ma. Hot, highly saline, dolomite-rich, evaporite-derived brines from the rift-phase Duruchaus Formation served as the lubricating medium for emplacement of the complex.

On the South American side of the Atlantic Ocean, the counterpart of the Kaoko Belt is thought to occur in southeastern Brazil and Uruguay. Several authors (e.g. Porada, 1979; Ebert et al., 1996; Machado et al., 1996; Heilbron and Machado, 2003; Will et al., 2004) consider the Ribeira Belt in southern Brazil as the likely extension of the Kaoko Belt into South America. Other authors favour a connection to the Dom Feliciano Belt in Uruguay and southern Brazil (Porada, 1989; Basei et al., 2000, 2005; Goscombe and Gray, 2007).

The ~650 Ma calc-alkaline, I-type granites in the Coastal Terrane of the Kaoko Belt may be compared with the 640–590 Ma Florianopolis-Pelotas-Aigua Batholith, a remnant of a magmatic arc system that follows almost the entire length of the Dom Feliciano Belt. The Ribeira and Dom Feliciano Belts experienced a constrictional deformation with a strong transpressional component at about the same time as the main deformation event $M_2$ occurred in the Kaoko Belt (Heilbron and Machado, 2003). In addition, the Ribeira and northern Dom Feliciano Belts are characterised by transpressive dextral deformation and west- to northwest-verging structures (Machado et al., 1996; Basei et al., 2000). Goscombe and Gray (2007) suggested that this observation, together with the documented east-verging structures and the sinistral displacement in the Kaoko Belt, implies that the magmatic arc represented by the calc-alkaline granites in the Dom Feliciano Belt and the Coastal Terrane of the Kaoko Belt moved southwards relative to the bounding Congo and Rio de la Plata Cratons. A similar southward-directed extrusion of the allochthonous part of the Western Kaoko Zone was already implied by Franz et al. (1999) and proposed by Will et al. (2004).

Two different tectonic settings have been suggested to explain the occurrence of the magmatic arc rocks. A west-directed subduction of Neoproterozoic oceanic crust (i.e. the Adamastor Ocean) underneath the Rio de la

Plata Craton was proposed by Porada (1979), Machado et al. (1996) and Masberg et al. (2005) on mainly geochemical grounds. Based on geochronological and geochemical data, including a recent sediment provenance study, Basei et al. (2000, 2005) argued for an east-directed subduction of ocean crust west of the Dom Feliciano Arc. In the former model the Three Palms Mylonite Zone must be a suture between South America and Africa at which oceanic crust was destroyed (Goscombe and Gray, 2007). However, as ophiolite fragments and high-pressure metamorphism are absent in the Kaoko Belt, these authors favour an easterly subduction direction. Furthermore, they argue that ocean closure was initiated at $\sim$650 Ma and that the Coastal Terrane formed as a volcanic arc on an attenuated Congo cratonic crust. Both the Three Palms Mylonite Zone and the Purros Mylonite Zone would be part of a broad suturing zone in a back-arc position between the magmatic arc and the Congo Craton. Continued subduction is interpreted by Goscombe and Gray (2007) to have led to the closure of the northern Adamastor Ocean between the Dom Feliciano and Kaoko Belts by $\sim$580 Ma and caused overthrusting of the Coastal Terrane and the magmatic arc of the Dom Feliciano Belt onto the Congo and the Rio de la Plata Cratons, respectively. The docking of the Coastal Terrane is associated with the $576 \pm 15$ Ma coeval peak metamorphism and the main transpressional deformation from 580 to 550 Ma in the Orogen Core and the Central Kaoko Zone of the Kaoko Belt and the reworking of the Coastal Terrane (Gray et al., 2006). As pointed out by Basei et al. (2000, 2005), this scenario implies that both, the Kaoko Belt and the Gariep Belt to the south, are essentially back-arc basins. The strain partitioning associated with the oblique collision and the transpressional deformation caused the observed sinistral shear zones and the ESE-vergent thrusts in the Kaoko Belt and the dextral shear zones and the W to NW-verging structures in the Dom Feliciano Belt. Thus, Goscombe and Gray (2007) interpret the Kaoko-Dom Feliciano Belts as a classic doubly vergent orogen separated by a magmatic arc in the centre.

In summary, Neoproterozoic basin closure was diachronous and started earlier in the north (present coordinates). If a Wilson Cycle model is applied to the coastal belts of southwestern Africa, subduction-related closure of the Adamastor Ocean would have begun in the north as indicated by calc-alkaline magmatism between 655 and 625 Ma in the Coastal Terrane of the Kaoko Belt (Goscombe et al., 2005a; Masberg et al., 2005) and between 625 and 580 Ma in the Araçuaí and Ribeira Belts as well as the Dom Feliciano Belt (Basei et al., 2000). At the same time, the southern Adamastor Ocean (Gariep Basin) would have been still opening. Alternatively, the Gariep Basin at that stage could be explained as a back-arc basin behind the Dom Feliciano arc (Basei et al., 2005). Arc-continent collision in the Kaoko Belt occurred prior to 580 Ma with peak metamorphism having been reached between 580 and 570 Ma. Accretion of oceanic crust took place at about the same time in the Gariep Belt, while subduction-related closure of the Khomas Ocean is indicated by mafic and calc-alkaline magmatism between 560 and 550 Ma. Closure of the southern Adamastor Ocean, the Gariep Basin, was eventually achieved between 550 and 540 Ma, probably more or less simultaneously with collisional deformation and metamorphism, as well as widespread magmatism in the Damara Belt (see Chapter 5.5) and the Lufilian Arc (John et al., 2003; John et al., 2004). This was followed by post-orogenic A-type granite emplacement, continued cooling and exhumation through 480 Ma.

While the Kaoko, Damara and Gariep Belts are the products of typical collisional orogenic activities, the Saldania Belt records accretionary orogeny that followed the final assembly of Gondwana at approximately 540 Ma. The former three belts resulted from the amalgamation of the Rio de la Plata, Congo and Kalahari Cratons around a collisional triple junction, whereas the Saldania Belt has always taken a position at the margin of the Kalahari Craton and eventually of Gondwana. It thus forms a continuation of the ca. 520–500 Ma Ross-Delamerian accretionary orogen, which was a forerunner to the Late Palaeozoic Gondwanides along the leading edge of the Gondwana supercontinental margin (Milani and De Wit, 2008).

## ACKNOWLEDGEMENTS

This work summarises results obtained in the course of many research projects that received support from a variety of institutions and funding agencies, especially the Geological Survey of Namibia, the South African National Research Foundation (NRF) and the German Research Foundation (DFG). A. Tankard and M. Okrusch are thanked for providing helpful reviews of the original manuscript.

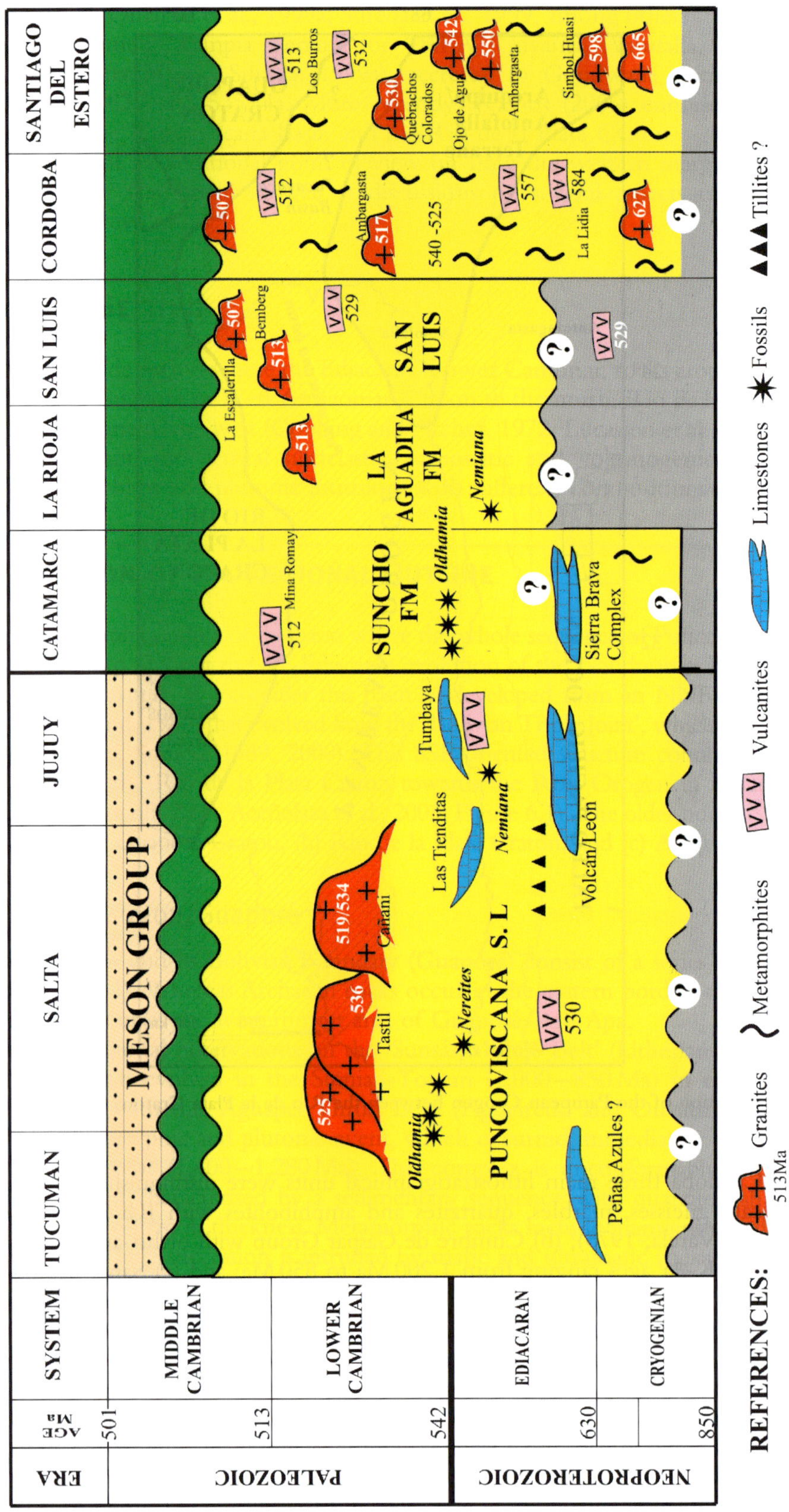


**Figure 6.2** Chronostratigraphy of the Pampean Orogen in western Gondwana.

1,900 to 1,200 Ma. Eastwards, in the Berenguela area, ages vary between 1,158 and 1,080 Ma. In the northern region of the Bolivian highlands (San Andrés de Huarina) a meta-granite was dated at 1,050 Ma, and interpreted as part of the Arequipa-Huarina Massif. Southwards, in Belén area of northern Chile, values range from 1,900 to 1,700 Ma. Finally in Urayani (Bolivia), gneisses and charnockites yielded 2,024 and 1,157 Ma while in

Cerro Choja between 1,024 and 1,200 Ma (Loewy et al., 2004). These rocks have also been metamorphosed in the Early Palaeozoic with radiometric ages for this event ranging from 650 to 360 Ma (Wörmer et al., 2000). The Cerro Chilla outcrop in the south of Titicaca Lake is composed of quarzites, shales and basalts, that are considered by several authors as belonging to the Ediacaran-Early Cambrian (Suarez Soruco, 2000). Further data from the Bolivian-Chilenian region date the high-grade metamorphosed rocks at 1,100 and 1,200 Ma (Wasteneys et al., 1995; Martignoli and Martelat, 2003).

The origin of the Arequipa terrane is a matter of debate as many authors have interpreted it to be a Laurentian fragment (micro-continent) placed on the Gondwanan margin during the Neoproterozoic (Ramos, 2008). This statement is confronted by Tosdal (1996), Lucassen and Thirwall (1998); Wörmer et al. (2000), Franz et al. (2000), who support a geological link between the rocks from the Arequipa Massif and those comprising the Amazonian Craton. According to these authors the latter would have been the source of the detrital material croping out in the former block. Antofalla was separate from the Arequipa terrane in Neoproterozoic times and reaccreted to the Gondwanan margin in the Late Ordovician (Ramos, 2008).

## 6.4. The Puncoviscana Basin

The Puncoviscana Basin displays a general NNE-SSW trend, and it was proposed that the unit was deposited in an aulacogenic structure developed in the western margin of Gondwana (Aceñolaza and Durand, 1986; Avila Salinas, 1992). The beginning of these aulacogenic structure has been named by Suarez Soruco (1989, 2000) as the 'Bolivian Triple Joint'. This structure was active in post-Stenian times (Tonian-Criogenian?), separating the Arequipa-Antofalla Terrane from Guaporé and Rio de La Plata cratons, allowing the opening of a marginal basin in the western border of Gondwana (Aceñolaza and Toselli, 2000; Figure 6.3).

In addition, it is necessary to mention the extensive sedimentary sequence affected by magmatic events (Figure 6.1). These events were assigned by Omarini et al. (1999a) to an intracontinental basin where a bimodal plutonic-volcanic event occurred, supporting the idea of an aulacogenic structure.

### 6.4.1. Lithostratigraphy

The sedimentary succession recognised in the basin is made up of fine siliciclastic rocks, carbonates, conglomerates and minor volcanogenic rocks. As previously noted, intense folding and faulting of the sequences prevent a clear understanding of the stratigraphy of the whole basin. Consequently, we will describe the lithostratigraphy in regional, basinal context.

### 6.4.2. Sedimentary rocks

The type area of the Puncoviscana Formation is placed in the homonymous locality in NW Argentina, near to the international border with Bolivia. Dark greyish, reddish and greenish slates, shales and quartzites are extensively exposed in the area. They generally occur as a heterolithic succession of dominantly fine-grained quarzites and dark shales. In some areas the succession is hundred of metres thick and laterally persistent. The same characteristics are observed southwards in Sierra de Zenta, with a frequent development of open folds with a NW general trend and some hundred metres of amplitude. Trace fossils are scarce and sedimentary structures, such as flute-casts, are frequent in the base of the quartzite levels. The same features are observed in the Tarija Department (southern Bolivia), where the unit is known as the San Cristóbal Formation.

In the many outcrops along Quebrada de Humahuaca, Quebrada del Toro, Lerma Valley and the Calchaquí mountains (NW Argentina), all described sedimentary lithotypes persist with very little modifications, notably the occurrence of lava flows, volcaniclastic levels and important limestone and dolostone bodies (Figure 6.4).

In the Tilcara-Purmamarca area (Jujuy), a heterolithic succession dominated by grey and greenish slates and sandstones is displayed with abundant primary structures (flute casts and ripple marks) and soft-body fossil impressions (*Nemiana simplex*). In the Cumbres de Lipán (Jujuy), dark brown and purple slates display large stratigraphic surfaces with *Helminthoida* and *Helminthoidichnites*. Nevertheless, in the area of Cuesta Muñano-La Polvorilla (Salta), the slates display fine-grained sandstone inliers. These are black, fissile and have provided an important ichnological association (*Oldhamia*, *Diplichnites*, *Dimorphichnus*; see also Chapter 9.3).

In Valle de Lerma (Salta), the sequence is represented by a succession of shales and sandstones which grade into conglomeratic-diamictitic levels (Sancha Formation/Los Noques Member), that are followed by limestones (Las Tienditas Formation). This series culminates with the typical rhythmite succession of ichnofossil-bearing shales

**Figure 6.3** Distribution of Pampean rocks in northwest Argentina.

and quarzites (Moya, 1998). The latter is widespread in the lower part of Quebrada del Toro and in the Calchaquí Valley (Cachi-Molinos). Besides, in Quebrada del Toro area, the presence of conglomeratic levels has also been recognised (Corralito Formation). Shaly levels appear randomly in the sequence, and have been regarded as sedimented on the apical areas of submarine fans or canyons, or corresponding to slumps in steep slope areas (Jezek, 1990). These lithotypes also crop out in Seclantás, Molinos and Tin-Tin areas (Salta). In some areas of the Quebrada del Toro, Escoipe, Molinos and Cachipampa (Salta), red and purple slates are abundant and have been interpreted by Jezek (1990) as the result of an hemipelagic sedimentation with high oxygen availability (Figure 6.5).

The westernmost outcrops are exposed in Sierra de Cobres (Salta and Jujuy), where greenish quartzite facies associated to greyish and yellowish slates have been described. These slates comprise abundant levels with trace fossils (*Oldhamia*, *Diplichnites*, *Helminthoidichnites* etc.) that are nicely preserved in the San Antonio de los Cobres area, allowing a fairly good correlation with the Abra Blanca-Muñano area to the east.

The sedimentary rocks are regionally distributed towards the south of Tucumán. Further south, metamorphosed slates are displayed in La Ovejería, Velasco and Famatina ranges of La Rioja province.

Most of the Cumbres Calchaquíes of Tucumán are represented by these rocks, and have been included in the Choromoro Belt by Mon and Salfity (1995). In the homonymous locality, the soft-bodied fossil *Selkirkia* sp. and the trace fossils *Treptichnus aequalternus* and *Ivesheadia lobata* have been described. A similar assemblage is recorded in

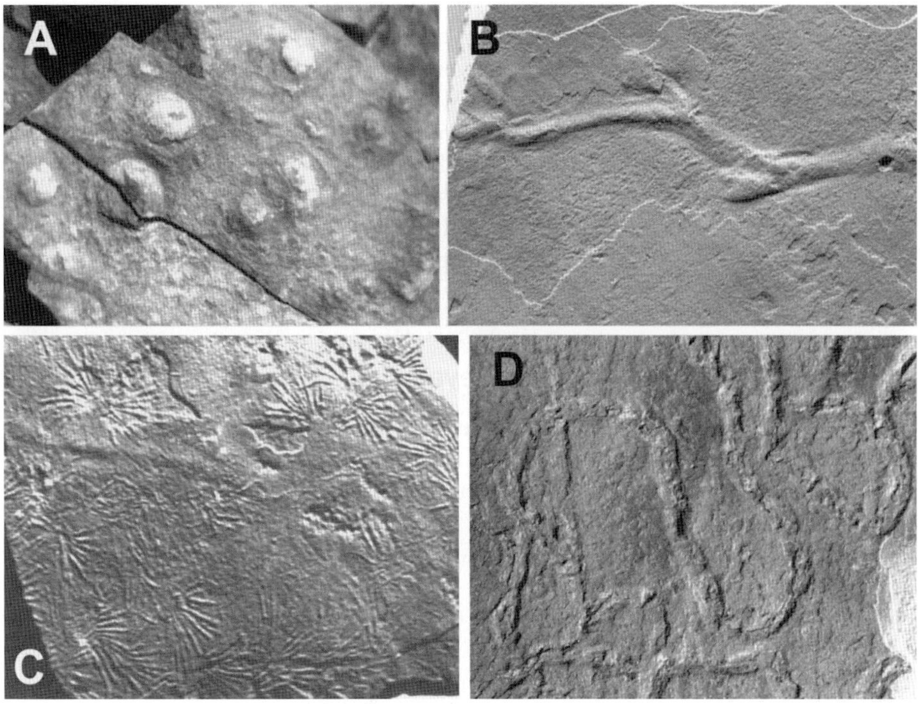


**Figure 6.4** Ediacaran-Lower Cambrian body and trace fossils. (A) *Nemiana simplex*, Purmamarca (Jujuy), (B) *Trepichnus inaequornatus*, Choromoro (Tucumán), (C) *Oldhamia* Abra Blanca (Salta) and (D) *Nereites saltensis*, Los Guachos (Salta).

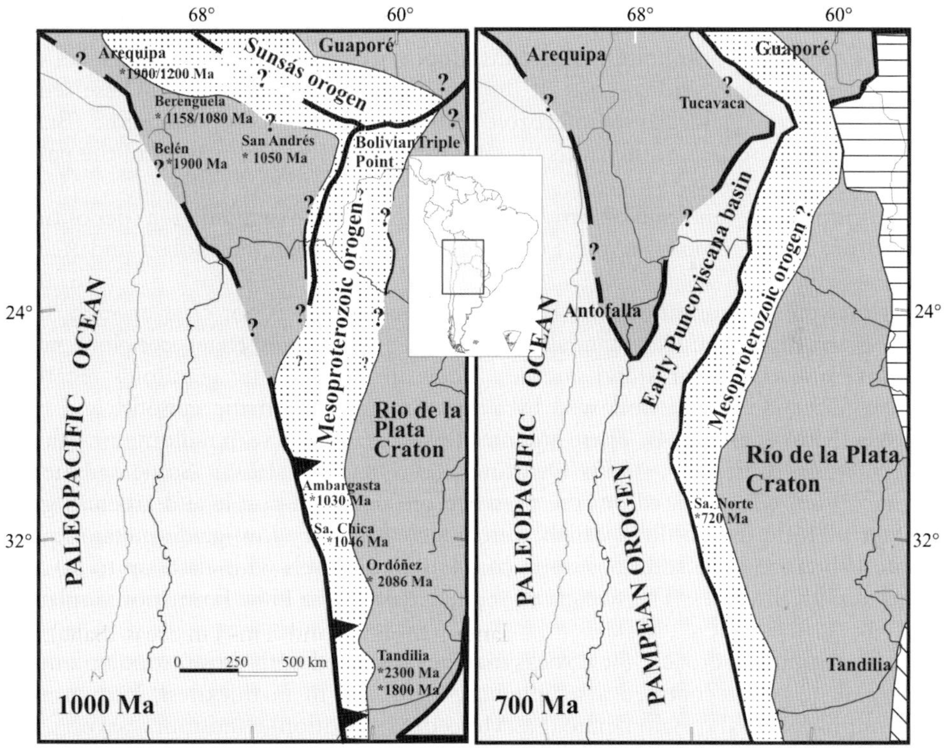


**Figure 6.5** Hypothetical position of the future Puncoviscana Basin at the end of the Sunsás Cycle at about 1,000 Ma (left). Early opening of the Puncoviscana Basin (700 Ma, right).

the Suncho Formation of the Sierra de la Ovejería (Catamarca), where shales and conglomerate display pavements with *Oldhamia* (Aceñolaza and Durand, 1986).

A southern extension of these facies has been proposed towards Ambato (Zimmermann and Van Staden, 2002), western Sierras de Cordoba (La Mermela Formation) and the central area of Sierra de San Luis (San Luis Formation; Baldo et al., 1996; Sato et al., 2002; Aceñolaza and Aceñolaza, 2007).

### 6.4.3. Provenance

According to sedimentological studies focusing the provenance of siliciclastic material of the Puncoviscana Basin in Salta and Tucumán, they display a main trend of E-W or SE-NW palaeocurrent directions, with a probable source area in the western part of Rio de la Plata Craton (Jezek, 1990). Based on several trace elements and geochemical data, Omarini and Do Campo (1993) and Do Campo and Ribeiro Guevara (2005) consider that provenance of the siliciclastic material of Puncoviscana represents a passive continental margin in western Gondwana during the Ediacaran-Lower Cambrian. On the other hand, and based on the moderate alteration of the sediments (CIA = 55 to 79) and certain geochemical patterns of trace element, Zimmermann (2005) assumed that Puncoviscana represents a foreland basin.

### 6.4.4. Sedimentation age of the siliciclastic levels

It becomes now possible to contrast ages of the sedimentary sequence deduced from fossils and detrital zircons (U-Pb ages) from sandstone levels.

According to fossils it can be established that the older levels may correspond to the occurrence of *Nemiana*, which is recognised in many places along the Quebrada de Humahuaca. This soft-bodied fossil has been found in the Ediacaran (Upper Proterozoic) around the world. Therefore, the Puncoviscana Formation *s.l.* cropping out in the northern area (Jujuy and Salta) is proposed to be older than those outcrops in the south (Salta—Tucumán-Catamarca) which contain *Oldhamia*, *Nereites* and *Treptichnus* deposited in the Lower Cambrian.

When considering detrital zircons ages from psammites of the Puncoviscana Formation *s.l.*, we interpret that they provide a maximum depositional age at the studied localities (Zimmermann, 2005; Figure 6.6). Detrital zircons were recovered from greywackes, and some of them may have been originated in coeval volcanic sources. Maximum depositional age constraints range from Late Cambrian to Late Neoproterozoic.

Detrital zircon age spectra suggests provenance from a continent with ages ranging from the Late Mesoproterozoic to the Early Neoproterozoic, between 1,080 and 850 Ma. Late Neoproterozoic ages between 650 and 542 Ma also occur. The morphology of zircons support a coeval volcanic provenance, which occurred in the Early Cambrian near the depositional areas of the Puncoviscana Formation *s.l.* These sediments may have been sourced in the Brazilian shield, Mesoproterozoic (Sunsás) and Late Neoproterozoic (Brasiliano), which is considered the main sediment source of the rivers flowing towards the passive margin of the palaeo-Pacific ocean (Adams et al., 2005, 2008a).

In addition, biostratigraphy and radiometric data are consistent, suggesting a depositional age for the Puncoviscana Formation *s.l.* encompassing the Ediacaran-Early Cambrian (Aceñolaza and Aceñolaza, 2005 and references therein).

Despite the fact that the relationship to the Rio de La Plata Craton is not clear enough, in Sierra Norte de Córdoba and in the Sierra de Ambargasta (Santiago del Estero) arkoses, sandstones, conglomerates and slates crop out, gradually turning into mica-schists and gneisses. They are intruded by granitoids recording calc-alkaline magmatism between 555 and 525 Ma, followed by a peraluminous magmatic episode, and by metamorphism dated at 525–505 Ma (Miró et al., 2004). These are intruded by granitoids with radiometric values ranging mostly between 500 and 600 Ma (Castellote, 1985; Quartino and Quartino, 1996). Some sedimentary rocks were described by Lucero (1969) who pointed out the presence of shales and purple and greenish greywackes on the western slope of Sierra Norte de Córdoba which were affected by very low-grade metamorphism and that are assigned to the Late Proterozoic. In addition, some conglomeratic levels with clasts up to 10 cm diameter were described interbedded with the slates. The siliciclastic sequence of the La Lidia Formation display ignimbrites with U-Pb ages of 584 Ma (Llambías et al., 2003).

Regarding a major regional approach, these rocks may be correlated with those cropping out in East Paraguay. There, Meinhold et al. (2000) recognised a sedimentary series 560 m in thickness, known as the Paso Pindó Group. The unit comprises a basal conglomerate (Las Mercedes Formation), sandstones, arkoses, greywackes (Paso Lima Formation) shales and tuffs (Cristo Redentor Formation). Meinhold et al. (2000) favour a Neoproterozoic age for the succession, because it is intruded by the Caapucú magmatic suite, made up of volcanic rocks and granitoids with an age of 531 Ma.

### 6.4.5. Age of the carbonate levels

#### 6.4.5.1. Cordillera Oriental

Thick limestones are interbedded in the Puncoviscana Formation *s.l.* in Salta and Jujuy. In Salta province, they crop out in Sierra del Castillejo and are included in a siliciclastic sequence. Whereas the basal siliciclastic level is

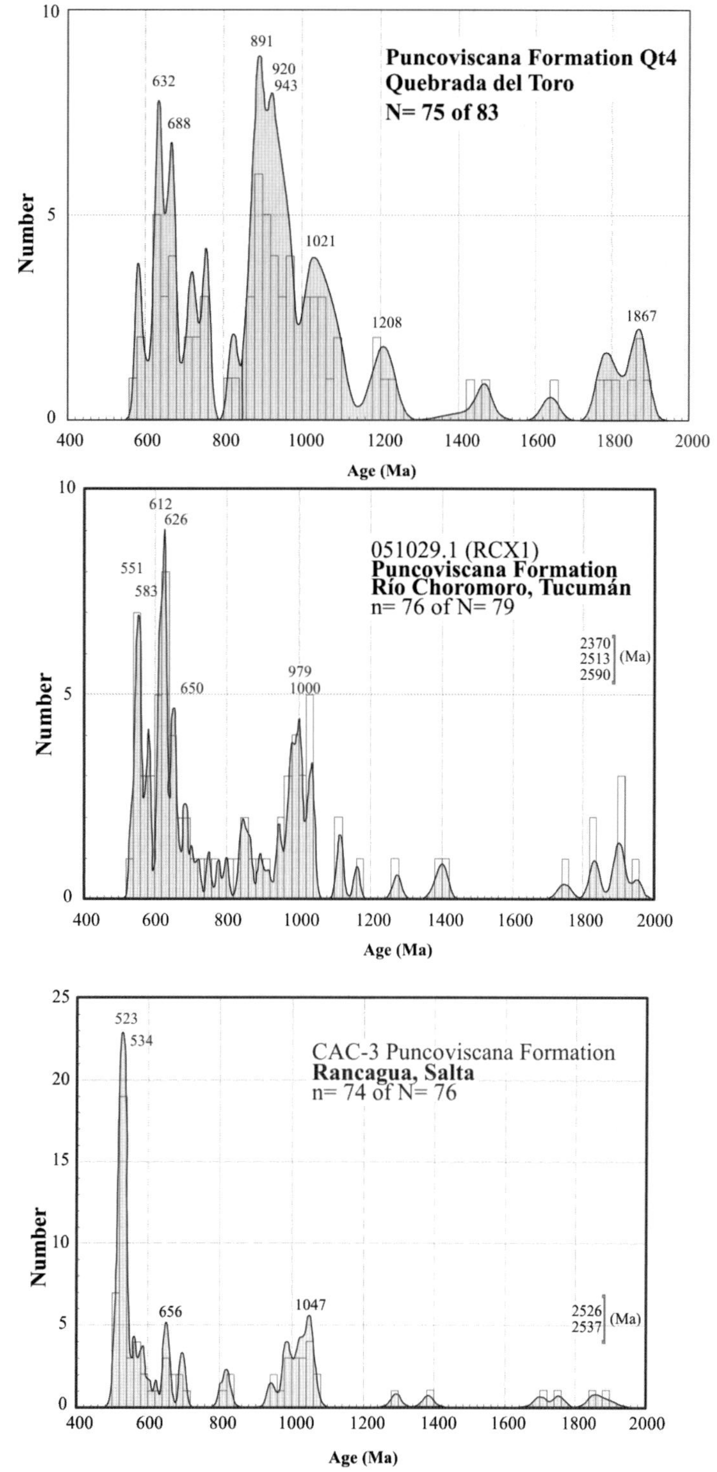


**Figure 6.6** Detrital zircon age spectra at different localities in the Puncoviscana Basin (modified from Adams et al., 2005, 2008a).

known as the Sancha Formation, the top is represented by the Puncoviscana Formation (Ortiz, 1962; Salfity et al., 1975). The base of the Sancha Formation is not exposed, and the succession includes red and green slates, greywackes, diamictites, conglomerates and turbidites, intercalated with basaltic flows. The Las Tienditas Formation (Ortiz, 1962) consists of a sequence of black, fine-grained limestones interbedded with marl beds and levels of brecciated, brownish-red limestones. The dark levels are rich in organic matter. The upper beds are covered by grey shales and green laminated sandstones and quartzites assigned to the Puncoviscana Formation *s.l.* (Baldis and Omarini, 1984; Figure 6.7).

In Jujuy province, carbonate outcrops are recognised in Quebrada de Humahuaca, between the León and Volcán localities as well as in Quebrada de Tumbaya. At Volcán and León the succession comprises black,

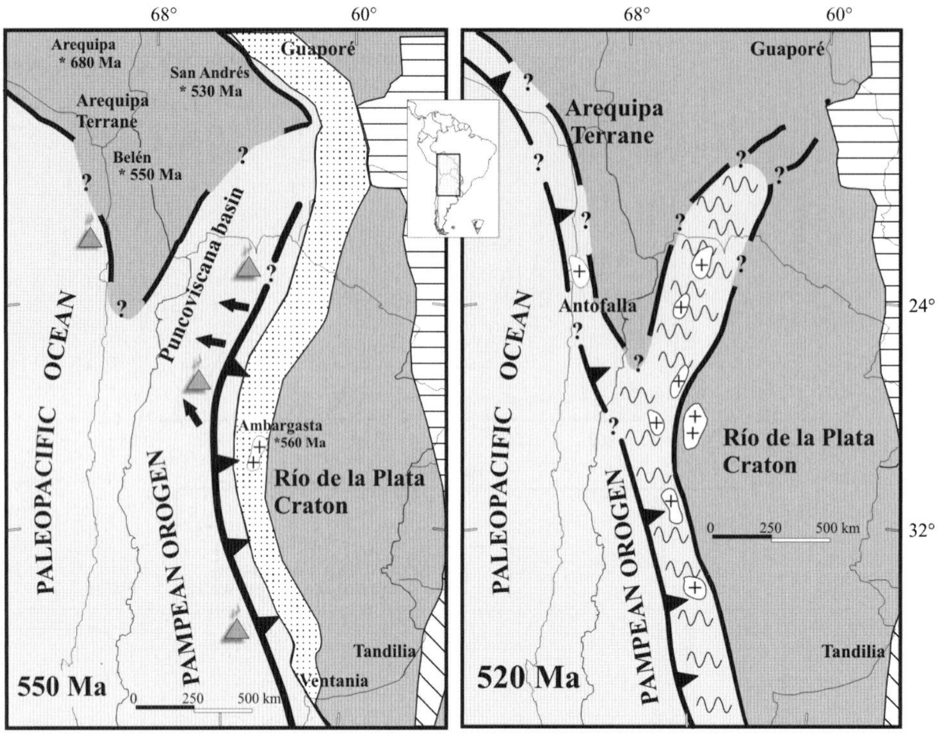


**Figure 6.7** Hypothetical evolution of the Puncoviscana Basin and the Pampean Orogen in Ediacaran-Lower Cambrian times. Left: Position at 550 Ma with illustration of volcanic activity, predominant palaeocurrents (arrows) and trench location. Right: Palaeogeography at 520 Ma, representative of the Tilcarian orogeny. Collision of the Arequipa Terrane closed and deformed the Puncoviscana Basin and the Pampean Orogen. The main intrusive granites are emplaced in this period.

fine-grained, micritic limestones, with white veins and strong folding; in some places they are interbedded with pelitic-psammitic levels (Loss and Giordana, 1952). Some clay–sand conglomeratic beds are interbedded with limestones, which together with alternated bands, are reminiscent of some glacial varves, and according to Loss and Giordana (1952) could be of glacial origin (tillites; Van Staden and Zimmermann, 2003). Scanavino and Guichón (1971) determined the maximum thickness of the calcareous sequence as 800 m, decreasing towards the north and south, wedging out within slates of the Puncoviscana Formation *s.l.*

In Cerro Agua Chica (Tumbaya Grande), the carbonate rocks comprise a sequence of grey, brown and black dolostones associated to pink limestones which overlie chloritised basaltic flows that locally show pillow structures (Camacho et al., 1993). Porto et al. (1990) estimated the total thickness as 85 m. The dolostone texture varies, with very fine-grained dolomite crystals in a glauconite matrix; sometimes showing oolites or a mosaic of subhedric to anhedric crystals filling gaps and cracks. Lamination is described in certain levels as being microbial in origin (cf., *Laminites*). The whole unit is strongly folded and faulted, preventing the determination of the stratigraphic sequence and its real thickness. Between the main dolostone level and the basalts, a heterolithic, varve-like sequence of shales and quartzites occurs.

$\delta^{13}C$ values vary between −1.6 and +3.4‰ Pee Dee belemnite (PDB) for carbonates of the Las Tienditas Formation. The highest positive records are near the base, decreasing upwards in the strata. In the uppermost sector of the curve a negative excursion has been interpreted as the probable record of the Precambrian-Cambrian transition. Tumbaya dolostones display a similar range (from −1.3 to +2.3‰ PDB), while Volcán-León limestones yielded values between +6.1 and +4.6‰ PDB.

The $\delta^{13}C$ values between −1.0 and +1.0‰ PDB are typical of the Early Cambrian while $\delta^{13}C$ values close to +5‰ PDB are common in Ediacaran carbonates (see Chapter 10). According to chemostratigraphic analyses, deposition of the Las Tienditas and Tumbaya dolostones may have started in the Ediacaran and culminated in the Early Lower Cambrian. Volcanic events are associated to the Tumbaya dolostones. A common element in both localities (Tumbaya and Las Tienditas) are volcanic rocks which are almost coeval with the carbonate sediments, an element not recorded in Volcán-León (Figure 6.8).

From the available data for the Puncoviscana Formation *s.l.* it may be deduced that Las Tienditas and Tumbaya carbonates may represent 'cap-carbonates' included on the same sedimentary cycle of the Puncoviscana Basin, which may include the post-Marinoan and post-Gaskiers glacial events, while the rocks from Volcán-León might be older.

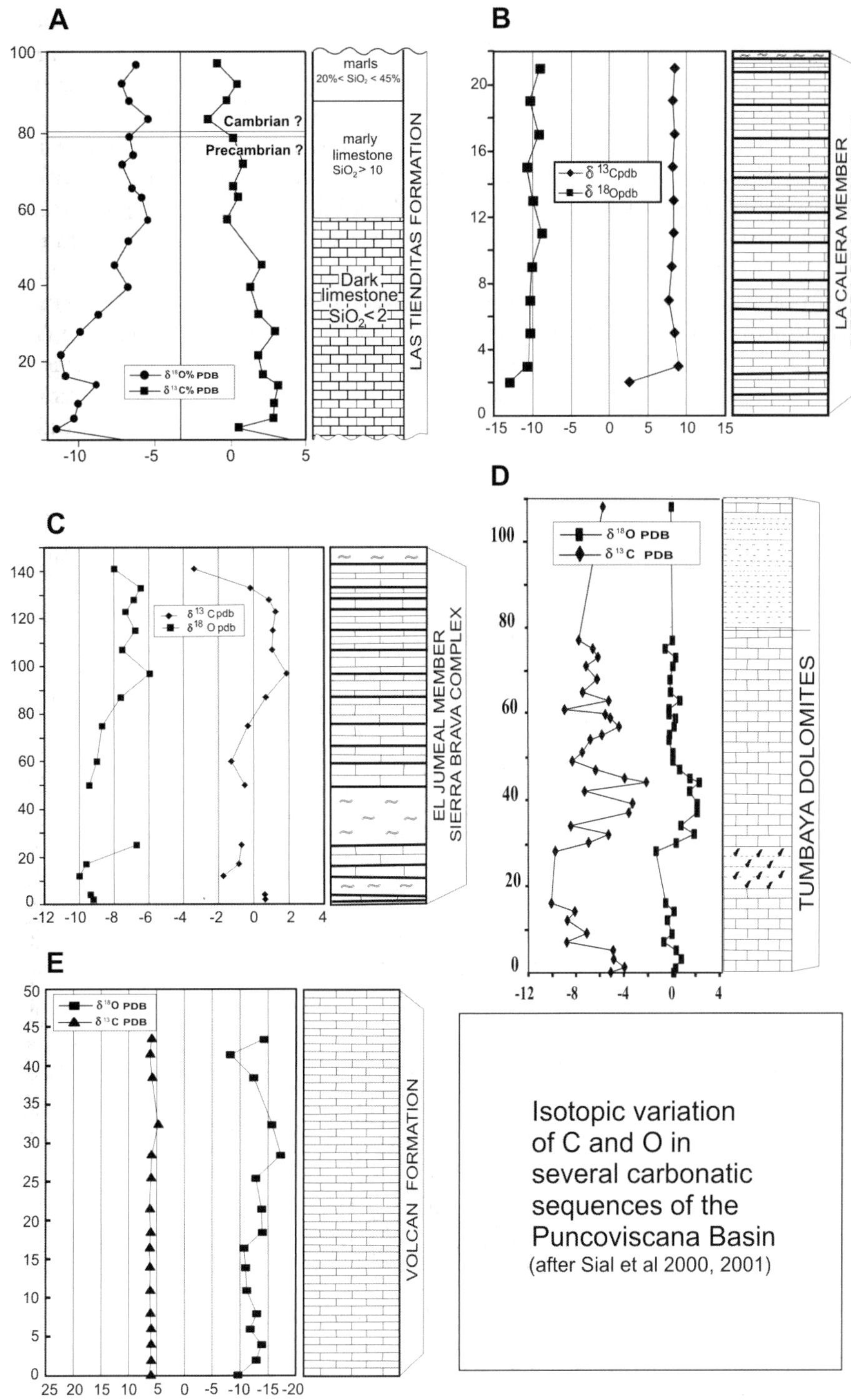


**Figure 6.8** C and O isotope chemostratigraphy for several carbonatic sequences of the Puncoviscana Basin. A. Las Tienditas Fm. (Salta); B. Sierra Brava Fm. La Calera, Sierra de Ancasti (Catamarca); C. Sierra Brava Fm. El Jumeal, Sierra de Ancasti (Catamarca); D. Tumbaya dolomites; and E. Volcan Fm. (Jujuy). (Modified from Sial et al., 2000b, 2001).

### 6.4.5.2. Sierras Pampeanas

In the Sierras Pampeanas and particularly in the Sierra de Ancasti, limestones and dolostones are widespread and included in the basement, being affected by low-to-medium metamorphic grade (Sierra Brava Complex). Its age, correlation and geological interpretation are still a matter of debate, C isotope chemostratigraphy being the tool which allowed the first correlations to be established. As an example, in El Jumeal Member, NW of Esquiú, marble levels are packages 5–10 m thick (maximum of 20 m), which are associated to basement gneisses, mica-schists and amphibolites. As a whole, marbles are granoblastic, medium-grained, white to grey or pinkish, and have been affected by dolomitisation with occurrence of tremolite. The mineral assemblage indicates amphibolite facies of metamorphism.

In La Calera Member, NW of Frías locality, outcrops are characterised by limestones interbedded with quartzites and schists. These fine-tomedium-grained carbonates range from white or grey to black in colour. The texture is granoblastic with quartz grains, magnetite and graphite. The association of calcite with quartz grains supports greenschist metamorphism facies. $\delta^{13}C$ values are +2.6‰ PDB, increasing up section to values of +8‰ PDB. $\delta^{18}O$ values at the base are −13‰ PDB and −10.0‰ PDB for the rest of the section.

The dolomitic limestones of El Jumeal Member display variable $\delta^{13}C$ values, starting in the lower part with +0.6‰ PDB, decreasing to −1.7 and −0.33‰ gradually turning positive again (+0.7 and +1.8‰), and culminating with a negative excursion to −3.4‰ PDB. This pattern is similar to $\delta^{18}O$, with negative values always below −6‰ PDB.

It is interesting to point out that a plateau observed for $\delta^{13}C$ (around +8‰ PDB) for la Calera Member is recognised worldwide for the 730−600 Ma time interval (Jacobsen and Kaufman, 1999; Melezhik et al., 2001). Dolostones of the El Jumeal Member show large fluctuation, as already explained, which makes it difficult to establish correlations with the different marble outcrops. Highly negative values at the top of the carbonate section has also been recognised in Las Tienditas limestones (Sial et al., 2001; Toselli et al., 2005).

A very useful lithologic element for sequence correlation is the presence of interbedded volcanic rocks or their metamorphic counterparts. In the Puncoviscana Formation *s.l.*, Las Tienditas limestones and Tumbaya dolostones are associated with volcanic rocks, while southwards the sequence of Volcán lack them completely. The same happens in Sierra de Ancasti, where the dolostones of El Jumeal Member are related to amphibolites, while the limestones of La Calera Member are related to siliciclastic metamorphic rocks.

The lithologic associations together with $\delta^{13}C$ values in the different sections allows the the separation of two different successions, in concordance to global $\delta^{13}C$ curves (Jacobsen and Kaufman 1999; Melezhik et al., 2001). The first assemblage is represented by carbonate rocks of the El Jumeal Member in Sierra de Ancasti and Las Tienditas-Tumbaya (Puncoviscana Formation *s.l.*) with $\delta^{13}C$ values ranging between −1.6 and +3.4‰ PDB, ending with a negative excursion which allows these rocks to be considered as Neoproterozoic to Middle Cambrian in age (Sial et al., 2001; Toselli et al., 2005, 2008). In Volcán (Puncoviscana Formation *s.l.*), and La Calera Member, $\delta^{13}C$ ranges between +4.6 and +8.9‰ PDB (Sial et al., 2001; Toselli et al., 2005, 2008), supporting a Neoproterozoic age, according to the secular variation curves of Jacobsen and Kaufman (1999), Melezhik et al. (2001), Hoffman et al. (1998) and Kha et al. (1999).

## 6.4.6. Volcanic events

The existence of volcanogenic levels interbedded in the Puncoviscana Formation *s.l.* was recognised at several localities of the Cordillera Oriental, Puna and Sierras Pampeanas.

In most cases, they are interbedded bodies usually less than 3 m in thickness. They are dark grey or greenish grey with porphyric texture. Petrographically they are mostly tholeiitic, transitional and alkaline basalts, latites, ankaramites, pyroxenic hornblendites and basanites. The studied outcrops are in the area of Tumbaya and Quebrada Las Chilcas, in Tinti Conchi (Río Iruya gulch), in Tipayoc, in Río Reyes and Alto de Minas (Jujuy) and in Quebrada El Vallecito (Salta; Coira et al., 1990). These volcanic contributions, although less consistent, can also be recorded in the north of Eastern Sierras Pampeanas (Koukharsky et al., 2003; Leal et al., 2003; Llambías et al., 2003) and are essentially made up of basaltic andesites, andesites, dacites and rhyolites.

### 6.4.6.1. Tectonic significance of the pre-Tilcarian volcanic rocks

In Ediacaran-Lower Cambrian times, different extrusive magmatic events coeval with the opening of Puncoviscana Basin are recognised. This material was represented by acid and basic tuffs and flows.

The petrographic and geochemical characteristics of the volcanic rocks have been analysed from the data published by Chayle and Coira (1987), Coira et al. (1990) and Omarini et al. (1999b), which reveal the transitional character of the magmatism in the Salta and Jujuy provinces. This magmatism might be interpreted as a consequence of several tectonic events and is recorded in three different groups of volcanic rocks. These rocks vary from intraplate alkaline basalts and transitional tholeiitic basalts (between continental and oceanic to island arc basalts emplaced next to the continent). This indicates a tectonic evolution from an intraplate setting (alkaline basalts) becoming extensional (continental and oceanic tholeiitic basalts) and finally turning into a compressional subduction phase typical of oceanic island arcs (oceanic tholeiitic basalts).

Ignimbrites cropping out in the Sierra Norte of Córdoba are worth a special analysis. They are interbedded with meta-conglomerates of the La Lidia Formation with conventional U-Pb age on zircons of 584 Ma. As regarded by Llambías et al. (2003), the Neoproterozoic silicic ignimbrites, together with the calc-alkaline batholith which succeeds them, suggest the existence of a magmatic arc associated to the western continental margin of

Gondwana. The ignimbrites could have been provided by strato-volcanoes, typical of a continental margin related to subduction, or they could also be associated to post-orogenic extensional tectonics.

Earlier, an andesitic-dacitic event occurred, corresponding to the Balbuena Formation in Sierra de Ambargasta, with a K-Ar age of 514 Ma. Koukharsky et al. (2003) interpreted it to be the typical example of a calc-alkaline series, with an acid extreme represented by the Oncán rhyodacite porphyry. In this area, the Ojo de Agua and Ambargasta formations are intruded by Middle to Late Cambrian dikes. Geochemical data support the existence of a magmatic arc which could have developed on continental crust.

Leal et al. (2003) describe granitoids of Late Proterozoic to Early Cambrian age, for the northernmost part of Sierras Pampeanas Orientales that are intruded by Los Burros dacite, which is in turn intruded by the Oncán rhyolite dikes. These volcanic rocks were emplaced after the deformational event and the exhumation of the batholith which happened during the Pampean orogeny. Los Burros dacite yielded a U-Pb SHRIMP zircon age of 512 Ma. According to these authors, the magmatic arc might have developed during the Late Precambrian to the Early Cambrian as the result of eastward subduction. This collision may have controlled extensive deformation and metamorphism at 530 Ma (Kraemer et al., 1995; Sims et al., 1998).

## 6.5. Tilcarian Orogeny

The deformation processes, metamorphism and associated plutonism which affected the Puncoviscana Basin occurred in the Early Cambrian. The orogeny developed towards the end of the Lower Cambrian and principally operated up to the Mid-Cambrian. These processes activated through the closure of the basin, probably related to the Arequipa-Antofalla Terrane colliding with the western Gondwanan margin and the activity of the Palaeo-Pacific Plate. Open to tight folding and axial plane cleavage was developed, which are mostly recognised in the pelitic facies. This folding was characterised by Willner and Miller (1986) and Willner (1990) as F1 and F2. The axial planes of these folds strike roughly NS, and vergence is to the east or to the west in accordance to the studied location. There are areas where chevron folds and flank slides are quite frequent.

Nevertheless, the metamorphic events north of Tucumán were of low grade with syn-kinematic and thermal characteristics that are particularly evident near the granitic plutons. Metamorphism usually generates a weak recrystallisation of the matrix and an oriented biotite growth. In some areas metamorphism reaches the chlorite zone and even higher in the Sierras Pampeanas south of Tucumán. In these geographical locations it is not easy to determine whether the metamorphic events correspond to the Tilcarian orogeny or to the following orogenic events in Ordovician times (Famatinian Cycle, *sensu* Aceñolaza and Toselli, 1976; Figure 6.9).

It is important to point out that from Tucumán to the north, the Tilcarian orogeny is older than the Middle to Late Cambrian Mesón Group, which unconformably overlies the Puncoviscana Formation *s.l.*

## 6.6. Plutonic Rocks

The granitic magmatism of the Pampean Cycle is classically restricted to Jujuy and Salta provinces, although rocks belonging to this cycle have also been recognised further south in the Ancasti, Ambargasta and Chica de Córdoba ranges.

The plutons described for the northern sector of Argentina are Cañaní, Tipayoc, Santa Rosa de Tastíl and Mojotoro, all of which are post-tectonic, and related to the diastrophic Tilcarian orogenic phase (Omarini et al., 1987a, 1987b).

The Cañaní Pluton (Turner, 1960) represents the northernmost intrusion in the Puncoviscana Formation in Argentina, and continues into Bolivia where it is named Cerro Condado granite (Avila Salinas, 1992). Omarini et al. (1987b) described it as a complex pluton formed by hornblende tonalites and porphyritic, two-mica granites, which show calc-alkaline affinity. Omarini et al. (1987b) suggest that it should be considered as an epizonal intrusive that was injected during the Tilcarian orogenic phase. Bachmann et al. (1987) determined, by conventional U-Pb method, the ages between 519 and 534 Ma for this granitc event.

In the Sierra de Santa Victoria, the Tipayoc pluton intrudes the Puncoviscana Formation *s.l.*, its petrographic composition being mainly granodioritic, trondhjemitic and quartz-dioritic (Omarini et al., 1996).

The Santa Rosa de Tastil batholith has been petrologically described by Kilmurray and Igarzabal (1971), Kilmurray et al. (1974) and Omarini et al. (1987a). It is formed by a grey biotite tonalite-granodiorite and a red biotite granite that constitute the pluton that grades into porphyry dacite. The granodiorite ages vary from 533 to 536 Ma, and the red granite aged between 517 and 514 Ma. The associated dacite porphyry yielded an age of 525 Ma (Omarini et al., 1985, 1987a, 1999b; Bachmann et al., 1987; Matteini et al., 2008). The batholith is

Therefore, most of the stations are located along main and secondary highways and when necessary, along roads, tracks and points of the Brazilian Fundamental Gravimetric Network (RGFB) and the Brazilian Institute of Geography and Statistics (IBGE) leveling network, once the gravimetric method requires that the altitude (a.s.l.) of the station be known.

To calculate the Bouguer anomaly, all measurements must be referred to the same Geodetic Reference System (GRS) and a density value must be defined to represent the portion of the upper crust that is above the geoid surface. For this study, more than 12,000 gravimetric stations were used, and the anomalies were calculated in relation to the 1967 GRS, assuming a density value of 2.67 g/cm$^3$ (Mantovani et al., 2005a; Figure 7.1.5). Topographic corrections were not applied in view of the survey scale.

The Bouguer anomaly thus contains several components that include structures at varying depths and scales, from those corresponding to more superficial bodies to those resulting from the lithospheric contribution. The components of interest are extracted using trend surfaces defined by a polynomial degree (Davis, 1986). In our study, the second-order polynomial trend was subtracted from the Bouguer anomaly in order to obtain the residual component of the Bouguer anomaly associated with crustal structures (Figure 7.1.5).

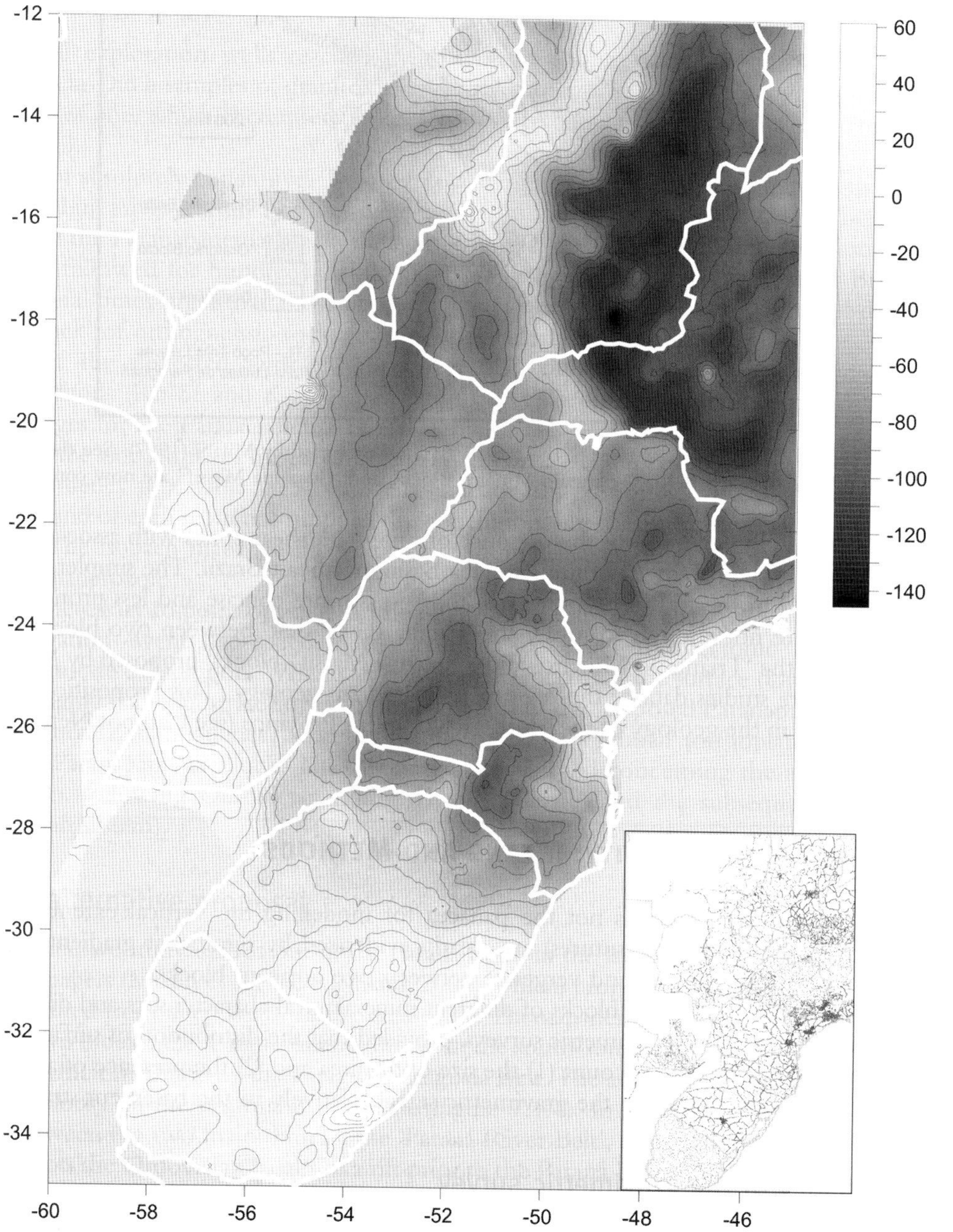


**Figure 7.1.5** Simple Bouguer anomaly for the whole surveyed area. Insert: distribution of gravimetric stations.

### 7.1.3.2. The presence of the Paraná Basin

As mentioned before, the Paraná Basin deposits act as a 'filter' on the gravimetric signal resulting from the basement structures. To individualise the component associated with the basement it is necessary to remove the gravimetric contribution of the Paraná Basin.

Deep-well data logs from oil exploration are available for the whole basin. This information plus seismic data helped to define the isopach maps of the sedimentary and volcanic (from basalt to rhyolite floods) formations and to determine several physical characteristics such as the average density of each formation (Zalán et al., 1990; Quintas et al., 1997).

From this information, the gravimetric contribution of the Paraná Basin was calculated and subtracted in order to obtain the Bouguer anomaly exclusively due to the crystalline basement structures (Quintas, 1995; Mantovani et al., 2005a; Figure 7.1.3). The methodology that takes into account the gravitational effect of the flexure caused by isostatic load compensation is described by Karner and Watts (1983) and the parameters used in the calculations are given by Quintas (1995) and Mantovani and Brito Neves, (2005).

The gravimetric configuration of the Paraná Basin framework evidences a structure bounded by marked gradients, and based on these boundaries it was named Paranapanema Block (Quintas, 1995; Mantovani et al., 2005a). The existence of an old lithospheric nucleus beneath the sediments, the Paraná Craton of Cordani et al. (1984), is attested by the positive gravimetric anomaly that stands out in the central portion of the basin framework.

Isopachs of all sedimentary formations, the flood basalt layer and sills (Paulipetro, 1982) were also used to map the main stresses that triggered and contributed to the formation of the basin. From the drill log records, the basement subsidence curves were obtained and used for a thermo-mechanical analysis of the Paraná Basin, based on the Royden and Keen (1980) extensional model (Quintas et al., 1997).

Mc Kenzie, 1978 and Royden and Keen (1980) created models for lithosphere stretching by pure shear. These authors took into account the isostatic compensation preserved during the whole process and the heat flow vertical component. The two models differ only in the discrimination of the response of the crust to stretching, the Royden and Keen (1980) model being a refinement of Mc Kenzie, 1978.

These authors consider that the lithosphere undergoes thinning during stretching which causes thermal perturbation, an increase in density by injection of asthenospheric material with a consequent subsidence by isostatic adjustment. In the Royden and Keen (1980) model, the crustal ($S_{brittle}$) and the subjacent lithospheric ($S_{ductile}$) subsidences are described. Quintas et al. (1999) and Mantovani et al. (2005a) take into account that $\beta$ is the subcrustal stretching factor and $\delta$ the crustal stretching factor. The attenuation $\varepsilon$ or total lithospheric stretching is the combination of both thinning events ($\delta$ and $\beta$) and is expressed by:

$$\varepsilon = \frac{a}{(t_c/\delta) + (a - t_c)/\beta}$$

which corresponds to McKenzie's '$\beta$ factor', widely used in the literature.

By means of the backstripping technique, two stretching events that contributed to the formation of the Paraná Basin were defined at 440 Ma and at 296 Ma, preceding the eruption of CFB (starting at 144 Ma). The backstripping process was successively applied from top to base of the sedimentary column at each deep well (Quintas et al., 1997, 1999). The Airy isostasy model was applied because it is adequate for the size of the Paraná Basin and for the values obtained for the elastic thickness Te $\leq$ 50 km (Vidotti, 1998; Mantovani et al., 2005b).

The application of the backstripping technique for 81 wells helped in determining the spatial distribution of stress (Quintas et al., 1997) and establishing the most probable sites of lithospheric and crustal thinning. Considerations on the mechanisms involved in the CFB emplacement can be drawn from this methodology (Figure 7.1.6).

The sites that correspond to the second stretching event (296 Ma) are superposed to those corresponding to the first (440 Ma) and are aligned with the present Paraná River waterway. Stresses parallel to the Ponta Grossa Arc match only with the second extensional event. Within the errors caused by the spatial distribution of the wells, the principal component of the first event is aligned with the western gravimetric gradient of the Paranapanema Block, which can be interpreted as a suture.

The geophysical observations corresponding to this limit and to the adjacent western portion are in agreement with the model of subduction of a continental margin of the Rio Apa Craton under the Paranapanema Block. Oceanic crust segments subjacent to the Paranapanema Block justify the dichotomy of the magmatism throughout the basin, as explained below.

The proposal of a collision between the Rio Apa Craton and the Paranapanema Block and the location of the gravimetric gradient, which is the western limit of the Paranapanema Block, implies the possibility of reactivation of old sutures which served as ducts for the large volume of magma that constitutes the Serra Geral Formation, among other possible mechanisms. This interpretation is supported by the CFB geochemistry and isotope geology.

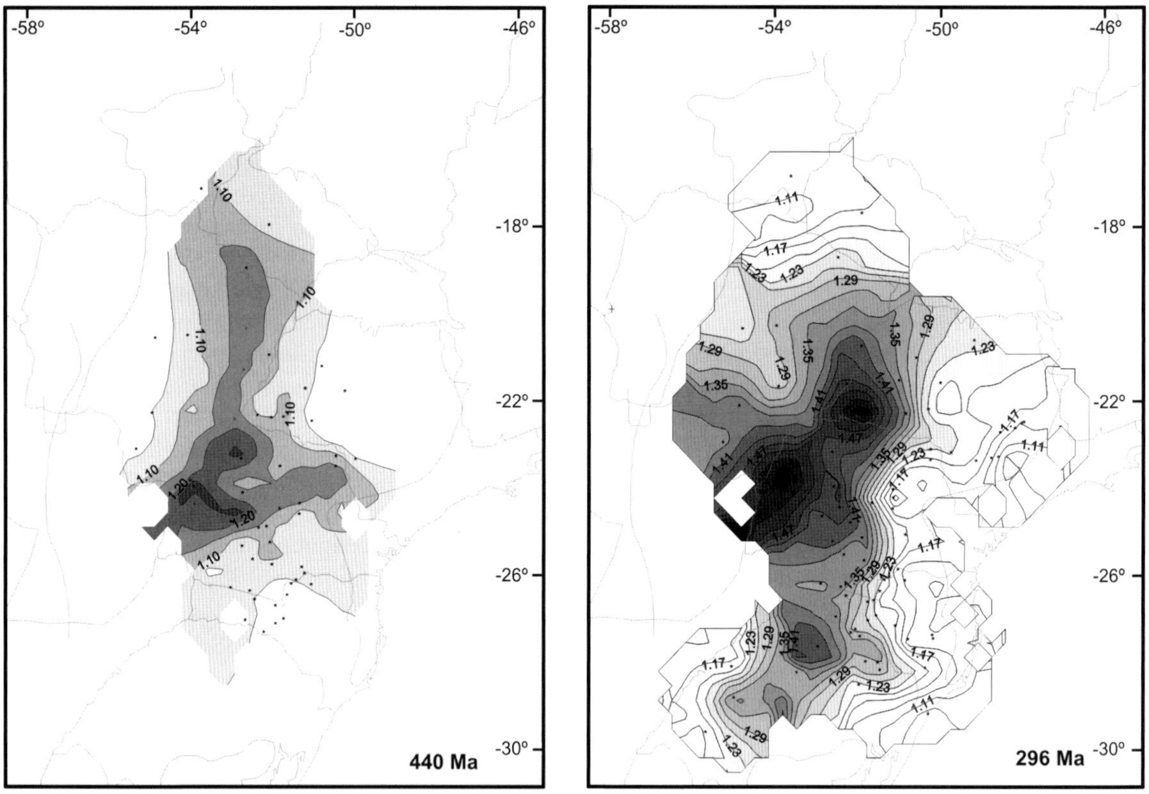


**Figure 7.1.6** Isopachs of the principal extensional stresses at the beginning (440 Ma) and during (296 Ma) the formation of the basin. The main depocenters and the slow variation of the stress field with time are identified (modified from Quintas et al. 1999).

## 7.1.4. Magnetotellurics and Magnetometry

### 7.1.4.1. Magnetotellurics

Two magnetotelluric sections, interpreted in association with the Pantanal Basin gravimetry (Woldemichael, 2003), show the presence of the western portion of the Paranapanema Block. The southern section yielded good quality data for 2D modelling (free of background, otherwise detected in the northern section). Five blocks have been characterised, three of them belonging to the basement. The Rio Apa Block ($R > 1{,}000\,\Omega$-m; $\Delta\rho = 2.715\,\mathrm{g/cm^3}$) and the Paraná Block ($R > 1{,}000\,\Omega$-m; $\Delta\rho = 2.703\,\mathrm{g/cm^3}$) which is equivalent to our Paranapanema Block, collided and dragged a lower crust chip and metamorphosed marginal sediments between them ($R = 5-50\,\Omega$-m; $\Delta\rho = 2.75\,\mathrm{g/cm^3}$). The three blocks that characterise the basement are located under the Paraguay Belt ($R = 100-1000\,\Omega$-m; $\Delta\rho = 2.70\,\mathrm{g/cm^3}$), which underlies the Pantanal Formation ($R = 5-150$ $\Omega$-m; $\Delta\rho = 2.54\,\mathrm{g/cm^3}$). The Rio Apa and Paraná cratonic blocks yield high resistivity.

### 7.1.4.2. Magnetometry

Aeromagnetic data from the Botucatu survey (Paulipetro, 1982) evidenced an important Lower Cretaceous dike swarm that corresponds to the Ponta Grossa Arc. Its N45W trend is that of the Guapiara, Curiúva, Rio Alonso and Rio Piquirí lineaments. Because the detail of the aerial surveys at a nominal altitude of 150 m, as is the case of the Botucatu aerogeophysical project, is not enough for direct inferences on the geometric characteristics, Ussami et al. (1991) carried out a 16-km ground magnetic section from the town of Sarutaiá to Fartura, for comparison.

Although dikes with weaker magnetisation were not observed from the aerial survey, sampling and ground measurements allowed to infer the lateral expansion along the Ponta Grossa Arc caused by the dikes. A rough calculation on the basis of the cited evidence suggests stretching between 9% and 13% perpendicular to NW, matching the model of Quintas et al. (1999).

## 7.1.5. Seismic Data

Global tomographic models which combine data from different seismic waves regarding variations of seismic wave velocities in the transition zone between 400 and 1,000 km depth (Ritsema et al., 2004 and references therein) have been proposed by several authors for different areas of the globe.

Van Decar et al. (1995) interpreted the low velocities observed in the South American Plate, particularly in the northern region of the Paraná Basin, as due to a thermal anomaly which was associated with a fossil mantle plume.

Later, An and Assumpção (2006), based on the gravimetric models of Ussami et al. (1993) and Vidotti et al. (1998) and using genetic algorithms, tried to identify the deep crustal structures that would attest to the proposed underplating model. The results led to an average Moho depth of 40–45 km revealing a thicker upper crust with lower S-waves velocity (3.5–3.6 km/s) down to 25–30 km; velocities in the lower crust reach, at most, 3.8 km/s. Finally, Feng et al. (2007) obtained S-waves anomalies of significantly low velocity under the Chaco basin and western portion of the Paraná Basin, suggesting a weak and/or thin lithosphere in that region.

Regarding the intracratonic basin, these authors used the new approach and identified higher velocities at 100–150 km depths which where associated with the 'Paraná Craton' otherwise known as Paranapanema Block (this work). Searching for new parameters, they attributed high values of flexural rigidity (Te) for the Paraná Basin structure. However, the results of Vidotti (1998), Mantovani et al. (2001, 2005b) and Tassara et al. (2007) do not characterise the rigidity of that area high enough to be considered as 'cratonic', even if Te values are higher than those obtained for Chaco and Pantanal basins.

## 7.1.6. Support from Geochemistry and Isotope Geology

Since the first geochemical studies, two trends were recognised for volcanic rocks of the Serra Geral Formation regarding $TiO_2$ contents and Sr isotopic ratios (Mantovani et al., 1985). The geographic distribution of the geochemical and isotopic data placed the high-$TiO_2$ basalts (HTi) that show a narrow interval of Sr isotopic ratio variations in the northern portion of the basin, geographically coinciding with the Paranapanema Block and adjacent structures. The low-$TiO_2$ basalts (LTi) that show large variation of Sr isotopic ratio are distributed in the southern portion, overlying the Rio de la Plata Craton, adjacent structures and those close to the present Atlantic Coast (Figure 7.1.7).

With the advance of Ar-Ar dating technology (MacDougall and Harrison, 1988; Kelley, 1995; Renne et al., 1992), variations in melting volumes and in source compositions during the evolution of this magmatic province could be detected. It is worth mentioning that the increase in the eruption rate can characterise melting by decompression associated with extensional processes that preceded the rupture of Gondwana. On the other hand, changes in geochemical characteristics suggest changes in magma sources. These characteristics were evidenced for the Paraná CFB by Turner et al. (1994) and the distribution of the ages for both trends can be seen in Figure 7.1.8. Geochronological data show that the eruption in the northern portion occurred in the 138–125 Ma interval, preceding those in the south which were dated 133–127 Ma.

The analysis of the vertical distribution using ditch samples identified geochemical subgroups in both provinces (Peate et al., 1988, 1992; Mantovani et al., 1988; Peate 1997): Pitanga, Paranapanema and Ribeira in the northern portion, and Gramado, Urubici and Esmeralda in the southern portion (Figure 7.1.9). From the three subgroups of the northern portion, the Ribeira (last effusion in the stratigraphic column of the Northern Province and SE of this province) has $TiO_2$ contents comparable to those of the LTi group, but it keeps the isotopic and geochemical characteristics of its two underlying subgroups. Ti depletion was attributed by Peate et al. (1992) to a low melting rate of the lithosphere or to a magma contribution from the proposed 'plume', while Gallagher and Hawkesworth (1992) ascribed Ti depletion to the presence of small percentages of volatiles in a heterogeneous lithosphere. More recent models (e.g. Korenaga, 2004; Anderson, 2005, 2007; Lustrino, 2005; Xu and He, 2007) attribute the variety of CFB magma compositions to lower crust delamination processes. In other words, when the continental crust becomes too thick, the dense eclogitic lower portion detaches from the rest causing uplift and melting by decompression.

Figure 7.1.7 illustrates the position of the Ponta Grossa Arc dikes in relation to the two magmatic provinces. The size of the magmatism exceeds the area of the basement blocks once the basalts flooded on a uniform sedimentary substrate, the topography of which responded to the subjacent thermomechanic state. In the border between the two provinces (LTi and HTi), the effusions mingled, which corresponds to the coeval final HTi– and the initial LTi–flood interface (Figure 7.1.9).

Hirano et al. (2006) observed volcanic manifestations in regions of plate flexure, where the stress regime is extensional, when studying an oceanic crust segment. Turner et al. (1994) associate the Ponta Grossa Arch and

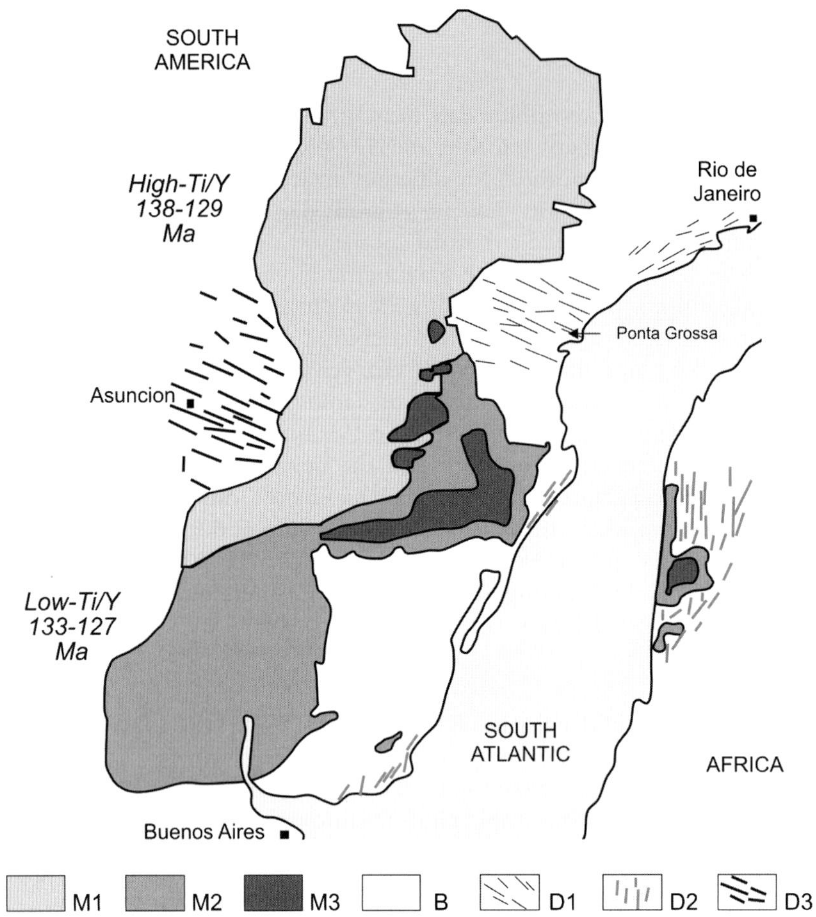


**Figure 7.1.7** Sketch map of the Paraná-Etendeka magmatic province just prior to the opening of the South Atlantic. The basaltic lavas are subdivided into broad high- and low-Ti/Y groups, and late-stage rhyolites (spaced diagonals). The high-Ti/Y lavas are associated with NW-SE trending dykes, and the low-Ti/Y with the N-S, more coast-parallel dykes. The Paraguay dyke swarm (i.e. in the area around Asuncion) is a major feature that is clearly evident from aeromagnetic surveys, and it includes both tholeiites of Paraná age and younger alkalic dykes. The coast-parallel dykes in the area of Rio de Janeiro also tend to be alkalic in composition and are unrelated to the Paraná-Etendeka CFB (from Hawkesworth et al., 2000).

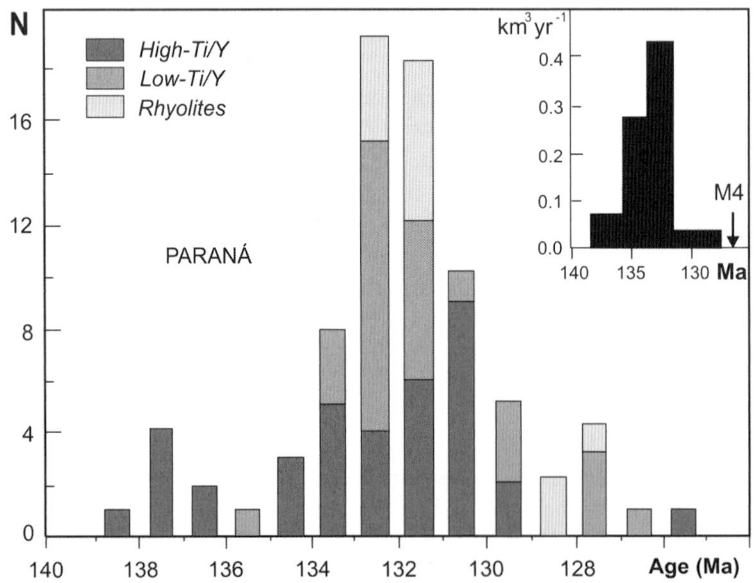


**Figure 7.1.8** Histogram of Ar-Ar ages from the Paraná-Etendeka province. The samples which record the older ages are all from the north and west of the main outcrop area. The inset is estimated melt generation rates, and the oldest oceanic basalts appear to be M4 equivalent to an age of 127 Ma (from Hawkesworth et al., 2000).

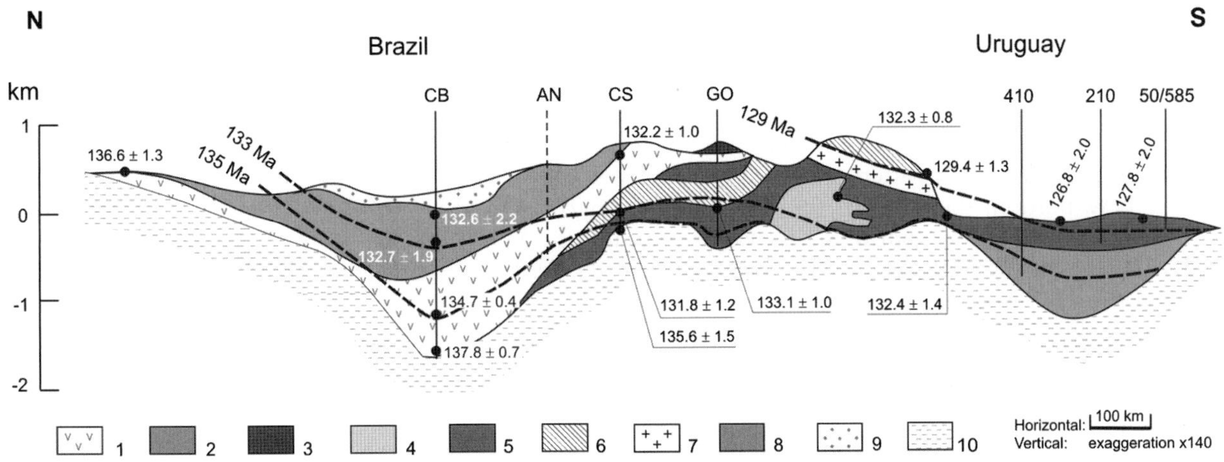


**Figure 7.1.9** N-S cross-section through the Paraná lavas after Turner et al. (1999). Magma types are marked as: 1: Pitanga, 2: Paranapanema, 3: Ribeira, 4: Urubici, 5: Gramado, 6: Esmeralda, 7: Rhyolites, 8: Lower southern volcanic layer, 9: Bauru Formation (sedimentary cover) and 10: Botucatu sandstone (basement). Note that Paranapanema lavas overlie Gramado-type lavas in the northern/central parts of the lava pile but underlie them in Uruguay indicating that the magma types are non-chronostratigraphic. Similarly, the dashed timelines for 135, 133 and 129 Ma, based on Ar-Ar age data from Turner et al. (1994) and Stewart et al. (1996), clearly cross-cut the magma types (CB, AN, CS, GO, 410, 210 and 50/585 are borehole locations from where samples were extracted).

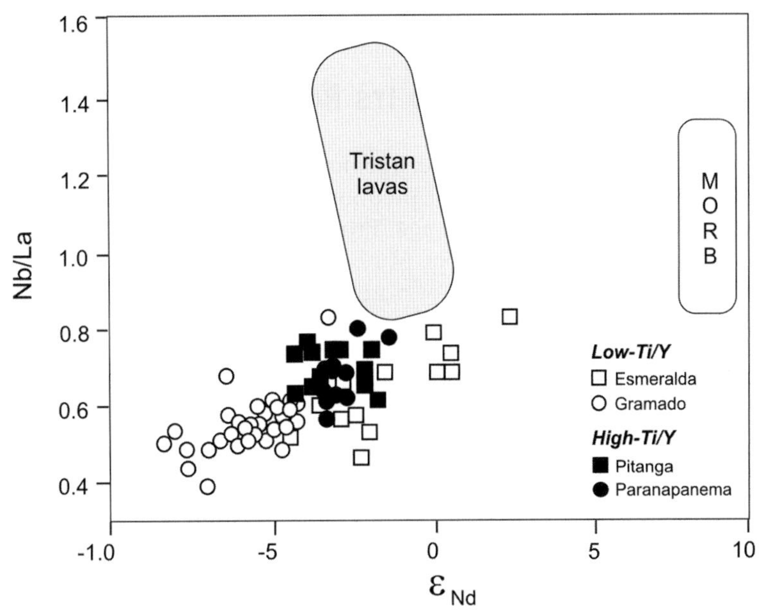


**Figure 7.1.10** $\varepsilon_{Nd}$-Nb/La for vocanic rocks of the Paraná Basin compared with those from the Tristan = Walvis plume and MORB. Neither $\varepsilon_{Nd}$ nor Nb/La change significantly during open system differentiation, and so both reflect the values in their source regions, and these are different from those sampled in oceanic areas (from Peate, 1997).

structures parallel to it (e.g. Salado Basin and Colorado Basin in the Argentinian coast) to extensional stresses associated with a clockwise rotation of the plate that, in the case of the Arch, superposes a flexure. The NW-SE dikes of the Ponta Grossa Arch yield the same compositions of the high-$TiO_2$ magmas of the northern magmatic province, suggesting that this extensional and flexural structure also served as a duct to the northern magmatism (Hawkesworth et al., 2000).

When compared to Oceanic Island basalts (OIB; Figure 7.1.10) Nd/La and initial Nd isotope ratios of Paraná CFB magmas are lower; therefore, they must have derived from old regions (Proteozoic nuclei). Its origin has been attributed to lithospheric mantle sources because it has trace element ratios different from those observed for the upper oceanic mantle (Erlank et al., 1984; Hawkesworth et al., 1986; Peate et al., 1992; Turner and Hawkesworth, 1995; Stewart et al., 1996). This proposal is corroborated by the major element composition of the Paraná CFB (more depleted in Fe and Na and enriched in Si) compared to the basalts generated by fertile peridotitic sources or directly associated with plumes (Hawkesworth et al., 2000). Thus, this observation is also consistent with the presence of an old lithospheric block underneath the Paraná Basin cover, that is the Paranapanema Block.

A summary of the present state of knowledge, obtained from indirect methods of investigation is as follows:

a. The geometry and boundaries of Paranapanema Block (PPM) are clearly defined by the marked gravity gradients of the Bouguer pattern obtained for the basement.
b. In the NW portion, the densities the Rio Apa Craton/Block (CRA) and PPM adopted in the gravimetric modelling correspond to high resistivity values measured by MT.
c. The CRA continental margin that subducted beneath the PPM could have contributed to the magmatism that formed the CFB northern province.
d. Crustal thickening derived from convergent (collisions and formation of belts and island arcs) and divergent (formation of the Paraná Basin) processes could have reached the conditions that culminated in delamination, which is one of the theories to explain the heterogeneity of the CFB. Decompression and increase in temperature would cause melting of the underplated material. In this case, high seismic velocities between and below 100 and 150 km would not be observed.
e. The alternative hypothesis is the presence of a plume, which would be the heat source to melt the base of the ($\sim$4% hydrated) lithosphere. It would consume part of the lithosphere base, and consequently lower seismic velocities and/or thicknesses would be observed when compared with cratonic regions.
f. The distribution of extensional stresses during the formation of the basin must have created and reactivated faults (PPM sutures), which would act together with the new fissures as ducts for the CFB magmatism.
g. Probable eastward dislocation of the continental rupture site and stress field variations is a consequence of the presence of PPM and is associated with crustal thickening resulting from the formation of the basin and the continuous change in topography due to loading forces (tilt).

### 7.1.7. The Paranapanema Block and Its Role in the Accretion of Gondwana

From a few direct geologic and many indirect (geologic, geochemical and isotopic) available data, the context envisaged for the Paranapanema Block is of predominant granitic nature (orthogneisses), confirmed by deep well sampling, and probable pre-Brasiliano in age. From the geotectonic point of view, it is a continental lithospheric segment that has first acted as an upper plate during subduction processes and later as a backland during interaction with neighbouring plates. Zones of plate interaction mark the external limits of all this portion of the basin (west, north-northeast and east) with the generation of continental magmatic arcs and post-collisional structures. Given a size estimated in ca. $10^5$ km$^2$ (from its inferred shape) and reworking (of unknown magnitude, but certainly high), an intermediate-sized plate is here proposed as the Paranapanema Block. The designation 'Paranapanema block' should be preserved in face of the present scenario, which is by far not the ideal one, and the demand for future surveys (e.g. deep wells and seismic surveys). Up to now there are not enough available data to state a cratonic crustal type. Requirements as "antiquity", stability, low thermal flow, thick lithosphere (keels), etc. are still necessary to ascertain. The form and size of this block stand in the list of unsolved problems, but this may be refined in the future.

A number of similar blocks were also involved in Gondwana amalgamation, such as the Goiás Massif (a microplate north of Paranapanema), the Pernambuco-Alagoas Massif, the Granja Massif and the Bangweulu Block in Africa. Whenever they are exposed, the common designation that Brazilian geotectonists would give to such structures would be 'massif' (a heritage from the soviet nomenclature).

The option for an intermediate-sized plate and backland for the neighbouring belts (Paraguay, southern Brasília and Apiaí-São Roque) is assumed; meanwhile, necessary complementary observations are being carried out.

Regarding the behaviour of the block, two aspects must be considered. The first is the comparison with what has been found in the interior of other major South American syneclises. The second is the feedback from the majority of researches which have outlined the branching orogen system of the Brasiliano-Panafrican collage by the end of the Proterozoic and beginning of the Palaeozoic. This fraction of the continental lithosphere has been used in the majority of studies of Gondwana components, that is fractions of the Rodinia supercontinent fission re-arranged in the Gondwana fusion.

Along the main axis, the Amazonas Syneclise (east of the Manaus parallel), there is a series of Cambrian (late-Brasiliano) mafic-ultramafic intrusives. In the interior of the Parnaíba Syneclise, several blocks are known and some others (Parnaíba Block) were identified by geophysical studies as being inter-related by a dense Brasiliano shearing network. South and southeast of the Paraná Basin, several small blocks (Curitiba, Luis Alves and Rio de La Plata) and mobile belts are present. It all indicates that precursor rifts, centred in Brasiliano structures, played an important role in supplying these major and other basins (e.g. Araripe and Parecis). In other words, very rarely

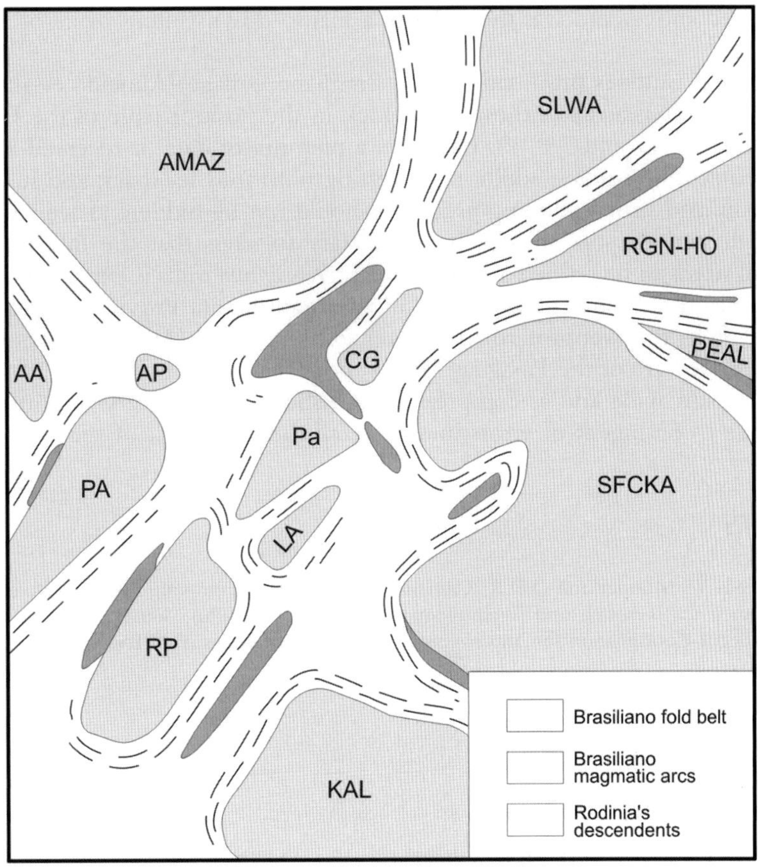


**Figure 7.1.11** The lithospheric descendants of Rodinia amalgamated by the Neoproterozoic Brasiliano branching system of orogens, as proposed by Cordani et al. (2003). Amaz: Amazonian; RGN-HO: Rio Grande do Norte-Hoggar; CG: Central Goiás; Pa: Paranpanema; PA: Pampia; LA: Luis Alves; RP: Rio de la Plata; AA: Arequipa-Antofalla and SFCKA: São Francisco/Congo/Kasai/Angola; PEAL: Pernambuco/Alagoas; KAL: Kalahari; SLWA: São Luis-West Africa.

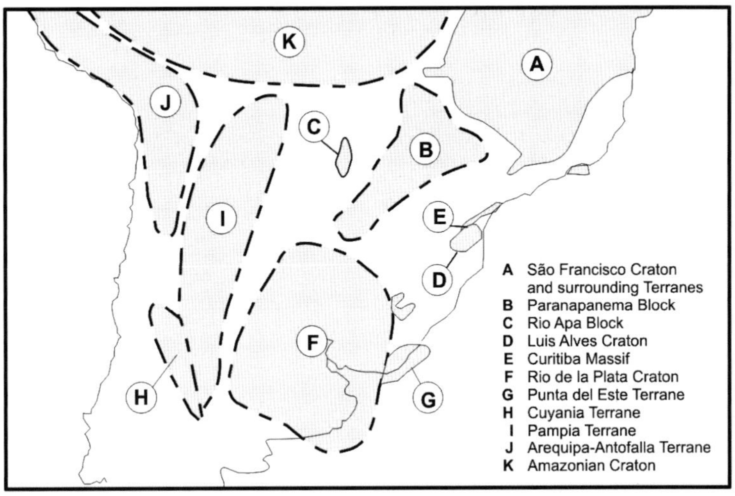


**Figure 7.1.12** The descendants of Rodinia in the South American continent. An excerpt of the map by Fuck et al. (2008).

these major Gondwanan basins are entirely situated on stable lithospheric cratonic contexts, especially considering their very large extent.

In the concert of pre-Brasiliano blocks (Rodinia descendants) articulated in the accretion of the Gondwana supercontinent at the Proterozoic-Phanerozoic Boundary (Fuck et al., 2008), smaller blocks (terranes, microcontinents and microplates less than $10^5$ km$^2$) underwent considerable reworking/tectonic regeneration, so that they cannot be considered cratonic entities. This has several theoretical explanations and is an issue in the agenda of more detailed geotectonic studies in other continents (the decratonisation phenomenon). We cannot expect that the Paranapanema Block behaved as a full cratonic domain, because it once was surrounded by several subduction zones (discussed above) and was the locus of a major 'downwelling' zone of asthenospheric currents during the Neoproterozoic.

In the majority of recent outlines proposed for the Brasiliano collage (Almeida et al., 2000; Cordani et al., 2003a, 2003b, Fuck et al., 2008), a rigid lithospheric block in the position/role of the Paranapanema has been acknowledged (see Figures 7.1.11 and 7.1.12). Based on a plethora of data, it received a practically unanimous approval from recent geotectonic studies, additionally getting notorious feedback and support.

Nevertheless, the many and varied pending problems should not be hidden, as well as the demand for more data. Because a thorough study is still missing, it is necessary to stress that the information on the proposed Paranapanema Block still needs to be refined and a certain *experimentum crucis* is expected from different sectors of the geosciences. The foreland for the central portion of the Mantiqueira Province (Apiaí-São Roque Domain) would be ca. 150 km far from the Neoproterozoic reference mobile belt (Apiaí-São Roque). *Idem*, the foreland (upper plate) for the presumed arc of the Paraguay Belt is more than 200 km distant. As the continental margins tend to be more than 100 km wide and as compression and shearing were very important in structuring these belts, there will always be a contingent of errors involved and there is a lack of accuracy in the interpretation.

## ACKNOWLEDGEMENTS

Comments and suggestions from W. Mooney and C.D.R. Carneiro contributed to improving the original version. Financial support was received from National Council for Research and Technology Development (CNPq), São Paulo State Foundation for Supporting Research (FAPESP), and National Coordination for Improvements in Graduate Studies (CAPES).

CHAPTER 7.2

# The Evolution and Tectonic Setting of the Luis Alves Microplate of Southeastern Brazil: An Exotic Terrane during the Assembly of Western Gondwana☆

Miguel A.S. Basei[1], Allen Nutman[2,3], Oswaldo SigaJúnior[1],
Cláudia R. Passarelli[1] *and* Cesar O. Drukas[1]

## Contents



## 7.2.1. Introduction

The structural framework of southeastern Brazil (central portion of the Mantiqueira Province) resulted from the juxtaposition of several crustal blocks, which were variably affected by Neoproterozoic subduction and collision events. The resulting assembly of continental masses led to the formation of western Gondwana (Brito Neves et al., 1999; Basei et al., 2000, 2008; Campos Neto, 2000; Cordani et al., 2000, 2003). Part of this crustal segment is the older domain to the west represented by the Río de La Plata and Paranapanema cratons (see Part 4 and Chapter 7.1), which are now extensively covered by the Palaeozoic sediments of the Paraná Basin. The eastern, accreted domain is formed by the Neoproterozoic Ribeira (to the north) and Dom Feliciano (to the south) fold belts, separated by migmatitic–granitic–gneissic terrains that constitute the basement of the Curitiba Microplate (to the north) and Luís Alves Microplate (LAM; to the south) (Figure 7.2.1). Further east is the Paranaguá Batholith, which is an accreted arc consisting of calc-alkaline granitoids. This compartmentalisation can be seen in Figure 7.2.1A and the spatial relationships between the main geotectonic units are sketched in the hypothetical section of Figure 7.2.1B.

☆Basei, M.A.S., Nutman, A., SigaJúnior, O., Passarelli, C.R., Drukas, C.O. 2009. The evolution and tectonic setting of the Luis Alves Microplate of Southeastern Brazil: an exotic terrane during the assembly of Western Gondwana. In: Gaucher, C., Sial, A.N., Halverson, G.P., Frimmel, H.E. (Eds): Neoproterozoic-Cambrian Tectonics, Global Change and Evolution: a focus on southwestern Gondwana. Developments in Precambrian Geology, 16, Elsevier, pp. 273–291.

[1] Centro de Pesquisas Geocronológicas, Instituto de Geociências, Universidade de São Paulo, Rua do Lago 562, CEP 05508-080 São Paulo, SP, Brazil.
[2] Research School of Earth Sciences, The Australian National University, Canberra, ACT 0200, Australia.
[3] Beijing SHRIMP Centre, Chinese Academy of Geological Sciences, 26 Baiwanzhuang Road, Beijing 100037, China.

*Developments in Precambrian Geology*, Volume 16
ISSN 0166-2635, DOI 10.1016/S0166-2635(09)01620-X

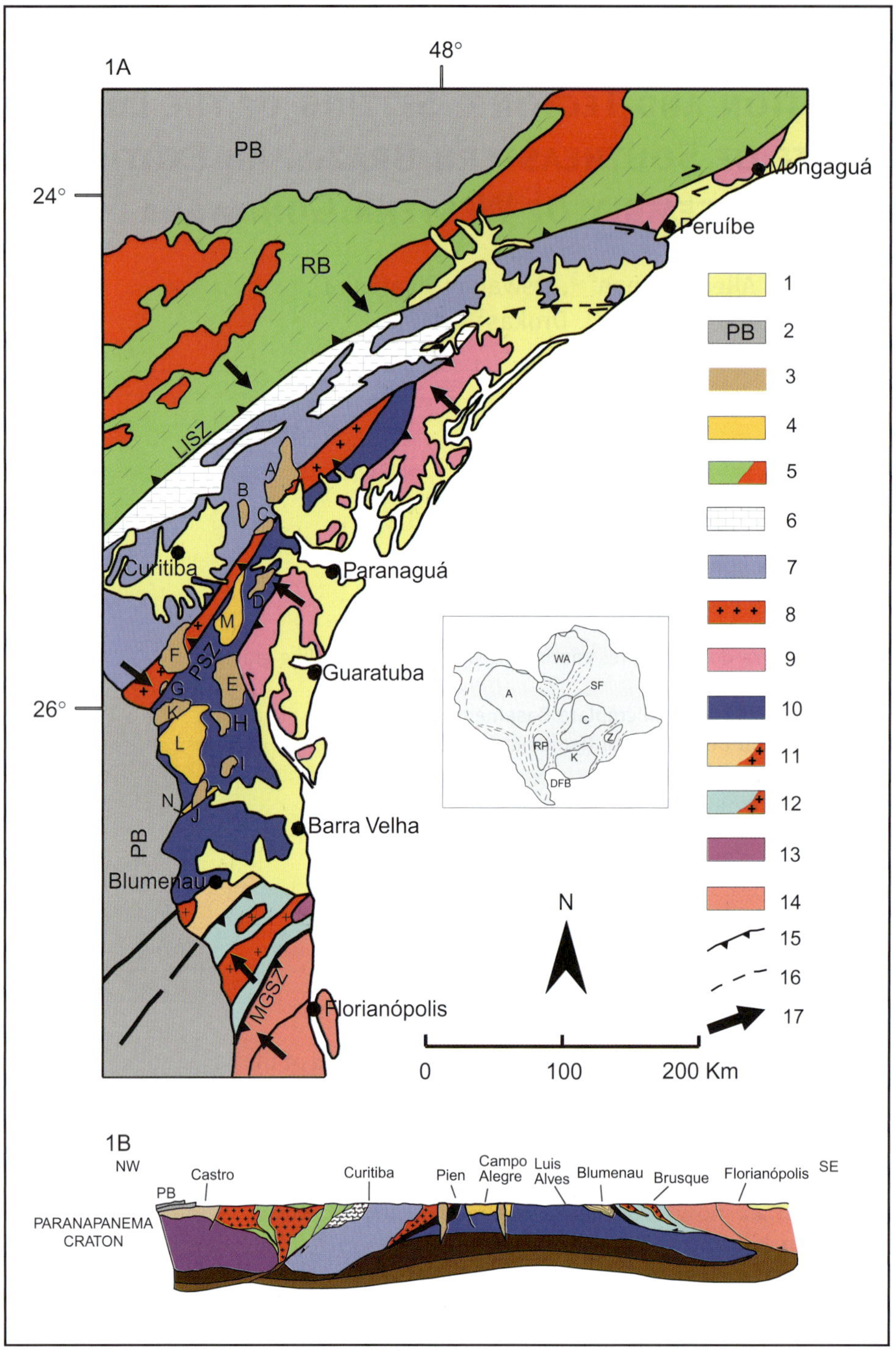


**Figure 7.2.1** Simplified geological map of the Brazilian southern portion between Curitiba (PR) and Florianópolis (SC): (A) 1, recent cover, 2, Paraná Basin sedimentary cover; *Serra do Mar Suite*: 3, A-type granitoids (a: Graciosa, b: Avanganhava, c: Marumbi, d: Serra da Igreja, e: Morro Redondo, f: Palermo, g: Agudos do Sul, h: Dona Francisca, i: Piraí, j: Corupá, k: Rio Negro); *Fini-Proterozoic basins*: 4, non-deformed volcano-sedimentary rocks (l: Campo Alegre, m: Guaratubinha, n: Corupá); *Southern Ribeira Belt*: 5, low-grade metasediments (green) and major calc-alkaline batholiths (red); *Curitiba Microplate*: 6, Capiru and Setuva metasedimentary covers, 7, Atuba Complex; *Pien Batholith*: 8, deformed arc-related calc-alkaline granitoids; *Paranaguá Batholith*: 9, calc-alkaline granitoids; *Luis Alves Microplate*: 10, high-grade metamorphic rocks; *Dom Feliciano Belt*: 11, Itajaí foreland basin and Subida granite, 12, Brusque schist belt and intrusive granitoids, 13, Basement Inlier (gneisses and migmatites of Camboriu Complex), 14, Florianópolis Batholith (arc-related granitoids); *Symbols*: 15, major thrust faults, 16, inferred geological contacts, 17, Neoproterozoic tectonic transport; (LISZ) Lancinha Itariri Suture Zone, (PSZ) Pien Suture Zone, (MGSZ) Major Gercino Suture Zone. (Inset) Gondwana reconstruction (modified from Unrug, 1996). The studied area is located between Rio de La Plata and Kalahari cratons. (B) Geological NW–SW section between Curitiba and Florianópolis, showing the spatial relations between the major geotectonic units, with special emphasis on the tectonic contacts between the Luís Alves Microplate and adjacent units.

This chapter examines the geology of the LAM and demonstrates that it is an allochthonous basement terrane, distinctively different in its history from the western Rio de La Plata and Paranapanema cratons. By examining its relationships with adjacent geological units, we show that it was emplaced from afar during the assembly of western Gondwana by plate tectonic processes.

## 7.2.2. Regional Geological Divisions

The Ribeira Belt (Almeida, 1967; Almeida et al., 1973) is a wide NE-trending orogenic belt subparallel to the southern Brazilian coast. The term Apiaí Fold Belt (Hasui et al., 1975) was proposed in order to subdivide the supracrustal sequences that constitute the southern part of the Ribeira Belt. Its N and NW limit is covered by sediments of the Paraná Basin and its southeastern termination is defined by the Lancinha Shear Zone (Figure 7.2.1A). The belt has been reconstructed, comprising from NW to SE a stromatolite-rich carbonate platform, a slope zone with abundant dolarenites and calcarenites, and a distal turbiditic sequence intercalated with metabasic rocks. This could represent a single passive margin. However, recent geochronologic data indicated that this belt contains Mesoproterozoic and Neoproterozoic supracrustal sequences juxtaposed by the end of the Neoproterozoic, thus revealing a polycyclic evolution (Campanha and Sadowski, 1999; Basei et al., 2003). Roots of Neoproterozoic magmatic arcs (Cunhaporanga, Três Córregos and Agudos Grandes) were recognised in the majority of the supracrustal sequences. In the south-southeastern segment, there are Palaeoproterozoic basement inliers composed of Rhyacian tonalitic gneisses (ca. 2.1 Ga) and deformed Statherian A-type granites 1.75 Ga in age (Cury et al., 2002).

South of the Lancinha Shear Zone, the Atuba Complex exemplifies the basement of the Curitiba Microplate. It is 60-km wide and 200-km long and is limited to SE by the gneisses that constitute the basement of the LAM (Siga et al., 1995; Basei et al., 1998a). The Atuba Complex is covered by low-grade Neoproterozoic metavolcano-sedimentary sequences (Capiru, Setuva and Turvo-Cajati) and has been intruded by anorogenic alkaline–peralkaline granitoids (Figure 7.2.1A) of the Serra do Mar Suite. Banded migmatitic gneisses predominate in the Atuba Complex. They consist predominantly of biotite–amphibole gneisses and tonalitic–trondhjemitic leucosome of Palaeoproterozoic age (2,100–2,200 Ma). Features related to a second Neoproterozoic migmatisation phase are also common; white or pink neosomes are usually concordant with the gneissic banding. The contact between the southern portion of the Curitiba Microplate with the LAM is marked by the heterogeneously deformed (ca. 615 ± 5 Ma) calc-alkaline granitoids that constitute the Rio Piên Batholith. Basic and ultrabasic rocks (630 Ma) also occur, and are possible remains of obducted oceanic floor (Basei et al., 1992; Machiavelli et al., 1993; Harara, 2001) between the microplates thought to be a Neoproterozoic active margin, where oceanic crust was consumed.

The LAM (Hartmann et al., 1979, 2000; Basei et al., 1992, 1998a; Siga, 1995; Harara et al., 2002, 2003) consists predominantly of banded gneisses which regionally carry hypersthene. Migmatitic and granitic rocks and to a lesser extent kinzigitic calc-silicatic gneisses also occur, together with iron formations and quartzites. A striking characteristic of this domain is the presence of basic and ultrabasic rocks (pyroxenites, metagabbros, amphibolites and magnesian schists), which in some places (e.g. Barra Velha) constitute the predominant rock type. The recognition of Palaeoproterozoic gneissic–granulitic and tonalitic rocks in the field with frequent bands of basic and ultrabasic nature is a unifying character of the LAM northeastwards, from Morretes to Serra Negra (Paraná State).

The Paranaguá Batholith occupies the eastern portion of the crustal segment shown in Figure 7.2.1. It is mostly made up of a Neoproterozoic igneous complex (ages from 616 to 590 Ma), which includes a great variety of granitic rocks distributed along a belt more than 100 km long and ca. 30 km wide (Basei et al., 1990; Siga, 1995; Passarelli et al., 2004). It is limited to the east by the present coastline and to the west by the thrusting of the granitoids onto gneissic–granulitic rocks of the LAM. The Morro Inglês Suite, composed of granitoids with calc-alkaline affinity, represents the main rock type, affected by the isotropic granites of alkaline affinity of the Rio do Poço and Estrela suites (Lopes, 1992; Siga, 1995). Gneisses (sometimes garnet-rich), mica-schists, quartzites and amphibolites, which were migmatised to different degrees, occur as roof pendants within these granitoids.

## 7.2.3. Basement Rocks of the Luís Alves Microplate

The LAM is floored by a Palaeoproterozoic migmatitic granitic–gneissic basement, the history of which goes back to the Archaean (Hartmann et al. 1979, 2000; Basei, 1985; Basei et al., 1998a). A Neoproterozoic cover composed of volcano-sedimentary basins plus A-type alkaline–peralkaline granitoids (Figure 7.2.1A and B) covers/intrudes the basement.

Good exposures of basement rocks, included in the Santa Catarina Granulitic Complex by Hartmann et al. (1979), occur across the LAM. Typical rocks are orthogneisses characterised by alternating millimetre- to

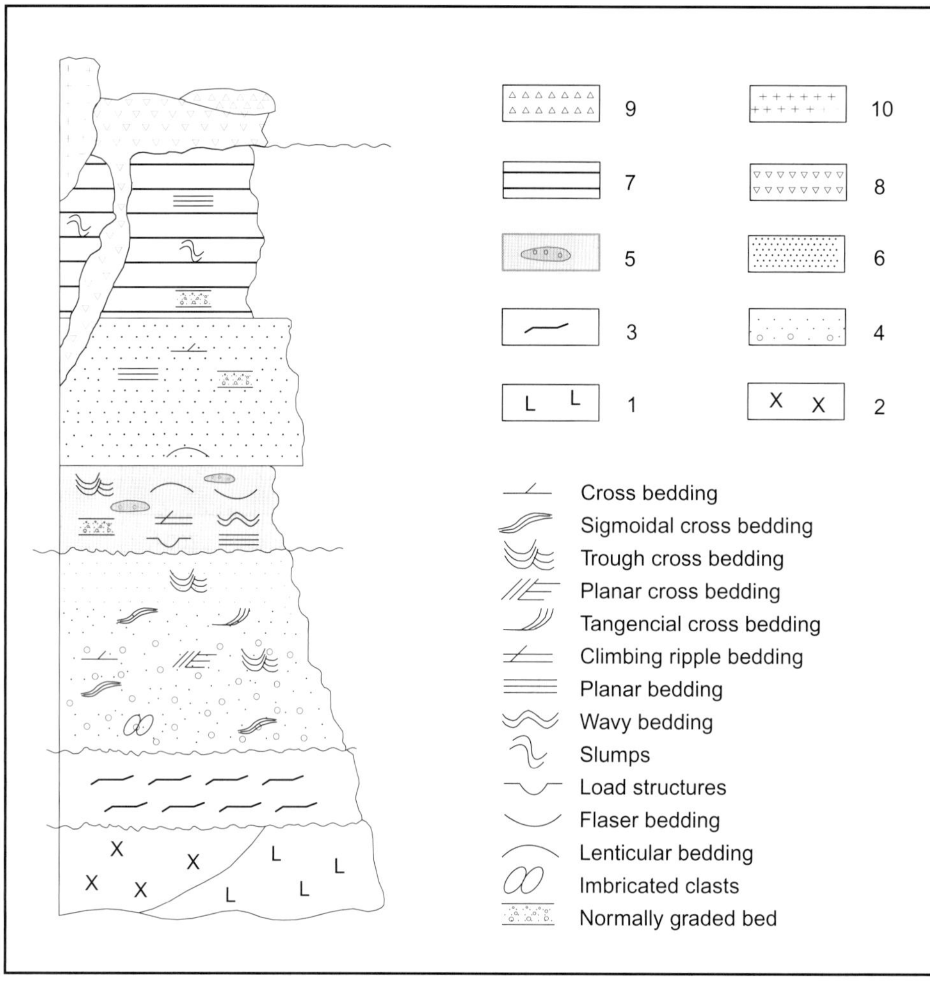


**Figure 7.2.5** Lithostratigraphic column for the Itajaí Basin. *Basement*: (1) Santa Catarina Granulite Complex, (2) São Miguel Gneissic–migmatite Complex, (3) Brusque Schist belt; *Itajaí Basin*: (4) conglomerates and arkosic sandstones, (5) rhythmites, (6) laminated to massif arkosic sandstones, (7) fine laminated siltstones, (8) Apiúna felsic volcanics, (9) talus with acidic volcanic clasts, (10) Subida granite.

reported the occurrence of ichnofossils and body fossils in low-density turbiditic levels of the lower part of the upper Itajaí Group. Body fossils were assigned to *Chancelloria*, which represents a Cambrian taxon restricted to the Lower and Middle Cambrian. Trace fossils also suggest a Cambrian age. This has been questioned by Basei et al. (1999, 2008), on the basis of a 565 Ma value obtained for the volcanic rocks that crosscut the sedimentary pile, placing deposition of the Itajaí Group at the end of the Neoproterozoic.

The sediments of the Itajaí Basin are intruded by the Subida granite of Cambrian age (529 Ma; Basei et al., 2008). It is an isotropic body with rare mafic minerals (amphibole) and of alkaline affinity. It produced a contact metamorphism aureole in the sediments of the basin. Xenoliths of sediments from the Itajaí Group are occasionally found in the Subida granitoid.

## 7.2.5. The Age of the Luís Alves Microplate Basement

Analytical methods and data of U-Pb zircon and monazite dating by the isotope dilution thermal ionisation mass spectrometry (ID-TIMS) and sensitive high resolution ion microprobe (SHRIMP) methods are presented in Appendix 1.

Two samples of the quartzo-feldspathic hypersthene gneiss that occurs in the northern domain of the microplate, near the Campo Alegre Basin (Dona Francisca highway), were analysed. The first sample (MJ 192) yielded, from three concordant fractions, a zircon U-Pb age of 2,200.2 $\pm$ 4.4 Ma (Figure 7.2.6A), which is a value similar to that obtained for prismatic and rosy crystals extracted from another granulitic gneiss sample (MJ 258), which yielded an age around 2,230 Ma. Despite the dispersion of points, brown zircons of the same sample, with rounded rims and edges, yielded an age of 2,361.0 $\pm$ 8.3 Ma (Basei et al., 1999). Harara (2001) presented several

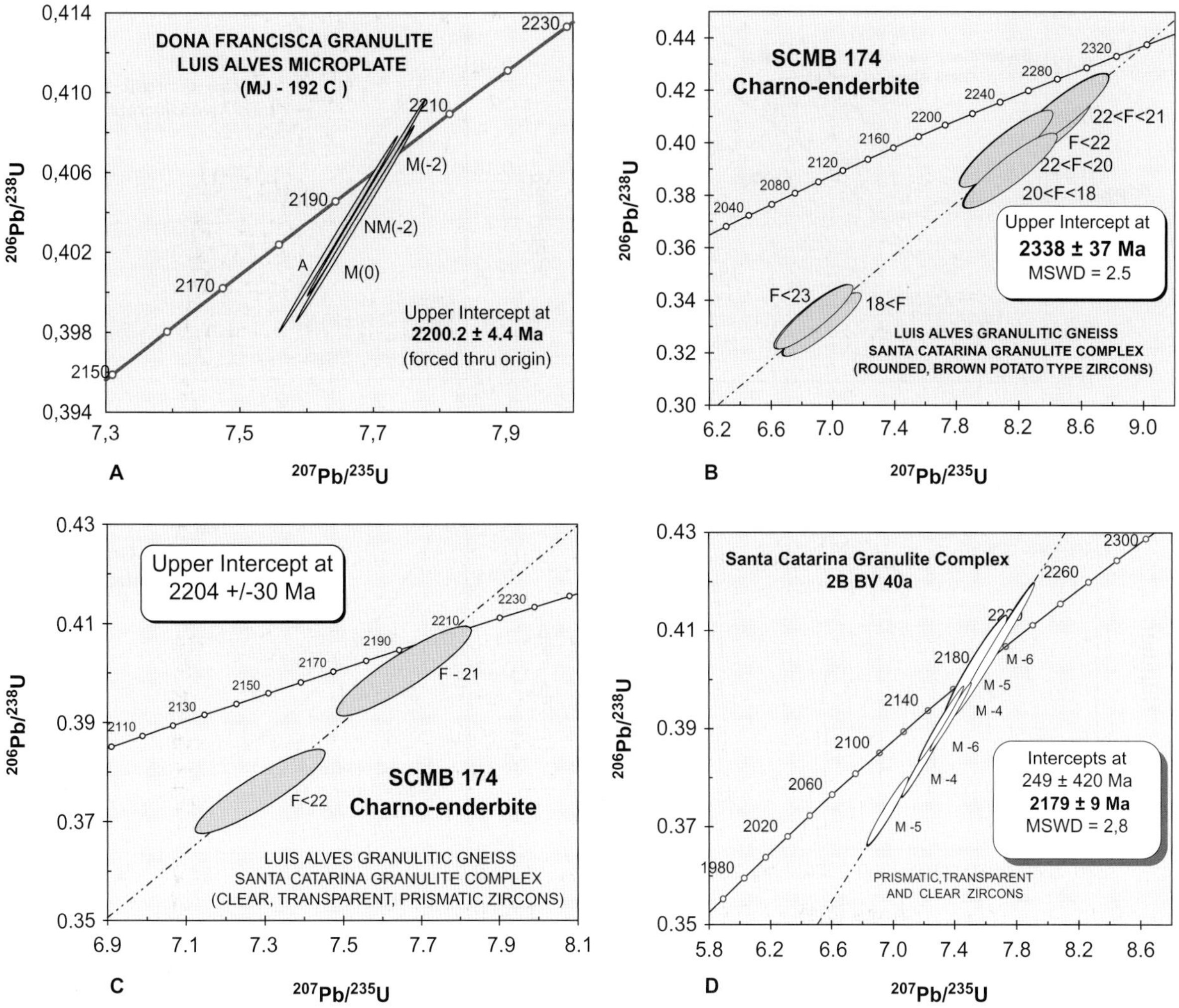


**Figure 7.2.6** (A) Concordia diagram for the zircons extracted from the MJ 192 granulitic gneiss. This value is interpreted as representative of the age of the second high-grade metamorphism that affected the Santa Catarina Granulitic Complex.
(B) Concordia diagram for the zircons extracted from the SCMB174 granulitic gneiss. The value of 2,338 ± 37 Ma obtained for oval crystals and crystals with rounded rims and edges is interpreted as indicative of the age of the first high-grade metamorphism that affected the Santa Catarina Granulitic Complex. (C) Concordia diagram for the zircons extracted from the SCMB174 granulitic gneiss. The value of 2,204 ± 30 Ma obtained for transparent, euhedral crystals with well-developed edges and faces is interpreted as indicative of the age of the second high-grade metamorphism that affected the Santa Catarina Granulitic Complex.
(D) Concordia diagram for transparent, euhedral zircons with well-developed edges and faces from paragneisses (2BBV40A).
The value of 2,179 ± Ma is interpreted as indicative of zircons formed during the second high-grade metamorphism.

zircon ages for granulitic rocks of the same region. The values ranged between 2,050 and 2,200 Ma. A Sm-Nd isochron obtained from mineral radiometric data yielded younger values, between 1,704 and 1,830 Ma, compatible with K-Ar mineral results (Harara, 2001).

Zircons extracted from charnockitic–enderbitic rocks of the central domain (sample SCMB 174 – quartzo-feldspathic hypersthene gneiss near Luís Alves) were analysed by ID-TIMS. Similarly to the study carried out in the northern portion, two zircon populations were characterised, one composed of crystals with rounded edges and well-developed pyramidal faces and the other constituted by well-formed, needle-shaped, transparent crystals. The ages obtained for the two populations were, respectively, 2,338 ± 37 and 2,204 ± 30 Ma (Figure 7.2.6B and C). Zircons from charnockitic–enderbitic rocks bearing ca. 5% hypersthene and blue quartz (SSAGI 28) from the central-northern region of the Complex were dated by the ID-TIMS method, yielding 2,372 ± 48 Ma (Basei et al., 1998a).

Quartzose paragneisses of the central domain (sample 2BBV 40a), intercalated with quartzitic bands, yielded a heterogeneous population of detrital zircons of varied shapes, sizes and colours. A set of translucent, brown crystals with well-developed pyramidal faces yielded, by means of discordant points, an upper concordia intercept Archaean age of around 2,800 Ma. Due to the high degree of discordance, these values must be taken only as an indication of the possible existence of Archaean rocks in the source area. Six fractions of well-formed, needle-shaped transparent crystals, similar to those extracted from sample SCMB 174, yielded an ID-TIMS age of

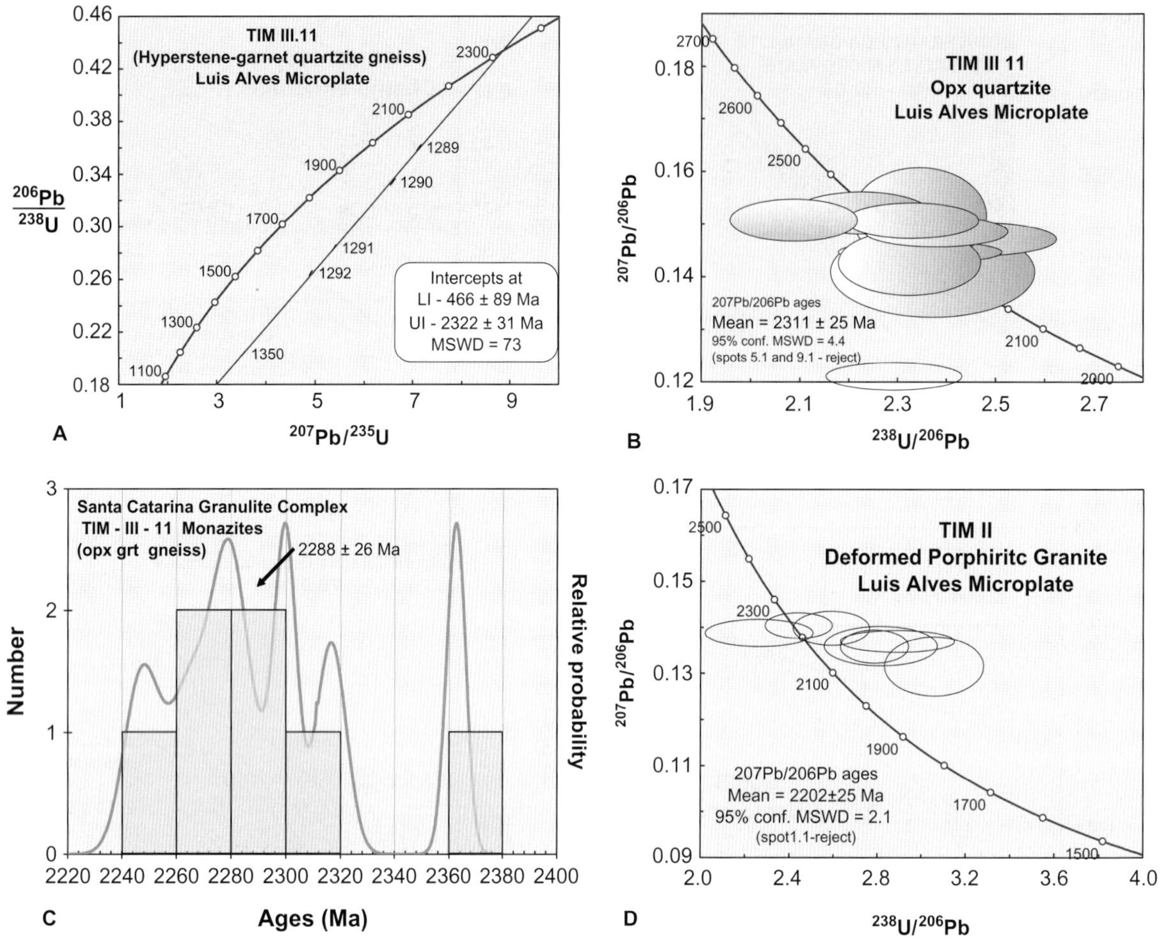


**Figure 7.2.7** (A) Concordia diagram for metamorphic oval zircons with rounded rims (TIM III 11). The ID-TIMS age (2,322 ± 32 Ma) results from the first high-grade metamorphism that affected these gneisses. (B) Concordia diagram with results for the zircons from the hypersthene–garnet gneiss (sample TIM III 11), showing the concordance of SHRIMP ages (2,311 ± 25 Ma) with the ID-TIMS result. (C) Histogram of monazite ages for sample TIM III 11. The age of 2,288 Ma is a little lower than the value for the zircons of the same sample, suggesting that the monazite and zircon ages result from the same metamorphic event. (D) U-Pb SHRIMP age for the porphyritic hornblende granitoid (TIM II 75) intrusive in granulitic gneisses around Blumenau. The age of 2,202 Ma is interpreted as related to zircon crystallisation and emplacement of the igneous body.

2,179 ± 9 Ma (Figure 7.2.6D). This result was confirmed by a SHRIMP age of 2,182 ± 9 Ma (Sato et al., 2008), obtained from well-formed transparent crystals. Cathodoluminescence (CL) images did not show complex internal structures, oscillatory zoning predominating in these zircons.

ID-TIMS and SHRIMP analyses of zircons from rocks affected by high-grade events (TIM III 11) of the southern portion of the Complex (north of the Itajaí Basin, near Timbó) have been carried out. Five zircon fractions extracted from a hypersthene–garnet gneiss yielded an age of 2,322 ± 32 Ma (Figure 7.2.7A). The U-Pb study by SHRIMP of the same sample indicated an age of 2,311 ± 25 Ma, similar to that obtained (Figure 7.2.7B). Monazites yielded a similar ID-TIMS mean age of 2,288 ± 26 Ma (Figure 7.2.7C). Two deformed granitoids, intrusive in the regional gneisses of the southern domain, have also been dated. By the SHRIMP method, zircon from a porphyroid granodioritic hornblende–biotite granitoid (sample TIM II 75) indicated an age of 2,202 ± 25 Ma (Figure 7.2.7D). This value is a little older than the age of 2,112 ± 21 Ma yielded by the zircons from a homogeneous (sample IB VII 23), finely foliated, mylonitic, pink leucogranite (Basei et al., 1998a).

CL images of TIM III 11 (hypersthene–garnet gneiss) zircons indicated the predominance of rounded crystals (metamorphic?) without internal structures while the zircons from the deformed granitoid of sample TIM II 75 show prismatic crystals (igneous) with internal oscillatory zoning (Figure 7.2.8).

Three different samples from the same outcrop of granulitic felsic gneisses crosscut by basic bodies (sample set MBAN 02) in the coastal region, south of the Barra Velha resort (Grant Beach), were dated. Good concordance was obtained from the analyses of zircons extracted from the basic and felsic bodies and from the zone of interaction between the two. Treated together, the data from all three samples indicate an age of 2,192 ± 14 Ma (Figure 7.2.9A).

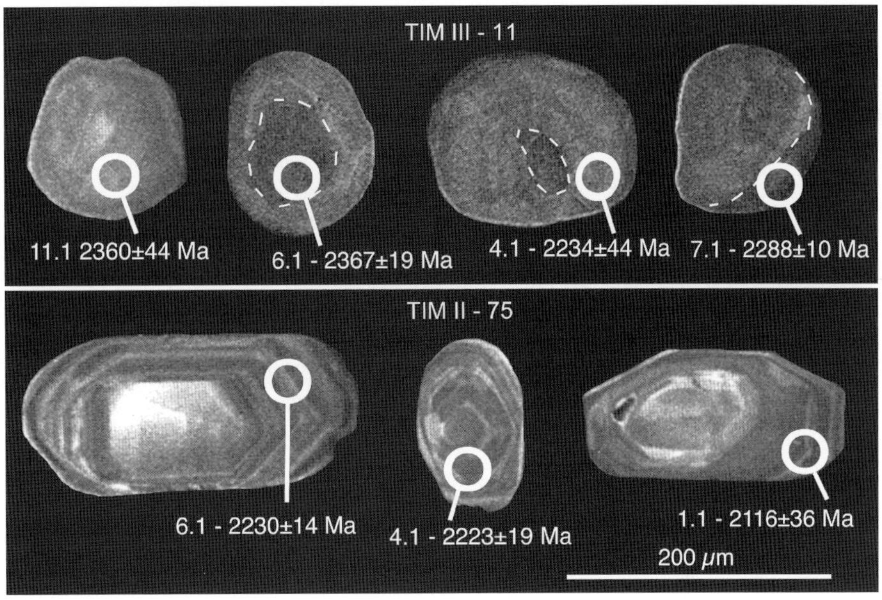


**Figure 7.2.8** Cathodoluminescence images for sample TIM III 11. Note the rounded and oval shape of the crystals and lack of inherited nuclei and complex internal structures.

## 7.2.6. Interpretation and Discussion of Radiometric Results

The first systematic geochronologic study of the Santa Catarina Granulitic Complex was by the K-Ar method by Minioli (1972), who analysed mafic–ultramafic rocks of the Barra Velha region of Santa Catarina State. This study indicated both Palaeoproterozoic and Archaean ages. The Archaean values were attributed to argon excess, whereas the Palaeoproterozoic ages seemed compatible with the regional cooling pattern obtained from analyses of biotite and amphibole from gneissic rocks (Figure 7.2.9B). In the surroundings of Barra Velha, the characterisation of a single Palaeoproterozoic U-Pb age for the zircons from basic rocks and tonalitic gneisses ruled out previous proposals that assigned an Archaean age for the mafic–ultramafic bodies, and at the same time they demonstrated the existence of a high-grade metamorphism that affected these rocks, younger than 2,200 Ma. Additionally, an age of ca. 2,250 Ma was obtained for the basic rocks by means of a Sm-Nd mineral isochron (Hartmann et al., 1999).

Ages around 2,350 Ma attributed to a high-grade metamorphic event were obtained in all domains of the complex, highlighting the importance of this event in the context of the LAM. This metamorphism preceded by ca. 200 million years a second high-grade metamorphism at around 2,180 Ma (Figure 7.2.9C). The existence of two high-grade metamorphic events was postulated by Basei et al. (1998a, 1999) and Hartmann et al. (2000), the latter confirming both events by the presence of simplectitic reactions involving two pyroxenes. The age of deformation and metamorphism of paragneisses of the LAM has always been a matter of debate. However, considering the presence of granulitic paragneisses with ages around 2,180 Ma, a certain period for erosion and deposition of sedimentary rocks between the two high-grade metamorphic events must be taken into account.

Archaean U-Pb values close to 2,700 Ma were obtained only from zircon cores studied by the SHRIMP method (Hartmann et al., 2000). However, despite the fact that Archaean rocks have not yet been recognised, similar Archaean values were observed in Rb-Sr and Sm-Nd whole-rock isochrons, which suggest that igneous Archaean rocks might be the protoliths of the Santa Catarina Granulitic Complex. At 1,800 Ma, the Santa Catarina Granulitic Complex had already cooled below the 450°C isotherm, as indicated by the K-Ar ages in biotite (Siga, 1995; Basei et al., 1998a).

The A-type Serra do Mar granitic magmatism and the coeval bimodal volcanic rocks of the volcano-sedimentary basins with ages around $588.5 \pm 4.7$ Ma intruded, and were deposited on, the Palaeoproterozoic high-grade metamorphic basement of the microplate.

Table 7.2.1 summarises the proposed tectonic evolution for the LAM.

## 7.2.7. Tectonic Implications

The LAM (Basei et al., 1992, 2005) represents the oldest continental fragment in southeastern Brazil to be sandwiched between Neoproterozoic supracrustal belts that characterise the Mantiqueira Province (Almeida, 1967).

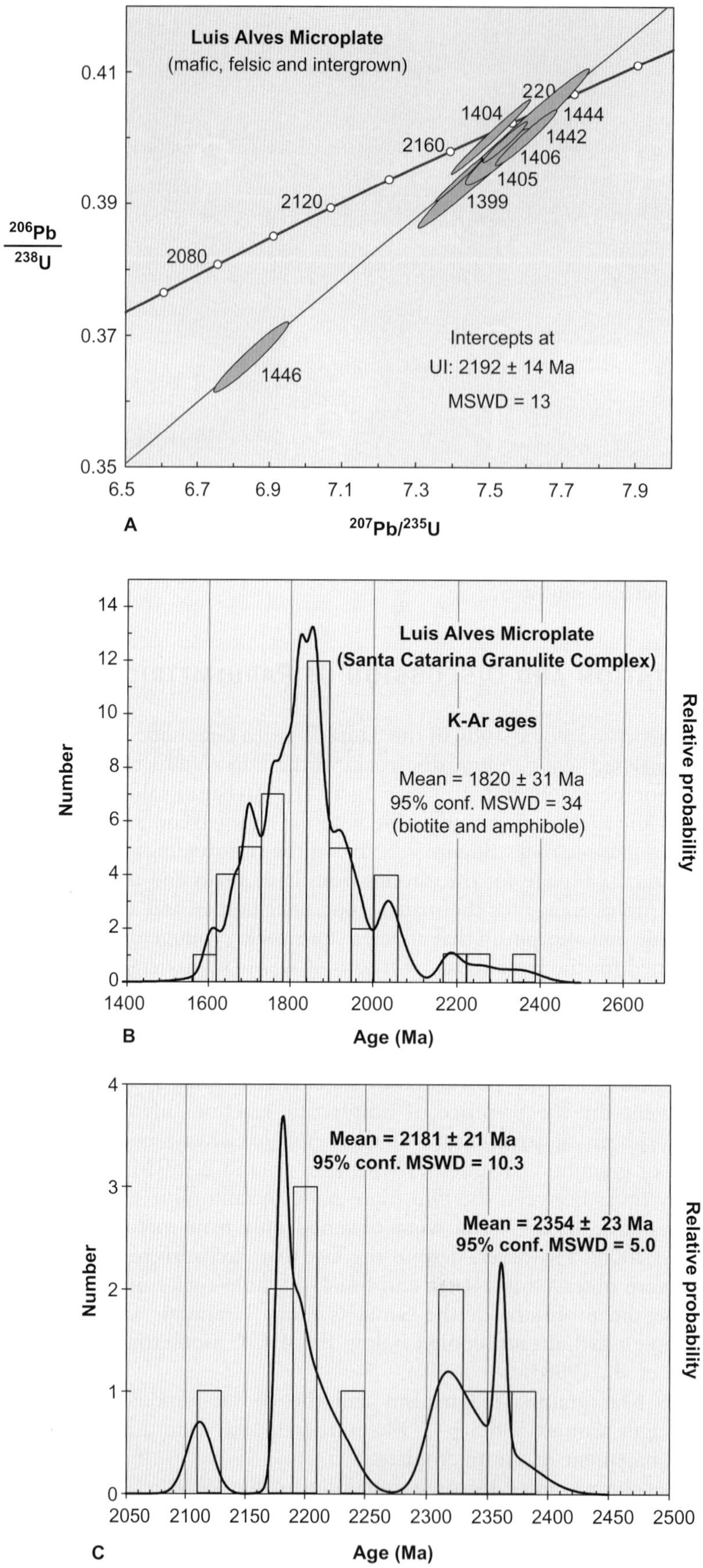


**Figure 7.2.9** (A) Concordia diagram for the basic dikes, felsic granulites and interaction region between these two rock types (MBAN 02 set sample). The age of 2,192 Ma common to the three rock types indicates that the bimodal magmatism that characterises the TTG magmatism of the Santa Catarina Granulitic Complex has an important Palaeoproterozoic component and is a little older than the second granulitic metamorphism that regionally affects the Complex. (B) Histogram with the density curve for K-Ar results yielded by biotite and amphibole from the Santa Catarina Granulitic Complex. Clearly, problematic results (due to excess or loss of argon), materials with low potassium contents (pyroxene) or low argon retention (K feldspar and whole rock) and ages affected by the Neoproterozoic superposition (at the microplate borders) were excluded from this histogram. (C) U-Pb ages of the main thermo-tectonic episodes related to high grade metamorphism of Santa Catarina Granulite Complex.

**Table 7.2.1** Summary of the proposed tectonic evolution for the Luis Alves Microplate.

| Unit | Main lithotype | U-Pb zircon ages (Ga) |
|---|---|---|
| *Neoproterozoic basins and granitoids* | | |
| Itajaí Basin | Sedimentary rocks with subordinate felsic volcanics | 0.58–0.56 |
| Serra do Mar Suite | A-type granites | 0.59–0.58 |
| Campo Alegre, Corupa and Guaratubinha basins | Felsic volcanics with subordinate sedimentary rocks | ~0.60–0.59 |
| *Paleoproterozoic Basement* | | |
| Santa Catarina Granulite Complex | Regional cooling | 1.8–1.7 |
| | Retrograde metamorphism | ~2.0–1.9 |
| | Second high-grade metamorphism | 2.1–2.0 |
| | Deformed granitoids | ~2.25–2.15 |
| | First high-grade metamorphism | ~2.35 |
| | TTG igneous activity | ~2.7 |

Despite the lack of Neoproterozoic deformation and metamorphism, this segment was affected by the events that, between 900 and 530 Ma, led to the fragmentation of Rodinia and the amalgamation of Gondwana. That Neoproterozoic tectonics affected the microplate can be demonstrated by its tectonic contacts with the adjacent Neoproterozoic domains (Brito Neves and Cordani, 1991; Brito Neves et al., 1999; Basei et al., 2000, 2008).

According to Condie's (1989) classification, this crustal segment, made up of Palaeoproterozoic rocks, is properly named microplate, because it is a geologically exotic unit showing no consanguinity with the other Precambrian domains of the southeastern portion of South America. The main evidences that the northern boundary of the LAM represents an active margin are the calc-alkaline granitoids of the Piên Batholith and the ophiolitic remains also found there (Machiavelli et al., 1993; Basei et al., 2000; Harara, 2001; Harara et al., 2003). Although the roots of these units are unknown, geophysical data (Hallinan et al., 1993) suggest the lack of lateral continuity with the Rio de La Plata and Paranapanema cratonic domains, supporting the surface geology interpretation of a suture between the two. In contrast to the LAM, the northern Curitiba Microplate shows a radically different, early Precambrian geological history with abundant Archaean rocks back to about 3,000 Ma (Sato et al., 2003) and a quite distinct late Precambrian thermo-tectonic history with all K-Ar data indicating Neoproterozoic ages. These differences also support the existence of a suture between these units.

The mosaic of blocks observed in southeastern Brazil was consolidated between 630 and 600 Ma, when the oceanic crust was consumed and several magmatic arcs were active (Brito Neves et al., 1999; Basei et al., 2005, 2008). Subsequently, Neoproterozoic fold belts formed between the main gneissic–migmatitic blocks (Figure 7.2.10).

Related to the same amalgamation process, but ca. 20 Myr after the accretion of the LAM to the early Precambrian western continental margin, the Itajaí foreland basin was established in its southern border, at ca. 580 Ma. The development of this basin lasted for ca. 15 Ma, with acidic volcanism culminating at ca. 565 Ma (Basei et al., 2008). Assuming that the basement was the source area for only the lower continental sequence of the Itajaí Basin (Basei, 1985; Basei et al., 2008) and that the Brusque Group (metavolcano-sedimentary sequence of the Dom Feliciano Belt) was the source area for the upper part of the sequence, the Brusque Group became proximal to the Itajaí Basin between 580 and 565 Ma. In this scenario, the Brusque Group became an elevated area, providing sediment to the upper part of the Itajaí Basin (Basei et al., 2000, 2008). In the Lower Cambrian, at ca. 535 Ma, deposition ceased when nappes consisting of the Brusque Group were emplaced over the Itajaí Basin fill, causing deformation and low-grade metamorphism. During the opening and later deformation of the Itajaí Basin in the southern border of the LAM, the juxtaposition of the Paranaguá Batholith to the ENE occurred. These granitoids evolved into a magmatic arc and were transported from E to W and tectonically juxtaposed to the LAM between 580 and 530 Ma. This lower Cambrian tectonism marked the complete closure of the oceanic basins that existed between the Rio de la Plata–Paranapanema–São Francisco and Congo–Kalahari cratons, resulting in the amalgamation of western Gondwana.

The information in the literature about the polar trajectory of the Rio de la Plata craton is mostly based on the good quality of the palaeomagnetic pole obtained in the Campo Alegre Basin (D'Agrella Filho and Pacca, 1978, 2007; Janikian et al., 2008). However, considering that this basin is situated on the LAM and not on the Rio de la Plata Craton, the palaeogeographic considerations regarding the latter are only valid after the juxtaposition of these two geotectonic units. The problem that persists in relation to the palaeomagnetic positioning of the group is the suggestion that the Campo Alegre Basin would be situated at ca. 590 Ma at the South Pole. There is no indication whatsoever of glacial deposits in this basin.

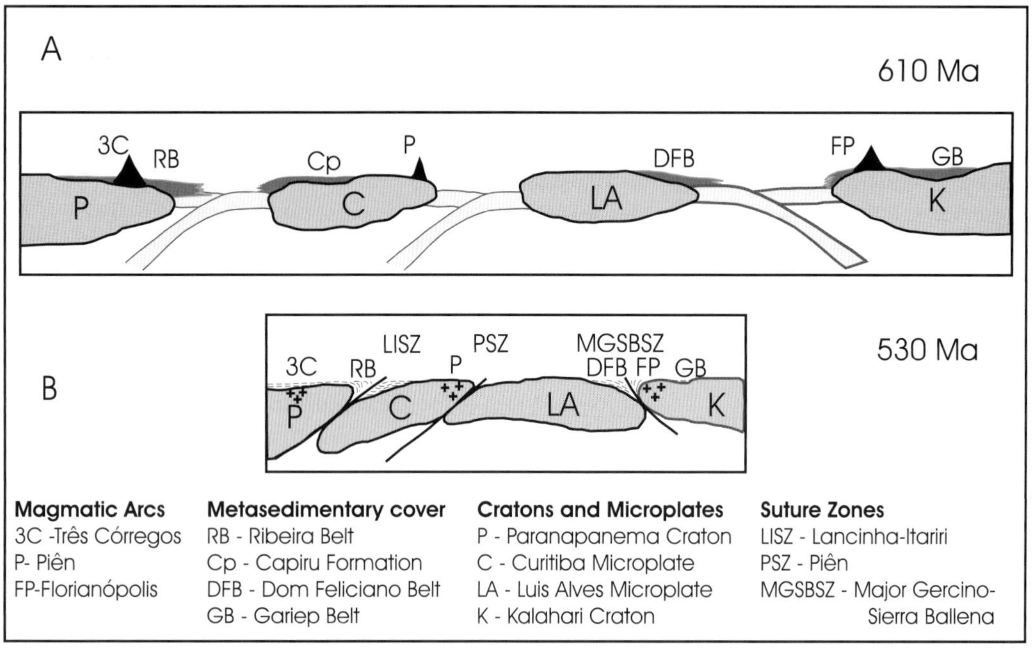


**Figure 7.2.10** (A and B) Schematic NW-SE section of the crustal blocks involved in the closing of the Neoproterozoic oceans existing between the South-American (Paranapanema and Rio de La Plata) and African (Kalahari and Congo) cratons (simplified from Basei et al., 2008).

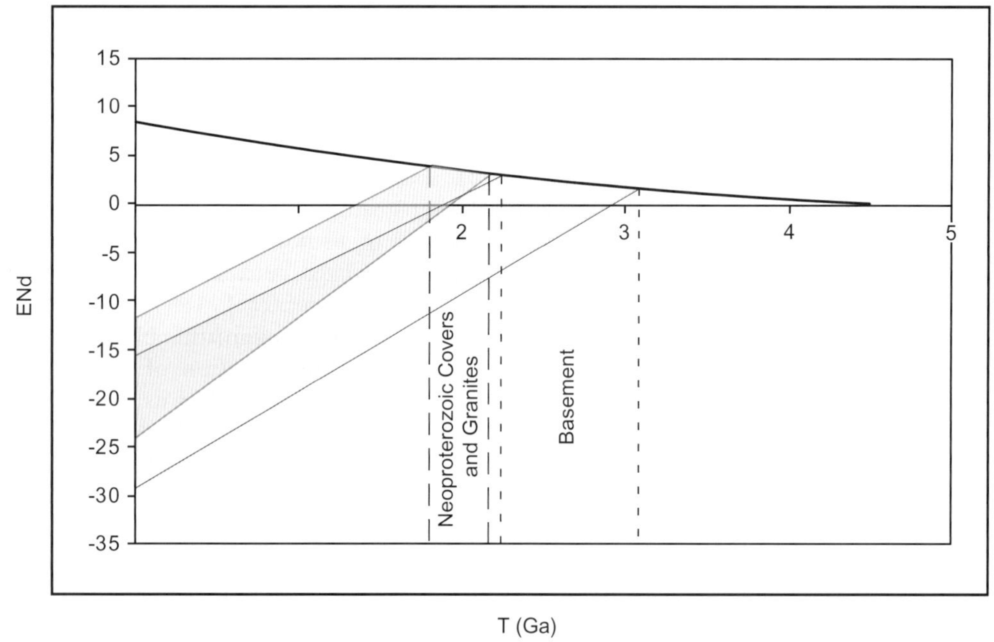


**Figure 7.2.11** Epsilon Nd versus time diagram showing the contrast observed between the Nd model ages ($T_{DM}$) for Neoproterozoic granitoids and Palaeoproterozoic gneisses.

## 7.2.8. Conclusion

The LAM is an exotic continental fragment in relation to the other geotectonic units of southeastern Brazil. It consists of migmatitic-gneisses at granulite facies or retrogressed under amphibolite facies conditions, that formed from a TTG association that since the end of the Palaeoproterozoic has been cold and stable, with cooling age around 1,800 Ma. Two high-grade metamorphic events, culminating around 2,354 $\pm$ 23 and 2,181 $\pm$ 21 Ma, can be characterised and are separated by a time gap of ca. 200 million years during which erosion and deposition of sedimentary rocks must have occurred.

Gravimetric data suggest that the thickness of the LAM reached ca. 45 km (Mantovani et al., 1989), making it impossible to consider that it is composed entirely of the outcropping granulitic gneisses. Therefore, it is suggested that crustal thickening occurred by the end of the Palaeoproterozoic, with the Santa Catarina Granulitic Complex thrust over a Palaeoproterozoic juvenile complex (Mantovani et al., 1989; Hallinan et al., 1993; Basei et al., 1998a). This hypothesis is also based on the contrast observed between Nd model ages ($T_{DM}$) that indicate mostly Palaeoproterozoic model ages for the intrusive granitoids and Neoarchean for the Santa Catarina Granulitic Complex gneisses (Figure 7.2.11). Despite the involvement of the LAM in the Neoproterozoic tectonics, its internal constitution was preserved from thermal-tectonic overprint. Neoproterozoic volcano-sedimentary basins and undeformed A-type granitoids, formed under an extensional regime between 600 and 588 Ma, represent the influence of the surrounding tectonic regime on the LAM.

Despite its Palaeoproterozoic basement, the LAM, as such, must be understood as an exotic unit that was involved in the Neoproterozoic tectonic events related to the closure of the oceans between large cratonic masses (Rio de la Plata, Paranapanema, Congo and Kalahari) that culminated in the formation of western Gondwana (Table 7.2.2).

## *ACKNOWLEDGEMENTS*

The authors wish to thank the research funding agency of São Paulo State—Fundação de Amparo a Pesquisa do Estado de São Paulo (FAPESP process 2005/58688-1) for the financial support for the SHRIMP and ID-TIMS analyses and field work. This is a contribution to IGCP 478.

## **APPENDIX 1**

Analytical methods for U-Pb zircon and monazite dating by the ID-TIMS and SHRIMP.

### ID-TIMS

U-Pb analytical routine used in the performed analysis was established at the Centro de Pesquisas Geocronológicas, IGc/USP circa15 years ago. Sample fragments are reduced to 100- and 250-mesh grain sizes in a disc mill. After being classified in a Wilfley table, the portion rich in heavy minerals is treated with bromoform ($d = 2.89\,g\,cm^{-3}$). The heavy concentrate is processed in a Frantz magnetic separator at 0.5 A. The non-magnetic fraction is treated with methyl iodide ($d = 3.3\,g\,cm^{-3}$), and the fraction containing the heavy minerals is once more processed in the Frantz separator at 1.0 and 1.5 A. When necessary, washing with nitric acid ($HNO_3$) is carried out for sulphide elimination. After washing, the material is once again processed in the Frantz separator and split in several zircon-rich magnetic fractions. The final purification of each fraction (preferably the least magnetic) is carried out by hand picking under the stereomicroscope. At this step, air-abrasion of a small quantity of zircon crystals is carried out using the same amount of pyrite for ca. 15 min, to remove common Pb adsorbed at the crystal surface. Then, the fraction is washed with hot $HNO_3$ (7N) to remove pyrite and then with milli-Q water in the ultrasound. The weight of each zircon fraction is obtained after the estimation, under the stereomicroscope, of the volume of crystals; using the volume/density ratio, the final weight of the fraction is calculated. After weighing, washing with $HNO_3$ (7N) and milli-Q water in the ultrasound is carried out for removal of organic material. The dissolution of zircon crystals is carried out with the addition of HF and $HNO_3$ in Teflon microbombs. A $^{205}Pb/^{235}U$ spike is also added. A set of 15 microbombs arranged in a metal jacket is left for three days in a stove at 200°C. Then, the HF is evaporated and HCl (6N) is added. The microbombs rest in a stove for 24 h. After the evaporation of HCl (6N), the residue is dissolved in HCl (3N). U and Pb are concentrated and purified by passing the solution through an anionic exchange resin column. The solution enriched in U and Pb is, after addition of phosphoric acid, evaporated until the formation of a micro-drop. The sample is deposited in a rhenium filament and the isotopic composition is determined with a Finnigan MAT 262 solid source mass spectrometer. After reduction of the data (PBDAT), the results are plotted in appropriate diagrams using the software Isoplot Ex (Ludwig, 2001). All ID-TIMS zircon analyses are shown in Table 7.2.1.

### SHRIMP

As the study aimed to reveal the general picture of the age of rocks, sample collection was concentrated in typical units for the Luis Alves basement. Zircon separation by standard gravimetric and isodynamic techniques and the mounting of selected zircons into epoxy resin discs were carried out at the Institute of Geological Sciences,

**Table 7.2.2** ID-TIMS zircon analysis.

| Sample | Laboratory number | 207/235[a] | Error (%) | 206/238[a] | Error (%) | 207/206 | Error (%) | 206/204[b] | Pb (ppm) | U (ppm) | Weight (mg) | Age (Ma) | | |
|---|---|---|---|---|---|---|---|---|---|---|---|---|---|---|
| | | | | | | | | | | | | 206/238 | 207/235 | 207/206 |
| TIM-III-11 Monazite | 851 | 8.2034 | 0.55 | 0.4146 | 0.55 | 0.1435 | 0.05 | 16482 | 6103.9 | 5576.4 | 0.0104 | 2236 | 2254 | 2270 |
| TIM-III-11 Monazite | 852 | 8.5803 | 0.52 | 0.4290 | 0.50 | 0.1450 | 0.13 | 760 | 8731.9 | 6101.1 | 0.0059 | 2301 | 2294 | 2288 |
| TIM-III-11 Monazite | 1253 | 7.7553 | 0.97 | 0.4053 | 0.96 | 0.1388 | 0.15 | 1983 | 1482.9 | 2364.4 | 0.0250 | 2194 | 2203 | 2212 |
| TIM-III-11 Monazite | 1156 | 8.1635 | 0.53 | 0.4165 | 0.52 | 0.1422 | 0.06 | 5701 | 12016.0 | 1017.0 | 0.0021 | 2245 | 2249 | 2254 |
| TIM-III-11 Monazite | 2044 | 8.4018 | 0.47 | 0.4224 | 0.46 | 0.1443 | 0.05 | 115103 | 7883.3 | 7176.9 | 0.0181 | 2271 | 2275 | 2279 |
| TIM-III-11 Monazite | 2045 | 8.1387 | 0.47 | 0.4167 | 0.47 | 0.1416 | 0.05 | 24483 | 7425.2 | 4744.6 | 0.0088 | 2246 | 2247 | 2247 |
| TIM-III-11 Monazite | 2046 | 8.3116 | 0.47 | 0.4213 | 0.46 | 0.1431 | 0.05 | 38760 | 8655.2 | 4481.8 | 0.0087 | 2266 | 2266 | 2265 |
| TIM-III-11 Monazite | 2047 | 9.3167 | 0.48 | 0.4461 | 0.48 | 0.1515 | 0.05 | 20142 | 6631.9 | 5611.9 | 0.0026 | 2378 | 2370 | 2363 |
| TIM-III-11 Monazite | 2371 | 8.5910 | 0.47 | 0.4267 | 0.46 | 0.1460 | 0.05 | 153741 | 6944.0 | 7505.2 | 0.0226 | 2291 | 2296 | 2300 |
| TIM-III-11 Monazite | 2372 | 8.7659 | 0.46 | 0.4312 | 0.46 | 0.1475 | 0.05 | 86934 | 11101.0 | 6269.7 | 0.0217 | 2311 | 2314 | 2317 |
| TIM-III-11 Monazite | 2373 | 8.4903 | 0.47 | 0.4254 | 0.47 | 0.1447 | 0.05 | 55970 | 8379.7 | 5692.5 | 0.0177 | 2285 | 2285 | 2285 |
| TIM III.11 Zircon | 1289 | 7.1257 | 0.57 | 0.3601 | 0.57 | 0.1435 | 0.09 | 797 | 67.0 | 174.6 | 0.0062 | 1982 | 2127 | 2270 |
| TIM III.11 Zircon | 1290 | 6.5916 | 0.64 | 0.3340 | 0.64 | 0.1431 | 0.06 | 4832 | 143.4 | 425.4 | 0.0070 | 1858 | 2058 | 2265 |
| TIM III.11 Zircon | 1291 | 5.3883 | 0.69 | 0.2837 | 0.68 | 0.1377 | 0.11 | 1173 | 187.2 | 631.4 | 0.0040 | 1610 | 1883 | 2199 |
| TIM III.11 Zircon | 1292 | 4.9162 | 0.63 | 0.2647 | 0.62 | 0.1347 | 0.08 | 4006 | 232.5 | 874.1 | 0.0030 | 1514 | 1805 | 2160 |
| TIM III.11 Zircon | 1350 | 3.5488 | 0.65 | 0.2032 | 0.64 | 0.1267 | 0.07 | 1908 | 64.1 | 314.5 | 0.0096 | 1192 | 1538 | 2053 |
| SCMB 174 Zircon | F<22 Brown | 8.3582 | 1.50 | 0.4032 | 1.50 | 0.1504 | 0.55 | 4810 | 458.7 | 1066.4 | N.A. | 2183 | 2271 | 2350 |
| SCMB 174 Zircon | 22<F<20 Brown | 8.1160 | 1.50 | 0.3979 | 1.50 | 0.1479 | 0.55 | 11100 | 228.2 | 543.3 | N.A. | 2159 | 2244 | 2322 |

| | | | | | | | | | | | | | | |
|---|---|---|---|---|---|---|---|---|---|---|---|---|---|---|
| SCMB 174 Zircon | 22<F<21 Brown | 8.4651 | 1.50 | 0.4116 | 1.50 | 0.1492 | 0.55 | 6914 | 258.1 | 590.6 | N.A. | 2222 | 2282 | 2336 |
| SCMB 174 Zircon | 18<F Brown | 6.9233 | 1.50 | 0.3307 | 1.50 | 0.1518 | 0.55 | 8074 | 200.5 | 570.8 | N.A. | 1842 | 2102 | 2367 |
| SCMB 174 Zircon | 20<F<18 Brown | 8.1383 | 1.50 | 0.3895 | 1.50 | 0.1516 | 0.55 | 6752 | 250.7 | 605.0 | N.A. | 2120 | 2246 | 2363 |
| SCMB 174 Zircon | F<23 Brown | 6.8705 | 1.50 | 0.3337 | 1.50 | 0.1493 | 0.55 | 1905 | 412.7 | 1138.8 | N.A. | 1856 | 2095 | 2338 |
| SCMB 174 Zircon | F<22 clear | 7.2881 | 1.50 | 0.3761 | 1.50 | 0.1405 | 0.55 | 4489 | 131.6 | 330.7 | N.A. | 2058 | 2147 | 2234 |
| SCMB 174 Zircon | 21 clear | 7.6572 | 1.50 | 0.4005 | 1.50 | N.A. | 0.55 | N.A. | N.A. | N.A. | N.A. | N.A. | N.A. | N.A. |
| MJ 192 zircon | NM(-2) | 7.6740 | 0.95 | 0.4034 | 1.00 | 0.1380 | 0.35 | 1235 | 38.0 | 72.0 | 0.5016 | 2185 | 2194 | 2202 |
| MJ 192 zircon | M(-2) | 7.6920 | 0.95 | 0.4047 | 1.00 | 0.1378 | 0.35 | 1667 | 38.0 | 70.0 | 0.5332 | 2191 | 2196 | 2200 |
| MJ 192 zircon | M(0) | 7.6490 | 0.95 | 0.4029 | 1.00 | 0.1377 | 0.35 | 2703 | 38.0 | 71.0 | 0.4237 | 2182 | 2191 | 2198 |
| 2BBV 40a zircon | 1025 M(-6)A | 7.5330 | 2.20 | 0.4027 | 2.13 | 0.1354 | 0.55 | 156 | 90.4 | 161.2 | 0.0105 | 2162 | 2165 | 2169 |
| 2BBV 40a zircon | 1028 M(-5)A | 7.6555 | 2.74 | 0.4063 | 2.72 | 0.1366 | 0.35 | 445 | 44.4 | 96.2 | 0.0109 | 2198 | 2191 | 2185 |
| 2BBV 40a zircon | 1031 M(-4)A | 7.2000 | 1.68 | 0.3837 | 1.65 | 0.1361 | 0.28 | 294 | 111.6 | 232.4 | 0.0113 | 2094 | 2137 | 2178 |
| 2BBV 40a zircon | 1026 M(-6)A | 7.3706 | 1.49 | 0.3924 | 1.47 | 0.1362 | 0.24 | 313 | 118.7 | 251.4 | 0.0106 | 2134 | 2157 | 2180 |
| 2BBV 40a zircon | 1029 M(-5)A | 6.9611 | 1.56 | 0.3730 | 1.52 | 0.1354 | 0.37 | 390 | 119.4 | 276.0 | 0.0099 | 2043 | 2107 | 2169 |
| 2BBV 40a zircon | 1032 M(-4)A | 7.3112 | 1.63 | 0.3910 | 1.61 | 0.1356 | 0.25 | 211 | 93.1 | 179.4 | 0.0132 | 2128 | 2150 | 2172 |
| 2BBV 40a zircon | 1027 M(-6)A | 6.1732 | 1.15 | 0.3223 | 1.13 | 0.1389 | 0.25 | 940 | 146.1 | 434.2 | 0.0112 | 1801 | 2001 | 2214 |
| 2BBV 40a zircon | 1030 M(-5)A | 7.4557 | 1.10 | 0.3546 | 1.03 | 0.1525 | 0.36 | 1157 | 262.5 | 699.2 | 0.0105 | 1957 | 2168 | 2374 |
| 2BBV 40a zircon | 1033 M(-4)A | 6.8400 | 0.77 | 0.3429 | 0.76 | 0.1447 | 0.12 | 3219 | 195.8 | 552.7 | 0.0160 | 1901 | 2091 | 2284 |
| MBAN zircon | 1399 2A C | 7.4156 | 0.70 | 0.3938 | 0.70 | 0.1366 | 0.09 | 842 | 25.2 | 48.3 | 0.1036 | 2140 | 2163 | 2184 |
| MBAN zircon | 1404 2C 4 | 7.4614 | 1.17 | 0.3988 | 1.16 | 0.1357 | 0.18 | 389 | 14.8 | 27.4 | 0.0529 | 2164 | 2168 | 2173 |
| MBAN zircon | 1405 2C 5 | 7.5213 | 0.95 | 0.3973 | 0.88 | 0.1373 | 0.35 | 1286 | 17.2 | 35.1 | 0.1047 | 2156 | 2176 | 2194 |
| MBAN zircon | 1406 2C 6 | 7.5387 | 0.65 | 0.3995 | 0.64 | 0.1369 | 0.15 | 701 | 35.4 | 65.4 | 0.0626 | 2167 | 2178 | 2188 |
| MBAN zircon | 1442 2B 2A | 7.5365 | 2.52 | 0.3985 | 2.49 | 0.1372 | 0.39 | 702 | 19.8 | 39.4 | 0.0830 | 2162 | 2177 | 2192 |
| MBAN zircon | 1444 2B 2B | 7.5956 | 0.90 | 0.4000 | 0.87 | 0.1377 | 0.21 | 2646 | 31.5 | 62.8 | 0.2005 | 2169 | 2184 | 2199 |
| MBAN zircon | 1446 2B 3B | 6.8510 | 1.23 | 0.3667 | 1.19 | 0.1355 | 0.33 | 450 | 12.6 | 27.2 | 0.0792 | 2014 | 2097 | 2171 |

SPU: laboratory number, magnetic fractions: numbers in parentheses indicate the tilt used on Frantz separator at 1.5 A current. N.A.: not available. Total U and Pb concentrations corrected for analytical blank. Ages: given in Ma using Ludwig Isoplot/Ex program (2002); decay constants recommended by Steiger and Jäger (1977).

[a]Radiogenic Pb corrected for blank and initial Pb, U corrected for blank.

[b]Not corrected for blank or non-radiogenic Pb.

**Table 7.2.3** SHRIMP zircon analyses.

| Labels | Grain type | U (ppm) | Th (ppm) | Th/U | Comm. ($^{206}$Pb%) | Error | $^{238}$U/$^{206}$Pb ratio | Error | $^{207}$Pb/$^{206}$Pb ratio | Error | Date (Ma) | Error (±1 s) | Disc (%) |
|---|---|---|---|---|---|---|---|---|---|---|---|---|---|
| TIM III 11 | | | | | | | | | | | | | |
| 1.1 | m, osc, eq | 268.4 | 64.2 | 0.24 | 0.015 | 0.025 | 2.4521 | 0.070 | 0.14686 | 0.001 | 2309.7 | 15.47 | −4.5 |
| 3.1 | e, h, eq | 545.4 | 71.0 | 0.13 | 0.006 | 0.021 | 2.3278 | 0.056 | 0.15038 | 0.001 | 2350.3 | 15.95 | −2 |
| 4.1 | e, osc, eq | 275.4 | 19.0 | 0.07 | 0.036 | 0.057 | 2.3701 | 0.086 | 0.14052 | 0.004 | 2233.7 | 44.29 | 1.6 |
| 5.1 | e, h, eq | 418.5 | 32.7 | 0.08 | 0.002 | 0.004 | 2.3223 | 0.060 | 0.1423 | 0.003 | 2255.4 | 31.28 | 2.4 |
| 6.1 | ?c, osc/h, eq | 635.3 | 71.9 | 0.11 | 0.035 | 0.017 | 2.2242 | 0.057 | 0.15188 | 0.002 | 2367.2 | 18.81 | 1.1 |
| 7.1 | og(±c?), h, eq | 625.6 | 15.9 | 0.03 | 0.015 | 0.015 | 2.3414 | 0.069 | 0.14502 | 0.001 | 2288 | 10.36 | 0.2 |
| 8.1 | e, osc, eq | 405.6 | 47.0 | 0.12 | 0.022 | 0.05 | 2.291 | 0.057 | 0.14411 | 0.001 | 2277.2 | 13.08 | 2.5 |
| 10.1 | e, osc, sp | 636.6 | 61.1 | 0.10 | 0.013 | 0.029 | 2.3767 | 0.060 | 0.14832 | 0.001 | 2326.6 | 13.98 | −2.7 |
| 11.1 | ?c, osc/h, eq | 560.8 | 94.6 | 0.17 | 0.015 | 0.015 | 2.3424 | 0.057 | 0.15127 | 0.004 | 2360.3 | 43.83 | −2.9 |
| 12.1 | ?c, osc/h, eq | 377.5 | 56.9 | 0.15 | 0.059 | 0.047 | 2.0845 | 0.053 | 0.15047 | 0.002 | 2351.3 | 18.92 | 7.4 |
| 2.1 | c, osc/h, eq | 546.9 | 21.7 | 0.04 | 0 | 0 | 2.7255 | 0.083 | 0.13761 | 0.003 | 2197.3 | 36.25 | −8.3 |
| 9.1 | e, h, eq | 431.7 | 30.4 | 0.07 | 0.015 | 0.025 | 2.4325 | 0.057 | 0.1363 | 0.001 | 2180.7 | 17.65 | 1.8 |
| TIM II 75 | | | | | | | | | | | | | |
| 1.1 | e, osc, p | 599.0 | 284.3 | 0.47 | 0.123 | 0.045 | 3.0561 | 0.092 | 0.13132 | 0.003 | 2115.7 | 35.98 | −13.7 |
| 2.1 | e, osc, p | 420.1 | 282.4 | 0.67 | 0.155 | 0.063 | 2.8127 | 0.105 | 0.13578 | 0.002 | 2174.1 | 22.99 | −9.8 |
| 3.1 | e, osc, sp | 637.3 | 401.1 | 0.63 | 0.035 | 0.028 | 2.7852 | 0.063 | 0.13563 | 0.001 | 2172.2 | 19.21 | −9 |
| 4.1 | e, osc, sp | 314.2 | 197.2 | 0.63 | 0.053 | 0.091 | 2.5883 | 0.071 | 0.13964 | 0.002 | 2222.8 | 18.99 | −5.3 |
| 5.1 | e, osc, p | 503.8 | 244.4 | 0.48 | 0.101 | 0.048 | 2.8889 | 0.106 | 0.13681 | 0.001 | 2187.2 | 12.54 | −12.4 |
| 6.1 | e, osc, p | 354.4 | 193.7 | 0.55 | 0.015 | 0.025 | 2.4403 | 0.062 | 0.14022 | 0.001 | 2229.9 | 14.08 | −0.7 |
| 7.1 | e, osc, p | 197.6 | 119.8 | 0.61 | 0.063 | 0.063 | 2.2629 | 0.099 | 0.13864 | 0.001 | 2210.3 | 15.51 | 6.7 |

Labels: (*x.y*) grain number followed by analysis number. Grain habit: (sp) stubby prism/equant grain (aspect ratio <2), (p) prism and (anh) anhedral. Site: (e) end or edge, (int) interior, (og) overgrowth and (c) core. CL image microstructure: (osc) oscillatory finescale zoning, (rex) recrystallised and (h) homogeneous (d: dark and b: bright). Corrected with model Pb of Cumming and Richards (1975) for likely age of rock. All errors are 1 s.

University of São Paulo. CL imaging and age determinations by SHRIMP took place in the Research School of Earth Sciences, The Australian National University, according to standard procedures (Compston et al., 1984; Stern, 1998; Williams, 1998; Sircombe, 2000). Choice of SHRIMP analytical sites was guided by CL imaging (McClaren et al., 1994). All SHRIMP zircon analyses are shown in Table 7.2.3. Most of the data yielded close to concordant ages. The data are portrayed in Concordia or Tera Wasserburg diagrams generated in the program Isoplot/Ex (Ludwig, 2001). The data were filtered prior to plotting, to remove analyses with the most disturbed radiogenic Pb-systematics and with $>2.5\%$ $^{206}Pb$ of common origin (calculated from measured $^{204}Pb$ using Cumming and Richards' (1975) Pb evolution curves for common Pb compositions).

# SECTION III:
# SYNTHESIS AND GLOBAL IMPLICATIONS

CHAPTER 8

# Tectonic Events and Palaeogeographic Evolution of Southwestern Gondwana in the Neoproterozoic and Cambrian ☆

Claudio Gaucher[1], Hartwig E. Frimmel[2] *and* Gerard J.B. Germs[3]

## Contents



## 8.1. Introduction

It is generally accepted that a supercontinent formed by amalgamation of small continents approximately 1,200–1,000 Ma ago, although its exact configuration remains speculative (Palaeopangea: Piper, 2000; Rodinia: McMenamin and McMenamin, 1990; Hoffman, 1991; Li et al., 2008). Subsequently, this supercontinent broke up into various fragments (ca. 850–550 Ma; Li et al., 2008). By collision some of these fragments may have formed larger blocks, whereas others continued to disperse and later took part in the amalgamation of Pangea (ca. 550–280 Ma, e.g. Dalziel, 1997). According to previous palaeogeographic reconstructions (e.g. Dalziel, 1997), the Adamastor Ocean formed between the Amazonia–Rio de la Plata cratons and the Kalahari Craton during the stages of Neoproterozoic continent dispersal. Grunow et al. (1996) suggested that continent-to-continent collision between the above-mentioned cratons and the resulting closure of the Adamastor Ocean were linked to the opening of the Iapetus Ocean between Laurentia and the South American cratons (ca. 550 Ma). The sequence of opening and closure of oceanic basins prior to Pan-African/Brasiliano orogeny is unlikely to be coeval in the various parts of Rodinia. As evident from the review by Li et al. (2008), Neoproterozoic supercontinental breakup was diachronous, with rifting events at approximately 825, 780, 750, 600 and 550 Ma, and may have been caused by a mantle superplume beneath the supercontinent. Gaucher and Germs (2002) proposed that oceanic basins

☆ Gaucher, C., Frimmel, H.E., Germs, G.J.B. 2009. Tectonic events and palaeogeographic evolution of southwestern Gondwana in the Neoproterozoic and Cambrian. In: Gaucher, C., Sial, A.N., Halverson, G.P., Frimmel, H.E. (Eds): Neoproterozoic-Cambrian Tectonics, Global Change and Evolution: a focus on southwestern Gondwana. Developments in Precambrian Geology, 16, Elsevier, pp. 295–316.

[1] Departamento de Geología, Instituto de Ciencias Geológicas, Facultad de Ciencias, Iguá 4225, 11400 Montevideo, Uruguay.
[2] Geodynamics & Geomaterials Research Division, University of Würzburg, Am Hubland, D-97074 Würzburg, Germany.
[3] University of Johannesburg, Private Bag X607, 6620 Oudtshoorn, South Africa.

*Developments in Precambrian Geology*, Volume 16
ISSN 0166-2635, DOI 10.1016/S0166-2635(09)01621-1

**Table 8.1** Fold belts and units in southwestern Gondwana and neighbouring regions, with age estimates for the opening and closure of their precursor oceans.

| Ocean | Belt/unit | Rift–drift (Ma) | Final closure (Ma) | References |
|---|---|---|---|---|
| Mozambique | Mozambique | 830 | 645 (initial) | Jöns and Schenk (2008), Hauzenberger et al. (2007) |
| Adamastor | Gariep | 771–740 | 550 | Frimmel et al. (2001a,b, 2002) |
| | Saldania | >650 | 550 | Scheepers and Armstrong (2002), Chapter 10 |
| | Kaoko | ca. 750 | 570 | Gray et al. (2006) |
| Brazilides | Paraguay | ca. 600 | 540–510 | Boggiani (1998), Gaucher et al. (2003), Ferreira et al. (2008) |
| | Araguaia | uncertain | 530 | Moura et al. (2008) |
| | Nico Pérez Terrane basins (A. Soldado) | 633–580 | 535 | Gaucher et al. (2008a), Bossi and Gaucher (2004) |
| | Cabo Frío Terrane | 620 | 525 | Schmitt et al. (2008a,b) |
| Pampean | Eastern Pampean | ca. 550 | 520–515 | Rapela et al. (1998a, 2007), Mulcahy et al. (2007) |
| Damara | Damara | 756–746 | 530 | Hoffman et al. (1996), Gray et al. (2006) |
| | Zambezi, Lufilian | 880 | 530 | Porada and Berhorst (2000) |
| Unnamed | Araçuaí | 880 | 560 | Silva et al. (2005, 2008) |
| | West Congo | 920 | 565 | Tack et al. (2001) |
| | Ribeira | 850 | 580–560 | Heilbron et al. (2008) |
| Goianides | Brasília | 900 | 630–620 | Valeriano et al. (2004, 2008) |
| Iapetus | Appalachians | 550 | 435 | Cawood et al. (2001), Van Staal et al. (1998) |

opened and closed sequentially from east to west in southwestern Gondwana, and that the palaeogeographic evolution from Rodinia to Gondwana could be described in terms of stepwise rifting of Rodinia and concomitant sequential closure of the thus-formed basins.

New insights from sedimentary successions in southwestern Gondwana, presented in Parts 2–7, as well as recently published data, will be reviewed in this chapter. Palaeocurrent analyses and provenance studies have shown that not two oceans (the Adamastor and Iapetus oceans; Grunow et al., 1996) but at least five oceans separated southwestern Gondwanan cratons during the late Neoproterozoic to early Palaeozoic. These oceans were, from east to west, the Damara, Adamastor, Brazilides, Pampean and Iapetus oceans (Table 8.1). Furthermore, the Mozambique Ocean separated East from West Gondwana, and in Early Palaeozoic times the Pacific Ocean extended to the south and west of Gondwana (Cawood, 2005). The Goianides Ocean, to the west of the Congo–São Francisco Craton, was consumed during the long-lived Brasília accretionary orogeny (Figure 8.1). The older (900–750 Ma) Charrúa Ocean (Fragoso-César, 1991) is represented by mafic rocks of the São Gabriel Block, tectonically attached to the Río de la Plata Craton (e.g. Saalmann et al., 2005) but of yet uncertain significance. In the light of new data, many of which have been presented in previous chapters of this book, we shall examine whether the relative timing of basin closure that eventually led to the amalgamation of southwestern Gondwana had an underlying common cause, and if yes, which one. The present contribution aims at providing a broader framework, to help explain the mechanisms behind Rodinia breakup and subsequent Gondwana amalgamation, as well as their profound environmental impacts.

## 8.2. Ocean Basins and Corresponding Sedimentary Successions

Figure 8.1 shows a reconstruction of Gondwana with the most relevant cratons, terranes and orogenic belts. The term "block" is used here to designate cratonic nuclei, which are still poorly known regarding their extent, boundaries and evolution (e.g. Paranapanema Block, see Chapter 7.1). The terrane concept is here applied in the sense of a "tectonostratigraphic terrane" (Coney et al., 1980; Howell, 1989), that is, a fault-bounded lithospheric block with a stratigraphy and geological evolution different from surrounding areas. Southwestern Gondwana was formed by the amalgamation of the Kalahari, Congo–São Francisco, Angola, Amazonia and Río de la Plata cratons, plus a number of smaller lithospheric blocks, such as the Pampean Block/Orogen, the Paranapanema Block, the Cuchilla Dionisio Terrane, Cabo Frío Terrane, Oriental Terrane, Coastal Terrane and Luis Alves

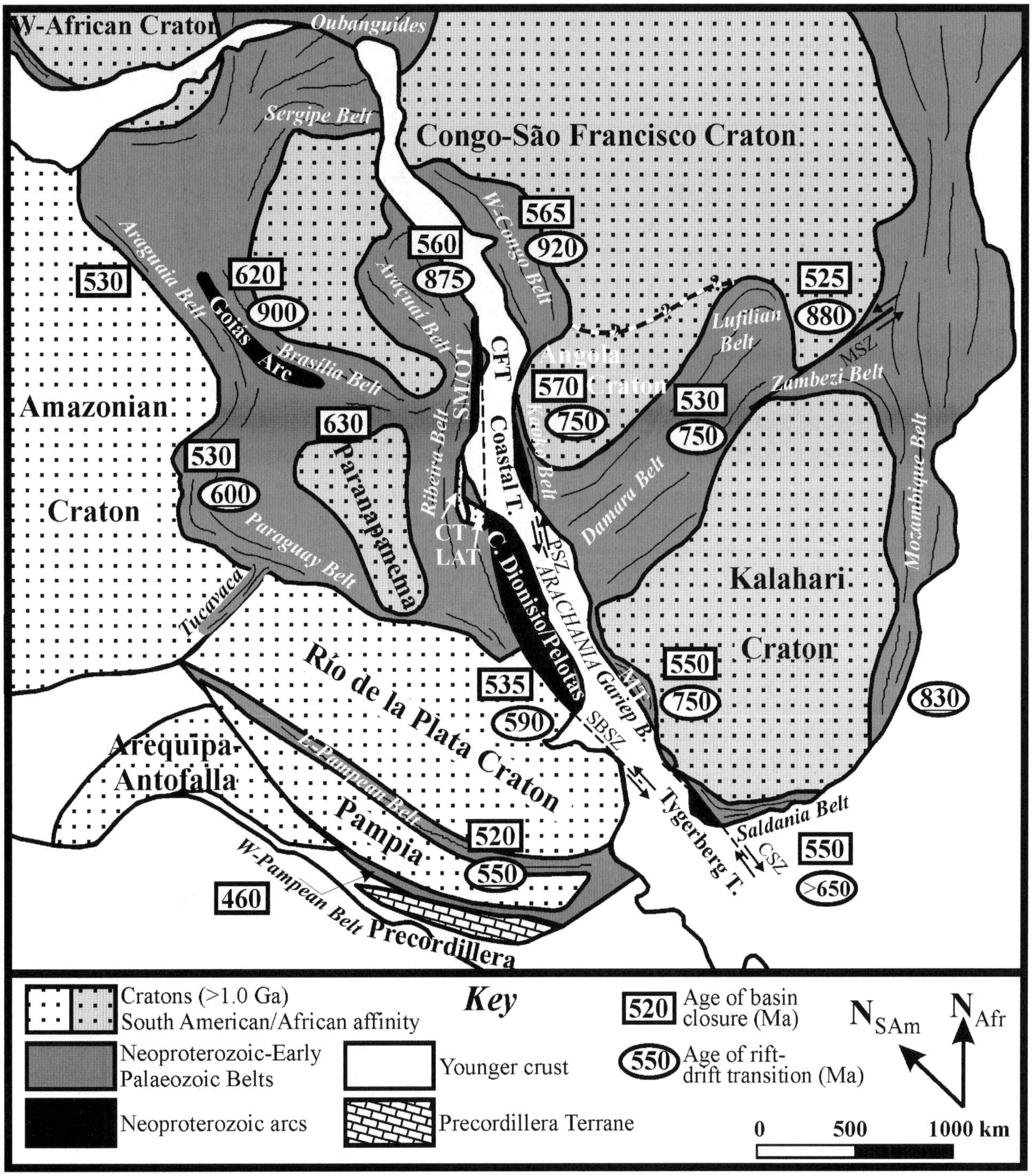


**Figure 8.1** Reconstruction of southwestern Gondwana and adjacent areas, showing intervening cratons and fold belts. Opening and closure of the precursor basins of each belt is indicated (underlined dates: estimates; sources of data: see Table 8.1). Note the central position in SW-Gondwana of Arachania, represented by the Cabo Frío, Coastal, Cuchilla Dionsio-Pelotas and Tygerberg terranes. CFT: Cabo Frío Terrane; CT: Curitiba Terrane; LAT: Luis Alves Terrane; MT: Marmora Terrane; SM/OT: Serra do Mar-Oriental Terrane; PSZ: Purros Shear Zone; SBSZ: Sierra Ballena Shear Zone; CSZ: Colenso Shear Zone. The "African" inheritance of the Paranapanema Block is tentative. Based on Porada (1989), Ramos (1988), Trompette (1994), Gaucher et al. (2003), Bossi and Gaucher (2004) and Parts 4 and 5 (this volume).

Block. Fringing these cratons/blocks, a number of orogenic belts mark the position of former oceanic basins (Figure 8.1 and Table 8.1), namely (from east to west):

(a) The Mozambique Belt, fringing the Kalahari and Congo cratons to the east, evolved from the *Mozambique Ocean*, and marks the boundary between East and West Gondwana. This basin will not be discussed here; it suffices to note that its opening probably occurred prior to 850 Ma, closure begun as early as 750 Ma (e.g. Veevers, 2003; Fitzsimons and Hulscher, 2005) and tectonometamorphic events continued into the Cambrian.

(b) The thick sedimentary succession of the Damara and Zambezi belts (see Part 5; John et al., 2003), separating the Congo and Kalahari cratons, was deposited in the *Damara Ocean* between ca. 750 and 540 Ma. Substantial evidence suggests that the Congo–São Francisco and Kalahari cratons were only welded together as a result of the *Pan-African Orogeny*, and were not adjacent in the supercontinent Rodinia (Cordani et al., 2003; De Waele et al., 2003). To avoid confusion, we use the term "Damara Ocean" for the entire oceanic basin that separated the Angola and Congo cratons from the Kalahari Craton, but note that the term "Khomas Ocean/Sea" has also been used in the literature for the southern section of that basin (see discussion in Part 5).

(c) The Gariep Belt (see Chapter 5.1) is associated with the evolution of the *Adamastor Ocean* (Hartnady et al., 1985), which opened between the western margin of the Kalahari Craton and the Cuchilla Dionisio-Pelotas Terrane (Bossi and Gaucher, 2004, senior synonim of "Punta del Este Terrane"), a block rifted off the Kalahari Craton between 770 and 740 Ma (Frimmel et al., 2001b, 2002; Basei et al., 2005). It is worth noting that the Rocha Formation in the Cuchilla Dionisio Terrane (Uruguay: Bossi and Gaucher, 2004), the Oranjemund Group in the Gariep Belt and the Tygerberg Formation of the homonymous terrane in the Saldania Belt (Gresse et al., 2006, and references therein; Figure 8.1) were also deposited in the Adamastor Ocean, in a deep-marine setting (Basei et al., 2005; see Section 8.3.2). As for the autochthonous units of the Saldania Belt (e.g. Cango Caves and Gamtoos groups), it is here assumed that they were deposited on the margins of the southern Adamastor Ocean. The lower Palaeozoic Kansa Group was deposited on the margins of the Pacific Ocean (*sensu* Cawood, 2005), after closure of the Adamastor Ocean.

(d) The Arroyo del Soldado, Sierras Bayas, Corumbá and Araras groups (Chapters 2 and 4.6), the latter units being part of the Paraguay Belt ("Brasilides" of Almeida, 1984), were deposited on the Río de la Plata and Amazonia cratons, in the western margin of the Brazilides Ocean (Dalziel, 1997). These units were deposited in basins that opened between 630 and 590 Ma, and closed around 530–520 Ma (Gaucher et al., 2008a,b; see Chapter 4.6). The blocks that marked the eastern margin of the *Brazilides Ocean* were the magmatic arc or arcs ("Arachania", see below) and associated terranes, comprising the Cuchilla Dionisio-Pelotas, Cabo Frío, Tygerberg and Coastal terranes, and later also the Congo São Francisco+Paranapanema Block+Goiás Arc and other minor terranes. Extensive Phanerozoic cover largely prevents the observation of the suture between these latter blocks and the Río de la Plata/Amazonia cratons.

(e) The western margin of the Río de la Plata Craton is marked by the Pampean Belt. The Puncoviscana Formation and its metamorphic correlates (e.g. Aceñolaza, 2003) were deposited in the *Pampean Ocean* that opened between the Río de la Plata Craton and Pampia (*sensu* Ramos, 2008). An equally likely hypothesis assumes that the Pampean Ocean opened between the Río de la Plata Craton and the Arequipa Terrane (see Chapter 6). Most lines of evidence suggest that this basin was short-lived, opening in the latest Ediacaran and closing in the Lower Cambrian (e.g. Rapela et al., 1998a; Casquet et al., 2008).

(f) In the Appalachians, the Labrador Group contains volcanic rocks (ca. 550 Ma) of the Skinner Cove Formation at its base, which probably accumulated as a result of the opening of the *Iapetus Ocean* (Cawood et al., 2001). Latest Ediacaran to Ordovician units in the Appalachians were deposited at the eastern margin of the Laurentian palaeocontinent, after breakup of Rodinia, which separated Laurentia from the South American cratons (Amazonia, Río de la Plata-Pampia). Although earlier rifting events have been noted, most authors agree that the rift–drift transition took place no earlier than 550 Ma (Van Staal et al., 1998; Cawood et al., 2001; Puffer, 2002).

All in all, six different oceans separated or surrounded the palaeocontinents that later formed southwestern Gondwana. This complexity of Neoproterozoic basins adds much to on-going problems in pre-Gondwana palaeogeographic reconstructions.

## 8.3. Key Blocks in Southwestern Gondwana Amalgamation

### 8.3.1. The role of the Congo–São Francisco Craton and the Angola Craton

Two key features shall be highlighted here, both of which require a complete revision of previous geodynamic models for southwestern Gondwana amalgamation. One of them is the timing of the collision between the Congo and the São Francisco cratons with the rest of southwestern Gondwana; the other involves the existence of an allochthonous crustal block to the west of the Kalahari Craton for which we propose the name Arachania (see below). In contrast to previous models, which assumed a pre-550 Ma assembly of the Congo–São Francisco Craton with southwestern Gondwana, a series of arguments, as listed in the following, speak for the amalgamation of these cratonic blocks with the rest of SW-Gondwana only in the Lower Cambrian:

(a) The Congo–São Francisco Craton was not part of Rodinia according to palaeomagnetic studies (Tohver et al., 2006) and geological data (Cordani et al., 2003; Fuck et al., 2008), showing a separate evolution until at least 550 Ma.

(b) The Mesoproterozoic tectonic evolution of the Irumide Belt in the southern Congo Craton is distinctively different to that of Mesoproterozoic belts in the Kalahari Craton, showing that accretion of the two blocks along the Zambezi Belt took place in the late Neoproterozoic (De Waele et al., 2003) and not earlier.
(c) Rifting at the margins of the Congo–São Francisco Craton occurred mostly between 920 and 875 Ma (Porada and Berhorst, 2000; Tack et al., 2001; Valeriano et al., 2004; Silva et al., 2008), thus 100–300 myr earlier than in any other SW-Gondwanan block. The southwestern corner of the Congo Craton as considered by most authors, where rift-related magmatism yielded ages between 756 and 746 Ma (Hoffman et al., 1996), is here regarded as a separate block (Angola Craton), in line with previous suggestions by Porada (1989) and Porada and Berhorst (2000; see below).
(d) The mobile belts peripheral to the western Congo–São Francisco Craton, such as the Brasília Belt, record long-lived convergence along an accretionary orogen. This convergence implies collision of the craton with intra-oceanic, juvenile magmatic arcs (Goiás Arc) and microplates (Goiás Archean Massif, Paranapanema Block) as early as 630 Ma (Pimentel and Fuck, 1992; Cordani et al., 2003; Valeriano et al., 2004, 2008, and references therein). Convergence with the larger cratons (Amazonia, Río de la Plata, Kalahari) took place only after the accretion of these arc terranes and microplates with the Congo–São Francisco Craton, along a second range of more peripheral belts, such as the Paraguay and Araguaia belts (see Figure 8.1). Thus, a large ocean (Goianides Ocean), which separated the Congo–São Francisco Craton from the Goiás Arc, Goiás Massif and Paranapanema Block, was consumed between ca. 900 and 630 Ma. A second ocean (Brazilides Ocean) was consumed between 600 and 530 Ma, before the blocks peripheral to the Congo–São Francisco Craton collided with the Amazonian and Río de la Plata cratons (Valeriano et al., 2008).
(e) Convergence of the Congo–São Francisco with the Kalahari and Angola cratons along the Damara, Lufilian and Zambezi belts was also long-lived, and lasted until 510 Ma (Porada and Berhorst, 2000; Gray et al., 2006, 2008).

Rifting at the southern margin of the Angola Craton took place ca. 150 myr later than in the Congo Craton, and more or less simultaneously to rifting in other areas of the Kalahari Craton further south (see Section 8.4.1). Moreover, the tectonic evolution of the Kaoko Belt, at the western margin of the Angola Craton, parallels that of the Gariep Belt, and is distinctively different to the older, longer lived West Congo, Araçuaí, Lufilian and Zambezi belts (see Sections 8.4.2 and 8.4.3). The suture of the Angola and Congo cratons is obscured by younger sedimentary covers, but as noted by Porada and Berhorst (2000) it may link up with the West Congo Belt to the NW and with the Lufilian Belt to the SE (Figure 8.1). Alternatively, it may be represented by the Luanda Shear Zone in Angola (Heilbron et al., 2008). Remnants of oceanic crust in the Araçuaí Belt show that an oceanic basin existed to the east of the São Francisco palaeopeninsula (Pedrosa-Soares et al., 2001, 2008). Finally, long-lived (190 myr) subduction of oceanic lithosphere recorded in the Ribeira Belt is better explained by the Angola and São Francisco–Congo cratons being independent plates and not a monolithic block (Heilbron et al., 2008). Thus, in our scheme, Damaran rifting separated the Kalahari and Angola cratons and not the Kalahari and Congo Cratons.

## 8.3.2. The Arachania arc

The existence of a large magmatic arc between the Río de la Plata and Kalahari cratons is of utmost importance for the understanding of the processes involved in SW-Gondwana amalgamation. Most previous models assumed a direct collision between the Kalahari and Río de la Plata cratons, with the resultant magmatic arc located mostly along the Atlantic coast of Uruguay and southern Brazil (Figure 8.1). This arc terrane (Cuchilla Dionisio Terrane: Bossi et al., 1998; Punta del Este Terrane: Preciozzi et al., 1999), however, was shown to be allochthonous to the Río de la Plata Craton, and accreted along the continental-scale Sierra Ballena-Canguçú-Major Gercino Shear Zone in the Cambrian (Bossi et al., 1998; Gaucher et al., 1998; Preciozzi et al., 1999; Basei et al., 2000, 2005; Bossi and Gaucher, 2004). Basei et al. (2005) suggested that the Rocha Formation in the Cuchilla Dionisio Terrane represents the western extension of the Oranjemund Group. The latter unit was thrust onto the western Kalahari Craton with the rest of the allochthonous Marmora Terrane at 545 Ma (Frimmel and Frank, 1998). Based on this, Gaucher and Germs (2002) recognised that one ocean separated the Cuchilla Dionisio-Pelotas Terrane from the Kalahari Craton (Adamastor Ocean), and a more westerly located, younger ocean (Brazilides Ocean) separated this terrane from the Río de la Plata Craton. We propose below the name "Arachania" for the block that comprises the Cuchilla Dionisio-Pelotas and correlative terranes, which likely represents a fragment of the Kalahari Craton that a later stage evolved into a magmatic arc.

The mode of closure of the Adamastor Ocean has remained controversial. Originally, west-directed subduction of the Adamastor Ocean as part of a Wilson cycle was assumed (Hartnady et al., 1985; Frimmel et al., 1996a) in which case the long-lived, 660–590 Ma calc-alkaline, largely granitic magmatism in the Cuchilla Dionisio-Pelotas arc (Granite Belt of the Dom Feliciano Belt; Philipp et al., 2000; Frantz et al., 2003; Silva et al., 2005) would have resulted from that subduction. Similarly, Gaucher and Germs (2002) envisaged west-directed subduction of the Adamastor Ocean beneath the Cuchilla Dionisio-Pelotas arc (650–550 Ma), collision of the arc with the Kalahari Craton at 545 Ma and east-directed subduction of the Brazilides Ocean beneath the Cuchilla Dionisio-Pelotas until 530–520 Ma (Gaucher et al., 2008b). In contrast, Basei et al. (2005) proposed that arc magmatism resulted from east-directed subduction of the Brazilides Ocean beneath the extended Kalahari palaeocontinent, with the Adamastor Ocean taking the role of a narrow back-arc basin near the western margin of the Kalahari Craton. Closure of this back-arc basin eventually led to the emplacement of the Marmora Terrane onto the para-autochthonous eastern, external part of the Gariep Belt at 545 Ma. A number of observations need to be accounted for, namely:

(a) Rift events at the eastern margin of the Río de la Plata and Amazonia cratons are no older than 630–590 Ma, with a rift–drift transition no older than 600 Ma (Gaucher et al., 2008a; see Chapter 4.6). On the other hand, continental extension-related magmatism and rifting along the western Kalahari Craton were dated between 833 and 741 Ma (Frimmel et al., 1996c, 2001b).
(b) Sedimentary successions at the eastern Río de la Plata and Amazonia cratons are exclusively Ediacaran to Cambrian in age; Cryogenian (850–635 Ma) deposits are notably absent.
(c) Cryogenian to Ediacaran shelf deposits occur in the western Kalahari Craton in the Gariep and Saldania belts.
(d) An extensive foreland basin represented by the Nama Group developed in the western Kalahari Craton between 550 and <530 Ma (i.e. Germs, 1995; Blanco, 2008).
(e) The Rocha Formation and Oranjemund Group are correlative units that once formed a continuous belt fringing the Cuchilla Dionisio-Pelotas arc to the east. They are characterised by low-grade metamorphism and a distinctive detrital zircon "fingerprint", with Namaquan, Eburnean and late Neoproterozoic sources (750–600 Ma), the latter showing that the source area was the arc itself (Basei et al., 2005).
(f) Obduction of oceanic crust onto the western Kalahari Craton took place at approximately 545 Ma (Marmora Terrane). No high-grade metamorphic rock and, more significantly, no high-pressure metamorphic belt related to this tectonic event has been recognised so far.

Considering all the previous findings above, we envisage that a combination of both models may best describe the geodynamic evolution of the Cuchilla Dionisio-Pelotas arc and associated units, as shown in Figure 8.3. According to our proposed new model, opening of the Adamastor Ocean would have begun at about 740 Ma, following continental rifting in that section of the combined Kalahari–Rio de la Plata cratons that had experienced intense crustal reworking and suturing during the late Mesoproterozoic (Namaqua event). By approximately 650 Ma, west-directed subduction of the Adamastor Ocean began while the Brazilides Ocean further west started to open. The two oceans were separated by a small block of older crust with typical Kalahari Craton-type isotopic make-up, Arachania, which became the host of arc magmatism between 650 and 570 Ma (Figure 8.3B). Collision of the arc with the Kalahari Craton at 580–570 Ma (Figure 8.3C) may account for a compressional event well documented in the northern Kalahari Craton (Southern Foreland: Gray et al., 2006) and in the Gariep Belt (Frimmel and Frank, 1998). Continuing spreading of the Brazilides Ocean may have resulted in east- or southeast-directed subduction beneath the arc, leading to the opening of a back-arc basin (Marmora Basin) around 570–560 Ma (Figure 8.3D). Evidence of extensional tectonics at the eastern side of Arachania is provided by geochemical studies of volcanic rocks of the Cerros de Aguirre Formation in Uruguay (Campal and Schipilov, 2005), which yielded a U–Pb SHRIMP zircon age of $571 \pm 8$ Ma (Hartmann et al., 2002a). This magmatism may be related to the opening of the Marmora Basin, as also shown by the weak deformation and metamorphism of the Cerros de Aguirre Formation (Campal and Schipilov, 2005), suggesting that major tectonometamorphic events on eastern Arachania were finished by 570 Ma. The Marmora Basin was floored by oceanic crust, and was the depositional locus of the Rocha Formation, Oranjemund Group and related units. Closure of this basin eventually led to the emplacement of the Marmora Terrane onto the para-autochthonous Port Nolloth Zone (Gariep Belt) at 545 Ma, involving obduction of young, buoyant oceanic crust and weak metamorphic overprint (Frimmel et al., 2002; Figure 8.3E).

We propose the name "Arachania" for the arc terrane that is made up of the Cuchilla Dionisio Terrane and other fragments of this block at both sides of the Atlantic, including the Rocha-Oranjemund belt. The name is derived from the tribe of the *Arachanes*, who inhabited the Atlantic coast of Uruguay and southern Brazil (the same area that is made up of the Cuchilla Dionisio-Pelotas Terrane) between 2000 B.C. and ca. 1700 A.D. (López Mazz, 2001). Apart from the Cuchilla Dionsio-Pelotas Terrane (*sensu* Bossi and Gaucher, 2004), other fragments of Arachania include (Figure 8.1):

(a) *Marmora Terrane*: This tectonostratigraphic terrane is bounded by the Schakalsberge Thrust and made up of mafic rocks with oceanic island and MORB geochemical signature (Frimmel et al., 1996a) and predominantly turbiditic siliciclastic units (Oranjemund Group; Frimmel et al., 2002). It differs from the adjacent, para-autochthonous Port Nolloth Zone not only with regard to stratigraphy and deformation, but also in its metamorphic history (Frimmel and Frank, 1998; Frimmel et al., 2002). Basei et al. (2005) obtained very similar detrital zircon age spectra for both the upper Oranjemund Group and the Rocha Formation, which led them to suggest that both units represent the same basin fill. The common features include – (i) prominent zircon age peaks at 1.0 Ga (Namaqua) and 1.8–2.0 Ga (Eburnean), consistent with provenance from a source related to the Kalahari Craton, and (ii) a large proportion (>30%) of zircon grains between 764 and 600 Ma in age. For the younger zircon population with ages between 700 and 600 Ma, there are no potential sources in the Kalahari Craton but they can be well explained by derivation from the magmatic arc to the west (Basei et al., 2005; Frimmel and Basei, 2006). Ediacaran sedimentary units sourced in the Río de la Plata Craton show distinctively different detrital zircon age spectra (Gaucher et al., 2008b), thus supporting the contention that the Rocha Formation and upper Oranjemund Group were not sourced in that palaeocontinent.

(b) *Tygerberg Terrane*: This block is separated from the Kalahari Craton by the Colenso Shear Zone (Rozendaal et al., 1999; Kisters et al., 2002). According to Kisters et al. (2002), a reversal in strike–slip motion from sinistral to dextral occurred in the Colenso Shear Zone at 540 Ma. The Tygerberg Formation (Malmesbury Group), to the west of the Colenso Shear Zone, comprises metapelites and turbiditic intercalations, and yielded U-Pb detrital zircon ages that define peaks at 1.0 Ga, 1.8–2.0 Ga and 750–575 Ma (Armstrong et al., 1998). This is similar to the pattern obtained for the upper Oranjemund Group and opens up the possibility that the Tygerberg Formation was also sourced in the Cuchilla Dionisio Terrane (Arachania). The age of deposition is constrained between 560 Ma (youngest detrital zircon) and 550–540 Ma (age of the intruding Cape granites; Armstrong et al., 1998), thus comparable to the Rocha Formation and Oranjemund Group (Basei et al., 2005, 2008). Granitic intrusions in the Tygerberg Terrane yielded U-Pb ages ranging between $552 \pm 4$ and $539 \pm 4$ Ma (U-Pb SHRIMP: Scheepers and Armstrong, 2002; Scheepers and Schoch, 2006), which are similar to late calc-alkaline granitic magmatism in the Cuchilla Dionisio Terrane around 550 Ma, such as the Santa Teresa Batholith ($555 \pm 7$, Rb-Sr: Umpierre and Halpern, 1971; Bossi et al., 1998; Muzio and Artur, 1999).

(c) *Cabo Frío Terrane*: This exotic block comprises a Palaeoproterozoic basement (1.8–2.0 Ga), high-grade Neoproterozoic mafic rocks and metasedimentary rocks known as the Búzios-Palmital successions (Schmitt et al., 2004, 2008a,b). U-Pb SHRIMP ages of detrital zircon grains from the Búzios-Palmital succession show peaks at 2.6, 1.9, 1.1–0.9 and 0.8–0.62 Ga, with the latter two being the most prominent peaks (Schmitt et al., 2004). Sm-Nd model ages range between 1.7 and 1.0 Ga. The age of the youngest detrital zircon in the Búzios-Palmital succession is reported as 620 Ma by Schmitt et al. (2004). These values are almost identical to those obtained from the Rocha Formation, Oranjemund Group and Tygerberg Formation (Armstrong et al., 1998; Basei et al., 2005, 2008), showing that the Búzios-Palmital successions were probably sourced in the Cuchilla Dionisio-Pelotas arc too. A source area in the Angola Craton is unlikely, because no Neoproterozoic zircon grains have been reported so far from sedimentary units of the Kaoko Belt sourced in that block (Goscombe et al., 2005). Palaeoproterozoic basement/protolith ages around 2.0 Ga were reported from both the Uruguayan (Hartmann et al., 2002a) and Brazilian (Silva et al., 2005) parts of the Cuchilla Dionisio-Pelotas Terrane. Therefore, the available evidence suggests that the Cabo Frío Terrane could represent the northern tip of Arachania. The adjacent Oriental or Serra do Mar Terrane (Heilbron and Machado, 2003; Heilbron et al., 2004b, 2008), a magmatic arc active between ca. 790–580 Ma, was probably accreted to the Congo–São Francisco palaeocontinent at 585 Ma during consumption of the vast ocean that separated that craton from the rest of SW-Gondwana (see below). Docking of the Cabo Frío Terrane took place ca. 50 myr after this event (Schmitt et al., 2008b), during final closure of the Brazilides Ocean. So far, no firm evidence exists of a pre-530 Ma connection between the Oriental and Cabo Frío terranes.

(d) *Coastal Terrane (Kaoko Belt)*: Despite the wealth of data available for this terrane in Namibia (e.g. Goscombe et al., 2005; Gray et al., 2006, 2008; Goscombe and Gray, 2007; see Chapters 5.1–5.8), its relationship with the surrounding blocks remains elusive. Whereas calc-alkaline granitic magmatism between 655 and 620 Ma, followed by post-tectonic granites between 575 and 555 Ma (Seth et al., 1998; Gray et al., 2006; Goscombe and Gray, 2007), makes a correlation with the Cuchilla Dionsio-Pelotas Terrane feasible, accretion of the Coastal Terrane to the Angola Craton occurred no later than 580 Ma according to Gray et al. (2006). Collision of the Marmora Terrane and the Cuchilla Dionsio-Pelotas Terrane with the Kalahari Craton occurred at 580–570 Ma, but final emplacement only was achieved by 545 Ma (Frimmel and Frank, 1998). If Arachania was attached or associated to the Kalahari Craton from 570 Ma on, it was separated from the Angola and Congo cratons by the Damara Ocean. Thus, the contradicting evidence regarding the Coastal Terrane and Arachania may be solved by (i) the Coastal Terrane representing a separate magmatic arc; (ii) postulating an earlier collision of Arachania with the Angola Craton in the north, followed by a later collision with the Kalahari Craton in the south (Gray et al., 2006); or (iii) displacement of the terrane along the Purros and/or Three Palms Shear Zone has been underestimated, and the alleged docking at 580 Ma (Goscombe and Gray, 2007) actually took place later by tangential tectonics. The striking proximity and

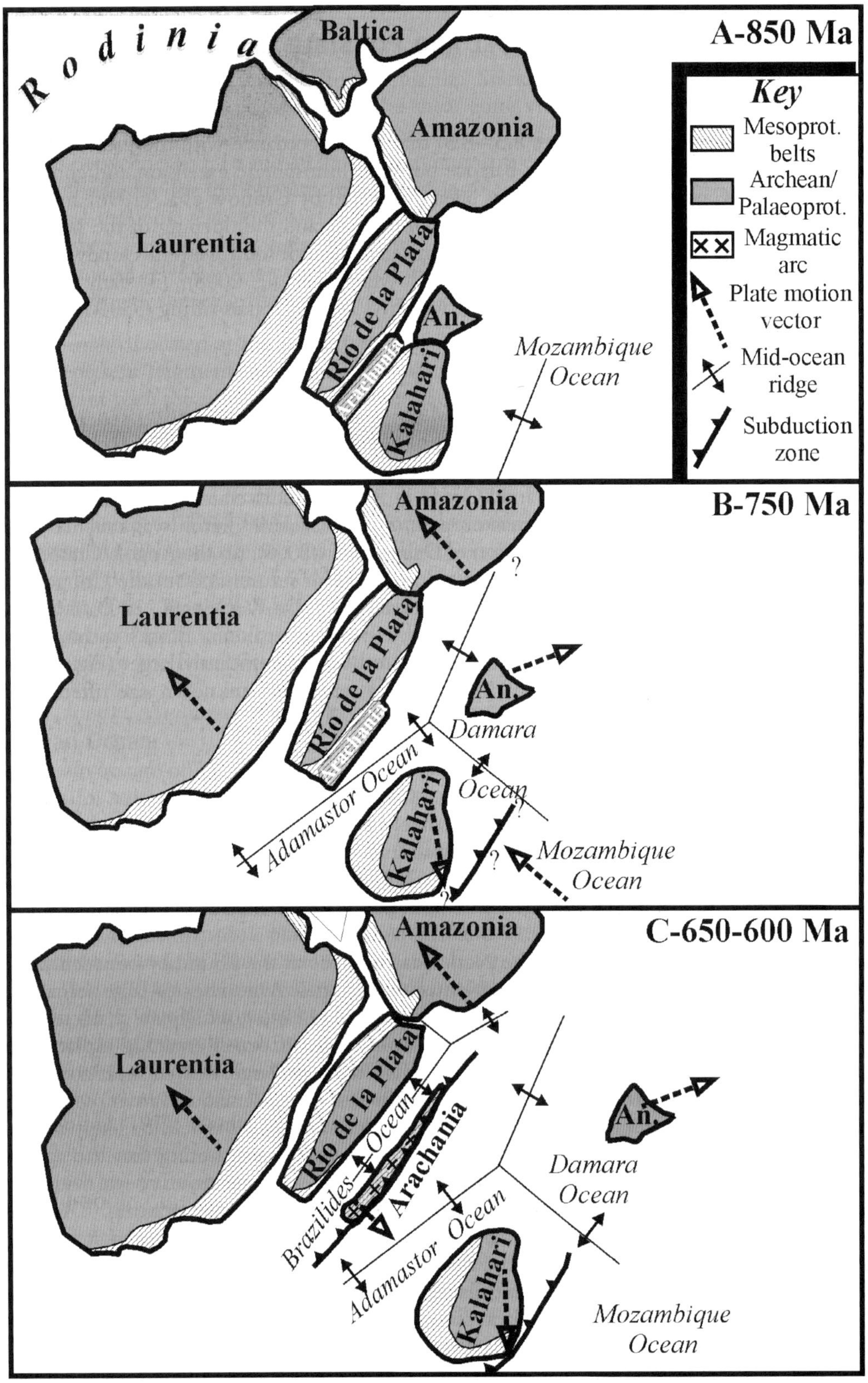


**Figure 8.4** (A–C) Palaeogeographic reconstruction of southwestern Gondwana and Laurentia between 850 Ma (fully assembled Rodinia) and 600 Ma. Note that a large cratonic block must have existed to the south of Laurentia, Río de la Plata and Kalahari in Rodinia configuration (not shown). Blocks to the west of Laurentia and to the east of Kalahari are not shown. An: Angola Craton. Position of Baltica according to Pisarevsky et al. (2008).

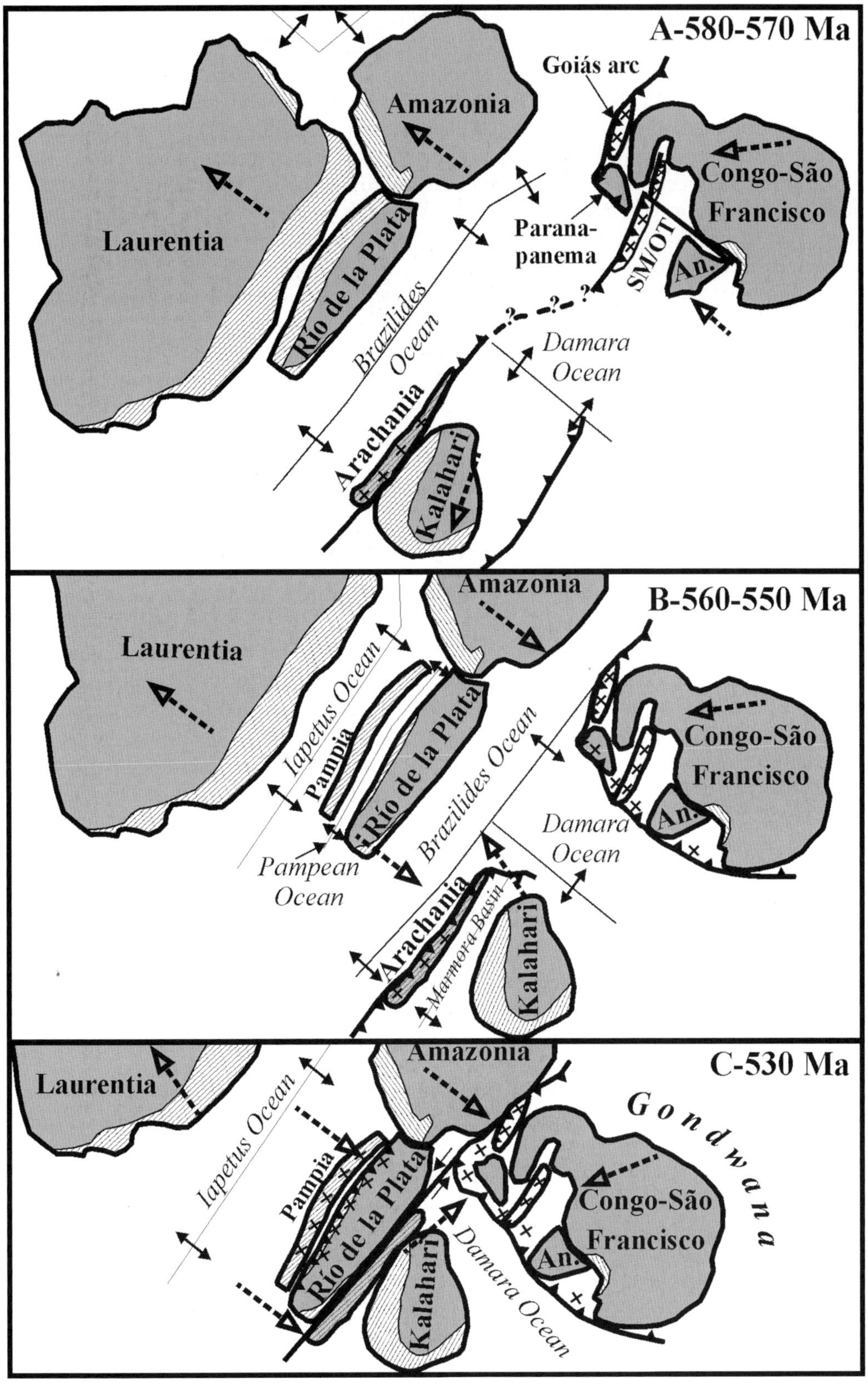


**Figure 8.5** (A–C) Palaeogeographic reconstruction of southwestern Gondwana and Laurentia between 580 and 530 Ma (final assembly of Gondwana). An: Angola Craton. SM/OT: Serra do Mar/Oriental Terrane. Smaller blocks related to the Congo–São Francisco and Paranapanema cratons (Curitiba, Luis Alves) not shown for clarity. Figure B mainly according to Gaucher et al. (2008b).

Consequently, and due to the onset of west-directed subduction at the eastern margin of Arachania, this block evolved into a magmatic arc (Figure 8.4C). This evolution explains the 650–550 Ma, arc-related granitic magmatism in the different Arachania fragments (Philipp et al., 2000; Frantz et al., 2003; Silva et al., 2005). According to the modified model presented in Figure 8.3, closure of the Adamastor Ocean was completed around 570 Ma with the collision of Arachania with the Kalahari Craton. Evidence of this compressional event has been reported from the Gariep, Kaoko and Damara (Southern Foreland) belts (Frimmel and Frank, 1998; Gray et al., 2006). An unconformity between the Nooitgedagt and Kombuis members (Frimmel et al., 2001a), of similar age, may be related to the collision of Arachania as well. We envisage that compression generated by continued spreading in the adjacent Brazilides Ocean was accommodated by a shift to east- or southeast-directed subduction at the western Arachania and northern Kalahari margins (Figure 8.5B).

It is worth noting that Pisarevsky et al. (2008) postulated a triple junction on top of a plume between Amazonia, Laurentia and Baltica, leading to separation of Baltica from the other two palaeocontinents between 615 and 600 Ma. This rifting event is probably associated to the coeval opening of the Brazilides Ocean.

By 630 Ma, most of the Goianides Ocean had been consumed, followed by collision of the juvenile Goiás Arc, the Goiás Massif and the Paranapanema Block with the Congo–São Francisco Craton (Valeriano et al., 2004, 2008). However, this craton was still far from the rest of SW-Gondwanan blocks (Figure 8.5A), including all other major cratons. A second ocean, possibly connected to the Brazilides and Damara oceans, separated these large continental masses (Figure 8.5A). Consumption of the ocean separating Amazonia and Congo–São Francisco has been linked by Pisarevsky et al. (2008) to rifting between Baltica and Amazonia, coeval to the opening of the Brazilides Ocean.

#### 8.5.1.4. Opening of the Pampean Ocean, closure of the Marmora Basin (560–545 Ma)

In the latest Ediacaran, rifting events shifted to the western Río de la Plata and Amazonian cratons and neighbouring Laurentia (Figure 8.5B). Opening of the Pampean Ocean is not well constrained in time, but was likely coeval to Iapetus opening. This might explain the short-lived nature of the Pampean Ocean, and the later obduction of young, buoyant oceanic crust in the Pampean Orogen. Meanwhile, and as a response to east-directed subduction of the Brazilides Ocean beneath Arachania, opening of the Marmora back-arc basin took place, in which the upper Oranjemund Group and the Rocha and Tygerberg formations were deposited. Final collision of Arachania with the Kalahari Craton took place at 545 Ma (Frimmel and Frank, 1998; Frimmel et al., 2002; Figure 8.5B), and was characterised by weak metamorphic overprint and obduction of oceanic crustal remnants (Marmora Terrane).

In the north, subduction of the Brazilides Ocean beneath the western Congo–São Francisco Craton and its peripheral terranes (Paranapanema block, Goiás Arc, Goiás Massif) was underway by 550 Ma (Figure 8.5B). Finally, closure of the Damara Ocean was initiated between the Kalahari Craton and the already assembled Congo–São Francisco and Angola cratons. Thus, the western and southern margins of the palaeocontinent represented by the Congo–São Francisco–Angola cratons and associated blocks was an active continental margin by latest Ediacaran times (Figure 8.5B).

#### 8.5.1.5. Opening of the Iapetus Ocean, final Gondwana assembly (530 Ma)

As noted by Grunow et al. (1996), the opening of the Iapetus Ocean (starting at ca. 550 Ma: Cawood et al., 2001) was a dramatic event leading to the "final crunch" between the three clusters of continental blocks (Figure 8.5C): Amazonia, Río de la Plata and Pampia in the west, Kalahari and Arachania in the south, and Congo, São Francisco, Angola, Paranapanema and peripheral terranes in the north. Opening of the Iapetus Ocean was not co-axial with Brazilides Ocean spreading, pushing Laurentia to the northwest (Van Staal et al., 1998) and Amazonia–Río de la Plata to the southeast. This resulted in closure of the mature Brazilides Ocean adjacent to the latter blocks by the following processes (Figure 8.5C): (a) the tangential collision of Kalahari and Arachania in the south and (b) the roughly orthogonal collision of the landmass comprising the Congo–São Francisco Craton and its peripheral, already amalgamated blocks in the north. The last oceanic basins remaining were the Damara Ocean and the central Brazilides Ocean (Figure 8.5C), which were closed at ca. 530 Ma (Schmitt et al., 2004, 2008b; Gray et al., 2006, 2008). Subduction possibly continued between Arachania and the Congo–São Francisco cluster well into the Cambrian, as indicated by the subduction-related Búzios Orogeny (Schmitt et al., 2004). Plate motion vectors changed from predominantly east–west to northeast–southwest at this stage, resulting in large transcurrent faults due to oblique or tangential collisions (Bossi and Gaucher, 2004; Gaucher et al., 2008b; Figure 8.5C). The Ribeira Belt is a complex puzzle of para-autochthonous basement terranes and tectonically interleaved Neoproterozoic units (Campanha et al., 2008), and is more properly viewed as a "shear belt". Arachania, in particular, was dissected and bounded by shear zones, and displaced to the north relative to the Río de la Plata and Kalahari cratons. The onset of dextral strike–slip motion recorded in the Colenso Shear Zone at 540 Ma (Kisters et al., 2002) is coeval to the sinistral Sierra Ballena Shear Zone (Bossi and Gaucher, 2004) at the other side of Arachania. In this context, the sinistral sense of the Purros and Three Palms shear zones is not easily explained, but may relate to the fact that the Western Kaoko Zone was interacting there with a different plate (Angola–Congo–São Francisco). A broad continental shelf between Africa and South America prevents the elucidation of the relationships between the Western Kaoko Zone and neighbouring areas in the Cuchilla Dionisio Terrane and Ribeira Belt.

The opening of the Rheic Ocean and concomitant closure of the Iapetus Ocean in the Late Cambrian to Silurian (Murphy et al., 2006) are the likely continuation of the process described above.

### 8.5.2. Possible causes of the westward migration of rifting

From the proposed evolution from Rodinia to southwestern Gondwana, a number of features and common patterns can be recognised. These are:

(a) Opening of the different oceans advanced in a westward direction over a time span of 200 myr, between 750 Ma (Adamastor and Damara oceans) and 550 Ma (Iapetus Ocean). If the Mozambique Ocean is included, this rifting process may have started already by 850 Ma at the eastern Kalahari Craton.
(b) The Congo–São Francisco Craton was probably far away from Rodinia and only joined other major southwestern Gondwana cratons during the latest Ediacaran.
(c) Closure of each ocean was related to the opening of the adjacent, more westerly located ocean. Closure of the Adamastor Ocean is linked to the opening and continued spreading of the Brazilides Ocean. The Pampean and Brazilides oceans, in turn, were closed as a result of the opening of the Iapetus Ocean. This is especially evident for the initial closure of oceanic basins, and not for their final closure.
(d) Collision-related orogenic events young from east to west: 570–545 Ma in the Gariep Belt, 535–530 Ma in the Paraguay Belt, 530–515 Ma in the Pampean Orogen and 435 Ma in the Appalachians. In the easternmost Mozambique Belt, onset of orogenic events by 750 Ma agrees with this pattern, although orogenic activity continued well into the Cambrian.

A mantle plume has been suggested by a number of authors as the cause of Rodinia breakup (Li et al., 1999, 2003, 2008; Frimmel et al., 2001b; Puffer, 2002; Wang et al., 2007), on the basis of: (a) the large heat source required for protracted trans-Rodinian rifting events, (b) occurrence of plume-related, high-temperature volcanic rocks (picrites, komatiites), (c) anorogenic and bimodal characteristics of rift-related rocks and (d) syn-rift lithospheric doming and unroofing. It is generally accepted that superplumes may be responsible for the breakup of supercontinents (e.g. Hill, 1991). Thus, breakup of Rodinia may be explained by the superplume hypothesis in a manner similar to that held responsible for the demise of the later Pangaea. We discuss below the nature and origin of the inferred post-Rodinian superplume in view of data from other regions.

## 8.6. Discussion and Environmental Consequences

### 8.6.1. Discussion

In the detailed palaeogeographic model of Rodinia presented by Li et al. (2008), the authors show that rifting and superplume activity migrated from the western margin (present coordinates) of Rodinia to its core at the western Laurentian margin between 825 and 720 Ma. According to these authors, Greater India, East Antarctica–Australia and South China progressively rifted away from the western margin of Laurentia. On the other hand, we have shown that rifting progressed from the eastern margin (present coordinates) of Rodinia to eastern Laurentia between ca. 800 and 550 Ma in a westward-migrating manner. One possible explanation would be to assume two different plume heads. However, if the plumes were static and Rodinia acted as a coherent plate, rifting at both Rodinian margins requires opposite plate motion vectors. This apparent inconsistency can be solved if the dominant mechanism was the generation of superplumes by circum-Rodinian "mantle avalanches" (Li et al., 2008), rather than a static superplume beneath a drifting supercontinent. Mantle avalanches (downwellings) are related to the subduction of cold oceanic lithosphere at the margins of supercontinents, triggering upwelling of hot mantle as the descending plates accumulate at the 660-km discontinuity and then catastrophically sink to the core–mantle boundary (Condie, 1998, 2000). There, "slab graveyards" accumulate and plumes are generated due to exothermic post-perovskite to perovskite transformation (Rino et al., 2008). Thus, as continental blocks rifted off the margins of Rodinia, the location of mantle upwellings must have shifted to a more central position, for two reasons: (a) continued subduction at the margins of the supercontinent would provide subducted slabs to fuel the mantle avalanche process and (b) the thermal isolation and shielding effects of continental crust would be greatest at the centre of the plate. As a result, a concentric pattern of Rodinia breakup

## 9.2.2. Calcareous Skeletons

### 9.2.2.1. *Cloudina*

As mentioned above, *Cloudina* was first described from the Nama Group in Namibia, and two species were erected: the larger *Cloudina hartmannae* (Figures 9.2.1A and 9.2.2G,H) and the smaller *Cloudina riemkeae* (Figures 9.2.1C, D and 9.2.2G; Germs, 1972a). The third currently accepted species is *Cloudina lucianoi* (Beurlen and Sommer) Zaine and Fairchild (1985; Figure 9.2.1B), first reported from the Corumbá Group in Brazil, and intermediate in size (diameter 0.2–3.8 mm) between *C. hartmannae* (diameter 2.5–6.5 mm) and *C. riemkeae* (0.3–1.5 mm). It is worth noting that *Cloudina waldei* Hahn and Pflug (1985) is a junior synonim of *C. lucianoi* (see discussion in Gaucher et al., 2003). All the species have been included by Hahn and Pflug (1985) in the family Cloudinidae.

#### 9.2.2.1.1. Anatomy and affinity

The genus *Cloudina* is characterised by a calcareous test of stacked cones or funnels which are eccentrically nested (Figure 9.2.1D), giving a 'cone-in-cone' structure (Germs, 1972a; Grant, 1990; Figures 9.2.1A–D and 9.2.2G,H). Axes of stacked funnels are often not parallel to each other (Figure 9.2.1B). Tubes are straight or sinuous (e.g. Figure 9.2.2G) and open at one end. Budding has been convincingly demonstrated for several occurrences (Germs, 1972a; Gaucher and Sprechmann, 1999; Hua et al., 2005; Figure 9.2.1C). The most likely modern analogous to *Cloudina*, considering its anatomy, calcareous shell and reproduction by budding are annelids, possibly serpulids (Germs, 1972a, Hua et al., 2005).

#### 9.2.2.1.2. Occurrences

*Cloudina* is a cosmopolitan genus and has been described from all continents. This may be related to *Cloudina* having planktic larvae, as modern serpulids do (e.g. Kupriyanova, 2003). In Africa it has been reported from the Nama Group, Namibia (Figures 9.2.1A,D; Germs, 1972a; Grotzinger et al., 1995, 2000; see Chapter 5.4). In South America it is known from the Arroyo del Soldado Group of Uruguay (Figure 9.2.1C, Gaucher and Sprechmann, 1999; Gaucher, 2000; Gaucher et al., 2003; see Chapter 4.3), from the Sierras Bayas Group of Argentina (Gaucher et al., 2005b; see Chapter 4.3), from the Corumbá Group and the Ribeira Belt in Brazil (Figure 9.2.1B, Hahn and Pflug, 1985; Zaine and Fairchild, 1985; Gaucher et al., 2003; Teixeira and Gaucher, 2004) and from the Itapucumí Group of Paraguay (Boggiani and Gaucher, 2004). In North America *Cloudina* occurs in the White Inyo Mountains, USA (Grant, 1990 and references therein; Hagadorn and Waggoner, 2000) in the La Ciénega Formation of México (Grant, 1990; Sour-Tovar et al., 2007) and in the Miette Group of Canada (Hofmann and Mountjoy, 2001). In Eurasia, *Cloudina* has been reported from the Ibor and Río Huso groups of Spain (Vidal et al., 1994; Jensen et al., 2007; Gamez Vintaned et al., 2008), from the Siberian Platform, Russia (Gamez Vintaned et al., 2008) from the Dengying Formation in southern China (Conway Morris et al., 1990; Bengtson and Zhao, 1992; Bengtson, 1994; Hua et al., 2003, 2005; Chen et al., 2008) and from the Ara Group in Oman (Conway Morris et al., 1990; Amthor et al., 2003). Finally, *Cloudina* also occurs in the Shackleton Glacier area in Antarctica (Yochelson and Stump, 1977; Grant, 1990).

#### 9.2.2.1.3. Stratigraphic range

The stratigraphic range of *Cloudina* is limited to the Late Ediacaran, which led Grant (1990) to propose *Cloudina* as an index fossil of that period. Almost two decades later Grant's (1990) proposal still holds, making a strong case for *Cloudina* as the index fossil of the – still unnamed – youngest series of the Ediacaran (see below). Whereas its extinction at the Precambrian-Cambrian boundary (542 Ma) is reasonably well documented (e.g. Amthor et al., 2003), its appearance datum is not so well constrained. In the Nama Group *Cloudina* occurs beneath ash beds that yielded an U-Pb zircon age of $549 \pm 1$ Ma (Grotzinger et al., 1995), thus the taxon appeared prior to 550 Ma. In the Dengying Formation, *Cloudina* occurs well above an ash bed dated by Condon et al. (2005; U-Pb zircon) at $551 \pm 0.7$ Ma (e.g. McFadden et al., 2008). In both cases, *Cloudina* appears at the top or immediately above the Shuram-Wonoka negative $\delta^{13}C$ anomaly (Grotzinger et al., 2000; McFadden et al., 2008), which is constrained between ca. 560 and $551 \pm 0.7$ Ma. In the Arroyo del Soldado Group in Uruguay, however, *Cloudina* appears within and below an expressive negative excursion possibly correlative of the Shuram-Wonoka anomaly (see Chapter 4.4; Gaucher and Sprechmann, 1999; Gaucher, 2000; Gaucher et al., 2004c). $^{87}Sr/^{86}Sr$ values between 0.7073 and 0.7085 for the host carbonates also suggest that the age of the Uruguayan occurrences may be significantly older than 550 Ma, possibly closer to 560–570 Ma (Jacobsen and Kaufman, 1999; Melezhik et al.,

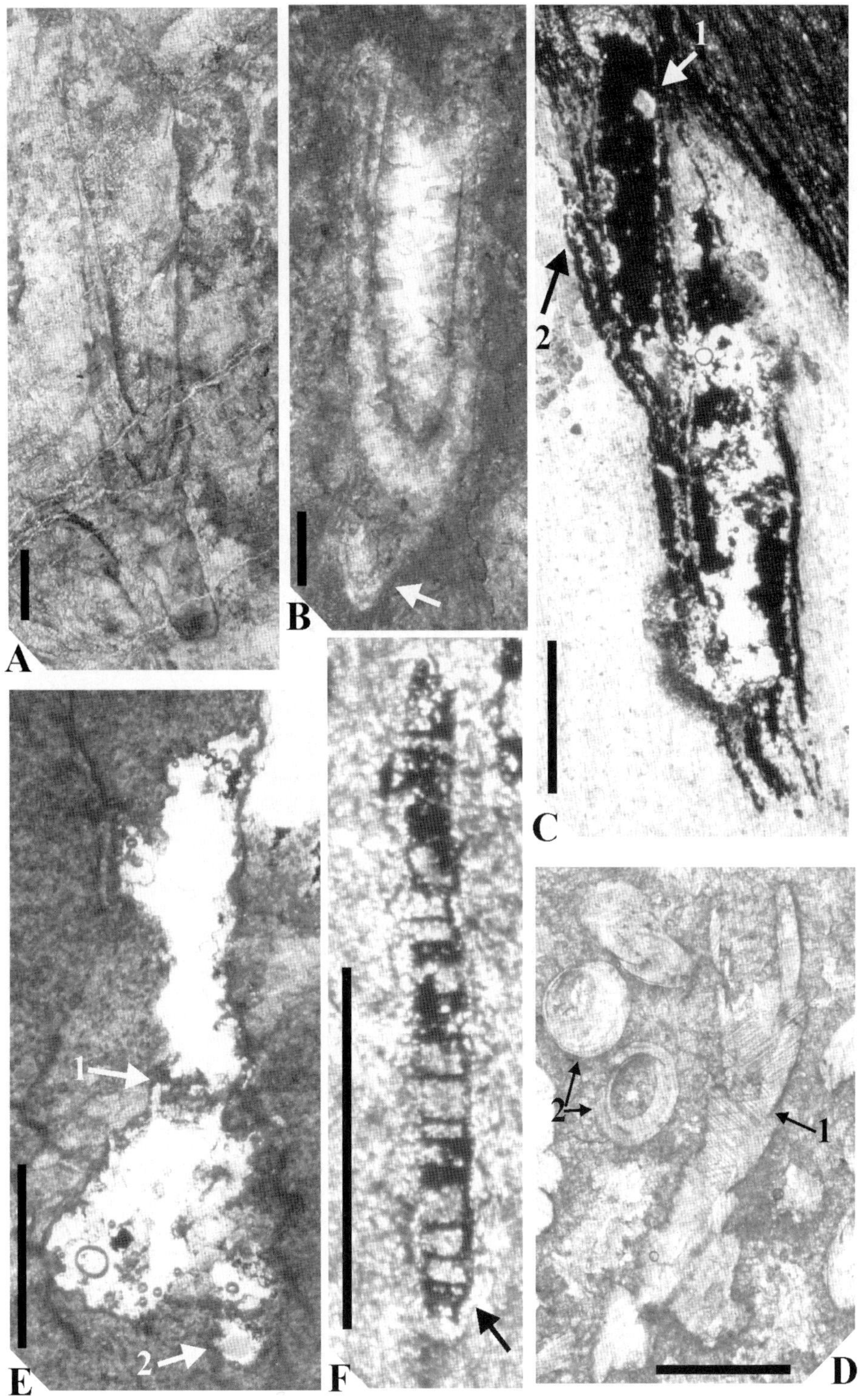


**Figure 9.2.1** *Cloudina*, *Titanotheca* and *Waltheria* in thin sections. A: *Cloudina hartmannae*, bioclastic carbonate, Omkyk Member at Driedoornvlakte reef (Nama Group). Note deeply stacked cones. B: *Cloudina lucianoi*, bioclastic limestone of the Tamengo Formation (Corumbá Group, Brazil). Note first cone (arrowed) not coaxial with the rest, suggesting that the organism was moved by currents and had to adjust its growth direction. C: *Cloudina riemkeae*, haematised, from banded siltstones of the Yerbal Formation (Arroyo del Soldado Group, Uruguay; Gaucher and Sprechmann, 1999). The boundary between two siltstone layers can be seen in the upper right. Note (1) bud growing from parent individual and (2) cone-in-cone structure. D: *Cloudina riemkeae* in limestone of the Nama Group. Note (1) longitudinal section with cone-in-cone structure and (2) cross sections showing excentrical nature of cones. E: *Titanotheca coimbrae*, holotype (Gaucher and Sprechmann, 1999). Note: (1) septum dividing both chambers and (2) small sphaeroidal bud. F: *Waltheria marburgensis* from siltstone of the Yerbal Formation (Arroyo del Soldado Group, Uruguay; Gaucher and Sprechmann, 1999), replaced by haematite. Note sphaerical chamber and cylindrical protuberance (arrowed). Scale bars represent 1 mm for all figures.

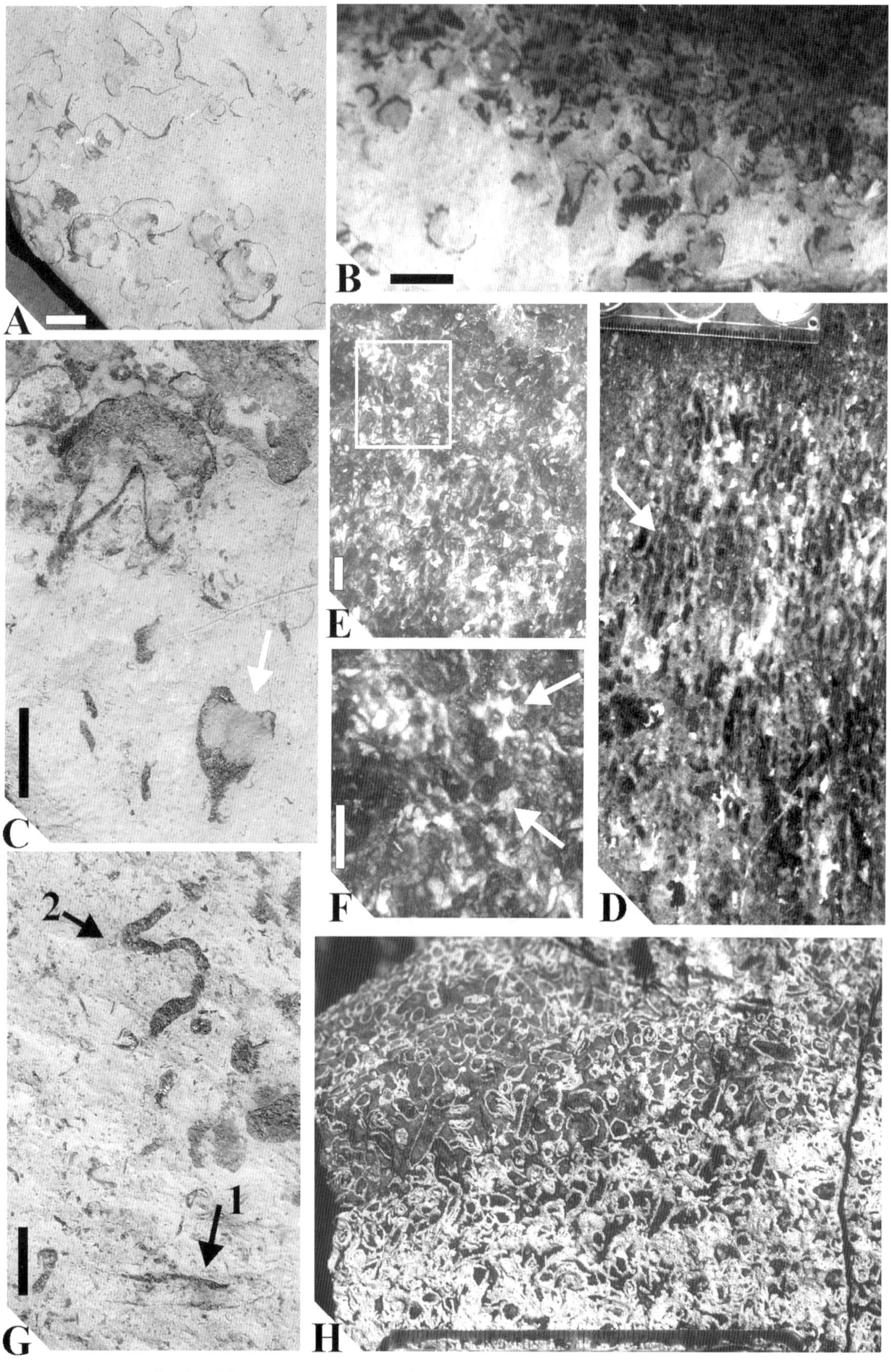


**Figure 9.2.2** Calcareous fossils of the Nama Group on rock surfaces. A–C: *Namacalathus hermanastes*, bioclastic limestones of the Omkyk Member at Farm Omkyk/Zwartmodder. Note specimen in C (arrowed) with stem and top opening showing inward-curving lip. D–F: *Namapoikia rietoogensis*, Omkyk Member at Driedoornvlakte reef. D: Longitudinal section (scale: 7 cm) showing that the colony comprises septate parallel tubes (arrowed) analogous to tabulate corals. E: Cross section. F: Enlargement of area indicated in subpart E, showing roughly hexagonal cross sections for some individuals (arrowed). G: *Cloudina hartmannae* (arrowed, 1) and sinuous *Cloudina riemkeae* (arrowed, 2) in limestones of the Omkyk Member at Farm Omkyk/Zwartmodder. H: *Cloudina hartmannae* event accumulation (bioclastic calcarenite), Omkyk Member at Driedoornvlakte build up (scale at the bottom: 8 cm). Scale bars represent 1 cm for all figures.

2001; Halverson et al., 2007a). An U-Pb SIMS zircon age for basement granite of $583 \pm 7$ Ma is consistent with the latter interpretation (Gaucher et al., 2008b). Thus, pending a more precise definition of the first appearance datum, we assume a range for *Cloudina* between ca. 560 and 542 Ma (see Section 9.2.6).

## 9.2.2.2. *Sinotubulites*

### 9.2.2.2.1. Anatomy

Although *Sinotubulites* was erected by Chen et al. (1981), a clear-cut differential diagnosis became available only very recently (Chen et al., 2008). *Sinotubulites* is very similar to *Cloudina*, but it shows several morphological differences which allow its separation as a different genus. The most important differences are (Chen et al., 2008): (a) *Sinotubulites* tubes are open at both extremes and often polygonal in cross section, unlike *Cloudina*; (b) the wall is multilayered, the layers being mostly concentrical ('tube-in-tube' structure); (c) the tubes show irregular annulations and longitudinal ridges on the surface formed by wrinkling of the outer shell layers and (d) up to 0.5% Sr in the shell suggests that it was probably composed of aragonite (Chen et al., 2008), contrasting with the possibly calcitic composition of *Cloudina*. Diameter of *Sinotubulites* ranges between 0.3 and 4.2 mm (Chen et al., 2008).

### 9.2.2.2.2. Occurrences, taxonomy and palaeoecology

Two species have been described. *Sinotubulites baimatuoensis* occurs in the latest Ediacaran Dengying Formation in China (Chen et al., 1981, 2008), and *Sinotubulites cienegensis* was described from the La Ciénega Formation (Sonora, México) and the Deep Spring Formation of the White Inyo Mountains, California and Nevada (USA; McMenamin, 1985; Signor et al., 1987). In all localities *Sinotubulites* co-occurs with *Cloudina*. However, unlike the latter, *Sinotubulites* was probably an epifaunal (reclined) organism. Its polygonal form and longitudinal ridges may have somewhat stabilised the fossil, preventing it from rolling (Chen et al., 2008).

## 9.2.2.3. *Namacalathus*

### 9.2.2.3.1. Occurrences and anatomy

Calcareous, goblet-shaped fossils were first reported from the Nama Group by Germs (1972b), but formally described as *Namacalathus hermanastes* by Grotzinger et al. (2000). The species is very abundant in the Nama Group (Figure 9.2.2A–C), and has been also described from the Miette Group of Canada (Hofmann and Mountjoy, 2001) and the Ara Group of Oman (Amthor et al., 2003). A few specimens occurring in the Itapucumí Group of Paraguay may be preliminarily assigned to *Namacalathus*. Although only four occurrences have been so far reported, they suggest a broad palaeogeographic distribution like *Cloudina*, with which it always co-occurs.

*Namacalathus* is characterised by a goblet-shaped, calcareous skeleton (Figure 9.2.2A–C) made up of hollow stem attached to a sphaeroidal cup, the latter showing a circular opening at the top (Figure 9.2.2C) and six or seven side holes (Grotzinger et al., 2000).

### 9.2.2.3.2. Stratigraphic range

In the Nama Group, the oldest *Namacalathus* occurrences are from the Kuibis Subgroup (Omkyk Member: Grotzinger et al., 2000; see Chapter 5.4), with an age $>549 \pm 1$ Ma (Grotzinger et al., 1995). Thus a minimum age of 550 Ma can be assumed for the first appearance of the genus. The extinction of *Namacalathus* at the Precambrian-Cambrian boundary is radiometrically well constrained in Oman by a U-Pb zircon age of $542 \pm 0.3$ Ma.

### 9.2.2.3.3. Palaeoecology and affinities

All *in situ Namacalathus* occurrences are associated to thrombolite and stromatolite build ups (e.g. Grotzinger et al., 2000; Hofmann and Mountjoy, 2001). Seilacher et al. (2003) interpreted the mode of life of *Namacalathus* as cemented to hard-substrates, which is consistent with its association to microbial reefs. Regarding the phylogenetic affinities of *Namacalathus*, its radial symmetry and sessile nature favours an association to anthozoan cnidarians (Grotzinger et al., 2000). A protozoan affinity, as suggested by Seilacher et al. (2003) is here considered less likely.

## 9.2.2.4. *Namapoikia*

*Namapoikia rietoogensis* Wood et al. (2002) has so far been described only from the Nama Group at Driedoornvlakte (Kuibis Subgroup, Omkyk Member), where it occurs associated with thrombolite–stromatolite

characterised by tubular, straight, septate shells which taper towards both extremes (Figure 9.2.1F), and often show branching (see Chapter 4.3). The shell is three-layered and uniserial (Figure 9.2.1F), but discoidal individuals with parenchymatous structure and bundles of individuals are common (Gaucher and Sprechmann, 1999; Gaucher, 2000; see Chapter 4.3). The initial chamber(s) form a bulbous structure with a cylindrical protrusion (Figure 9.2.1F). Maximum diameter of tubular *Waltheria* ranges between 10 μm and 1 mm, length varying between 0.8 and 10 mm. The haematised shells contain up to 5 wt% P, corresponding to 25–30% fluor-apatite, which has been interpreted by Gaucher and Sprechmann (1999) as a relic of the original shell mineralogy. In support of this is the scarcity of phosphates in the host siltstones (0.10–0.12 wt%), suggesting that phosphatisation was not likely responsible for the apatite in the shell walls of *Waltheria*. It is worth noting that co-occurring *Cloudina* is often haematised as well (Gaucher and Sprechmann, 1999; Gaucher, 2000; Gaucher et al., 2003; Figure 9.2.1C), showing that ferrification and not phosphatisation was the main preservation mechanism. *Waltheria* is often found perpendicular to bedding (Gaucher, 2000), which is inferred to be its life position ('mat sticker': Seilacher, 1999). Thin 'shell beds' formed by current-reworked *Waltheria* are also common (Gaucher, 2000). Fossils may be very abundant in certain beds, and concentrations of up to 1,500 individuals/cm$^3$ of rock were measured by Gaucher (2000). The palaeobiological affinity of *Waltheria* is still unclear, but it may represent a primitive metazoan, considering its complex anatomy and phosphatic shell. If this is confirmed, *Waltheria* may represent the oldest phosphatic fossils, and possibly the first metazoans with such a shell. The age of the Yerbal Formation is constrained by C and Sr chemostratigraphy and available radiometric ages of basement rocks and intrusive granites between 570 and 560 Ma (Gaucher et al., 2004c, 2007c, 2008b, see Chapters 4.3 and 4.4; Figure 9.2.4).

Another important occurrence of Neoproterozoic metazoans with possible phosphatic shells is represented by *Vendoconularia triradiata* Ivantsov and Fedonkin (2002). The fossil consists of a conical test with six identical faces limited by corner sulci, each face showing a midline flanked by chevron-like ribs (Ivantsov and Fedonkin, 2002; Van Iten et al., 2005). Although the test has a six-fold symmetry, three internal longitudinal bands reduce the overall symmetry to three-fold (Ivantsov and Fedonkin, 2002). *Vendoconularia* was probably weakly mineralised, and in view of its clear affinity to the Conulata (Van Iten et al., 2005), a phosphatic mineralisation is assumed (Ivantsov and Fedonkin, 2002), although no chemical analyses of the shell are available. The fossils occur in the lower Ust'-Pinega Formation, which is constrained to be older than 555 ± 0.3 Ma, and close to 558 Ma (U-Pb zircon ages of ash beds: Martin et al., 2000; Fedonkin et al., 2007a; Figure 9.2.4).

Both examples cited above show that by ca. 560 Ma metazoans had likely evolved the ability to secrete phosphate skeletons, long thought to have arisen only in the earliest Cambrian (Figure 9.2.4).

## 9.2.5. Agglutinated Skeletons

Agglutinated skeletons are common in Early Cambrian successions, such as the foraminifer *Platysolenites* (McIlroy et al., 2001), and the agmatan *Volborthella* (Yochelson and Kisselev, 2003; Figure 9.2.4). In the Neoproterozoic, however, very few agglutinated organisms have been so far reported.

*Titanotheca coimbrae* Gaucher and Sprechmann (1999) was first described from the Yerbal Formation (Arroyo del Soldado Group, Uruguay), co-occurring with *Cloudina* and *Waltheria*. The fossils are sphaeroidal, vase- or flask-shaped (Figure 9.2.3) and have one or two chambers separated by a septum (Figure 9.2.1E; see Chapter 4.3). The shell is made up of tiny, agglutinated rutile grains (Figure 9.2.3; Gaucher and Sprechmann, 1999; Gaucher, 2000), much like modern foraminiferans of the genus *Bathysiphon* (Lowenstam and Weiner, 1989). On this basis, and considering the anatomy of *Titanotheca*, Gaucher and Sprechmann (1999) assigned the species to the Foraminiferida (Suborder: Textulariina), representing the oldest record of this important protists worldwide. Maximum diameter of test varies between 0.1 and 1.5 mm, and length between 0.1 and 5 mm. The gregarious nature of this species is demonstrated by samples of the upper Yerbal Formation, where a number of attached specimens have been observed (see Chapter 4.3). Apart from the Arroyo del Soldado Group, *Titanotheca* has been described from the Corumbá Group (Gaucher et al., 2003), Pico do Itapeva, Cajamar and Eleutério basins in Brazil (Teixeira and Gaucher, 2004). In southern Africa, *Titanotheca* was been reported by Gaucher et al. (2005a) from the upper Port Nolloth Group (Holgat Formation) of southern Namibia. All occurrences are Late Ediacaran in age, between ca. 570 and 555 Ma (Gaucher et al., 2003, 2005a, 2008b; Teixeira and Gaucher, 2004; Figure 9.2.4).

More recently, Seilacher et al. (2003) and Seilacher (2007) suggested that the Vendobionta may represent giant agglutinated foraminifera (*Xenophyophoria*), capable also of engulfing sand grains within the protoplasm. The occurrence of *Titanotheca* in coeval strata shows that agglutinated foraminifers were indeed extant in Late Ediacaran times.

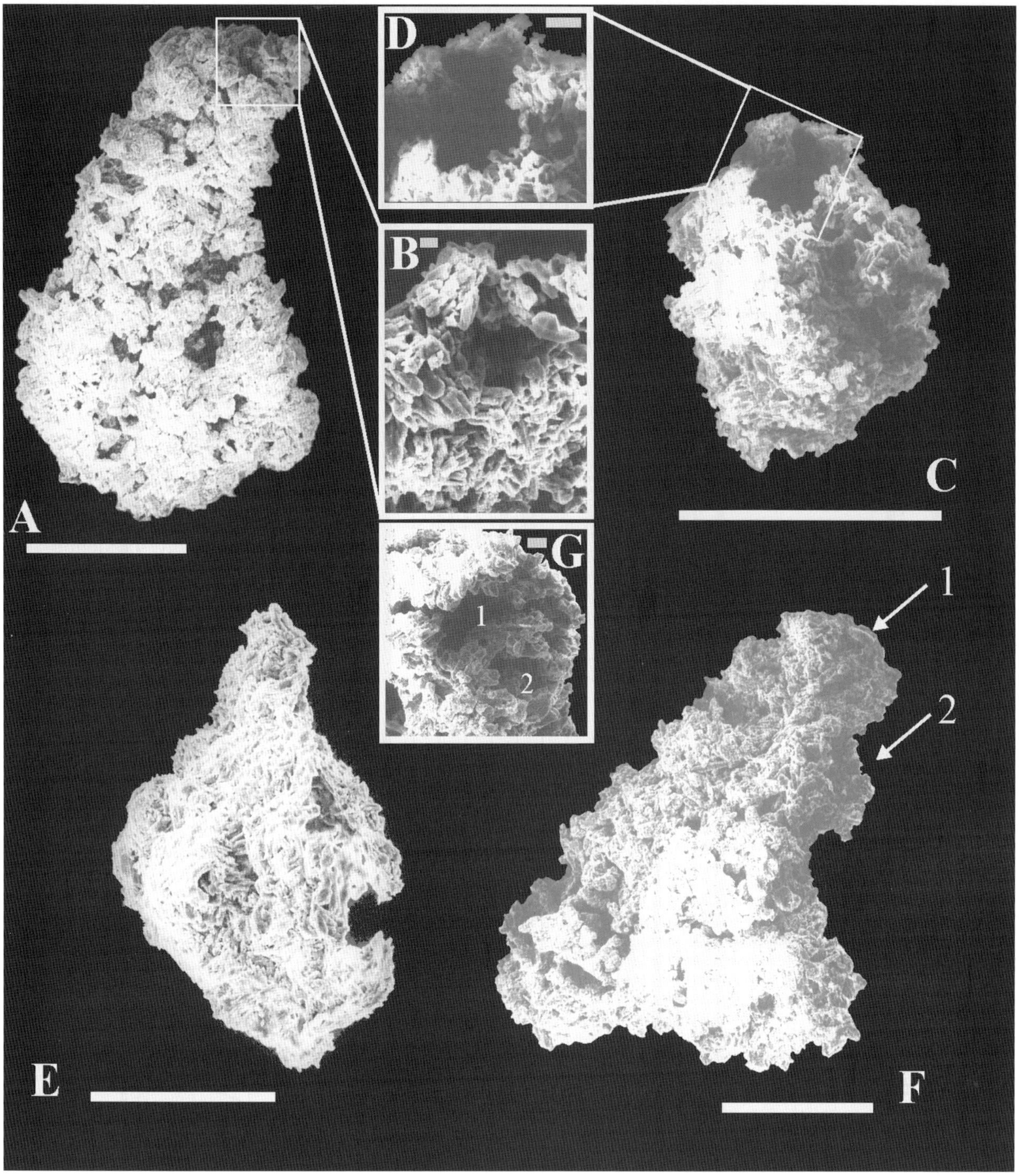


**Figure 9.2.3** SEM images of *Titanotheca coimbrae* specimens isolated from siltstones of the Yerbal Formation (Arroyo del Soldado Group, Uruguay). A–B: Paratype, with long neck and detail (subpart B) of small opening and agglutinated rutile crystals (Gaucher and Sprechmann, 1999). C–D: Paratype of juvenile specimen, without neck and larger opening (Gaucher and Sprechmann, 1999). E: Specimen with a short, incipient neck. F–G: Two attached individuals, one with curved neck. Note individual openings (1 and 2), shown in detail in G (Gaucher and Sprechmann, 1999). White scale bars equal 100 μm, and grey scale bars represent 10 μm.

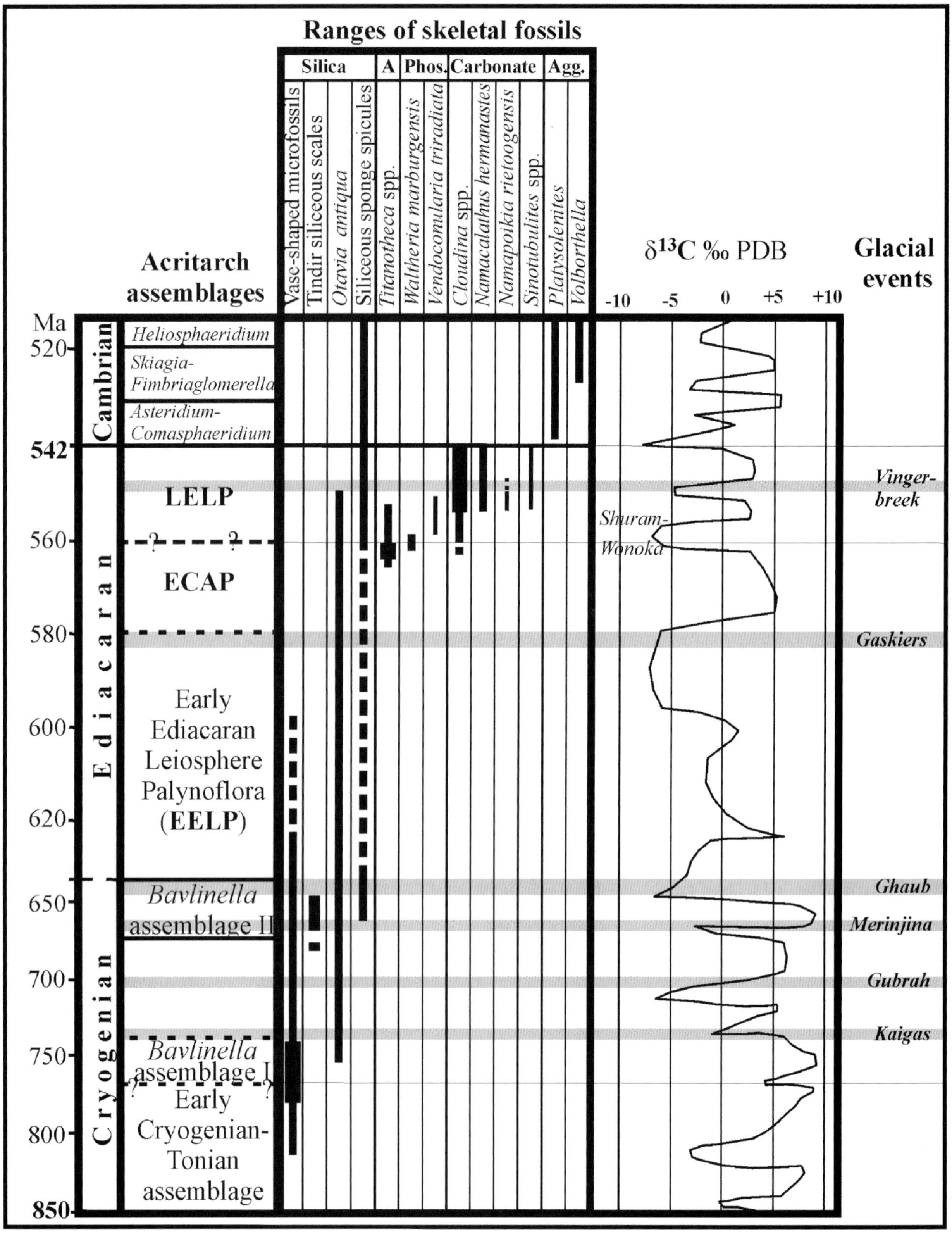


**Figure 9.2.4** Stratigraphic chart showing stratigraphic ranges of Neoproterozoic and selected Early Cambrian skeletal fossils (width of bars represents for each taxon relative abundance and/or diversity). Note change in absolute age scale at 635 Ma. Ranges of *Platysolenites* and *Volborthella* according to McIlroy et al. (2001) and Yochelson and Kisselev (2003). Global seawater $\delta^{13}C$ curve modified from Halverson et al. (2005) and Maloof et al. (2005). For acritarch assemblages, see Chapter 9.1. Biotic and glacial events are also indicated. Cambrian acritarch biozones after Moczydłowska (1991). LELP: Late Ediacaran Leiosphere Palynoflora; ECAP: Ediacaran Complex Acanthomorph Palynoflora (Grey, 2005). A/Agg.: Agglutinated, Phos.: phosphate. Ages of glacial events according (from base to top) to Frimmel et al. (1996), Allen et al. (2002), Fanning (2006), Hoffmann et al. (2004), Bowring et al. (2003b) and Germs (1995).

## 9.2.6. Discussion and Conclusions

### 9.2.6.1. Biostratigraphy

Many Neoproterozoic shelly fossils are useful biostratigraphic markers. Their higher fossilisation potential, often wide palaeogeographic distribution and relatively short stratigraphic range allow them to be used as index fossils.

VSMs are the oldest, potentially useful skeletal fossils for Neoproterozoic biostratigraphy. They are palaeogeographically widespread, may be very abundant in specific sections and individual species could have short stratigraphic range (e.g. Porter and Knoll, 2000; Porter et al., 2003). Both scale-bearing and organic-walled VSMs may be used for Cryogenian subdivision (Figure 9.2.4). Peak diversity of VSMs is recorded in pre-glacial Cryogenian successions constrained between ca. 800 and 740 Ma, such as the Chuar Group (Porter and Knoll, 2000; Porter et al., 2003; Knoll et al., 2006). In these time period other useful stratigraphic markers do occur, such as the acritarch *Cerebrosphaera buickii* Butterfield et al. (1994), which became extinct prior to the oldest Cryogenian glaciations (Hill et al., 2000; Grey, 2005; Moczydłowska, 2008; see Chapter 9.1). Thus, a combination of acritarch- and VSM-biozones offers the best prospects for the biostratigraphic definition of the base of the Cryogenian System (Figure 9.2.4), currently being considered by the Neoproterozoic Subcommission of the International Commission on Stratigraphy (ICS). The obvious advantage is that a Global Boundary Stratotype Section and Point (GSSP: Salvador, 1994) located at this stratigraphic level would leave most Neoproterozoic glacial events in the Cryogenian, including the most severe ones (see Chapters 11.1 and 11.2).

Grant (1990) proposed *Cloudina* as a potential index fossil for the Late Vendian-Ediacaran. We formally propose here the *Cloudina* Range Zone, which is defined by the lowest and highest stratigraphic occurrence of the genus *Cloudina*. It is a taxon-range zone according to the classification of biozones of the International Stratigraphic Guide (Salvador, 1994). Whereas the extinction of *Cloudina* has been shown to coincide with the Ediacaran-Cambrian boundary at 542 Ma (Grotzinger et al., 1995; Amthor et al., 2003), its first appearance datum is less well constrained but older than 550 Ma (Figure 9.2.4). As reference sections of the *Cloudina* Range Zone we designate the following:

(a) In the Kuibis and Schwarzrand Subgroups of the Nama Group (southern Africa), *Cloudina* (*C. riemkeae* and *C. hartmannae*) occurs between the lower Dabis Formation (Mara Member) and the uppermost Urusis Formation (Germs, 1972a, 1995; Grotzinger et al., 1995, 2000). Three ash beds within the biozone were dated by U-Pb on zircon between $549\pm1$ and $543\pm1$ Ma (Grotzinger et al., 1995). $\delta^{13}C$ values of host carbonates are negative at the base (Mara and lower Omkyk members) and positive, up to +5‰ PDB in the rest of the biozone (Kaufman et al., 1991; Grotzinger et al., 1995, 2000).

(b) In the Ara Group (Huqf Supergroup, Oman), *Cloudina* (*C. hartmannae*) occurs in units A1 to A3 (Conway Morris et al., 1990; Amthor et al., 2003). Its highest stratigraphic occurrence there coincides with an ash bed dated by U-Pb on zircon at $542.0\pm0.3$ Ma and a negative $\delta^{13}C$ excursion down to −5‰ PDB considered to represent the Ediacaran-Cambrian boundary (Amthor et al., 2003).

(c) In the Corumbá Group, *Cloudina* (*C. lucianoi*) occurs in the Tamengo Formation (Zaine and Fairchild, 1985; Boggiani, 1998; Gaucher et al., 2003). U-Pb SHRIMP zircon ages between $545\pm6$ and $543\pm2$ Ma were obtained for ash beds in the upper Tamengo Formation (Babinski et al., 2006, 2008). $\delta^{13}C$ values of host carbonates is mostly positive (up to +5.5‰ PDB), and possibly extending into an underlying negative excursion (Boggiani, 1998; Gaucher et al., 2003; Misi et al., 2007). $^{87}Sr/^{86}Sr$ ratios of 0.7085 have been reported for *Cloudina*-bearing carbonates (Boggiani, 1998; Gaucher et al., 2003).

(d) In the Arroyo del Soldado Group, *Cloudina* (*C. riemkeae*) occurs in the upper Yerbal and middle Polanco formations (Gaucher and Sprechmann, 1999; Gaucher, 2000; Gaucher et al., 2003; see Chapter 4.3). $\delta^{13}C$ values are moderately positive at the lowest stratigraphic occurrence (+2‰ PDB), up to +5.5‰ PDB in the middle of the biozone, and negative (−3‰ PDB) at the highest stratigraphic occurrence (Gaucher et al., 2004c; see Chapter 4.4). If the latter negative excursion is correlated with the Shuram-Wonoka anomaly and the basal Kuibis negative excursion in the Nama Group, the lowest stratigraphic occurrence of *Cloudina* in the Arroyo del Soldado Group may be one of the oldest so far reported. Corresponding $^{87}Sr/^{86}Sr$ ratios range between 0.7073 and 0.7085 (see Chapter 4.4).

The *Cloudina* Range Zone as defined above may be useful for the subdivision of the Ediacaran into different series, and could be used to characterise the youngest, still unnamed series.

*Titanotheca* is another potential index fossil of the Late Ediacaran, not least because of its unusually high preservation potential. The genus has a wide palaeogeographic distribution (three palaeocontinents), relatively short stratigraphic range (Figure 9.2.4), it is very abundant at the known occurrences and is not facies-dependent, occurring in siltstone, carbonate and phosphorite (Gaucher and Sprechmann, 1999; Gaucher, 2000; see Chapter 4.3).

from the Spitskop Member of the Urusis Formation (middle Schwarzrand Subgroup) (Jensen and Runnegar, 2005) (Figure 9.3.2).

Based on the radiometric date of 543 ± 1 Ma of a silicified ash bed in the Spitskop Member and the occurrence of the body fossil *Pteridinium carolinaense* in this member, a latest Ediacaran age has been assigned to the Spitskop Member (Grotzinger et al., 1995). However, the occurrence of *Streptichnus narbonnei* may be an indication that a part of the Spitskop Member may be Cambrian in age and that the Ediacaran-Early Cambrian boundary may be within the Spitskop Member and not at the base of the Nomtsas Formation (Jensen and Runnegar, 2005).

#### 9.3.3.2. Early Cambrian (upper Nama Group) trace fossils (ca. 540–530 Ma)

The trace fossils of the Nomtsas Formation (uppermost Schwarzrand Subgroup) and Fish River Subgroup are generally larger in size, more complex and slightly more diversified than the trace fossils occurring in the underlying Ediacaran sediments. '*Phycodes*' *pedum* occurs widespread and some complex trace fossils occur in these successions especially in the Fish River Subgroup (Crimes and Germs, 1982; Geyer and Uchman, 1995; Geyer, 2005; Table 5.4.1). Most of the trace fossils thus far found in the Fish River Subgroup have been found in the Rosenhof Member (Gross Aub Formation). The trace fossils in the Rosenhof Member are *Trichophycus pedum*, *Trichophycus pollardi*, *Paleophycus*, *Gordia*, *Skolithos* and *Enigmatichnus* (Crimes and Germs, 1982; Geyer and Uchman, 1995; Geyer, 2005). Geyer and Uchman (1995) and Geyer (2005) consider some trace fossils like *Phycodes* to belong to *Trichophycus* whereas, for example Jensen and Runnegar (2005) and Buatois et al. (2007) consider them to belong to *Treptichnus*. There is much dispute about the taxonomical status of *Trichophycus*, *Phycodes* and *Treptichnus*; and discussions are focused on fine morphological characters and the nature of the burrow (Erdogan et al., 2004; Aceñolaza and Aceñolaza, 2007).

The diverse ichnoassemblages of the Fish River Subgroup are indicative of an earliest Cambrian pre-trilobite age (Geyer, 2005). They comprise traces of deposit feeders as well as others that were probably created by suspension feeders and agrichnial farmers (Geyer, 2005). The trace maker of *Enigmatichnus* was probably an arthropod (Crimes and Germs, 1982).

### 9.3.4. Trace Fossils of the Vanrhynsdorp Group (ca. 550–535 Ma)

The stratigraphy of the Vanrhynsdorp Group and its depositional environments are discussed in Chapter 5.4 and shown in Figure 5.4.8. The Ediacaran-Cambrian boundary has not yet been pinpointed in the Vanrhynsdorp Group. The boundary may occur at the base of the Arondegas Formation or more likely at the base of the Besonderheid Formation of the Knersvlakte Subgroup. Future dating of silicified volcanic ash beds in the Gannabos and Besonderheid Formations most probably will make it possible to determine where this important boundary is stratigraphically located in the Vanrhynsdorp Group.

The Vanrhynsdorp Group is emerging as a key succession for documenting ichnodiversity, ichnostratigraphy, body size and burrowing behaviour among infaunal metazoans across the Ediacaran-Cambrian boundary (Buatois et al., 2007; Almond et al., 2008). The trace fossils thus far described from the Vanrhynsdorp area are shown in Figures 5.4.8 and 9.3.3. Much progress has recently been made with the study of trace fossils by Buatois et al. (2007) and Almond et al. (2008). The Vanrhynsdorp Group is considered to be a slightly deeper-water unit that correlates to the Nama Group (Germs and Gresse, 1991; Gresse and Germs, 1993; Figure 5.4.3). Even though, precise correlation of the Nama and Vanrhynsdorp successions remains disputable.

No trace fossils have yet been found in the basal Flaminkberg Formation. Narrow straight to curved horizontal burrows of *Helminthopsis* are associated with microbial dark mudstones of the younger Hoedberg Formation (Kwanous Subgroup) (Gresse, 1992; Buatois et al., 2007; Almond et al., 2008; Figure 9.3.3A). According to Buatois et al. (2007) and Almond et al. (2008) this trace fossil is characteristic of inshore as well as deeper water in the Ediacaran Period. More data are required to ascertain whether these *Helminthopsis*-bearing successions are latest Ediacaran or Early Cambrian in age.

Large 1–2.3 cm-wide horizontal burrows occur in the heterolithic upper Gannabos Formation of the Knersvlakte Subgroup (Gresse, 1992; Buatois et al., 2007; Almond et al., 2008; Figure 9.3.3B). This unidentified type of trace fossil (*Planolites*?) is also associated with microbial wrinkle marks and is characteristic for an offshore storm-influenced marine environment.

Buatois et al. (2007) and Almond et al. (2008) provisionally assigned an Ediacaran age to the Hoedberg, Arondegas and Gannabos formations based on the apparent absence of *Treptichnus*-type burrows which characterise comparable storm dominated shelf-facies in the Early Cambrian. If correct, this implies that infaunal large-bodied metazoans appeared before the end of the Ediacaran (Almond et al., 2008).

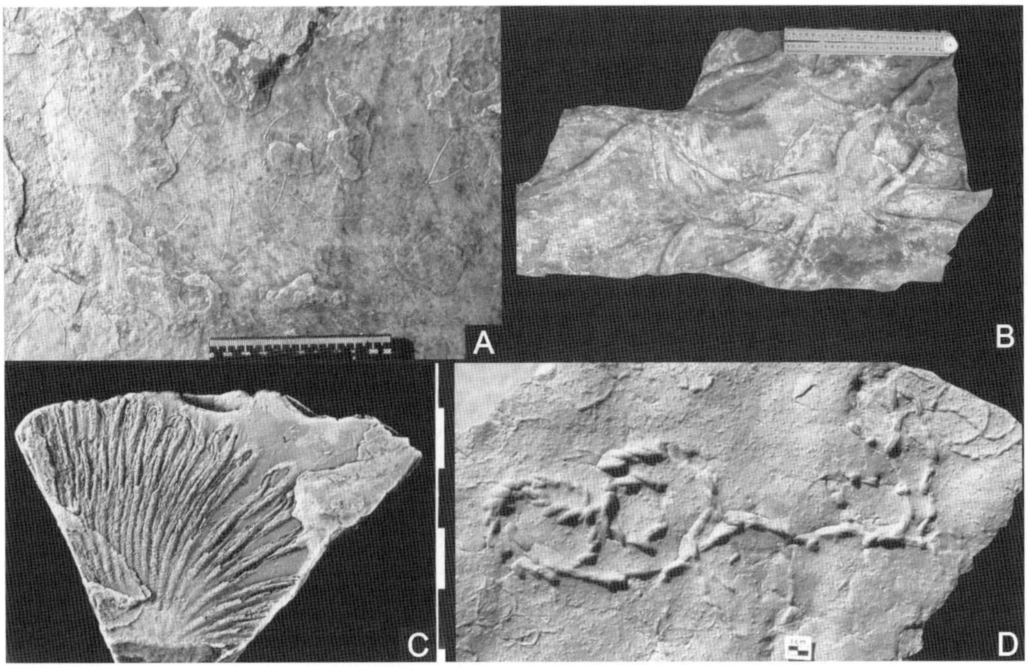


**Figure 9.3.3** Ediacaran-Early Cambrian trace fossils from South Africa. A- *Helminthopsis* ichnoguild, horizontal burrows, Hoedberg Formation, Vanrhynsdorp Group. B- *Planolites* isp., Gannabos Formation, Vanrhynsdorp Group. C- *Oldhamia geniculata*, Besonderheid Formation, Vanrhynsdorp Group (scale segments to the right are 1 cm). D- *Treptichnus pedum*, Kalkgat Formation, Vanrhynsdorp Group (Photographs supplied by J.E. Almond).

Complex sub-horizontal burrow systems of the distinctive Cambrian ichnofossil *Oldhamia geniculata* (Figure 9.3.3C) occur in the distal, finely laminated mudrocks of the Besonderheid Formation of the middle Knersvlakte Subgroup (Gresse, 1992; Buatois et al., 2007; Almond et al., 2008). This trace fossil is also known from the Early Cambrian of Argentina (Seilacher et al., 2005). The Vanrhynsdorp *Oldhamia* systems are associated with wrinkle marks and show a sophisticated foraging behaviour of the trace maker. The successions of the overlying upper Knersvlakte and the Klipbak subgroups contain prolific low diversity trace fossil assemblages, dominated by *Treptichnus*-type trace fossils (Germs, 1983; Gresse, 1992; Almond et al, 2008). The presence of *Treptichnus pedum* (Figure 9.3.3D), as well as the absence of unequivocal arthropod scratch burrows such as *Rusophycus/Cruziana*, suggests an earliest Cambrian age for the Vanrhynsdorp successions overlying the Besonderheid Formation (Buatois et al., 2007; Almond et al., 2008).

The Vanrhynsdorp treptichnids display considerable preservational, morphological and hence behaviour, plasticity with frequent intergradations between *Treptichnus pedum* and other treptichnid ichnospecies within the same burrow system (Buatois et al., 2007; Almond et al., 2008). Transitions with *Curvolithus—Saerichnites—Arthrophycus* and *Cruziana* morphs have been observed (Buatois et al., 2007; Almond et al., 2008). Buatois et al. (2007) and Almond et al. (2008) consider the identity and arthropod origin of *Monomorphichnus* reported from the Klipbak Formation controversial (Germs, 1983).

## 9.3.5. Conclusions

Thus far, definitive Ediacaran fossils have not yet been found in the Puncoviscana Formation, even though geochronological data includes indisputable Ediacaran strata in the basin (Aceñolaza and Aceñolaza, 2005, 2007). This is in contrast to the Nama and Vanrhynsdorp groups, where such deposits are well exposed and contain trace fossils (Germs, 1972b, 1983; Crimes and Germs, 1982; Almond et al., 2008).

The Late Ediacaran trace fossil assemblages of the lower Nama and lower Vanrhynsdorp groups are similar to Ediacaran trace fossils elsewhere in the world: generally sparse, very low in diversity and dominated by small, fairly simple horizontal burrows generated in a shallow marine to offshore depositional environment (Crimes and Germs, 1982; Buatois et al., 2007; Almond et al., 2008; Table 5.4.1; Figure 9.3.3A). Unnamed, up to 2.5 cm-wide trace fossils occurring in the probable Ediacaran, Gannabos Formation (Vanrhynsdorp Group; Figure 9.3.3B) suggest that large metazoans appeared prior to the Cambrian (Buatois et al., 2007; Almond et al., 2008).

In contrast, the latest Ediacaran middle and upper Nama Group contains complex *Treptichnus*-type trace fossils which occur in the Nasep Member and lower Spitskop Member of the Urusis Formation (Schwarzrand Subgroup)

(Germs, 1972b; Geyer and Uchman, 1995; Geyer, 2005; Jensen et al., 2000). In other words, as shown by the Nama Group, treptichnid-type trace fossils already occur in ca. 546 Ma old successions, that is, prior to the Cambrian (Figure 9.3.2). It is not yet certain if the newly described *Treptichnus*-type *Streptichnus narbonnei* from the Spitskop Member is Ediacaran (Grotzinger et al., 1995) or Early Cambrian in age (Jensen and Runnegar, 2005; Figure 9.3.2).

The earliest Cambrian trace fossil assemblages of the uppermost successions of the Nama and Vanrhynsdorp groups display a higher, but still relatively low, diversity of trace fossils dominated by *Treptichnus* (Crimes and Germs, 1982; Almond et al., 2008) (Table 5.4.1, Figures 9.3.2, 9.3.3D and 5.4.8). The treptichnid-dominated assemblages of the uppermost Nama and Vanrhynsdorp groups are predominantly associated with reddish shallow marine sandy deposits. *Diplichnites* and ?*Monomorphichnus* sparsely occur in more distally deposited finer clastic sediments of these uppermost successions (Crimes and Germs, 1982; Germs, 1983).

From the ichnofossil assemblages found in the Ediacaran Nama (Table 5.4.1) and Vanrhynsdorp (Figure 5.4.8) groups, it becomes evident that the increase in trace fossil diversity towards the Cambrian occurs more gradually than generally accepted. This has also been noted by Gehling et al. (2001). This more 'gradual' increase in trace fossil diversity from the Ediacaran to the Early Cambrian appears to apply also to body fossil diversity (Jensen et al., 1998). The relatively low diversity and apparent absence of trilobite tracks and trails probably indicate that the upper Nama and upper Vanrhynsdorp groups are of pre-trilobite age (Geyer, 2005; Almond et al., 2008), and that these successions accumulated prior to the acme of the 'agronomic revolution' and the 'Cambrian explosion', and that these events occurred gradually. Vertical bioturbation might have taken place during deposition of the Ediacaran lower and middle Nama groups which probably confirms the gradual evolution of these events.

The trace fossil diversity of the Cambrian Puncoviscana Formation is higher than that of the earliest Cambrian Nama and Vanrhynsdorp groups. Arthropod trace fossils as *Tasmanadia*, *Diplichnites*, *Monomorphichnus* and worm-like tracks and trails are quite common in the Puncoviscana Formation. Vertical bioturbation occurs both in the eastern shallow and western deeper part of the Puncoviscana Basin. The frequent occurrence of these trace fossils indicates that these fossiliferous levels of the Puncoviscana Formation may be younger than the uppermost lower Cambrian (pre-trilobite) successions of the Nama and Vanrhynsdorp groups and that the Puncoviscana Formation most likely accumulated during a time interval including 'trilobite' time. The trace fossil assemblages of the Puncoviscana Formation reflect the development of the true 'agronomic revolution' (Seilacher and Pflüger, 1994) or the 'Cambrian substrate revolution' (Bottjer et al., 2000).

The trace makers of the ichnofossils occurring in the Cambrian Puncoviscana Formation and in the Ediacaran and Cambrian Nama and Vanrhynsdorp groups, as elsewhere on Earth during that time, predominantly fed on or beneath the surface of biomats which occurred abundantly in the Puncoviscana, Nama and Vanrhynsdorp basins (Aceñolaza and Aceñolaza, 2001, 2007; Noffke et al., 2002a,b; Buatois and Mángano, 2004; Buatois et al., 2007; Almond et al., 2008). The biomats were formed by microbial communities and were a key element of the ecological structure of shallow and deep-water settings. They are characterised by wrinkle structures and elephant skin among other forms (e.g. Droser et al., 2002; Noffke et al., 2002b; McIlroy et al., 2005; Bailey et al., 2006). The occurrence of the trace fossil *Nereites* in shallow marine deposits of the Vingerbreek Member (Nudaus Formation, Schwarzrand Subgroup, Nama Group; Table 5.4.1, Chapter 5.4) and in the eastern shallow facies of the Puncoviscana Formation indicates that its trace maker had not yet moved to an offshore environment. The timing of this event is uncertain. More studies of the turbiditic Besonderheid Formation (Vanrhynsdorp Group) and deeper marine facies of the Puncoviscana Formation are needed to establish when this event took place.

The occurrence of *Oldhamia geniculata* (Figure 9.3.3C) in the Besonderheid Formation (Vanrhynsdorp Group) probably indicates a Cambrian age for this formation. Its occurrence in distal pro-delta deposits (probably distal turbidites) represents a step in the ecological onshore–offshore shift of its trace maker (Seilacher et al., 2005). The sparse occurrence of only one *Oldhamia* species (i.e. *Oldhamia geniculata*) in the Besonderheid Formation and the low trace fossil diversity of this formation contrasts with the higher trace fossil diversity of the Puncoviscana Formation which contains various *Oldhamia* ichnospecies (Figure 9.3.1D) associated with arthropod traces and probable early network systems comparable to *Thalassinoides* (Aceñolaza and Aceñolaza, 2005; Seilacher et al., 2005). The difference in diversity can most probably be ascribed to a difference in age, with a Puncoviscana Formation younger in age than the uppermost Cambrian successions of the Nama and Vanrhynsdorp groups.

It may also be explained by different environmental conditions. The general depositional environments of the Nama and Vanrhynsdorp groups are rather well known (Figures 5.4.5 and 5.4.8), but not those of the Puncoviscana Formation. More work is needed to identify the various depositional environments of the unit. A basic requirement for identifying those depositional environments is to have a good knowledge of the stratigraphy of the Puncoviscana Formation. Due to complex structural deformation and lack of chronological data the stratigraphy of the Puncoviscana Formation has not yet been unravelled. The Early Cambrian Guachos Formation is included in the Puncoviscana Basin, and represents a local environmentally controlled facies within the framework of the Ediacaran-Early Cambrian successions in northwest Argentina (Aceñolaza and Aceñolaza, 2005).

The eastern facies of the Puncoviscana Formation accumulated in a shallow marine setting, while there is no full consensus regarding the depositional setting of the western facies of the Puncoviscana Formation (Borrello, 1969; Aceñolaza, 2005). The generally accepted deep-water western facies of this formation is challenged in this publication. It is proposed herein that the Puncoviscana Formation was deposited in an overall shallow marine environment in a basin deepening towards the west. The occurrence of relatively deep-ocean flysch-like shales and sandstones indicate that the western Puncoviscana facies (García-Bellido and Aceñolaza, 2005) was generally deposited in deeper water than the Nama and Vanrhynsdorp (with the exception of the Besonderheid Formation) groups.

In the Vanrhynsdorp Group the multiple intergradations between *Treptichnus pedum* and different ichogenera, including other *Treptichnus* forms, cast serious doubts about the reliability of using *Treptichnus* ichnospecies, such as *Treptichnus pedum*, in biostratigraphic zonations of Ediacaran and lowermost Cambrian sediments (Buatois et al., 2007; Almond et al., 2008).

Finally, a re-examination of Ediacaran to Early Cambrian trace fossils is urgently needed from well-dated reference sections around the world, especially considering *Treptichnus pedum* and its evolutionary history.

## ACKNOWLEDGEMENTS

The authors are indebted to J.C. Gutiérrez-Marco, D. García-Bellido and S. Jensen for the revision of the manuscript, and to J.E. Almond for the graphical assistance with photographs. The manuscript benefited from field discussions with P. Vickers-Rich and M. Fedonkin. This contribution was supported by the Universidad Nacional de Tucumán (CIUNT).

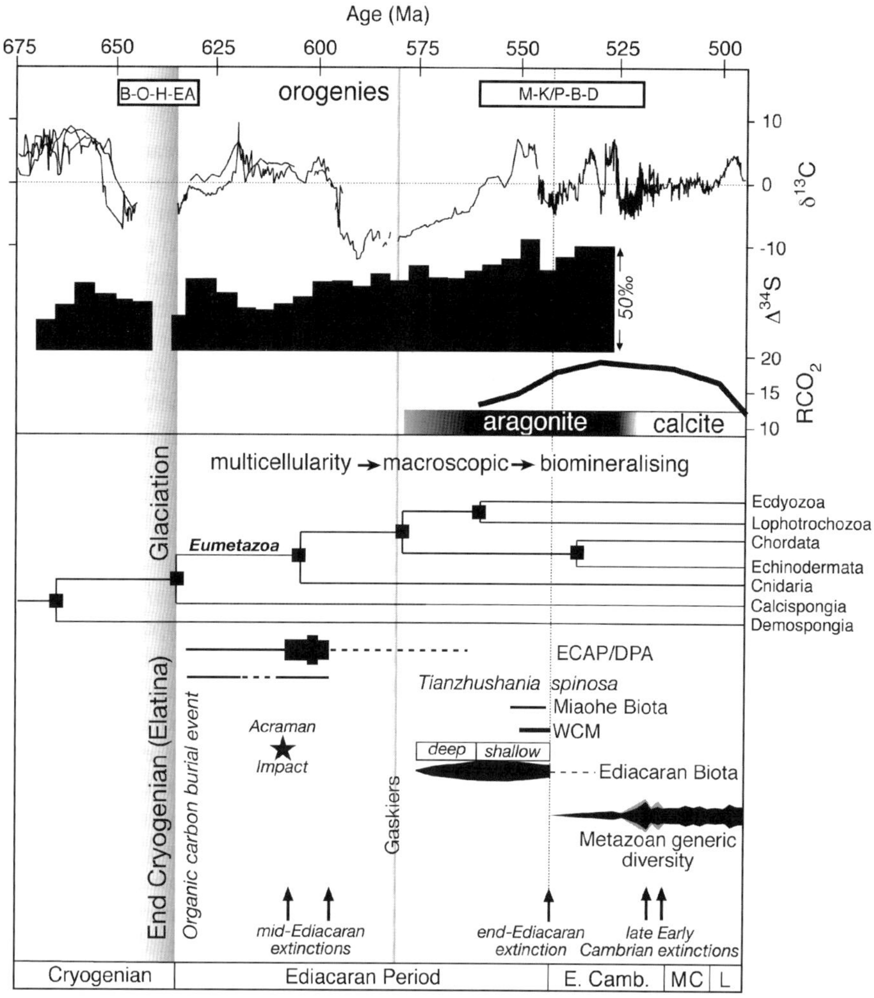


**Figure 10.5** Summary of tectonic events and geochemical and biological change spanning the Precambrian-Cambrian boundary. Orogenic events, as summarised in the text: B-O-H-EA stands for the Brazilia, Oubengides, Hoggar, East African orogenies and M-K/P-B-D stands for the Malagasy, Kuunga/Pinjarra, Buzios, and Damara orogenies. $\delta^{13}C$ data are from Figure 10.2. $\delta^{34}S$ [$\delta^{34}S_{sulph}(avg)-\delta^{34}S_{pyr}(avg)$] is approximated for 5 m.y. bins from the data set in Figure 10.4, but taking into consideration large amplitude fluctuations in $\delta^{34}S$ values in the earliest Ediacaran (from Hurtgen et al., 2006) that result in an overestimate of $\delta^{34}S$ computed in this way. $RCO_2$ ($CO_2$ concentration normalised against pre-industrial levels) is from Berner (2006). The switch from aragonite to calcite seas in the Early Cambrian is from Stanley and Hardie (1998), Porter (2007), and Zhuravlev and Wood (2008). The three-stage Ediacaran-Cambrian biological evolution is as described in the text (Section 6). Early animal diversification is modified from Peterson et al. (2007) with molecular clock estimates of major diversifications from Peterson and Butterfield (2005). ECAP/DPA: Ediacaran complex acanthomorphic palynoflora/Doushantuo—Pertatataka acritarch assemblage; relative species diversity from Grey et al. (2003). Note that while the timing of the extinction of the ECAP/DPA is reasonably constrained by its apparent link to the onset of the Shuram-Wonoka anomaly (Grey and Calver, 2007; McFadden et al., 2008), the timing of an earlier extinction of the leiospheric acritarchs and radiation of the ECAP/DPS in South Australia (linked to the Acraman impact; Grey et al., 2003) is not well constrained. Distribution of *Tianzhushania spinosa* is from Yin et al. (2007) and McFadden et al. (2008). Distribution of the Miaohe Biota is modified from Xiao et al. (2002). Distribution and relative specific diversity of the Ediacaran Biota and weakly calcifying Metazoa (WCM) are from Narbonne (2005). Metazoan generic diversity, including archaeocyaths (in grey), is from Zhuravlev (2001).

multicellularity in a number of eukaryotic clades, the first abundant macroscopic body fossils do not appear until ca. 575 Ma, after Gaskiers glaciation. These are the classical 'Ediacara Biota,' found in numerous localities around the world and characterised by unusual three-dimensional soft-bodied preservation in sandstones (Gehling, 1999; Narbonne, 2005). Older Ediacara assemblages (575–560 Ma) are found only in deep water slope to basinal environments and are dominated by rangeomorphs, frond-like organisms with fractal architecture that presumably lived as suspension feeders attached to the substrate (Clapham et al., 2003; Narbonne, 2005). Like most members of the classical Ediacara Biota, their affinities are enigmatic, though they are generally regarded as basal Metazoa (Narbonne, 2005). Younger Ediacara assemblages (560–542 Ma) are typically found in shallower environments and are of markedly different taxonomic composition, with relatively few rangeomorphs and a diversity of likely bilaterians (e.g., Fedonkin and Wagonner, 1997; Narbonne, 2005) and other enigmatic forms that might not be metazoan (*cf.* Peterson et al., 2003; Seilacher et al., 2003). These younger assemblages also record the first evidence of bilaterian mobility, with presumed bilaterian trace fossils found in rocks >555 Ma (Jensen, 2003; Droser et al., 2005). In addition to the classical Ediacara fossils, carbonaceous compression fossils from the Miaohe Biota (uppermost Doushantuo Formation, ca. 551 Ma in age; Condon et al., 2005) record macroscopic multicellularity in a number of clades, including red, brown, and green algae, and animals (Xiao et al., 2002). Recent discovery of the eight-armed Ediacara fossil *Eoandremeda octobrachiata* as impressions in both the Miaohe black shales of South China and as casts and moulds in South Australia provides an important link between the Ediacara and Miaohe biotas, promising a possible integration of the heretofore rich but enigmatically distinct fossil records between these two key regions (Zhu et al., 2008).

Another important milestone in animal evolution occurred in latest Ediacaran time, ca. 550 Ma, when biomineralised animal skeletons first appear (Grant, 1990; Grotzinger et al., 2000). These first, lightly mineralised skeletons – the massive framework of *Namapoikia*, the goblets of *Namacalathus*, and the flanged tubes of cloudiniids (the weakly calcifying Metazoa, or WCM) – are sufficiently different from each other to suggest they evolved independently, perhaps in response to the selective pressure of macrophagous predation, which also appeared at this time (Bengston and Yue, 1992; Hua et al., 2003). These mineralised taxa are relatively short-lived however; they disappear from the record at the end of the Ediacaran Period, perhaps in an extinction event that also wiped out the classical Ediacara Biota (Knoll and Carroll, 1999; Amthor et al., 2003; Wille et al., 2008). Latest Ediacaran acritarch assemblages are low diversity (Vidal and Moczydlowska-Vidal, 1997; Knoll et al., 2006), perhaps hinting at inimical environmental conditions at this time (e.g., Knoll and Carroll, 1999; Kimura and Watanabe, 2001).

The Early Cambrian fossil record is characterised by a dramatic diversification of metazoans, especially bilaterians. This is particularly evident from a comparison of late Ediacaran and early Cambrian soft-bodied lagerstätte (Xiao et al., 2002): whereas the ~550 Ma Miaohe Biota includes only *Eoandromeda* (Zhu et al., 2008) and a few tubular forms that might be metazoan, the ~520 Ma Chengjiang Biota includes at least 150 metazoan genera distributed among more than 10 phyla, including arthropods, priapulids, ctenophores, cnidarians, nematomorphs, chaetognaths, and chordates (Zhang et al., 2001; Chen et al., 2002; Hou et al., 2004; Vannier et al., 2007). Early Cambrian skeletal fossils provide a similar view of metazoan diversification. The earliest skeletal assemblages are relatively low diversity and – with the exception of molluscan shells and chaetognath spines (Szaniawski, 1982; Runnegar, 1996) – include mostly problematic forms that are likely stem-groups of one or more bilaterian clades (e.g., Skovsted et al., 2008). By ~520 Ma, however, the diversity of skeletal taxa had reached its Cambrian peak (Zhuravlev, 2001), and nearly all of the major biomineralising animal clades had appeared (Bengtson and Conway Morris, 1992). This diversification is accompanied by, and probably linked to, another dramatic biological event: the widespread and independent acquisition of biomineralised skeletons among numerous clades – not only metazoan, but protistan as well (Bengtson and Conway Morris, 1992). Two extinctions in the middle and late Botomian resulted in a rapid decline in metazoan diversity, and nearly complete disappearance of archaeocyaths (Zhuravlev and Wood, 1996; Zhuravlev, 2001). Four additional extinction events during the Middle and Late Cambrian kept metazoan diversity at relatively low levels for the remainder of Cambrian time (Zhuravlev, 2001).

At the risk of overgeneralisation and pending refinement of age constraints and correlations between successions (such as South China and South Australia), Ediacaran-Cambrian biological evolution can be divided into three phases: (1) an initial phase, ca. 635–580 Ma, where multicellularity was present in several clades, possibly including Eumetazoa, but organisms remained, for the most part, microscopic; (2) a second phase, ca. 580–550 Ma, when multicellular organisms evolved macroscopic size; and (3) a third phase, ca. 550–520 Ma, when biomineralised skeletons appeared. The advances that characterise these phases were not limited to single clades, that is, they evolved convergently in several groups. Both multicellularity and macroscopic size evolved convergently in animals, fungi, red algae, green algae, and brown algae, all of which were present (and multicellular) by early Ediacaran time, and, with the possible exception of fungi, macroscopic by late Ediacaran time (see discussion above).

Mineralised skeletons evolved independently not only in numerous eukaryotic clades, but also in numerous metazoan lineages. Evidence for the latter is based on the difference in structure and mode of growth among metazoan skeletons (Bengtson and Conway Morris, 1992), as well as on molecular clock and body fossil evidence for significant divergence within crown-group bilaterians during Ediacaran time (Peterson et al., 2008; Figure 10.5), long before mineralised skeletons appear. It is particularly noteworthy that in the case of these last two characters – macroscopic size and biomineralisation – episodes of convergent evolution are closely clustered in time. This suggests that extrinsic factors – a changing physical or ecological environment or both – drove these innovations (although see Butterfield (2009) for an alternative hypothesis that intrinsic eukaryotic evolution triggered the apparent changes in the Ediacaran-Cambrian environment). Indeed, a hint that Ediacaran evolution was intimately linked to environmental change is suggested by the appearance of macroscopic organisms on the heels of Gaskiers glaciation, in association with deep-ocean oxygenation (Canfield et al., 2007; Shen et al., 2008c).

## 10.7. Late Neoproterozoic-Cambrian Ocean Redox

A growing body of data supports Canfield's (1998) hypothesis that at least a significant percentage of the global deep ocean was sulphidic through most of the middle part of the Proterozoic Eon (e.g., Scott et al., 2008, and references therein). Insofar as this hypothesis is correct, then the Neoproterozoic must have experienced a transition to a more oxic marine environment, with accompanying modifications to the abundance of various redox-sensitive elements, such as Mo, U, and key bioessential transition metals (Anbar, 2008). While it has therefore been generally assumed that the late Precambrian environment experienced some sort of oxygenation event (e.g., Catling and Claire, 2005), only recently have high-resolution data sets of redox-sensitive proxies begun to bear the hypothesis out. While Canfield and Teske (1996) argued for oxygenation some time in the latter half of the Neoproterozoic based on $\delta^{34}S$ variations in the sedimentary pyrite record, more recently, coupled sulphate and pyrite records, showing a rise in $\Delta^{34}S$ to near 40‰, have been used to argue more precisely for a mid-Ediacaran oxygenation event (Hurtgen et al., 2005; Fike et al., 2006; Halverson and Hurtgen, 2007; McFadden et al., 2008). Fike et al. (2006) further argued that this oxygenation event is recorded in the shift from uncorrelated to complementary $\delta^{13}C_{org}-\delta^{13}C_{carb}$ data at ca. 550 Ma (Condon et al., 2005) as a result of the progressive oxidation of a large deep-ocean DOC reservoir. These models appear to be corroborated by iron-speciation data. For example, a sharp decrease in the relative abundance of highly reactive iron (FeHR/FeTot) at about 580 Ma in deep water sediments of the Conception Group on the Avalon Peninsula, Newfoundland, suggests at least local deep-ocean oxygenation occurring shortly after Gaskiers glaciation and before the first appearance (Narbonnne and Gehling, 2003) of Ediacaran fossils in the same succession (Canfield et al., 2007). Likewise, FeHR/FeTot data from the Sheepbed Formation in northwest Canada and the Doushantuo Formation in South China suggest mid-Ediacaran oxygenation of open continental shelf bottom waters (Shen et al., 2008c). In the case of northwest Canada, this oxygenation appears to have been abrupt and preceded the sharp downturn in $\delta^{34}S_{pyr}$ characteristic of the middle Ediacaran Period. This record of ocean oxygenation insinuates a tantalising, if not unambiguous, link with the origin of the Eumetazoa (Figure 10.5).

More recently, Canfield et al. (2008) used an expanded iron-speciation data set from multiple basins to argue that while the surface ocean was generally oxygenated in the latter half of the Neoproterozoic, the deep ocean was dominantly iron-rich, shifting transiently to a euxinic deep ocean at the PCB. This hypothesis has important implications for the biogeochemical evolution of the deep oceans and biospheric evolution. Sulphidic waters are both much more depleted in the trace nutrients Fe and Mo (Anbar and Knoll, 2002) and much more prone to triggering biotic crises (Meyer and Kump, 2008). Furthermore, a switch to iron-rich deep oceans in the latter half of the Neoproterozoic would hint to a return to Archaen-like conditions. Thus, although deposition of banded iron-formation (BIF) in the Neoproterozoic is restricted to glacial intervals, casting some doubt on the ubiquity of iron-rich deep waters, this hypothesis deserves serious consideration and needs to be tested by examining in more detail multiple late Neoproterozoic sedimentary basins.

The sulphur isotope record (Figure 10.4) suggests a somewhat more complex evolution in ocean redox, with large fluctuations in average $\Delta^{34}S$ and excursions in both $\delta^{34}S_{sulph}$ and $\delta^{34}S_{pyr}$. The first expansion in $\Delta^{34}S$ coincides with the Bitter Springs negative $\delta^{13}C$ stage (Halverson et al., 2007b), and may record ventilation of the deep ocean related to changing ocean circulation and nutrient supply attending a large-scale true polar wander event (Maloof et al., 2006). Subsequent increases in $\Delta^{34}S$ occur between the Bitter Springs stage and the onset of the earliest Cryogenian glaciation, and in the interglacial period prior to the end-Cryogenian (i.e., Elatina) glaciation (Figure 10.3). In all three cases, $\Delta^{34}S$ then abruptly decreases and $\delta^{34}S_{pyr}$ returns to values near 0‰. The late Cryogenian decrease in $\Delta^{34}S$ coincides with the Trezona negative carbon isotope anomaly (*cf.* Halverson et al., 2002), which presaged the onset of end-Cryogenian glaciation, as well as a transient increase in $\delta^{34}S_{sulph}$ and reduction in CAS concentrations to near nil (Halverson and Hurtgen, 2007). Thus, it appears that this event entailed the depletion of the marine sulphate reservoir via extensive BSR of a $^{13}C$-depleted reservoir of reduced carbon, consistent with (if not proof of) a massive and prolonged flux of methane from sedimentary clathrate reservoirs (Schrag et al., 2002). Low sulphate conditions appear to have continued through end-Cryogenian glaciation and the earliest part of the Ediacaran Period (Hurtgen et al., 2006; Shen et al., 2008c).

### 10.7.1. An early Ediacaran oxygen pulse?

Irrespective of the precise mechanism for the successive depletions in the marine sulphate reservoir through the late Tonian and Cryogenian, it is clear that both expansions and contractions in marine sulphate, and implicitly ocean redox, correspond closely with fluctuations in marine $\delta^{13}C$ and glacial intervals. This coupling is perhaps most impressive in the middle Ediacaran Period, which is punctuated by the Shuram-Wonoka anomaly that appears to have commenced prior to the ca. 580 Ma onset of the Gaskiers glaciation (Halverson et al., 2005). The sulphur isotope record suggests that seawater sulphate concentrations increased dramatically prior to this anomaly, and this sulphate may have played an important role in both the generation of the excursion and maintaining $O_2$ levels. The carbon isotope anomaly itself implies that a huge supply of oxidising power must have already been available in the oceans (Bristow and Kennedy, 2008) to mineralise the $^{13}C$-depleted, reduced carbon responsible for the anomaly (be it sedimentary OM, DOC, or methane). And because BSR, coupled to pyrite burial, actually releases oxidising power (i.e., $2Fe_2O_3+8SO_4^{2-}+16H^+ \rightarrow 15O_2+4FeS_2+8H_2O$; Berner, 2006), the sulphate reservoir (along with iron oxides) may have acted as an oxygen capacitor, maintaining $O_2$ in the atmosphere–ocean system at high enough levels to permit the evolution of large Ediacaran animals by 575 Ma, even as total oxidising power in the ocean was being reduced. At the same time, BSR at this scale may well have driven the deep oceans to euxinic conditions. Likely not coincidentally, the onset of the Shuram-Wonoka anomaly appears to correspond with an abrupt disappearance of many acanthomorphic acritarch species (Grey and Calver, 2007; McFadden et al., 2008), as well as *Tianzhushania* and other putative metazoan embryo fossils (Zhou et al., 2007; Figure 10.5).

If oxygen and sulphate levels rose sharply in the early Ediacaran, what was the trigger for this immense flux of oxidising power? The extraordinary conditions in the aftermath of the end-Cryogenian snowball glaciation would have interacted to produce the ideal environment for high rates of organic carbon production and export to sediments in the deep ocean. The extremely high $CO_2$ levels (>200 mbar; Hoffman et al., 1998; Pierrehumbert, 2004) and consequent elevated rates of nutrient delivery (through silicate weathering) to the surface ocean would have driven high OM production (Kirschvink and Raub, 2003; Elie et al., 2007). At the same time, the rapid melting of continental glaciers and sea ice would have flooded the surface ocean, which would have then warmed rapidly under the ultragreenhouse atmosphere. The result would have been a highly density stratified ocean, with a hot brackish surface layer capping cold, saline deep waters. An anoxic deep ocean would have been inevitable given the huge oxygen demand imposed by sinking particulate organic matter (POM), and the accumulated OM would have been balanced by a high $O_2$ flux to the atmosphere. The physical stratification of the post-snowball ocean likely would have decayed within a few tens of thousands of years, but because the timescale for drawdown of $CO_2$ after a snowball glaciation might have been on the order of several million years (Le Hir et al., 2009) and positive feedbacks under the warm conditions would have been strong (Meyer and Kump, 2008), anoxia could have been maintained, at least in basins with focused nutrient supply, for millions of years. Importantly, this organic carbon burial event is not marked by high $\delta^{13}C_{carb}$ (Figure 10.5), but this is easily explained by the leverage of the immense ocean–atmosphere carbon reservoir at the time, which would have damped $^{13}C$-enrichment due to OM burial. The physical evidence for this organic carbon burial event may be found in widespread and thick black shales comprising the bulk of many early Ediacaran cap carbonate sequences, most notably the Sheepbed Formation in NW Canada, the Dracoisen Formation in Svalbard and its equivalent, the Canyon Formation, in East Greenland, the Pertatataka Formation in central Australia, the Masirah Bay Formation in Oman, and condensed Doushantuo Formation in South China.

### 10.7.2. Transition to the Palaeozoic

The purported middle Ediacaran extinction events (Figure 10.5) may have been the first to affect early animals, but end-Ediacaran mass extinction was the first major extinction to affect extant animal phyla. Both the Ediacaran Biota and the WCM (Grotzinger et al., 1995; Amthor et al., 2003) disappeared abruptly in pace with a prominent and short-lived negative $\delta^{13}C$ anomaly, an abrupt increase in $\delta^{34}S_{pyr}$ (Figures 10.4 and 10.5), and high FeHR/FeTot ratios in black shales (Canfield et al., 2008), which, based on previous arguments, likely herald a reprise of anoxic conditions in the deep oceans at this time. Trace elements (Kimura and Watanabe, 2001; Schröder and Grotzinger, 2007) and Mo isotope data (Wille et al., 2008) support this hypothesis and further indicate widespread euxinia across the PCB.

The redox evolution of the Early Cambrian ocean is poorly constrained, but iron-speciation data from the Yangtze Platform, South China, suggest euxinic bottom waters in the early Tommotian (Goldberg et al., 2007), while widespread black shale deposition in the Botomian indicates pervasive ocean anoxia at this time (Zhuravlev and Wood, 1996). Whether or not euxinic conditions were continuous from the PCB is impossible to establish at this point, but it is noteworthy that large fluctuations in $\delta^{13}C$, similar in magnitude but at a much higher frequency than those observed throughout the Neoproterozoic, continued through the Nemakit-Daldyn and Tommotian, only beginning to stabilise around 0‰ at the Tommotian–Adtabanian boundary (Figure 10.5; Maloof et al., 2005). This stabilisation in $\delta^{13}C_{carb}$ coincides with the expansion of archaeocyaths and onset of the Cambrian radiation (Maloof et al., 2005). A possibly analogous scenario occurred in the Early Triassic, following the end-Permian mass extinction, when large $\delta^{13}C$ fluctuations coincided with a protracted period of low biotic diversity and evidence for deep-ocean anoxia (Payne et al., 2004; Knoll et al., 2007). Only after $\delta^{13}C$ stabilised in the Middle Triassic did the biosphere fully recover. Like the Early Cambrian, geochemical evidence suggests that the anoxic deep oceans spanning the P–T boundary were also euxinic (Grice et al., 2005). If the Permo-Triassic superanoxic event (Isozaki, 1997) and mass extinction is analogous to the Early Cambrian, then perhaps the pattern of unstable $\delta^{13}C_{carb}$ spanning from the latest Ediacaran Period to the Adtabanian reflects the prevalence of euxinic deep oceans, which in turn contributed to the extinction of the Ediacaran Biota and WCM and the protracted onset and regularly interrupted Cambrian radiation.

## 10.8. Reorganisation of the Marine Carbon Cycle

Many factors would have contributed to significant adjustments to the global carbon cycle during the transition from the Proterozoic to the Palaeozoic, but the dominant driver of this reorganisation would have been the origin of the Metazoa. First, as Logan et al. (1995) argued, the origin of pelagic animals would have altered the transport of POM in the oceans by repackaging much of it as faecal pellets, which would have been able to penetrate deeper into the water column prior to remineralisation. Thus, the origin of animals would have enhanced the biological pump of DIC from surface to deep waters while reducing the contribution of sedimentary OM derived from heterotrophic bacteria. Furthermore, bioturbation, coupled with higher dissolved $O_2$ and $SO_4$ concentrations in seawater, would have also strongly diminished the preservation potential of that component of OM that reached the sediment and increased the role of BSR and subsequent oxidative recycling during early diagenesis (Canfield and Teske, 1996).

Second, the advent of metazoan biomineralisation would have decreased carbonate saturation in the surface oceans (Ridgwell and Zeebe, 2005). Since the early calcifying animals lived in the shallow marine environment, biogenic carbonate precipitation was concentrated on the continental shelves. However, carbonate precipitation in the late Precambrian was already dominantly occurring in the shelf environment, being largely mediated by cyanobacterial mats (Grotzinger and James, 2000). The latest Neoproterozoic origin of the WCM and subsequent appearance of the Cambrian small shelly fauna and the Archeocyatha would have accelerated a long-term decrease in the carbonate saturation of seawater spanning the Proterozoic-Palaeozoic boundary that is evident in unidirectional changes in the mode of carbonate precipitation (Knoll et al., 1993). For example, giant ooids, a hallmark feature of the Neoproterozoic, did not persist into the Cambrian (Grotzinger and James, 2000), while calcified microbes come into prominence across the PCB; both changes were likely the consequence of a transition to lower carbonate supersaturation in the surface oceans (Knoll et al., 1993). This shift to a lower saturation state is also evinced in a change from aragonite to calcite seas in the middle Early Cambrian (Figure 10.1), which may also have exerted an influence on the early evolution of calcifying metazoans (Porter, 2007; Zhuravlev and Wood, 2008).

It has been argued that a gradual decrease in the mass of the marine DIC reservoir could account for the increased frequency and greater magnitude of $\delta^{13}C$ fluctuations in the Neoproterozoic-Early Palaeozoic oceans (Bartley and Kah, 2004). While it seems likely that this phenomenon played a role in modulating the $\delta^{13}C$ of seawater and would have also affected carbonate saturation, the very high $CO_2$ levels hypothesised for the Early Cambrian (Figure 10.5) suggest that the story may be more complicated. Shields (2007b) has argued for high weathering rates, and thus high $CO_2$ outgassing rates, spanning the PCB to account for trends in the Sr and S isotope records (Figure 10.3 and 10.4). The standard mechanism for increased $CO_2$ outgassing is elevated rates of carbonate metamorphism (Edmond and Huh, 2003). But this mechanism, rather than simply requiring orogenesis, which was widespread across Gondwana at this time (Collins and Pisarevsky, 2005), requires subduction of carbonates. Because deep-water carbonates at this time would have largely been restricted to redeposited gravity flows derived from the margins of carbonate platforms and to carbonatised oceanic crust, it is not clear, based on any available evidence, that the late Neoproterozoic-Cambrian tectonic environment would in fact have favoured unusually high carbonate metamorphism and $CO_2$ outgassing rates.

Other factors may have influenced carbonate chemistry and $p_{CO_2}$ in the Cambrian. For example, decreased carbonate saturation in seawater, resulting from biomineralisation and elevated Ca/Mg in seawater, would have tilted carbonate equilibrium towards higher $CO_2$ levels (Ridgwell and Zeebe, 2005). So also would have the decreased burial efficiency of OM (evinced in lower average $\delta^{13}C_{carb}$). On the other hand, an externally imposed increase in $p_{CO_2}$ would also have decreased carbonate saturation, favoured a shift towards calcite seas (Stanley and Hardie, 1998), and would help to account for widespread anoxia and the abundance of black shales in the Early Cambrian. Thus, we are left with a chicken-and-egg scenario, where we can speculate that carbonate saturation states dropped sharply across the PCB, but whether this was externally forced by the exogenic carbon cycle or internally driven by changes in carbonate precipitation in the oceans and seawater chemistry remains unresolved. What is evident is that perturbations in the global carbon cycle, as recorded in $\delta^{13}C$ in carbonates, are closely allied with evolution of the biosphere, and that the modernised biosphere played a stronger regulatory (negative feedback) role in modulating the carbon, oxygen, and sulphur cycles.

## 10.9. Conclusions

A picture of biogeochemical change across the Proterozoic-Phanerozoic transition is emerging, stimulated by a surge in research into the extreme climates, tectonic upheaval, and biospheric change that make it such a compelling chapter in Earth's history. The marine carbon isotope record highlights this transition, showing a drop in average values from +5‰ to 0‰ between the Cryogenian and Ediacaran, and increased frequency and gradual damping in the amplitude of $\delta^{13}C$ fluctuations in the Cambrian (Figures 10.2 and 10.5). This wholesale shift in the mode of organic carbon burial was accompanied by changes in carbonate chemistry (Ridgwell et al., 2003), with the origin of biomineralisation in the latest Ediacaran and Early Cambrian switch from aragonite seas (high Mg/Ca) to calcite seas (low Mg/Ca) (Figure 10.5) decreasing carbonate saturation in the surface oceans.

While the chronological framework for the PCB interval is not yet radiometrically well calibrated, stratigraphic, palaeontological, and geochemical data sets are increasingly integrated, permitting more critical assessments of the interconnections between the evolving redox of the ocean–atmosphere system, the origin and diversification of animals and other eukaryotic clades, climatic catastrophes, and reorganisation of the continents. A consensus seems to be emerging that Earth experienced a major oxygenation event in the Ediacaran Period and that it was intimately linked to the first appearance of animals. Impressively, molecular clock data (Peterson and Butterfield, 2005) and palaeontological data (Yin et al., 2007; Zhou et al., 2007) converge to suggest that the first, microscopic animals appeared within a few million years of the end of the last snowball glaciation. Geological and geochemical data, along with theoretical considerations, indicate a major pulse in carbon burial and oxygen production at this time. A central and unresolved question is whether, as commonly assumed, this increased oxygenation triggered early animal evolution, or if instead biological innovations associated with the origin of animals drove oxygenation and other environmental changes (Butterfield, 2009).

Significant fluctuations in ocean redox, detected mainly in stable isotope and iron-speciation data, appear to be a regular feature of the Cryogenian to Cambrian world (Figure 10.5; Canfield et al., 2008). These redox changes are closely linked to episodes of glaciation in the Neoproterozoic, and, at the coarse biostratigraphic resolution available, seem to correlate with major biotic events, such as a middle Ediacaran extinction of acanthomorphic acritarchs, first appearance of macroscopic animals at ca. 575 Ma, and extinction of the Ediacaran and weakly calcifying biota at the Precambrian-Cambrian boundary (Figure 10.5). Thus, it is clear that a more detailed interrogation of the history of biogeochemical changes at this time will require closely allied stratigraphic, geochemical, and palaeontological studies.

## *Acknowledgements*

This contribution benefits from the first author's recent discussions with M. Ader, N. Butterfield, C. Gaucher, J. Gehling, P. Hoffman, D. Johnston, F. Macdonald, D. McKirdy, N. Swanson-Hysell, W. Preiss, and R. Trindade, and past discussions with colleagues too numerous to list. G. Shields and S. Xiao provided constructive reviews of the manuscript. This chapter is a contribution to IGCP projects 478 and 512. GPH's and ASC's contribution forms TRaX record No. 1.

# PART 11 Neoproterozoic-Cambrian Palaeoclimatic Events

# Neoproterozoic to Cambrian Palaeoclimatic Events in Southwestern Gondwana☆

A.J. Kaufman[1], A.N. Sial[2], H.E. Frimmel[3] *and* A. Misi[4]

## Contents



## 11.1.1. Constructing a Global Record of Neoproterozoic Palaeoclimatic Variations

Since publication of the 'Snowball Earth' hypothesis – initially by Kirschvink (1992a) based on an apparently robust equatorial palaeolatitude for glacial strata in the Neoproterozoic Elatina Formation of Australia, and later by Hoffman et al. (1998) based on their high-resolution stratigraphic investigation of the carbonate-dominated Otavi Group in Namibia – there has been an exponential growth in the number of publications focusing on the oscillating record of extreme climate change in the Cryogenian interval (generally constrained between 750 and 635 Ma) and through the Ediacaran and Cambrian periods. Many of these studies have revealed remarkable litho- and chemostratigraphic similarities in broadly equivalent, but widely separated successions (e.g. Halverson et al., 2005a).

The lack of a rigorous biostratigraphic framework through much of this time interval and the general absence of reliable radiometric age constraints, however, has led to the use of temporal carbon isotope trends (often constructed from single and likely incomplete sections) as a means of dating Neoproterozoic sedimentary successions through

☆Kaufman, A.J., Sial, A.N., Frimmel, H.E., Misi, A., 2009. Neoproterozoic to Cambrian palaeoclimatic events in southwestern Gondwana. In: Gaucher, C., Sial, A.N., Halverson, G.P., Frimmel, H.E. (Eds.): Neoproterozoic-Cambrian Tectonics, Global Change and Evolution: a focus on southwestern Gondwana. Developments in Precambrian Geology, 16, Elsevier, pp. 369–388.

[1] Departments of Geology and ESSIC, University of Maryland, College Park, MD 20742-4211, USA.
[2] NEG-LABISE, Departamento de Geologia, Universidade Federal de Pernambuco, C.P. 7852, Recife 50670-000, Brazil.
[3] Department of Geological Sciences, University of Cape Town, Rondebosch 7700, South Africa.
[4] Instituto de Geociências, Universidade Federal da Bahia, CEP 40170-290 Salvador-Bahia, Brazil.

*Developments in Precambrian Geology*, Volume 16
ISSN 0166-2635, DOI 10.1016/S0166-2635(09)01626-0

comparison with composite curves (e.g. Jacobsen and Kaufman, 1999). This is problematic insofar as carbon isotope variations, especially across glacial divides, are inherently oscillatory, the magnitude of $\delta^{13}C$ anomalies can vary between facies, and the age models used for the composites require periodic tuning (cf. Melezhik et al., 1999). Furthermore, chemostratigraphic correlations are complicated by the discovery of multiple diamictite/cap carbonate couplets in individual sections, the incomplete nature of the stratigraphic record in glaciated marginal marine environments and tectonic factors that either omit or duplicate strata. Further complicating the interpretation of Cryogenian to Cambrian palaeoclimatic variations are suggestions that the negative $\delta^{13}C$ excursions are diagenetic artefacts, or are the result of strong surface-to-deep carbon isotope gradients in seawater (Jiang et al., 2007; Swart, 2008). If the latter is true, the profound isotopic variations documented through the strata in any of the marginal marine successions may have more spatial than temporal significance.

Our ability to reconstruct a Neoproterozoic palaeoclimatic history from southwestern Gondwana depends fundamentally on accurate regional and global correlation of Cryogenian glacial diamictites. For example, in Namibia the correlation of diamictites between the Congo and Kalahari cratons has been a long standing geological problem because of the tectonic fragmentation and reassembly of the Damaran Orogen into several tectonostratigraphic zones (Martin, 1965; Miller, 1983; Hoffmann, 1989; see Part 5).

In the absence of temporally significant fossils and radiometric constraints, carbon isotope stratigraphy was specifically applied to the correlation of carbonates above and below the Namibian glacial deposits (Kaufman et al., 1991). These early results revealed profound swings in $^{13}C$ abundances across the glacial divides with a stratigraphic pattern of strongly negative values in a 'cap carbonate' above diamictites on the Congo Craton in the Otavi Mountainland. The sub-Maieberg diamictite was previously thought to be associated with the Chuos Formation, but later regional correlation led to a revised subdivision of the Otavi Group that included two discrete levels of glacial deposits and the re-assignment of the diamictite/cap carbonate couplet studied in 1991 as the Ghaub and Maieberg formations (Hoffmann and Prave, 1996). Assuming the strongly negative $\delta^{13}C$ anomalies were temporally significant and reflected unusual post-glacial ocean chemistry, the isotopic results predicted the stratigraphic position of a second ice age in the Witvlei Group (Bildah and LaFraque members of the Buschmannsklippe Formation: Hegenberger, 1993) on the Kalahari Craton to the south (Kaufman et al., 1991; Saylor et al., 1998).

While the two glacial events in the Otavi Group were originally ascribed to the Sturtian epoch (Kaufman et al., 1997; Hoffman et al., 1998), the stratigraphically higher Ghaub diamictite below the Maieberg Formation cap carbonate was later re-classified as a Marinoan archetype (Hoffman and Schrag, 2002) with the carbonate containing a suite of unusual textural, sedimentologic and isotopic characteristics of supposedly temporal significance. In this context there is a general assumption that the two levels of glacial diamictite found in each of the post-rift carbonate platform successions on the Congo and Kalahari cratons are equivalent, representing Sturtian (Chuos, Blaubeker and Kaigas diamictites) and Marinoan (Ghaub, Blässkranz and Numees diamictites) events.

Tests of these lithologic and carbon isotope correlations in Namibia, and evaluation of the significance of sedimentary features as temporal markers require independent isotopic constraints and stratigraphically significant radiometric dates. Furthermore, it is important to correlate the low grade successions on the foreland platforms with metamorphosed equivalents in the fold and thrust belts of the Damara Orogen. In these internal mobile zones (Outjo, Swakop and Hakos) two levels of glacial diamictite have been recognised, but the scale and magnitude of thrusting and nappe displacement is poorly constrained, thus making the direct equivalence of diamictites in the metamorphosed portion of the Damara with unmetamorphosed units in the external zones problematic (Hoffmann, 1989).

Correlation problems in Namibia and elsewhere are now further complicated by the recognition of diamictites from a post-Marinoan Ediacaran Period glacial epoch referred to as the Gaskiers (ca. $582 \pm 0.4$ Ma: Bowring et al., 2003b); these are best expressed in Newfoundland where they are overlain locally by a thin cap carbonate (Myrow and Kaufman, 1999) with $\delta^{13}C$ values as low as −8‰. Notably, this couplet has been tentatively correlated around the world to a profound negative $\delta^{13}C$ anomaly (to a nadir of −10‰ and lower) in carbonates preserved on several continents – called the Shuram event (Fike et al., 2006) – but besides the Gaskiers itself underlying glacial diamictites are currently known from only one example in Brazil (Alvarenga et al., 2007) and potentially one in Norway and the north Atlantic region (Halverson et al., 2005a) and possibly also in South Africa (Gaucher et al., 2005a).

In this review, we re-evaluate the correlation of palaeoclimatic events in the Neoproterozoic of southwestern Gondwana, including the Congo and Kalahari cratons in Africa, in light of new radiometric ages, field observations and chemostratigraphy in those regions, as well as those from the São Francisco, Amazon and Rio de la Plata cratons in South America. Combined with temporal constraints from Cryogenian successions external to southwestern Gondwana and recent observations of stratigraphic anomalies, this compilation of palaeoclimatic events provides insight to the number and duration of ice ages during the Neoproterozoic glacial epoch.

## 11.1.2. Age Constraints for Cryogenian Glacial Deposits in Southwestern Gondwana

The global distribution of Cryogenian diamictites (Harland, 1964) and their stratigraphic position within individual basins suggests that there were two main Neoproterozoic glacial epochs, which have come to be known as the Sturtian and Marinoan ice ages. When new age constraints on the Gaskiers diamictite in Newfoundland came to light, a third discrete glacial epoch was added to the Cryogenian inventory (Figure 11.1.1). Snowball Earth-like conditions have been proposed for the Sturtian and Marinoan events, but the narrow extent of Gaskiers-age glacial deposits suggests that this Ediacaran Period ice age may have been more regional in scope.

### 11.1.2.1. Sturtian

In Namibia on the Congo Craton, the Sturtian Chuos diamictite is constrained to be <746 Ma based on the age of a feeder dyke to a porphyry lava upon which the glaciogenic Chuos deposits in the Sumas Mountains rests (Hoffman et al., 1996).

The correlative Kaigas diamictite in the Port Nolloth Terrain is associated with felsic rift-related volcanic rocks (Rosh Pinah Formation) from which single zircon Pb-Pb and U-Pb ages of 741 ± 6 (Frimmel et al., 1996c) and 752 ± 6 Ma (Borg et al., 2003) have been obtained. The felsic volcanic rocks of the Rosh Pinah Formation rest above the Kaigas Formation (Frimmel, 2008) and thus provide a minimum age constraint for the Kaigas diamictite. The maximum age of sedimentation in the Gariep Basin is given by the youngest age obtained on

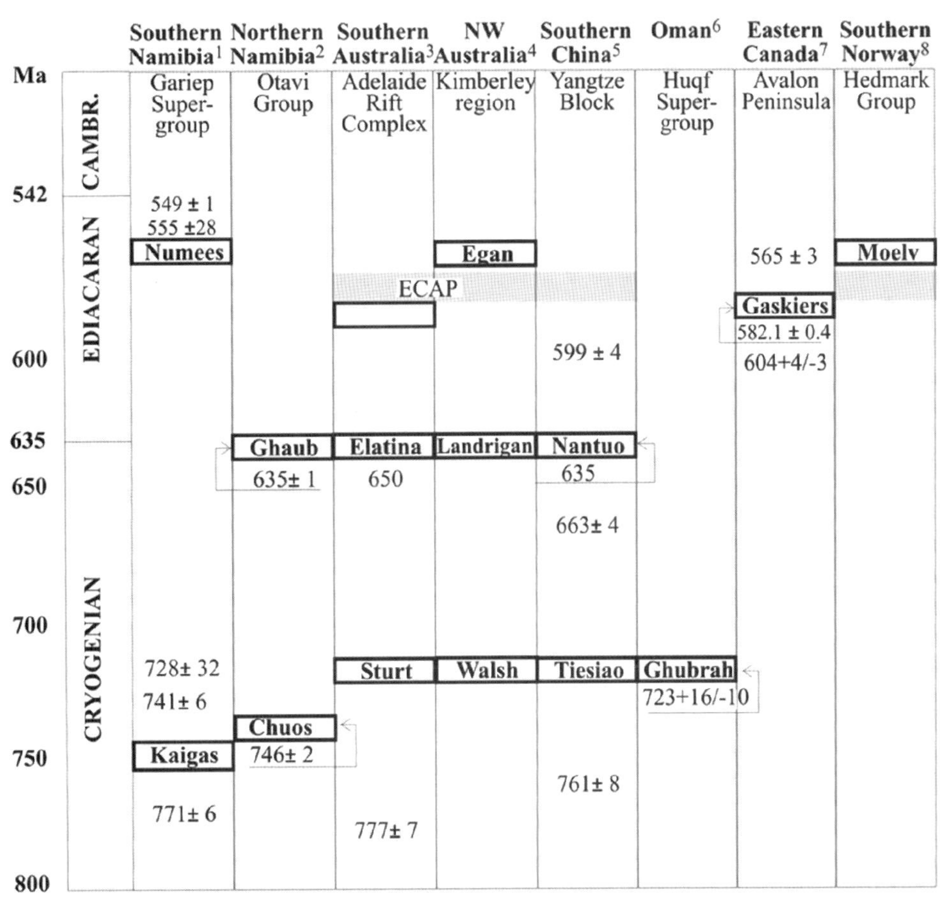


**Figure 11.1.1** Stratigraphic subdivision of the proposed Cryogenian and Ediacaran periods. Glacial units within selected late Neoproterozoic succession and their radiometric age constraints are shown to scale (Gaucher et al., 2005a). Except for Ghubrah, Ghaub and Gaskiers that have been directly dated (Brasier et al., 2000; Bowring et al., 2003b; Hoffmann et al., 2004), correlations for all other glacial units is interpretative (modified from Gaucher et al., 2005a and references therein). ECAP, Ediacaran Complex Acanthomorph Palynoflora (Grey et al., 2003). *Sources*: (1) Frimmel et al. (1996c); Fölling et al. (2000); Grotzinger et al. (1995); (2) Hoffmann et al. (2004); (3) Ireland et al. (1998), Preiss (2000); (4) Grey and Corkeron (1998); (5) Zhou et al. (2004a); (6) Brasier et al. (2000); (7) Myrow and Kaufman (1999); Bowring et al. (2003b); (8) Vidal and Nyestuen (1990).

## 11.1.3.4. Lithologic and isotopic observations of Sturtian ice ages in southwestern Gondwana

### 11.1.3.4.1. Congo Craton

The Sturtian Chuos Formation in Namibia is typically developed above continental rift-related siliciclastic deposits of the Nosib Group or pre-Damaran basement and is present in the Central and Eastern Kaoko Zone, the Northern Platform as well as the Northern and southern Central Zone of the Damara Belt (see Part 5). It has not been found to date in the southern parts of the Damara Belt. The thickness is variable and in places, a fluvioglacial origin is indicated. A characteristic feature of this formation is locally well-developed beds or lenses of iron-formation or iron-cemented diamictite with dropstones. Where present, the iron-formation is typically found in the lower parts of the unit, whereas Fe-enrichment at the top is related to secondary enrichment on palaeosurfaces. Thin intercalated beds of massive, white, very fine-grained dolomicrite reflect breaks in glacial deposition. In the western part of the southern Central Zone of the Damara Belt, the Chuos Formation partly consists of graded mass-flow to turbiditic arenaceous beds with dropstones in the fine-grained tops. The formation oversteps the underlying Rössing Formation (see Part 5) in a southerly and easterly direction. From this onlap it is concluded that the northern rift shoulder of the southern rift graben in the Damara Belt was still exposed during Chuos times.

In northern Namibia glacial deposits of the Sturtian Chuos Formation are sandwiched by carbonates that preserve the noted positive-to-negative $\delta^{13}C$ anomaly (Figure 11.1.2; Hoffman et al., 1998; Hoffman and Halverson, 2008). The post-glacial Rasthof cap carbonate is characterised by deep water rhythmites composed of authigenic limestone cements (with $^{87}Sr/^{86}Sr$ compositions as low as 0.7066) and allodapic grainstones overlain by a dark microbial reef with characteristic roll up structures. The Rasthof carbonates begin with $\delta^{13}C$ values near −5‰ at the base and quickly rise through 0 to a plateau of +5‰ in as few as 15 m of section, associated with shallowing of the basin and deposition of microbialaminite and stromatolites facies. It is likely that the rise in $\delta^{13}C$ compositions and wholesale precipitation of carbonate may be directly related to seafloor shallowing into the photic zone and the sudden bloom of photosynthetic mats. By rapidly drawing $^{12}CO_2$ out of seawater and pumping reduced carbon into sediments, the activity of the benthic mats would have increased pH and supersaturated oceanic alkalinity. This allowed for carbonate precipitation from evolving seawater with progressively more enriched $^{13}C$ compositions.

The Naos Formation in the Southern Marginal Zone (Hakos Terrain) of the Damara Belt, which is composed of metadiamictite interbedded with micaschist, amphibolite and quartzite is currently regarded as correlative of the Ghaub Formation. Earlier lithostratigraphic correlations, however, equated this metadiamictite with supposedly younger diamictites of the Blässkranz Formation in the nearby Naukluft Nappe Complex (Hoffmann, 1989). To date isotopic studies of overlying carbonates in the overlying Melrose/Samara formations have not been conducted. Locally, volcano-exhalative siliceous iron formation occurs in proximity to some of the amphibolite bodies. The entire succession in the SMZ has been interpreted as incipient passive continental margin deposits laid down at the end of continental break-up in the Southern Zone (see Chapter 5.3).

Other likely Sturtian diamictites on the Congo Craton include the Grand Conglomerat of the Nguba Group (Lufilian Arc in the Democratic Republic of Congo and Zambia), which is overlain by the Calcaires du Kokontwe limestone. The diamictite consists of fine-grained argillitic matrix, with scattered quartz and lithic clasts, including carbonate, biotite, clay and black shale. A U-Pb SHRIMP zircon age of zircons in Ngumba Group volcanic unit indicates that the Grand Conglomerat is younger than $760 \pm 5$ Ma (Key et al., 2001). Carbonate of the Calcaires du Kokontwe above the diamictite in two cores show upsection depletions in $^{13}C$ ranging from −2.4 to −4.7‰ and −3.6 to −6.2‰, respectively (Bodiselitisch et al., 2005; Bodiselitisch, personal communication, 2007). The lowest $^{87}Sr/^{86}Sr$ isotopic ratios of equivalent post-glacial (Bas Congo mixtite inférieuer) limestone from the 'Haut Shiloango' Subgroup in the West Congo foreland are ~0.7068 (Frimmel et al., 2006; Poidevin, 2007), which match well with those in the Rasthof Formation of northern Namibia.

### 11.1.3.4.2. Kalahari Craton

The Blaubeker Formation in the Southern Foreland of the Damara Belt on the shores of the Kalahari Craton may similarly be a correlative of the Chuos Formation but it lacks iron-formation. A dark finely laminated dolomicrite near the base of the Court Formation and above the diamictite preserves moderately negative $\delta^{13}C$ compositions (Kaufman et al., 1991; Saylor et al., 1998); it is considered a likely cap carbonate lithofacies.

In the Gariep Belt, the older diamictite is known as the Kaigas Formation, which takes a similar stratigraphic position as the Chuos Formation. Lithologically the Kaigas Formation differs from the Chuos by lacking iron-formation and carbonate intercalations. Most of the formation is not a typical tillite but a poorly sorted conglomerate, in places with upwards fining graded bedding and has a fluvioglacial signature. Whether the Kaigas deposits reflect a further, older, global glacial event, remains uncertain. As the Kaigas Formation diamictite is not developed in open-marine palaeoenvironments but only locally, at near-shore river mouths and/or entry points of land-based glaciers, it is possible that this formation was formed as consequence of only regional glaciation. Carbon isotope analyses of the thin (~2 m) dark-blue to grey, finely laminated dolomicrite of the Pickelhaube cap carbonate, as well as minor limestone and dolomite intercalated with arenite and shale, reveal continuously negative $\delta^{13}C$ values for ~100 m above the glacial contact (Fölling and Frimmel, 2002). The lowest $^{87}Sr/^{86}Sr$ values recorded in two sections of the Pickelhaube Formation are ~0.7071 and ~0.7073.

#### 11.1.3.4.3. São Francisco Craton

On the opposite shore of the Adamastor Ocean, Sturtian equivalent diamictites – variably known as the Bebedouro, Jequitaí, Carandaí or Carrancas formations (Karfunkel and Hoppe, 1988; Martins-Neto et al., 2001) – floor the flat-lying Bambuí and Una groups on the São Francisco Craton in Brazil. In the Irecê Basin, the iron-rich Bebedouro diamictite (Figure 11.1.3A) is overlain by a thinly laminated reddish dolomicrite at the base of the Salitre Formation, an equivalent of the Sete Lagoas Formation in the Bambuí Basin to the south. It is likely that the very first isotopic analysis of post-glacial cap carbonate anywhere came from the study of this deposit (Torquanto and Misi, 1977), although these authors then interpreted the −5‰ carbon isotope values as reflecting a lacustrine environment of deposition. Above the basal cap level, $\delta^{13}C$ values of carbonates in the Salitre and the equivalent organic-rich Sete Lagoas Formation rise sharply to near +8‰ in organic-rich microbialaminites (Figure 11.1.3B; Iyer et al., 1995; Misi and Veizer, 1998; Santos et al., 2000).

### 11.1.3.5. Lithologic and isotopic observations of Marinoan ice ages in southwestern Gondwana

#### 11.1.3.5.1. São Francisco Craton

Lithologic and geochemical evidence for a second Bambuí Group ice age is found stratigraphically higher near the top of the Sete Lagoas Formation. Associated with this ice age is a fall in $\delta^{13}C$ of marine limestones by as much as 15‰ to values as low as −5‰ in the Pedro Leopoldo facies (cf. Peryt et al., 1990; Iyer et al., 1995; Babinski et al., 2007; Misi et al., 2007), a remarkable outcrop of neomorphosed aragonitic seafloor cements (akin to those in the Keilberg Member of the Maieberg Formation: Figure 11.1.3C) and limestone rhythmites (Figure 11.1.3D). In the type locality at Sambra Quarry the Pedro Leopoldo facies sits on a basement high, but in nearby Inhaúma a basement clast dominated diamictite was recently discovered beneath a thick clay interval and a thinly laminated pink cap dolostone. This post-glacial carbonate contains precipitate fabrics and an upward trend to more negative $\delta^{13}C$ values starting around 0‰. In contrast, fans and draping micrite at the base of the Sambra Quarry exposure start with $\delta^{13}C$ values near −5‰ and then rise rapidly to near +1‰ along with evidence of sudden shallowing at 10.5 m above the quarry floor (Misi et al., 2007).

A notable negative carbon isotope excursion is also recognised in the correlative Salitre Formation associated with the informal B1 member (Misi and Veizer, 1998) and the deposition of phosphate-rich stromatolites described as *Jurussania krilov* (Figure 11.1.3E; Misi and Kyle, 1994). Biostromes of similar but somewhat larger plumb stromatolites are preserved as organic-rich limestone (Figure 11.1.3F) in the overlying Serra de Santa Helena Formation. These structures are indistinguishable from those near the base of the Maieberg Formation cap carbonate, but have $\delta^{13}C$ values near to +10‰. Analyses of well-preserved high Sr samples show $^{87}Sr/^{86}Sr$ values tightly bunched around 0.7072–0.7074 in the ex-aragonite seafloor precipitates at Sambra Quarry and in the equivalent horizon in Irecê.

Based on remarkable textural similarities (i.e. seafloor aragonite precipitates) and an exact match of Sr isotopes (~0.7073) in the high Sr precipitate fabrics, the pink cap carbonates at Inhuama above the diamictite and the crystal fans at Sambra Quarry are most likely to be a direct equivalent to the Maieberg Formation. The Pb-Pb age constraint (740±22 Ma: Babinski et al., 2007) for the aragonite fans at Sambra Quarry, however, would place the two units convincingly in the Sturtian epoch, and not the Marinoan (cf. Hoffman and Schrag, 2002). While this eleven point Pb-Pb isochron with notably low MSWD is compelling, the age should be considered tentative at present given the lack of complementary U-Pb analyses on the same samples.

#### 11.1.3.5.2. Congo Craton

In Namibia, one of the most precisely dated examples of Marinoan glacial deposits occurs in the southern Central Zone of the Damara Belt. Hoffmann et al. (2004) obtained a precise U-Pb zircon age of 635.5±1 Ma for a thin ash bed near the top of a metadiamictite unit that these authors correlate with the well-developed glaciomarine Ghaub Formation that occurs all along the Northern Platform, including the Eastern Kaoko Zone, Northern Marginal Zone, Northern Zone and Central Zone of the Damara Belt (see Chapter 5.3). Based on this correlation the above age is typically referred to as the most precise constraint on the age of the Ghaub Formation in southwestern Africa.

The Ghaub glaciation led to a major drop in sea level, estimated to be as much as 400 m, which resulted in exposure and widespread karsting of the Northern Platform. Consequently, the Ghaub Formation diamicite is not

**Figure 11.1.3** Photographs from the Neoproterozoic Bambuí Group and equivalents, Brazil. A: Iron cemented diamictite of the Bebedouro Formation; B: organic-rich microbialaminite limestone of the Sete Lagoas Formation with synaeresis (?) cracks; C: tufted ex-aragonite precipitates and draping micrite from the Pedro Leopoldo facies at Sambra Quarry; D: limestone rhythmite from the uppermost Sete Lagoas Formation at Paraíso Quarry; E: phosphate-rich stromatolites classified as *Jurussania krilov* from the B1 member of the Salitre Formation, Una Group (Misi and Kyle, 1994); F: limestone tubestone stromatolites from the Serra de Santa Helena Formation.

developed everywhere but is either very thin or missing over large areas. As the Ghaub glaciation took place during a tectonically less active period compared to the Chuos glaciation, the Ghaub Formation is lithologically not as variable (Hoffman and Halverson, 2008) being composed primarily of carbonate clasts. Iron enrichment is notable only in the dropstone-laden rhythmite immediately beneath the Maieberg cap, which is distinctly pale cream to pink, lean in organic carbon, and has characteristic sheet crack cements, graded laminations of clastic carbonate, plumb stromatolites, megaripples (Allen and Hoffman, 2005), barite and ex-aragonite seafloor precipitates.

The negative biogeochemical anomaly in the Maieberg cap carbonate contrasts with the Rasthof anomaly in both its trend and apparent duration. Basal Maieberg dolomicrites begin at −3‰ and then fall to −5‰ in overlying limestone rhythmites before rising over hundreds of metres upsection to near 0‰, as the shelf shallowed and was finally exposed (Hoffman et al., 1998; Hoffman and Schrag, 2002). Strontium isotope compositions of ex-aragonite limestone and rhythmite from deep water Maieberg lithofacies yield values of ca. 0.7073 (Kaufman et al., 1997; Halverson et al., 2007a), which are an exact match with the identical facies of the Sete Lagoas Formation at Sambra Quarry (and the high Sr limestones from the Pickelhaube cap carbonate in southern Namibia) supporting their direct equivalence.

Finally, the Petit Conglomerat of the Kundelungu Group in the northeastern Congo Craton and its equivalents in the Democratic Republic of Congo and Zambia is considered a Marinoan glacial diamictite (Bodiselitisch et al., 2005). The Petit Conglomerat overlies the Nguba Group with an erosional unconformity (Wendorff, 2003) and contains abundant faceted and striated clasts of both intrabasinal and extrabasinal origin. The diamicite is overlain by a 10 m thick fine-grained cap dolostone known as the Calcaire Rose. Carbon isotopes in this cap progressive decrease upsection from ca. −1.8 to −4.1‰ (Bodiselitisch et al., 2005; personal communication, 2007), but in the West Congo Belt the equivalent carbonates of the Schisto-Calcaire Subgroup reveal extremely variable $\delta^{13}C$ ratios ranging between −3 and +9‰ and a lowest $^{87}Sr/^{86}Sr$ value from Sr-rich limestone of ~0.7074 (Frimmel et al., 2006; Poidevin, 2007).

### 11.1.3.5.3. Kalahari Craton

Other potential Marinoan diamictites in Namibia include the Blässkranz Formation preserved in the Naukluft Nappe Complex on the Kalahari Craton. In the Naukluft the diamictite is overlain by the Tsabisis cap carbonate, which is a dark rhythmic limestone with negative $\delta^{13}C$ compositions and lowest $^{87}Sr/^{86}Sr$ values of ~0.7081. This value is identical to those in high Sr precipitate fabrics from the correlated LaFraque Member of the Buschmannsklippe Formation in the nearby Witvlei Group (Hegenberger, 1993; Saylor et al., 1998). It is also an exact match for high Sr rhythmites in transgressive limestone above the basal Doushantuo cap dolostone (Jiang et al., 2007). Analogous to stratigraphic relations seen in the Maieberg cap carbonate, plumb stromatolites in the Bildah Member underly ex-aragonite crystal fans of the LaFraque Member. Notably, the negative $\delta^{13}C$ values continue for a considerable stratigraphic throw (also like the Maieberg) continuing throughout the Buschmannsklippe Formation and stepping over a significant unconformity into the fossiliferous Nama Group (Kaufman et al., 1991; Saylor et al., 1998).

### 11.1.3.5.4. Amazon Craton

On the Amazon Craton in the northern Paraguay Belt of Brazil, glacial diamictite of the Puga Formation is generally believed to be Marinoan in age, which is consistent with a new Pb-Pb carbonate age of 633±25 Ma for the Mirassol d'Oeste cap (see Chapter 2). The glacial deposit comprises thick sheets of pebbly siltstone and diamictite that accumulated in a glaciomarine setting (Alvarenga and Trompette, 1992). The Puga is overlain by transgressive carbonates of the Araras Group (Nogueira et al., 2003, 2007b; Riccomini et al., 2007). The contact between the Puga and overlying Mirassol d'Oeste cap carbonate is highly irregular (Figure 11.1.4A), in contrast to the knife-sharp surface recognised across glacial terminations elsewhere, suggesting that at the time of emplacement of the cap the underlying diamictite was still soft (Nogueira et al., 2003). This observation supports the view of a rapid icehouse to greenhouse transition, and has similarly been noted at the contact between the Jequitaí diamictite and basal Sete Lagoas cap carbonate near Belo Horizonte in south central Brazil on the São Francisco Craton (Figure 11.1.4B).

The Mirassol d'Oeste cap dolostone and lowermost Guia Formation limestone atop the Puga diamictite contains many of the same suite of lithologic and textural features found in the Maieberg (sheet crack cements, plumb stromatolites and megaripples in the dolostone, and seafloor crystal fans in the organic-rich limestone facies) (Nogueira et al., 2003; Trinidade et al., 2003); barite was recently discovered in a horizon between the stromatolites and fans as predicted by their position relative to these features in similar cap lithofacies in Namibia and arctic Canada interpreted to be characteristic of Marinoan (~635 Ma) glaciation (Hoffman and Schrag, 2002).

A high-resolution carbon isotope study of the Mirassol d'Oeste cap dolostone reveals highly negative $\delta^{13}C$ values at the base of the unit (ca. −9‰) that rises to a plateau of −4 to −5‰ in the plumb stromatolite dominated interval (Font et al., 2006). Similar values characterise the precipitate interval at the base of the Guia Formation. The lowest $^{87}Sr/^{86}Sr$ value recorded in high Sr precipitate limestone from this unit is 0.7074 (Nogueira et al., 2003; Trinidade et al., 2003) supporting its equivalence with similar precipitates in the Sete Lagoas and Maieberg cap carbonates on the São Francisco and Congo cratons, respectively.

### 11.1.3.5.5. Rio de la Plata Craton

In the southern Paraguay fold belt of Brazil (state of Mato Grosso do Sul) on what some interpret to be the northern margin of the Rio de la Plata Craton (Gaucher et al., 2003, 2008b) the Puga diamictite in its type section and a thin cap dolostone are overlain by siliciclastics and carbonates of the Corumbá Group. Other researchers

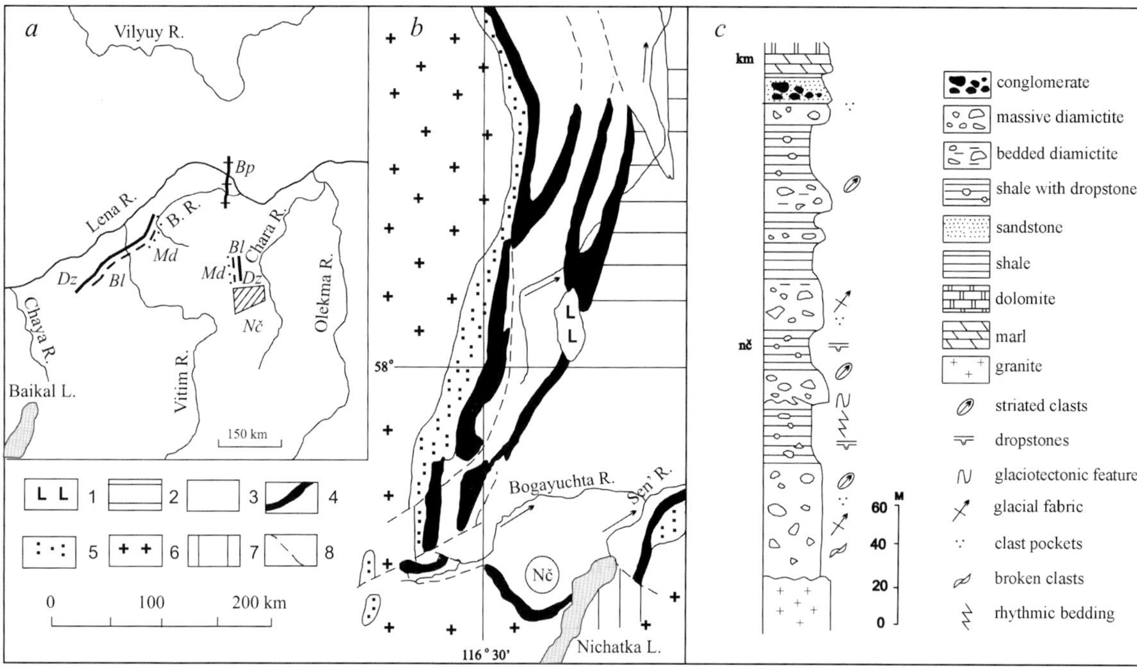


**Figure 11.2.2** (a) Occurrences of Neoproterozoic diamictites in the North Baikal and Patom highlands: Bp — Bol'hoy Patom Fm.; Dz — Dzhemkukan Fm.; Bl — Ballaganakh Group; Md — Medvezhevka Fm.; Nč — Nichatka Fm.; B.R. — Bol'shoy Patom River. (b) Geological scheme of the South Centrocline of Berezovskaya basin: (1) Mesozoic syenite-porphyry; (2) Tinnaaya, Zherba, Chencha and Nikol'skoe Fms; (3) Sen' and Kumakh-Ulakh Fms; (4) Nichatka Fm.; (5) pre-Nichatka Neoproterozoic deposits; (6) Palaeoproterozoic and Archaean granites; (7) Archaean gneisses and granites; (8) main thrusts and upthrusts. Nč — type Nichatka Fm. section. (c) The type section of the Nichatka Fm.

#### 11.2.2.3.2. India

Slightly metamorphosed Neoproterozoic deposits (in ascending order) of the Blaini and Infra Krol formations and the Krol Group are exposed in the Lesser Himalayas. The diamictite-bearing Blaini Formation unconformably overlies metamorphosed deposits of the Simla Group, traditionally referred to as the Tonian or lower Cryogenian. Diamictites occur at the base and in the upper part of the Blaini Formation. They are separated by a thick package of shales and sandstones. The lower diamictite (few to 400 m in thickness) consists of three or four, lenticular, massive diamictite beds and orthoconglomerates. Locally, diamictite beds bear striae and abrasion marks. At a few localities diamictites contain facetted and striated clasts and elsewhere distinct glacial fabrics of elongated clasts (Jain et al., 1981). Using sequence-stratigraphic analysis, Jiang et al. (2003) correlated the lower diamictitic member of the Blaini Formation to the middle Cryogenian Jiangkou and Gucheng formations of South China.

#### 11.2.2.3.3. Oman

Diamictites of the Ghubrah Formation occur at the base of the Abu Mahara Group in northern Oman. Ages of $712 \pm 1.6$ and $723+16/-10$ Ma (U-Pb on zircons; Rieu et al., 2007), obtained for ash beds within the diamictite, indicate a middle Cryogenian age. In southern Oman, the basal Ayn Formation of the Mirbat Group contains diamictites which yield a maximum, detrital zircon age of ca. 710 Ma. The diamictites unconformably overlie basement rocks dated as 700–750 Ma. It is thus inferred that diamictites of the Ayn and Ghubrah formations are more or less similar in age (Rieu et al., 2007). At some localities the Ayn diamictites overlie striated pavements and contain clasts denoting glacial abrasion. They alternate with bedded deposits bearing dropstones, and are in turn overlain by cap dolomites. The unit is interpreted as glacial, or proximal and distal glaciomarine deposits (Kellerhals and Matter, 2003).

## 11.2.3. Late Cryogenian Glaciations

### 11.2.3.1. East Asia

#### 11.2.3.1.1. South China

In the Yangtze Platform Ediacaran-late Cryogenian successions are well exposed and well studied. They can serve as a real standard for this stratigraphic interval (in contrast to the formally recognised Australian sections).

Upper Cryogenian (Lower Vendian) glacial deposits of the Nantuo Formation are well known regarding its stratigraphic position, isotopic and facies composition, and palaeogeographic evolution (Zhu et al., 2007, and references therein). The formation overlies with erosional unconformity the Datangpo Formation, dated between 654 ± 3.8 Ma (U-Pb on zircon; Zhang et al., 2008b) and 663 ± 4 Ma (U-Pb on tuff zircons; Zhou et al., 2004b). An ash bed in the lower Nantuo Formation yielded a U-Pb zircon age of 636.3 ± 4.9 Ma (Zhang et al., 2008b). The unit is composed of massive and, less frequently, bedded diamictites with subordinate, thin mudstone intercalations. Clasts comprise sedimentary, metamorphic and igneous rocks, and are variously rounded, frequently bullet-shaped, facetted and bear glacial striations. The diamictites are mostly basal tillites (Dobrzinski and Bahlburg, 2007), and are in turn overlain by cap dolomites which form the basal member of the predominantly carbonatic Doushantuo Formation. Ash beds in the lower part of the formation yielded U-Pb zircon ages of 621 ± 7 Ma (Zhang et al., 2005) and 635 ± 1.5 Ma (Condon et al., 2005). Large, acanthomorphic acritarchs of the Pertatataka-Doushantuo assemblage first appear at this level, which become more diverse and abundant up section, where non-skeletal Metazoa and abundant bioturbation are preserved (see Chapter 9.1). The overlying Dengying Formation contains Cloudiniids and is overlain by the Zhujiaqing Formation, which contains at the base a small shelly fauna similar to that of the Nemakit-Daldyn Horizon in Siberia (Steiner et al., 2007; Zhu et al., 2007). Background positive $\delta^{13}C$ values (2–7‰) are punctuated by three significant negative anomalies in the Doushantuo and Dengying formations (Zhu et al., 2007). The first anomaly, named Cance and down to −5‰ PDB, is confined to the cap dolomite. The second and most significant anomaly, the Dounce (down to −11‰), occurs in the upper part of the Doushantuo Formation. The third anomaly, the Bace (−5‰ to −7‰), is at the level of the Nemakit-Daldyn Horizon. Usually, although without sufficient evidence, the Nantuo Formation is correlated to the Elatina Subgroup of Australia and therefore considered to be deposited during the Marinoan Glaciation of the latest Cryogenian.

### 11.2.3.2. North Eurasia

#### 11.2.3.2.1. Middle Siberia

Glacial deposits of inferred late Cryogenian age occur in the southern part of the Siberian Craton and its margins (Figure 11.2.2a). The Bol'shoy Patom Formation is the basal unit of the middle Dal'nyaya Tayga Group of the thick (over 5 km) Patom Supergroup (Ura Uplift; Figures 11.2.1 and 11.2.2a) in the northeastern edge of the Baikal-Patom folded arc (Chumakov et al., 2007). The Bol'shoy Patom Formation comprises an alternation of massive and bedded diamictites, graded-bedded diamictites, sandstones and thin-bedded turbidites bearing dropstones and till pellets. The formation is about 1 km in thickness. Elongate clasts of massive diamictites show glacial orientation. This allows the diamictites to be classified as tillites. Besides, the Bol'shoy Patom Formation includes debris flow and glaciomarine deposits. The formation unconformably overlies older deposits of the Patom Supergroup, and is conformably overlain by cap dolomites (5 m in thickness) and carbonate-siliciclastic deposits of the Barakun Formation. The latter is in turn overlain by the Ura Formation, which contains a rich assemblage of acanthomorphic acritarchs, including the genera *Ericiasphaera*, *Tanarium*, *Variomargosphaeridium*, *Appendisphaera*, *Dicrospinasphaera*, *Alicesphaeridium*, *Meghistrichosphaeridium* and *Sinosphaera* (Vorob'eva et al., 2008; see Chapter 9.1). This assemblage is very similar to early-middle Ediacaran microbiotas of postglacial formations, such as Doushantuo of South China, Pertatataka of Central Australia and Infra Krol of the Lesser Himalayas (see Chapter 9.1). The Ura microbiota is closest to the assemblage of the second palynozone of Grey (2005). The assignment of the upper part of Dal'naya Tayga Group to the Ediacaran (Vendian) has been confirmed by maximum ages of 600 ± 10 Ma provided by detrital zircon dating (U-Th-Pb, LA-ICPMS; Meffre et al., 2008) from the upper part of Khomolkho Formation, which is a stratigraphic equivalent of the Ura Formation (e.g. Zhadnova, 1961; Keller et al., 1967; Ivanov et al., 1995; Khomentovsky and Postnikov, 2001). Chemostratigraphic data (Pokrovsky et al., 2006) and the occurrence of Ediacaran acanthomorphic acritarchs (Vorob'eva et al., 2008) are consistent with this age.

Carbonates of the Dal'nyaya Tayga Group (excluding the cap dolomites) are characterised by positive $\delta^{13}C$ values (3–8‰) and increasing $^{87}Sr/^{86}Sr$ ratios from 0.7073 to 0.7077 up section. The overlying Zhuya Group yielded highly negative $\delta^{13}C$ values (−5‰ to −13‰) and higher $^{87}Sr/^{86}Sr$ ratios of about 0.7079 (Pokrovsky et al., 2006). Approximately 750 m above the Zhuya Group, small shelly fossils characteristic of the Nemakit-Daldynian Stage were found, which are immediately followed by the Tommotian Stage fauna of the Lower Cambrian (Khomentovsky et al., 2004). According to both palaeontological and isotopic data the Bol'shoy Patom Formation was assigned to the upper Cryogenian (Lower Vendian). In fossil content and isotopic data the Vendian (late Cryogenian-Ediacaran) and Lower Cambrian section of the Ura Uplift is similar to the successions of the Nantuo, Doushantuo, Dengying formations and the basal Cambrian of South China (Steiner et al., 2007; Zhu et al., 2007). The Dal'nyaya Tayga Group can be correlated to the Nantuo Formation and the lower part of the Doushantuo Formation, which is characterised by predominantly positive $\delta^{13}C$ values and Pertatataka-type acritarchs. The negative $\delta^{13}C$ anomaly of the Zhuya Group can be correlated to the Dounce negative $\delta^{13}C$ anomaly in the upper part of the Doushantuo Formation, and the Bace negative $\delta^{13}C$ anomaly at the base of the Zhujiaqing Formation may match the negative $\delta^{13}C$ anomaly at the base of the Nemakit-Daldyn Horizon.

To the southeast of the Ura Uplift on the western slope of the Aldan Shield, the Patom Group is noticeably reduced in thickness. The Bol'shoy Patom Formation is replaced by the Nichatka Formation (130–145 m thick) consisting of red diamictites, conglomerates, sandstones and thin-bedded mudstones with dropstones (Figure 11.2.2b and c). The diamictites contain erratic, numerous facetted and striated clasts, and elongate pebbles showing glacial orientation. The base of some diamictite beds is deformed by glaciotectonics. Sandstones form fans structures with large boulder conglomerates at the top. All these facts suggest that the Nichatka Formation is mainly composed of moraine tillites, glaciofluvial and glaciolacustrine deposits. The formation is conformably overlain by cap dolomite (up to 10 m thick) made up of thin intercalations of reddish-brown marls, marly dolomites and pink dolomites. The cap dolomite is the basal member of the Kumakh-Ulakh Formation. The Nichatka Formation was correlated to Bol'shoy Patom Formation of the Ura Uplift on the basis of geological mapping and was assigned to the Lower Vendian (late Cryogenian) as well (Chumakov et al., 2007).

To the southwest of the Ura Uplift, glacial deposits are exposed with some interruptions along the western margin of the Baikal-Patom folded zone, where they are named the Dzhemkukan Formation of the Patom Group and the Buguldeyka Member of the Baikal Group. A stratigraphic similarity of the Baikal Group and the upper part of the Patom Group was evidenced by Sr isotopic records (Kuznetsov and Letnikova, 2005) and Pb-Pb dating of limestones (550 ± 40 Ma; Kuznetsov et al., 2006). Geological and sequence-stratigraphic data favour correlation of the Buguldeyka Member with the glaciogenic Marnya Formation of the Sayan Mountain region, South Siberia (Sovetov and Komlev, 2005).

All glacial units of Middle Siberia listed above were grouped into the single Middle Siberian Glacial Horizon (Chumakov, 1993). Clast composition shows that the Siberian Craton was the main source for the glacial deposits of the Middle Siberian Glacial Horizon. These data and the great extension of glacial deposits along the margins of the Siberian Craton (2,000 km) suggest that the main glaciation centres occurred in the craton, and at least all of its southern part was covered by Middle Siberian ice sheets.

#### 11.2.3.2.2. The West Urals

In the Middle Urals upper Cryogenian glacial deposits are represented by the Tany Formation belonging to the base of the Serebryanka Group and, probably, by the overlying Koyva Formation (Figure 11.2.3b). These units show stratigraphic and facies similarity and are separated by a turbidite sequence (the Garevka Formation), which is noticeably reduced in thickness at some sections (e.g. at the Koyva River). The Tany Formation is composed of two massive diamictite layers separated by sandstones and shales, with a combined thickness of 800 m. The lower, thicker diamictite layer is overlain by 'cap dolomites' 8 m in thickness. The diamictites are made up of predominantly erratic clasts which increase in number and size to the southwest, towards the Russian Craton (Ablizin et al., 1982). Clasts include granite-gneiss blocks up to 3.5 m in diameter; the shale interbeds contain 'dropstones'. Diamictites of the Koyva Formation occur only to the north of the Sylvitsa River. In the Us'va section they are associated with hematitic ores. In the Koyva type section the diamictites are overlain by a 6-m thick cap dolomite, the latter characterised by negative $\delta^{13}C$ values (−3‰ to −5‰ PDB). Volcanic rocks of the Koyva Formation yielded ages of 569 ± 42 and 559 ± 16 Ma (Sm-Nd on clinopyroxene and Rb-Sr whole rock, respectively). The Sm-Nd age has a large error and the Rb-Sr age seems to be rejuvenated, because 1.5 km up section the upper half of the Sylvitsa Group contains abundant non-skeletal Metazoa and yielded an age of 557 ± 13 Ma (U-Pb SHRIMP dating of tuff zircon; Maslov et al., 2007). Somewhat lower in the Sylvitsa Group a fossil assemblage occurs, which is similar to the Miaohe biota of the uppermost Doushantuo Formation of South China (Grazhdankin et al., 2007). It should also be taken into consideration that the basal unit of the Sylvitsa Group is the glaciogenic Starye Pechi Subformation, which may be equivalent to the glaciogenic Gaskiers Formation of middle Ediacaran age (see Section 11.2.4.1). The Tany Formation unconformably overlies the Basegy Group, which yielded an age of 671 ± 24 Ma in its middle part (Rb-Sr and U-Pb zircon dating of trachytes; Petrov et al., 2005; Ronkin et al., 2007). Thus, the Tany Formation must be significantly older than 560 Ma and younger than 670 Ma, therefore it can be regarded as late Cryogenian and tentatively correlated to the Nantuo Formation of South China.

Possible stratigraphic equivalents of the Tany and Koyva formations in the Poludov Ridge, North Urals, are the Churochnaya Formation (Figure 11.2.3a) composed of typical glacial continental and glaciomarine shelf deposits (Chumakov, 1992) and the glaciomarine Lower Kurgashlya Subformation in the South Urals (Chumakov, 1998).

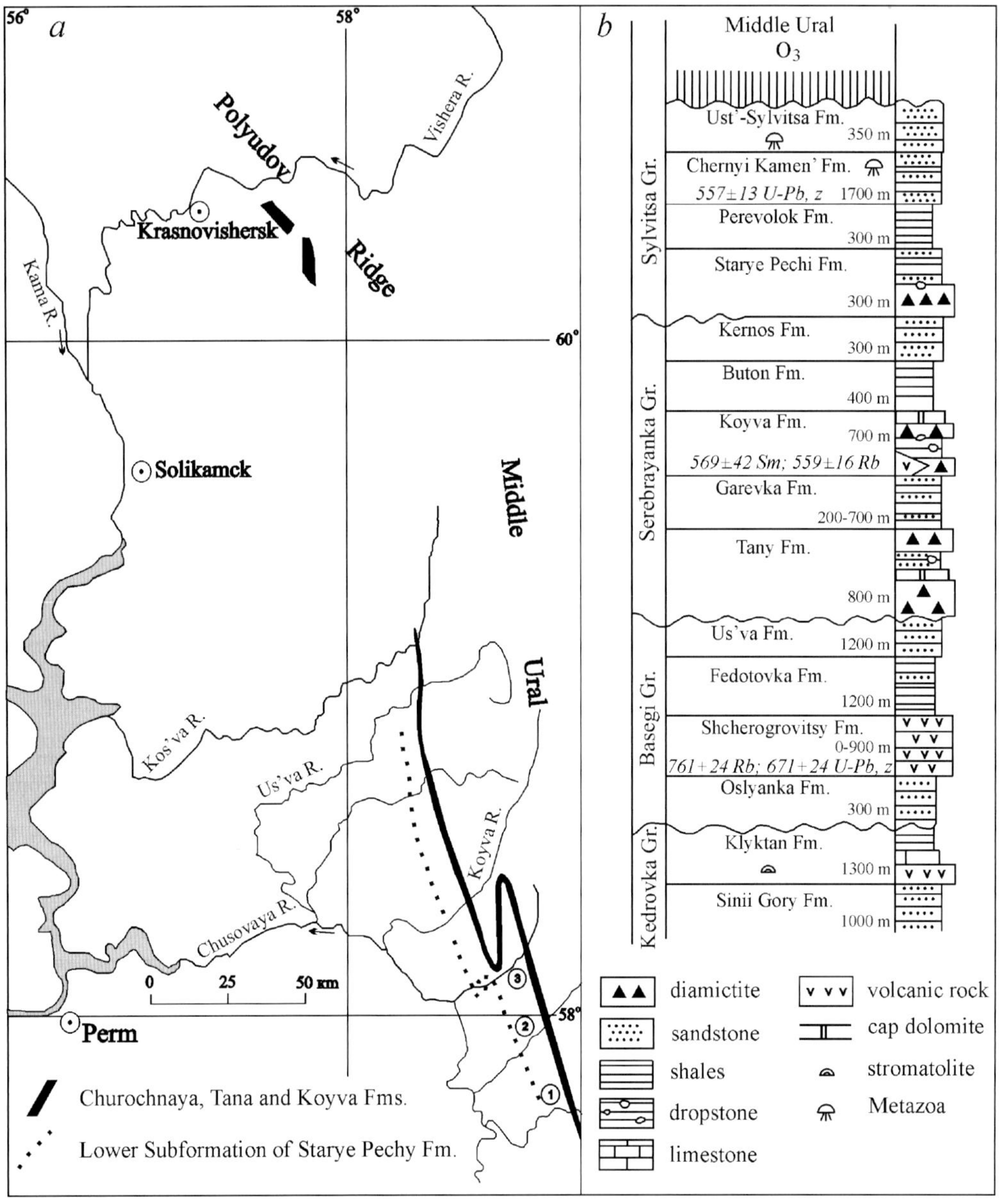


**Figure 11.2.3** (a) Occurrences of Neoproterozoic glacial diamictites in the North (Poludov Ridge) and Middle Urals. (1) Mezhevaya-Utka Rv.; (2) Serebryanka Rv.; (3) Sylvitsa Rv. (b) Composite stratigraphic section of the Middle Ural section. U-Pb, z — U-Pb age on zircon from tuffs; Rb — Rb-Sr age on volcanics, whole rock; Sm — Sm-Nd age.

### 11.2.3.2.3. East Europe

The composition of the Uralian diamictite clasts evidences that the crystalline basement and sedimentary cover of the Russian Plate were the main sources of the Churochnaya, Tany and Koyva formations. Extensive inland glacier sheets in the northeastern part of the East European Craton are evidenced by glacial units of Early Vendian (late Cryogenian) age buried in the Riphean-Early Vendian aulacogens. In the Orsha aulacogen of Byelorussia the Vil'chitsy Group contains two glacial units: the Lower Blon' Subformation below and the Glusk Formation above (Figure 11.2.4). They are separated by intervening sands of the Upper Blon' Subformation and by a deep erosional unconformity. Both glacial units are represented by continental facies, that is tills and tillites, thin-bedded glaciolacustrine varve-like muds with dropstones and glaciofluvial sands (Bessonova and Chumakov, 1969; Chumakov, 1981, 1992). The Vil'chitsy Group unconformably overlies the Riphean Poles'e Group (ca. 1,000 Ma) and is in turn unconformably overlain by the Volyn' volcanosedimentary succession (Figure 11.2.4). U-Pb ages of zircon from interbedded tuffs (Compston et al., 1995) and basalts (Shumlyanskii et al., 2006) of the Volyn' Group are $551 \pm 4$ and $549 \pm 29$ Ma, respectively. The Volyn' Group is unconformably overlain by the Mogilev-Podolsk Group containing abundant non-skeletal Metazoa (Velikanov et al., 1983). Whereas the stratigraphic position of the Lower Blon' Subformation is similar to that of late Cryogenian glacigenic deposits of

Babinski, M., Van Schmus, W.R. and Chemale, F., Jr., 1999. Pb–Pb dating and Pb isotope geochemistry of Neoproterozoic carbonate rocks from the São Francisco Basin, Brazil: implications for the mobility of Pb isotopes during tectonism and metamorphism. Chemical Geology, 160: 175–199.

Babinski, M., Vieira, L.C. and Trindade, R.I.F., 2007. Direct dating of the Sete Lagoas cap carbonate (Bambuí Group, Brazil) and implications for the Neoproterozoic glacial events. Terra Nova, 19: 401–406.

Bachmann, G., Grauert, B., Kramm, U., Lork, A. and Miller, H., 1987. El magmatismo del Cámbrico medio-Cámbrico superior en el basamento del noroeste argentino: investigaciones isotópicas y geocronológicas sobre los granitoides de los complejos intrusivos de Santa Rosa de Tastil y Cañaní. 10° Congreso Geológico Argentino, Tucumán, Acta IV, pp. 125–127.

Badenhorst, F.P., 1988. The lithostratigraphy of the Chuos mixtite in part of the southern Central Zone of the Damara Orogen, South West Africa. Communications of the Geological Survey of Namibia, 4: 103–110.

Badenhorst, F.P., 1992. The lithostratigraphy of area 2115 B and D in the Central Zone of the Damara Orogen, Namibia with emphasis on facies changes and regional correlation. Unpublished MSc Thesis, University of Port Elizabeth, Port Elizabeth, 124 pp.

Bailey, J.V., Corsetti, F.A., Bottjer, D.J. and Marenco, K.N., 2006. Microbially-mediated environmental influences on Metazoan colonization of matground ecosystems: evidence from the Lower Cambrian Harkless Formation. Palaios, 21: 215–226.

Baldis, B. and Omarini, R., 1984. El grupo Lerma (Precámbrico–Cámbrico) en la comarca central Salteña y su posición en el borde Pacífico Americano. Actas 9° Congreso Geológico Argentino 1, pp. 64–78.

Baldo, E., Demange, M. and Martino, R., 1996. Evolution of the Sierras de Córdoba, Argentina. Tectonophysics, 267: 121–142.

Bao, H., Fairchild, I.J., Wynn, P.M. and Spötl, C., 2009. Stretching the envelope of past surface environments: Neoproterozoic glacial lakes from Svalbard. Science, 323: 119–122.

Baldwin, J.A. and Brown, M., 2008. Age and duration of ultrahigh-temperature metamorphism in the Anápolis–Itauçu Complex, Southern Brasília Belt, central Brazil – constraints from U–Pb geochronology, mineral rare earth element chemistry and trace-element thermometry. Journal of Metamorphic Geology, 26: 213–233.

Bao, H., Lyons, J.R. and Zhou, C., 2008. Triple oxygen isotope evidence for elevated $CO_2$ levels after a Neoproterozoic glaciation. Nature, 453: 504–506.

Baptista, M.C., 2004. Estratigrafia e evolução geológica da região de Lagoa Formosa (MG). Master of Science Thesis, Federal University of Minas Gerais, Belo Horizonte, 104p.

Barfod, G.H., Albarede, F., Knoll, A.H., Xiao, S., Telouk, P., Frei, R. and Baker, J., 2002. New Lu-Hf and Pb-Pb age constraints on the earliest animal fossils. Earth and Planetary Science Letters, 201: 203–212.

Barnes, C.R., 1999. Paleoceanography and paleoclimatology: an Earth system perspective. Chemical Geology, 161: 17–35.

Barnes, S.-J., 1982. Serpentinites in central South West Africa/Namibia: a reconnaissance study. Memoirs of the Geological Survey of S.W. Africa/Namibia, 8: 1–90.

Barnes, S.-J., 1983. Pan-African serpentinites in central South West Africa/Namibia and the chemical classification of serpentinites. In: Miller, R.M. (Ed.), Evolution of the Damara Orogen of South West Africa/Namibia. Geological Society of South Africa, Johannesburg, pp. 147–155.

Barnett, W., Armstrong, R.A. and de Wit, M.J., 1997. Stratigraphy of the upper Neoproterozoic Kango and lower Palaeozoic Table Mountain Groups of the Cape Fold Belt revisited. South African Journal of Geology, 100: 237–250.

Barrio, C., Poiré, D.G. and Iñiguez, A.M., 1991. El contacto entre la Formación Loma Negra (Grupo Sierras Bayas) y la Formación Cerro Negro, un ejemplo de paleokarst, Olavarría, provincia de Buenos Aires. Asociación Geológica Argentina Revista, 46: 69–76.

Bartley, J.K. and Kah, L.C., 2004. Marine carbon reservoir, $C_{org}$–$C_{carb}$ coupling, and the evolution of the Proterozoic carbon cycle. Geology, 32: 129–132.

Bartley, J.K., Kah, L., McWilliams, J.L. and Stagner, A.F., 2007. Carbon isotope chemostratigraphy of the Middle Riphean type section (Avzyan Formation, Southern Urals, Russia): signal recovery in a fold-and-thrust belt. Chemical Geology, 237: 211–232.

Bartley, J.K., Pope, M., Knoll, A.H., Semikhatov, M.A. and Petrov, P.Yu., 1998. A Vendian-Cambrian boundary succession from the northwestern margin of the Siberian Platform; stratigraphy, palaeontology, chemostratigraphy and correlation. Geological Magazine, 135: 473–494.

Bartley, J.K., Semikhatov, M.A., Kaufman, A.J., Knoll, A.H., Pope, M.C. and Jacobsen, S.B., 2001. Global events across the Mesoproterozoic-Neoproterozoic boundary: C and Sr isotopic evidence from Siberia. Precambrian Research, 111: 165–202.

Basei, M.A.S., 1985. O Cinturão Dom Feliciano em Santa Catarina. PhD Thesis, University of São Paulo, São Paulo, Brazil, 191 pp.

Basei, M.A.S., Citroni, S.B. and Siga, O., Jr. 1998b. Stratigraphy and age of Fini-Proterozoic basins of Paraná and Santa Catarina states, southern Brazil. Bol. IG USP, Série Científica no. 29, pp. 195–216.

Basei, M.A.S., Frimmel, H.E., Nutman, A. and Preciozzi, F., 2006. Provenance and depositional age of the Dom Feliciano belt supracrustal units, Brazil-Uruguay. Correlation with SW Africa. V South American Symposium on Isotope Geology, Short Papers, Punta del Este, pp. 45–48.

Basei, M.A.S., Frimmel, H.E., Nutman, A.P. and Preciozzi, F., 2008. West Gondwana amalgamation based on detrital zircon ages from Neoproterozoic Ribeira and Dom Feliciano belts of South America and comparison with coeval sequences from SW Africa. In: Pankhurst, R.J., Trouw, R.A.J., de Brito Neves, B.B. and De Wit, M.J. (Eds.), West Gondwana: Pre-Cenozoic Correlations Across the South Atlantic Region. Geological Society, London, Special Publications, Vol. 294, pp. 239–256.

Basei, M.A.S., Frimmel, H.E., Nutman, A.P., Preciozzi, F. and Jacob, J., 2005. A connection between the Neoproterozoic Dom Feliciano (Brazil/Uruguay) and Gariep (Namibia/South Africa) orogenic belts – evidence from a reconnaissance provenance study. Precambrian Research, 139: 195–221.

Basei, M.A.S., Kawashita, K. and Siga, O., Jr. 1987. Idade, características litoestratigráficas e estruturais do Grupo Itajaí, Santa Catarina. Simpósio Sul Brasileiro de Geologia, 3, Curitiba-PR. Atas. 1: 93–106.

Basei, M.A.S., McReath, I. and Siga, O., Jr., 1998a. The Santa Catarina granulite complex of southern Brazil, a review. Gondwana Research, 1: 383–391.

Basei, M.A.S., Siga, O., Jr., Kaulfuss, G., Cordeiro, H., Tickyj, H., Cury, L.F., Prazeres Filho, C.R., Reis Neto, J.M. and Weber, W., 2003. Geochronology and isotope geochemistry of Votuverava and Perau Mesoproterozoic Basins, Southern Ribeira Belt, Brazil. Short Papers of the IV South American Symposium on Isotope Geology, Salvador, BA, Brazil, CBPM, IRD, Vol. 2, pp. 501–504.

Basei, M.A.S., Siga, O., Jr., Machiavelli, A. and Mancini, F., 1992. Evolução tectônica dos terrenos entre os Cinturões Ribeira e Dom Feliciano (PR-SC). Revista Brasileira de Geociências, 22: 216–221.

Basei, M., Siga, Jr., O., Masquelin, H., Harara, O., Reis Neto, J. and Preciozzi, F., 2000. The Dom Feliciano Belt of Brazil and Uruguay and its foreland domain, the Rio de la Plata Craton. In: Cordani, U.G., Milani, E.J., Thomaz Filho, A. and Campos, D.A. (Eds.), Tectonic evolution of South America, 31st International Geological Congress, Rio Janeiro, pp. 311–334.

Basei, M.A.S., Siga, O., Jr. and Reis Neto, J.M., 1990. O Batólito Paranaguá: proposição de idade, considerações petrogenéticas e implicações tectônicas. XXXVI Congresso Brasileiro de Geologia – SBG, Natal – RN, Vol. 4, pp. 2711–2722.

Basei, M.A.S., Siga, O., Jr., Reis Neto, J.M., Passarelli, C.R., Prazeres, H.J., Kaulfuss, G., Sato, K. and Lima, P.S., 1999. Paleoproterozoic granulitic belts of the Brazilian southern region (PR-SC) In: II South American Symposium on Isotope Geology, Cordoba, Argentina, Short Papers, 1, pp. 291–294.

Bassi, H.G.L., 1952. Los depósitos de ilmenita y magnetite titanífera de la Mina Podestá (ex-Romay). Dto. El Alto (Catamarca). Buenos Aires, Dirección Nacional de Geología y Minería. Boletín, 77: 5–25.

Bassi, H.G.L., 1977. Distrito Icaño y manifestaciones cupríferas Irma e Iniciación. Buenos Aires, Serv. Nac. Min. y Geología (unpublished).

Basu, A., Patranabis-Deb, S., Schieber, J. and Dhang, P.C., 2008. Stratigraphic position of the ~1000 Ma Sukhda Tuff (Chhatisgarh Supergroup, India) and the 500 Ma question. Precambrian Research, 167: 383–388.

Batten Hender, K.L. and Dix, G.R., 2008. Facies development of a Late Ordovician mixed carbonatesiliciclastic ramp proximal to the developing Taconic orogen: Lourdes Formation, Newfoundland, Canada. Facies, 54: 121–149.

Beder, R., 1925. Algunas observaciones sobre el yacimiento de mineral de hierro de la Mina Romay, cerca de Albigasta, provincia de Catamarca. Buenos Aires, Publ. Dirección General de Minas. Geología e Hidrología, 11: 1–12.

Behr, H.J., Ahrendt, H., Porada, H., Röhrs, J. and Weber, K., 1983. Upper Proterozoic playa and sabkha deposits in the Damara Orogen, SWA/Namibia. In: Miller, R.M. (Ed.), Evolution of the Damara Orogen of South West Africa/Namibia. Geological Society of South Africa, Special Publication, Vol. 11, pp. 1–20.

Behr, H.J., Ahrendt, H., Schmidt, A. and Weber, K., 1981. Saline horizons acting as thrust planes along the southern margin of the Damara Orogen (Namibia/SW-Africa). Geological Society, London, Special Publication. Vol. 9, pp. 167–172.

Bekker, A. and Kaufman, A.J., 2007. Oxidative forcing of global climate change: a biogeochemical record across the oldest Proterozoic ice age in North America. Earth and Planetary Science Letters, 258: 486–499.

Belcher, R.W. and Kisters, A.F.M., 2003. Lithostratigraphic correlations in the western branch of the Pan-African Saldania belt, South Africa: the Malmesbury Group revisited. South African Journal of Geology, 106: 327–342.

Bell, C.M., 1980. Deformation of the Table Mountain Group in the Cape Fold Belt south of Port Elizabeth. Transactions of the Geological Society of South Africa, 83: 115–124.

Bengtson, S., 1992. Proterozoic and earliest Cambrian skeletal metazoans. In: Schopf, J.W. and Klein, C. (Eds.), The Proterozoic biosphere – a multidisciplinary study. Cambridge University Press, Cambridge, pp. 397–411.

Bengtson, S., 1994. The advent of animal skeletons. In: Bengtson, S. (Ed.), Early life on earth, Nobel Symposium No. 84, Columbia University Press, New York, pp. 439–449.

Bengtson, S., 2002. Origins and early evolution of predation. Paleontological Society Papers, 8: 289–317.

Bengtson, S. and Morris, S.C., 1992. Early radiation of biomineralizing phyla. In: Lipps, J.H. and Signor, P.W. (Eds.), Origin and early evolution of the Metazoa. Plenum Press, New York, pp. 447–481.

Bengtson, S. and Zhao, Y., 1992. Predatorial borings in Late Precambrian mineralized exoskeletons. Science, 257: 367–369.

Benton, M., 1982. Trace fossils from Lower Paleozoic ocean-floor sediments of the Southern Uplands of Scotland. Transactions of the Royal Society of Edinburg Earth Sciences, 73: 67–87.

Berner, R.A., 2004. A model for calcium, magnesium and sulfate in seawater over Phanerozoic time. American Journal of Science, 304: 438–453.

Berner, R.A., 2006. GEOCARBSULF: a combined model for Phanerozoic atmospheric $O_2$ and $CO_2$. Geochimica et Cosmochimica Acta, 70: 5653–5664.

Berning, J., 1986. The Rössing uranium deposit, South West Africa/Namibia. In: Anhaeusser, C.R. and Maske, S. (Eds.), Mineral deposits of Southern Africa. Geological Society of South Africa 2, Johannesburg, pp. 1819–1832.

Bertrand-Sarfati, J., Moussine-Pouchkine, A., Amard, B. and Aït Kaci Ahmed, A., 1995. First Ediacaran fauna found in western Africa and evidence for an Early Cambrian glaciation. Geology, 23: 133–136.

Bessonova, V.Ya. and Chumakov, N.M., 1969. Upper Precambrian glacial deposits of western regions of USSR. Lithology and Mineral Resources, 2: 73–89.

Bettencourt, J.S., Monteiro, L.V.S., Bello, R.M.S., Oliveira, T.F. and Juliani, C., 2001. Metalogênese do Zinco e chumbo na região de Vazante-Paracatu, Minas Gerais. In: Pinto, C.P. and Martins-Neto, M.A. (Eds.), Bacia do São Francisco: geologia e recursos minerais. SBG-Núcleo de Minas Gerais, Belo Horizonte, pp. 161–198.

Beukes, N.J., 1973. Precambrian iron-formations of southern Africa. Economic Geology, 68: 960–1004.

Beukes, N.J. and Klein, C., 1992. Models for iron-formation deposition. In: Schopf, J.W. and Klein, C. (Eds.), The Proterozoic biosphere: a multidisciplinary study. Cambridge University Press, Cambridge, pp. 147–151.

Beurlen, K. and Sommer, F.W., 1957. Observações estratigráficas e paleontológicas sôbre o calcário Corumbá. Boletim Divisão Geologia e Mineralogia/DNPM, 168: 1–47.

Bjerrum, C.J. and Canfield, D.E., 2004. New insights into the burial history of organic carbon on the early Earth. Geochemistry, Geophysics, Geosystems, 5: 1–9.

Blaine, J.L., 1977. Tectonic evolution of the Waldau Ridge structure and the Okahandja Lineament in part of the Central Damara Orogen, west of Okahandja, South West Africa. Precambrian Research Unit, University of Cape Town, Bulletin 21, 99 pp.

Blanco, G., 2008. Provenance analysis of the Neoproterozoic-Cambrian Nama Group (Namibia) and the Arroyo del Soldado Group (Uruguay): implications for the palaeogeographic reconstruction of SW Gondwana. PhD Thesis, University of Johannesburg, 300 pp.

Blanco, G. and Gaucher, C., 2004. Estratigrafía, paleontología y edad de la Formación Las Ventanas (Neoproterozoico, Uruguay). IV Congreso Uruguayo de Geología, Actas (CD-ROM), Montevideo.

Blanco, G. and Gaucher, C., 2005. Estratigrafia, paleontologia y edad de la Formacion Las Ventanas (Neoproterozoico, Uruguay). Latin American Journal of Sedimentology and Basin Analysis, 12: 109–124.

Blanco, G., Rajesh, H.M., Gaucher, C., Germs, G.J.B. and Chemale, F., Jr., 2009. Provenance of the Arroyo del Soldado Group (Ediacaran to Cambrian, Uruguay): Implications for the paleogeographic evolution of southwestern Gondwana. Precambrian Research, 171: 57–73.
Blanco, G., Zimmermann, U., Gaucher, C., Chemale, Jr., F. and Germs, G.J.B., 2007. Provenance of the Arroyo del Soldado Group (Ediacaran-Lower Cambrian, Uruguay): detrital zircon U-Pb, Sm-Nd isotopes and geochemical data. V. Congreso Uruguayo de Geología, Montevideo (CD-ROM).
Blaxland, A., Gohn, E., Haack, U. and Hoffer, E., 1979. Rb/Sr ages of late-tectonic granites in the Damara Orogen, Southwest Africa/ Namibia. Neues Jahrbuch fur Mineralogie, Monatshefte, 11: 498–508.
Bloeser, B., Schopf, J.W., Hododyski, R.J. and Breed, W.J., 1977. Chitinozoans from the Late Precambrian Chuar Group of the Grand Canyon, Arizona. Science, 195: 676–679.
Bodiselitsch, B., Koeberl, C., Master, S. and Reimold, W.U., 2005. Estimating duration and intensity of Neoproterozoic snowball glaciations from Ir anomalies. Science, 308: 239–242.
Boggiani, P.C., 1998. Análise estratigráfica da Bacia Corumbá (Neoproterozóico) – Mato Grosso Do Sul. PhD Thesis, Instituto de Geociências, São Paulo University, 181 pp.
Boggiani, P.C. and Alvarenga, C.J.S. de, 2004. Faixa Paraguai. In: Mantesso-Neto, V., Bartorelli, A., Carneiro, C.D.R. and Brito-Neves, B.B. (Eds.), Geologia do Continente Sul-Americano: Evolução da Obra de Fernando Flávio Marques de Almeida. Editora Beca, São Paulo, pp. 113–120.
Boggiani, P.C., Fairchild, T.R. and Coimbra, A.M., 1993. O Grupo Corumbá (Neoproterozóico-Cambriano) na região central da Serra da Bodoquena (Faixa Paraguai), Mato Grosso do Sul. Revista Brasileira de Geociências, 23: 301–305.
Boggiani, P.C., Fairchild, T.R. and Riccomini, C., 2004. New level of diamictites in the Corumbá Group (Ediacaran), Paraguay Belt, South America. 1st Symposium Neoproterozoic-Early Palaeozoic Events in SW-Gondwana, Extended Abstracts, São Paulo, pp. 10–12.
Boggiani, P.C., Ferreira, V.P., Sial, A.N., Babinski, M., Trindade, R.I.F., Aceñolaza, G., Toselli, A.J. and Parada, M.A., 2003. The cap carbonate of the Puga Hill (central South America) in the context of the post-Varanger glaciation. IV South American Symposium on Isotope Geology, 24–27 August 2003, Salvador, Brazil, Short Papers, pp. 324–327.
Boggiani, P.C. and Gaucher, C., 2004. *Cloudina* from the Itapucumí Group (Vendian, Paraguay): age and correlations. In: 1st Symposium Neoproterozoic-Early Palaeozoic Events in SW-Gondwana, Extended Abstracts, São Paulo, pp. 13–15.
Bonhomme, M.G., 1976. Mineralogie des fractions fines et datations Rubidium–Strontium dans le Groupe Bambuí, MG, Brésil. Revista Brasileira de Geociências, 6: 211–220.
Bonhomme, M. and Cingolani, C., 1980. Mineralogía y geocronología Rb-Sr y K-Ar de fracciones finas de la "Formación La Tinta", provincia de Buenos Aires. Revista Asociación Geológica Argentina, 35: 519–538.
Bonhomme, M.G., Cordani, U.G., Kawashita, K., Macedo, M.H.F. and Thomaz Filho, A., 1982. Radiochronological age and correlation of Proterozoic sediments in Brazil. Precambriam Research, 18: 103–118.
Boni, M., Terracciano, R., Evans, N.J., Laukamp, C., Schneider, J. and Bechstädt, T., 2007. Genesis of vanadium ores in the Otavi Mountainland, Namibia. Economic Geology, 102: 441–469.
Booker, J.R., Favetto, A. and Pomposiello, M.C., 2004. Low electrical resistivity associated with plunging of the Nazca flat slab beneath Argentina. Nature, 429: 399–403.
Borba, A.W., Mizusaki, A.M.P., Santos, J.O.S., McNaughton, N.J., Onoe, A.T. and Hartmann, L.A., 2008. U-Pb zircon and $^{40}Ar$-$^{39}Ar$ K-feldspar dating of syn-sedimentary volcanism of the Neoproterozoic Maricá Formation: constraining the age of foreland basin inception and inversion in the Camaquã Basin of southern Brazil. Basin Research, 20: 359–375.
Borg, G., Kärner, K., Buxton, M., Armstrong, R. and Van der Merwe, S.W., 2003. Geology of the Skorpion zinc deposit, southern Namibia. Economic Geology, 98: 749–771.
Borrello, A.V., 1966. Trazas, restos tubiformes y cuerpos fósiles problemáticos de la Formación La Tinta, Sierras Septentrionales de la Provincia de Buenos Aires. Comisión de Investigaciones Científicas de la Provincia de Buenos Aires. Paleontografía Bonaerense, 5: 1–42.
Borrello, A., 1972. La estructura assyntica de la Argentina. Revista Brasileira de Geociencias, San Pablo, 2 (2): 71–84.
Borrello, A.V., 1969. Los geosinclinales de la Argentina. Anales de la Dirección Nacional de Geología y Minería, 14: 1–188.
Bossi, J., 1966. Geología del Uruguay. Departamento Publicaciones Universidad de la República, Montevideo, Uruguay.
Bossi, J., 1983. Breve reseña sobre el conocimiento geológico del Escudo Predevoniano en Uruguay, Sudamérica. Zentralblatt für Geologie und Paläontologie, 1: 417–429.
Bossi, J. and Campal, N., 1992. Magmatismo y tectónica transcurrente durante el Paleozoico Inferior en Uruguay. In: Gutierrez-Marco, J.G., Saavedra, J. and Rabano, I. (Eds.), Paleozoico Inferior de Iberoamérica. Universidad de Extremadura, Mérida, pp. 343–356.
Bossi, J., Campal, N., Civetta, L., Demarchi, G., Girardi, V., Mazzucchelli, M., Negrini, L., Rivalenti, G., Fragoso Cesar, A., Sinigoi, S., Texexeira, W., Piccirillo, E. and Molesini, M., 1993a. Early Proterozoic dike swarms from western Uruguay: geochemistry, Sr-Nd isotopes and petrogenesis. Chemical Geology, 106: 263–277.
Bossi, J., Cingolani, C., Llambías, E., Varela, R. and Campal, N., 1993b. Características del magmatismo post-orogénico finibrasiliano en el Uruguay: formaciones Sierra de Ríos y Sierra de Animas. Revista Brasileira de Geociências, 23: 282–288.
Bossi, J., Fernandez, A. and Elizalde, G., 1965. Predevoniano en el Uruguay. Boletín Facultad de Agronomía, 78: 1–49.
Bossi, J. and Ferrando, L., 2001. Carta Geológica del Uruguay a escala 1/500.000, versión digital 2.0. Facultad de Agronomía, Montevideo.
Bossi, J., Ferrando, L., Montaña, J., Campal, N., Morales, H., Gancio, F., Schipilov, A., Piñeyro, D., and Sprechmann, P., 1998. Carta geológica del Uruguay. Escala 1:500.000. Geoeditores, Montevideo.
Bossi, J. and Gaucher, C., 2004. The Cuchilla Dionisio Terrane, Uruguay: an allochthonous block accreted in the Cambrian to SW-Gondwana. Gondwana Research, 7 (3): 661–674.
Bossi, J., Gaucher, C., Chiglino, L., Sial, A.N., Navarro, R. and Piñeyro, D., 2008. Parque UTE Group: Mesoproterozoic of the Río de la Plata Craton (Nico Pérez Terrane, Uruguay). 44 Congresso Brasileiro de Geologia, Anais, Curitiba, p. 67.
Bossi, J., Gaucher, C., Navarro, R., Piñeyro, D. and Chiglino, L., 2007. Escama tectónica Carapé: litoestratigrafia de una pieza importante del rompecabezas Neoproterozoico–Cámbrico en el Uruguay. V Congreso Uruguayo de Geología, Montevideo (CD-ROM).
Bossi, J. and Navarro, R., 1991. Geología del Uruguay. Vol. 1, Universidad de la República, Montevideo, 453 pp.

Bossi, J. and Navarro, R., 2001. Grupo Carapé: su reivindicación. Revista Sociedad Uruguaya Geología, 8: 2–12.

Bossi, J., Navarro, R. and Gaucher, C., 2002. Aspectos geológicos de las rocas metavolcánicas y metasedimentarias del Grupo Lavalleja, sudeste de Uruguay. Discussao. Revista Brasileira de Geociências, 32: 598–601.

Bossi, J., Pineyro, D. and Cingolani, C.A., 2005. El límite sur del Terreno Piedra Alta (Uruguay). Importancia de la faja milonítica sinistral de Colonia. Actas XVI Congreso Geológico Argentino 1, pp. 173–180.

Bossi, J., Preciozzi, F. and Campal, N., 1993c. Predevoniano del Uruguay. Parte 1: Terreno Piedra Alta. DINAMIGE, Montevideo, pp. 1–50.

Bossi, J. and Schipilov, A., 2007. Rocas ígneas básicas del Uruguay. Facultad de Agronomía, Montevideo, 364 pp.

Bottjer, D., Hagadorn, J.W. and Dornbos, S., 2000. The Cambrian substrate revolution. GSA Today, 10 (9): 2–7.

Bouougri, E.H. and Porada, H., 2007. Siliciclastic biolaminites indicative of widespread microbial mats in the Neoproterozoic Nama Group of Namibia. Journal of African Earth Sciences, 48: 38–48.

Bowring, S.A., Grotzinger, J.P., Condon, D.J., Ramezani, J., Newall, M.J. and Allen, P.A., 2007. Geochronological constraints on the chronostratigraphic framework of the Neoproterozoic Huqf Superroup, Sultanate of Oman. American Journal of Science, 307: 1097–1145.

Bowring S.A., Myrow P., Landing E., and Ramenzani J., 2003a. Geochronological constraints on terminal Neoproterozoic events and the rise of Metazoans. NASA Astrobiology Institute (NAI Gen. Mtg. Abstr.), pp. 113–114.

Bowring, S., Myrow, P., Landing, E., Ramezani, J. and Grotzinger, J., 2003b. Geochronological constraints on terminal Neoproterozoic events and the rise of metazoans. Geophysical Research Abstracts, 5: 13219.

Bowring, S.A. and Schmitz, M.D., 2003. High precision U-Pb geochronology an the stratigraphic record. In: Hanchar, J.M. and Hoskin, W.O. (Eds.), Zircons, Reviews in Mineralogy and Geochemistry Vol. 53, pp. 305–326.

Bradley, D.C., 2008. Passive margins through Earth history. Earth-Science Reviews, 91: 1–26.

Brain, C.K., Hoffmann, K.-H., Prave, A.R., Fallick, A.E., Coetzee, J. and Botha, A.J., 2001. Interpretive problems in a search for micro-invertebrate fossils from a Neoproterozoic limestone in Namibia. Palaeontologia Africana, 37: 1–12.

Brain, C.K., Prave, A.R., Fallick A.E. and Hoffmann, K.-H., 2003. Sponge-like microfossils from Neoproterozoic intertillite limestones of the Otavi Group in northern Namibia. In: Frimmel, H.E. (Ed.), III International Colloquium Vendian-Cambrian of W-Gondwana, Programme and Extended Abstracts, Cape Town, pp. 19–22.

Brain, C.K., P rave, A.R., Hoffmann, K-H., Fallick A.E., Botha, A.J., Condon, D., Herd, D., Young, I. and Sturrock, C.J., 2008. Progress in the evaluation of sponge-like fossils from Neoproterozoic limestones of Namibia. 15th Biennial Conference of the Palaeontological Society of Southern Africa, Conference Programme and Abstracts, Matjiesfontein, pp. 56–57.

Branco, J.J.R., 1957. O conglomerado do Samburá, Minas Gerais. Anais Academia Brasileira de Ciências, 28 (3): 295–301.

Brasier, M., Cowie, J. and Taylor, M., 1994a. Decision on the Precambrian-Cambrian boundary stratotype. Episodes, 17: 3–8.

Brasier, M., Green, O. and Shields, G., 1997. Ediacarian sponge spicule clusters from southwestern Mongolia and the origins of the Cambrian fauna. Geology, 25: 303–306.

Brasier, M.D., Corfield, R., Luo, H., Wu, X., Ouyang, L., Jiang, Z. and Hamdi, B., 1990. The carbon- and oxygen-isotope record of the Precambrian-Cambrian boundary interval in China and Iran and their correlation. Geological Magazine, 127: 319–332.

Brasier, M.D., Corfield, R.M., Derry, L.A., Rozanov, A.Y. and Zhuravlev, A.Y., 1994b. Multiple $\delta^{13}C$ excursions spanning the Cambrian explosion to the Botomian crisis in Siberia. Geology, 22: 455–458.

Brasier, M.D. and Lindsay, J.F., 1995. A billion years of environmental stability and the emergence of eukaryotes: new data from northern Australia. Geology, 26: 555–558.

Brasier, M.D., McCarron, G., Tucker, R., Leather, J., Allen, P.A. and Shields, G.A., 2000. New U-Pb zircon dates for the Neoproterozoic Ghubrah glaciation and for the top of the Huqf Supergroup, Oman. Geology, 28: 175–178.

Brasier, M.D. and Sukhov, S.S., 1998. The falling amplitude of carbon isotopic oscillation through the Lower and Middle Cambrian. Canadian Journal of Earth Sciences, 35: 353–373.

Braun, A., Chen, J., Waloszek, D. and Maas, A., 2008. First Early Cambrian Radiolaria. In: Vickers-Rich, P. and Komarower, P. (Eds.), The rise and fall of the Ediacaran biota. Geological Society of London, Special Publications, Vol. 286, pp. 143–149.

Breitkopf, J.H., 1988. Iron formations related to mafic volcanism and ensialic rifting in the southern margin zone of the Damara Orogen, Namibia. Precambrian Research, 38: 111–130.

Breitkopf, J.H. and Maiden, K.J., 1988. Tectonic setting of the Matchless Belt pyritic copper deposits, Namibia. Economic Geology, 83: 710–723.

Briqueu, L., Lancelot, J.P., Valois, J.P. and Walgenwitz, F., 1980. Géochronologie U-Pb et genèse d'un type de minéralisation uranifère: les alaskitesde Goanikontes (Namibie) et leur encaissant. Bulletin des Centres de Recherches Exploration-Production Elf-Aquitaine, 4: 759–811.

Bristow, T.F. and Kennedy, M.J., 2008. Carbon isotope excursions and the oxidant budget of the Ediacaran atmosphere and ocean. Geology, 36: 863–866.

Brito Neves, B.B.de., 1983. O mapa geológico do Nordeste Oriental, escala 1:1.000.000. Tese de Livre-Docência, Instituto de Geociências, Universidade de São Paulo, São Paulo, 177 pp.

Brito Neves, B.B. and Cordani, U.C., 1991. Tectonic evolution of South America during Late Proterozoic. Precambrian Research, 33: 23–40.

Brito Neves, B.B.de, Campos Neto, M.C. and Fuck, R., 1999. From Rodinia to eastern Gondwana: an approach to the Brasiliano–Pan African Cycle and orogenic collage. Episodes, 22 (3): 155–166.

Brito Neves, B.B.de, Fuck, R.A., Cordani, U.G. and Thomaz Filho, A., 1984. Influence of basement structures on the evolution of the major sedimentary basins of Brazil. Journal of Geodynamics, 1: 495–510.

Brito Neves, B.B.de, Sial, A.N. and Albuquerque, J.P.T., 1977. Vergência centrífiga residual no Sistema de Dobramentos Sergipano. Revista Brasilera de Geociências, 7: 102–114.

Brito Neves, B.B.de, Sial, A.N., Rand, H.M. and Manso, V.V., 1982. The Pernambuco-Alagoas massif, northeast Brazil. Revista Brasileira de Geociências, 12: 240–250.

Brito Neves, B.B.de, Santos, E.J. and Van Schmus, W.R., 2000. Tectonic history of the Borborema province. In: Cordani, U.G., Milani, E.J., Thomaz Filho, A. and Campos, D.A. (Eds.), Tectonic evolution of South America. 31st International Geological Congress, Rio de Janeiro, pp. 151–182.

Grotzinger, J.P., Adams, E.W. and Schröder, S., 2005. Microbial – metazoan reefs of the terminal Proterozoic Nama Group (ca. 550-543 Ma), Namibia. Geological Magazine, 142: 499–517.

Grotzinger, J.P., Bowring, S.A., Saylor, B.Z. and Kaufman, A.J., 1995. Biostratigraphic and geochronologic constraints on early animal evolution. Science, 270: 598–604.

Grotzinger, J.P. and James, N.P., 2000. Precambrian carbonates: evolution of understanding. In: Grotzinger, J.P. and James, N.P. (Eds.), Carbonate sedimentation and diagenesis in the evolving Precambrian world, SEPM Special Publication 67. SEPM, Tulsa, pp. 3–23.

Grotzinger, J.P. and Knoll, A.H., 1995. Anomalous carbonate precipitates: is the Precambrian the key to the Permian? PALAIOS, 10: 578–596.

Grotzinger, J.P., Watters, W.A. and Knoll, A.H., 2000. Calcified metazoans in thrombolite-stromatolite reefs of the terminal Proterozoic Nama Group, Namibia. Paleobiology, 26: 334–359.

Gruner, B.B., 2000. Metamorphoseentwicklung im Kaokogürtel, NW-Namibia: phasenpetrologische und geothermobarometrische Untersuchungen panafrikanischer Metapelite. Freiberger Forschungshefte, Bergakademie Freiberg, Germany, C486, 221pp.

Grunow, A., Hanson, R. and Wilson, T., 1996. Were aspects of Pan-African deformation linked to Iapetus opening? Geology, 24: 1063–1066.

Gualda, G.A.R., 2001. Evolução petrográfica e mineralógica das associações alcalina e aluminosa dos Granitos Tipo-A da Graciosa, PR. MSc Dissertation, Institute of Geosciences, University of São Paulo, São Paulo, 250 pp.

Gualda, GA.R. and Vlach, S.R.F., 2007a. The Serra da Graciosa A-type granites and syenites, southern Brazil. Part 2: petrographic and mineralogical evolution of the alkaline and aluminous associations. Lithos, 93: 310–327.

Gualda, G.A.R. and Vlach, S.R.F., 2007b. The Serra da Graciosa A-type granites and syenites, southern Brazil. Part 3: magmatic evolution and post-magmatic breakdown of amphiboles of the alkaline association. Lithos, 93: 328–339.

Guan, B., Ruitang, W.U., Hambrey, M.J. and Geng, Wu., 1986. Glacial sediments and erosional pavement near the Cambrian-Precambrian boundary in western Henan Province, China. Journal of Geological Society, London, 143: 311–323.

Guimarães, E.M., 1997. Estudos de Proveniência e diagênese com ênfase na caracterização dos filossilicatos dos grupos Paranoá e Bambuí na região de Bezerra-Cabeceiras, GO. Tese de doutorado, Instituto de Geociencias, Universidade de Brasília, Brasilia, 269 pp.

Guimarães, J.T., 1996. A Formação Bebedouro no Estado da Bahia; faciologia, estratigrafia e ambientes de sedimentação. MSc Thesis, Instituto de Geociências, Universidade Federal da Bahia, Salvador, Bahia, 146 pp.

Guj, P., 1970. The Damara mobile belt in the south-western Kaokoveld, South West Africa. Precambrian Research Unit, University of Cape Town, Bulletin, 10: 168 pp.

Guj, P., 1974. A revision of the Damara stratigraphy along the southern margin of the Kamanjab inlier, South West Africa. Precambrian Research Unit, University of Cape Town, Bulletin, 15: 167–176.

Haack, U. and Gohn, E., 1988. Rb-Sr data on some pegmatites in the Damara Orogen, Namibia. Communications of the Geological Survey of Namibia, 4: 13–17.

Haack, U., Gohn, E. and Klein, J.A., 1980. Rb/Sr ages of granitic rocks along the middle reaches of the Omaruru River and the timing of orogenetic events in the Damara Belt (Namibia). Contributions to Mineralogy and Petrology, 74: 349–360.

Habicht, K.S. and Canfield, D.E., 1997. Sulfur isotope fractionation during bacterial sulfate reduction in organic-rich sediments. Geochimica Cosmochimica Acta, 61: 5351–5361.

Habicht, K.S. and Canfield, D.E., 2001. Isotope fractionation by sulfate-reducing natural populations and the isotopic composition of sulfide in marine sediments. Geology, 29: 555–558.

Habicht, K.S., Gade, M., Thamdrup, B., Berg, P. and Canfield, D.E., 2002. Calibration of sulfate levels in the Archean ocean. Science, 298: 2372–2374.

Hagadorn, J.W. and Bottjer, D.J., 1997. Wrinkle structures: microbially mediated sedimentary structures common in subtidal siliciclastic settings at the Proterozoic-Phanerozoic transition. Geology, 25: 1047–1050.

Hagadorn, J.W. and Bottjer, D.J., 1999. Restriction of a late Neoproterozoic biotope: suspects microbial structures and trace fossils at the Vendian-Cambrian transition. Palaios, 14: 73–85.

Hagadorn, J.W., Fedo, C.W. and Waggoner, B.M., 2000. Lower Cambrian Ediacaran fossils from the Great Basin, U.S.A.. Journal of Paleontology, 74: 731–740.

Hagadorn, J.W. and Waggoner, B., 2000. Ediacaran fossils from the southwestern Great Basin, United States. Journal of Paleontology, 74: 349–359.

Hagadorn, J.W., Xiao, S., Donoghue, P.C.J., Bengston, S., Gostling, N.J., Pawlowska, M., Raff, E.C., Raff, R.A., Turner, F.R., Yin, C., Zhou, C., Yuan, X., McFeely, M.B., Stampanoni, M. and Nealson, K.H., 2006. Cellular and subcellular structure of Neoproterozoic embryos. Science, 314: 291–294.

Hahn, G., Hahn, R., Leonardos, O.H., Pflug, H.D. and Walde, D.H.G., 1982. Körperlich erhaltene Scyphozoen-Reste aus dem Jungprakambrium Brasiliens. Geologica et Palaeontologica, 16: 1–18.

Hahn, G. and Pflug, H.D., 1985. Die Cloudinidae n. fam., Kalk- Röhren aus dem Vendium und Unter-Kambrium. Senckenbergiana lethaea, 65: 413–431.

Hälbich, I.W., 1964. Observations on primary features in the Fish River Series and Dwyka Series in South West Africa. Transactions of the Geological Society of South Africa, 67: 95–100.

Hälbich, I.W. and Alchin, D.J., 1995. The Gariep belt: stratigraphic-structural evidence for obliquely transformed grabens and back-folded thrust stacks in a combined thick-skin thin-skin structural setting. Journal of African Earth Sciences, 21: 9–33.

Hälbich, I.W. and Freyer, E.E., 1985. Structure and metamorphism of Damaran rocks in the Ugab Profile. Progress Report for 1982/83. Communications of the Geological Survey of Namibia, 1: 97.

Hallinan, S.E., Mantovani, M.S.M., Shukowsky, W. and Braggion, I., Jr., 1993. Estrutura do Escudo Sul-Brasileiro: Uma revisão através de dados gravimétricos e magnetométricos. Revista Brasileira de Geociências, 23: 201–214.

Halls, H.C., Campal, N., Davis, D.W. and Bossi, J., 2001. Magnetic studies and U-Pb geochronology of the Uruguayan dyke swarm, Rio de la Plata craton, Uruguay: paleomagnetic and economic implications. Journal of South American Earth Sciences, 14: 349–361.

Halverson, G.P., 2006. A Neoproterozoic chronology. In: Xiao, S. and Kaufman, A. (Eds.), Neoproterozoic geobiology and paleobiology. Springer, Dordrecht, pp. 231–271.

Halverson, G.P., Dudás, F.Ö., Maloof, A. and Bowring, S.A., 2007a. Evolution of the $^{87}Sr/^{86}Sr$ composition of Neoproterozoic seawater. Palaeogeography, Palaeoclimatology, Palaeoecology, 256: 103–129.

Halverson, G.P., Hoffman, P.F., Schrag, D.P. and Kaufman, A.J., 2002. A major perturbation of the carbon cycle before the Ghaub glaciation (Neoproterozoic) in Namibia: prelude to snowball Earth? Geochemistry, Geophysics, and Geosystems, 3: 1–24.
Halverson, G.P., Hoffman, P.F., Schrag, D.P., Maloof, A.C. and Rice, A.H.N., 2005. Towards a Neoproterozoic composite carbon isotope record. Geological Society of America Bulletin, 117: 1181–1207.
Halverson, G.P. and Hurtgen, M.T., 2007. Ediacaran growth of the marine sulfate reservoir. Earth and Planetary Science Letters, 263: 32–44.
Halverson, G.P., Maloof, A.C. and Hoffman, P.F., 2004. The Marinoan glaciation (Neoproterozoic) in northeast Svalbard. Basin Research, 16: 297–324.
Halverson, G.P., Maloof, A.C., Schrag, D.P., Dudas, F.O. and Hurtgen, M.T., 2007b. Stratigraphy and geochemistry of a ca. 800 Ma negative carbon isotope interval in northeastern Svalbard. Chemical Geology, 237: 5–27.
Hambrey, M.J. and Harland, W.B. (Eds.), 1981. Earth's pre-Pleistocene glacial record. Cambridge University Press, Cambridge, 1004 pp.
Hanson, R.E., Wilson, T.J. and Munyanyiwa, H., 1994. Geological evolution of the Neoproterozoic Zambezi orogenic belt in Zambia. Journal of African Earth Sciences, 18: 135–150.
Haralyi, N.L.E. and Walde, D.H.G., 1986. Os minerios de ferro e manganês da região de Urucum, Corumbá, Mato Grosso do Sul. In: Schobbenhaus, C. and Coelho, C.E.S. (Eds.), Principais depósitos minerais do Brasil. Vol. II, Ferro e metais da industria do aço, Ministério das Minas e Energia, Brazil, pp. 127–144.
Harara, O.M.M., 1996. Análise estrutural, petrológica e geocronológica dos litotipos da região de Piên (PR) e adjacências. Unpublished MSc Dissertation, Institute of Geosciences, University of São Paulo, São Paulo, 196 pp.
Harara, O.M.M., 2001. Mapeamento e investigação petrológica e geocronológica dos litotipos da região do Alto Rio Negro (PR-SC): Um exemplo de sucessivas e distintas atividades magmáticas durante o Neoproterozóico. Unpublished PhD Thesis, Institute of Geosciences, University of São Paulo, São Paulo, 206 pp.
Harara, O.M.M., Basei, M.A.S. and Siga, O., Jr. 2002. From subduction to late and post-collision settings: a record from Neoproterozoic successive magmatic in the upper Rio Negro region. XXXXI Congresso Brasileiro de Geologia, João Pessoa – PB, Vol. 1, pp. 310.
Harara, O.M.M., Basei, M.A.S., Siga, O., Jr., Campos Neto, M.C. and Prazeres Filho, H.J., 2003. Dating of high-grade metamorphism by U–Pb, Sm–Nd and K–Ar isotopic systems: Paleoproterozoic I-type granulites from the northern border of the Luis Alves gneiss-granulite terrain, southern Brazil. IV SSAGI, Salvador – BA, Vol. 2, pp. 568–571.
Harland, W.B., 1964. Evidence of Late Precambrian glaciation and its significance. Problems in palaeoclimatology. Proceedings of the NATO Palaeoclimates Conference, pp. 119–149.
Harland, W.B., Hambrey, M.J. and Waddams, P., 1993. Vendian geology of Svalbard. Norsk Polarinstitutt, Oslo, 150 pp.
Hart, S., 1966. Radiometric ages in Uruguay and Argentina and their implications concerning continental drift. Geological Society of America, Annual Meeting, San Francisco, USA.
Hartmann, L.A., Campal, N., Santos, J.O.S., McNaughton, N.J., Bossi, J., Schipilov, A. and Lafon, J.M., 2001. Archean crust in the Río de la Plata Craton, Uruguay-SHRIMP U-Pb zircon reconnaissance geochronology. Journal of South American Earth Sciences, 14: 557–570.
Hartmann, L.A., Leite, J.A.D., McNaughton, N.J. and Santos, J.O.S., 1999. Deepest exposed crust of Brazil – SHRIMP establishes three events. Geology, 27: 947–950.
Hartmann, L.A., Santos, J.O.S., Bossi, J., Campal, N., Schipilov, A. and McNaughton, N.J., 2002a. Zircon and titanite U-Pb SHRIMP geochronology of Neoproterozoic felsic magmatism on the eastern border of the Rio de la Plata Craton, Uruguay. Journal South American Earth Sciences, 15: 229–236.
Hartmann, L.A., Santos, J.O.S., Cingolani, C.A. and McNaughton, N.J., 2002b. Two Paleoproterozoic orogenies in the evolution of the Tandilia Belt, Buenos Aires, as evidenced by zircon U-Pb SHRIMP geochronology. International Geology Review, 44: 528–543.
Hartmann, L.A., Santos, J.O.S., McNaughton, N.J., Vasconcellos, M.A.Z. and Silva, L.C., 2000. SHRIMP dates recurrent granulite facies metamorphism in the Santa Catarina granulites, southern Brazil. Anais Academia Brasileira de Ciências, 72: 559–572.
Hartmann, L.A., Silva, L.C. and Orlandi Filho, V., 1979. O Complexo Granulítico de Santa Catarina. Descrição e implicações genéticas. Acta Geológica Leopoldensia, 3: 93–112.
Hartmann, O., Hoffer, E. and Haack, U., 1983. Regional metamorphism in the Damara Orogen: interaction of crustal motion and heat transfer. In: Miller, R.M. (Ed.), Evolution of the Damara Orogen of South West Africa/Namibia. Geological Society of South Africa, Special Publication, Marshalltown, pp. 233–241.
Hartnady, C., Joubert, P. and Stowe, C., 1985. Proterozoic crustal evolution in southwestern Africa. Episodes, 8 (4): 236–244.
Hartnady, C.J.H., 1969. Structural analysis of some pre-Cape formations in the Western Province. Precambrian Research Unit, University of Cape Town, Bulletin, 6: 70 pp.
Hartnady, C.J.H., Newton, A.R. and Theron, J.N., 1974. The stratigraphy and structure of the Malmesbury Group in the southwestern Cape. Precambrian Research Unit, University of Cape Town, Bulletin, 15: 193–213.
Hartt, C.F., 1941.Geologia e Geografia Física do Brasil. Companhia Editora Nacional, São Paulo, 649 pp.
Hasui, Y., Carneiro, C.D.R. and Coimbra, A.M., 1975. The Ribeira Folded Belt. Revista Brasileira de Geociências, 5: 257–266.
Haughton, S.H., Frommurze, H.F. and Visser, D.J.L., 1937. The geology of a portion of the coastal belt near the Gamtoos Valley, Cape Province. Explanation Sheet 151 Gamtoos River. Geological Survey of South Africa, Pretoria.
Häussinger, H. and Okrusch, M., 1993. Geochemistry of premetamorphic hydrothermal alteration of metasedimentary rocks associated with the Gorob massive sulfide prospect, Damara Orogen, Namibia. Economic Geology, 88: 72–90.
Hauzenberger, C.A., Sommer, H., Fritz, H., Bauernhofer, A., Kröner, A., Hoinkes, G., Wallbrecher, E. and Thöni, M., 2007. SHRIMP U–Pb zircon and Sm–Nd garnet ages from the granulite-facies basement of SE Kenya: evidence for Neoproterozoic polycyclic assembly of the Mozambique Belt. Journal of the Geological Society, 164: 189–201.
Hawkesworth, C.J., Gallagher, K., Kirstein, L., Mantovani, M.S.M., Peate, D.W. and Turner, S.P., 2000. Tectonic controls on magmatism associated with cotinenta break-up: an example from the Paraná Etendeka Province. Earth and Planetary Science Letters, 179: 335–349.
Hawkesworth, C.J., Gledhill, A.R., Roddick, J.C., Miller, R.M. and Kröner, A., 1983. Rb-Sr and $^{40}Ar/^{39}Ar$ studies bearing on models for the thermal evolution of the Damara Belt, Namibia. In: Miller, R.M. (Ed.), Evolution of the Damara Orogen of South West Africa/ Namibia. Geological Society of South Africa, Special Publication, Vol. 11, pp. 323–338.

Hawkesworth, C.J., Mantovani, M.S.M., Taylor, P.N. and Palacz, Z., 1986. Coupled crust-mantle systems: evidence from the Paraná of South Brazil. Nature, 322: 356–359.
Hayes, J.M., Strauss, H. and Kaufman, A.J., 1999. The abundance of C in marine organic carbon and isotopic fractionation in the global biogeochemical cycle of carbon during the past 800 Ma. Chemical Geology, 161: 103–125.
Hayes, J.M. and Waldbauer, J.R., 2006. The carbon cycle and associated redox processes through time. Philosophical Transactions of the Royal Society of London, 361: 931–950.
Hedberg, R.M., 1979. Stratigraphy of the Ovamboland Basin, South West Africa. Precambrian Research Unit, University of Cape Town, Bulletin, 24: 1–325.
Hegenberger, W., 1987. Gas escape structures in Precambrian peritidal carbonate rocks. Communications of the Geological Survey of Namibia, 3: 49–55.
Hegenberger, W., 1993. Stratigraphy and sedimentology of the Late Precambrian Witvlei and Nama Groups, east of Windhoek. Memoirs of the Geological Survey of Namibia, 17: 1–82.
Heilbron, M. and Machado, N., 2003. Timing of terrane accretion in the Neoproterozoic-Eopaleozoic Ribeira belt, SE Brazil. Precambrian Research, 125: 87–112.
Heilbron, M., Pedrosa-Soares, A.C., Campos-Neto, M.C., Silva, L.C., Trouw, R. and Janasi, V.A., 2004a. Brasiliano orogens in southeast and south Brazil. Journal of Virtual Explorer, 17: 4.
Heilbron, M., Pedrosa-Soares, A.C., Campos Neto, M.C., Silva, L.C., Trouw, R.A.J. and Janasi, V.A., 2004b. Província Mantiqueira. In: Mantesso-Neto, V., Bartorelli, A., Carneiro, C.D.R. and Brito Neves, B.B.de (Eds.), Geologia do Continente Sul-Americano: evoluç ão da obra de F.F.M. de Almeida, Beca, São Paulo, pp. 203–234.
Heilbron, M., Valeriano, C.M., Tassinari, C.C.G., Almeida, J., Tupinamba, M., Siga, O. and Trouw, R.A.J., 2008. Correlation of Neoproterozoic terranes between the Ribeira Belt, SE Brazil and its African counterpart: comparative tectonic evolution and open questions. In: Pankhurst, R.J., Trouw, R.A.J., Brito-Neves, B.B. and De Wit, M.J. (Eds.), West Gondwana: Pre-Cenozoic correlations across the South Atlantic region. Geological Society of London, Special Publication 294, pp. 211–237.
Heilbron, M.L., Valeriano, C., Zimbres, E., Chrispim, S.J. and Tupinambá, M., 1987. O Contato Basal do Grupo Canastra entre Itaú de Minas e Carmo do Rio Claro. IV Simpósio de Geologia de Minas Gerais, Sociedade Brasileira de Geologia, Núcleo Minas Gerais, Belo Horizonte, pp. 178–198.
Hennies, W.T., 1966. Geologia do centro-oeste matogrossense. PhD Thesis, Instituto de Geociências, São Paulo University, 65 pp.
Henry, G., 1992. The sedimentary evolution of the Damara Sequence in the Lower Khan River Valley, Namibia. University of the Witwatersrand, Johannesburg, 217 pp.
Higgins, J.A. and Schrag, D.P., 2003. Aftermath of a snowball Earth. Geochemistry, Geophysics, Geosystems, 431: doi: 10.1029/2002GC000403.
Higgins, J.A. and Schrag, D.P., 2006. Beyond methane: towards a theory for the Paleocene–Eocene thermal maximum. Earth and Planetary Science Letters, 245: 523–537.
Hill, A.C., Cotter, K.L. and Grey, K., 2000. Mid-Neoproterozoic biostratigraphy and isotope stratigraphy in Australia. Precambrian Research, 100: 281–298.
Hill, R., 1991. Starting plumes and continental break-up. Earth and Planetary Science letters, 104: 398–416.
Hill, R.S. and Nolte, C.C., 1989. Gamtoos Group. In: Johnson, M.R. (Ed.), Catalogue of South African lithostratigraphic sheets. South African Committee for Stratigraphy, Council for Geoscience, Pretoria.
Hirano, N., Takahashi, E., Yamamoto, J., Abe, N., Ingle, S.P., Kaneoka, I., Hirata, T., Kimura, J., Ishii, T., Ogawa, Y., Machida, S. and Suyehiro, K., 2006. Volcanism in response to plate. Flexure Science, 313: 1426–1428.
Hoffer, E., 1983. Compositional variations of minerals in metapelites involved in low to medium-grade isograd reactions in the Southern Damara Orogen, South West Africa/Namibia. In: Martin, H. and Eder, F.W. (Eds.), Intracontinental fold belts: case studies in the Variscan Belt of Europe and the Damara Belt in Namibia. Springer, Heidelberg, pp. 745–765.
Hoffman, P.F., 1991. Did the breakout of Laurentia turn Gondwanaland inside-out? Science, 252: 1409–1412.
Hoffman, P.F., 1999. The break-up of Rodinia, birth of Gondwana, true polar wander and the snowball Earth. Journal of American Earth Science, 28: 17–33.
Hoffman, P.F., 2005. On Cryogenian (Neoproterozoic) ice-sheet dynamics and the limitations of the glacial sedimentary record. South African Journal of Earth Sciences, 108: 557–576.
Hoffman, P.F. and Halverson, G.P., 2008. The Otavi Group of the western Northern Platform. The Eastern Kaoko Zone and the western Northern Margin Zone. In: Miller, R.M. (Ed.), The geology of Namibia. Geological Survey of Namibia, Windhoek, Vol. 2, pp. 13-70–13-137.
Hoffman, P.F., Halverson, G.P., Domack, E.W., Husson, J.M., Higgins, J.A. and Schrag, D.P., 2007. Are basal Ediacaran (635 Ma) post-glacial "cap dolostones" diachronous? Earth and Planetary Science Letters, 358: 114–131.
Hoffman, P.F., Hawkins, D.P., Isachsen, C.E. and Bowring, S.A., 1996. Precise U-Pb zircon ages for early Damaran magmatism in the Summas Mountains and Welwitschia inlier, northern Damara belt, Namibia. Communications of the Geological Survey of Namibia, 11: 47–52.
Hoffman, P.F., Kaufman, A.J. and Halverson, G.P., 1998a. Comings and goings of global glaciations on a Neoproterozoic tropical platform in Namibia. GSA Today, 8 (5): 1–9.
Hoffman, P.F., Kaufman, A.J., Halverson, G.P. and Schrag, D.P., 1998b. A Neoproterozoic snowball Earth. Science, 281: 1342–1346.
Hoffman, P.F. and Schrag, D.P., 2002. The snowball Earth hypothesis: testing the limits of global change. Terra Nova, 14: 129–155.
Hoffmann, K.-H., 1983. Lithostratigraphy and facies of the Swakop Group of the southern Damara Belt, SWA/Namibia. Special Publication of the Geological Society of South Africa, 11: 43–63.
Hoffmann, K.H., 1989. New aspects of lithostratigraphic subdivision and correlation of late Proterozoic to early Cambrian rocks of the southern Damara Belt and their correlation with the central and northern Damara Belt and the Gariep Belt. Communications of the Geological Survey of Namibia, 5: 59–67.
Hoffmann, K.H., Condon, D.J., Bowring, S.A. and Crowley, J.L., 2004. A U-Pb zircon date from the Neoproterozoic Ghaub Formation, Namibia: constraints on Marinoan glaciation. Geology, 32: 817–820.
Hoffmann, K.H. and Prave, A.R., 1996. A preliminary note on a revised subdivision and regional correlation of the Otavi Group based on glaciogenic diamictites and associated cap dolostones. Communications of the Geological Survey of Namibia, 11: 77–82.

Hofmann, H.J. and Mountjoy, E.W., 2001. *Namacalathus-Cloudina* assemblage in Neoproterozoic Miette Group (Byng Formation), British Columbia: Canada's oldest shelly fossils. Geology, 29: 1091–1094.

Holland, H., 2006. The oxygenation of the atmosphere and oceans. Philosophical Transactions of the Royal Society B, 361: 903–915.

Holser, W.T. and Kaplan, I.R., 1966. Isotope geochemistry of sedimentary sulfates. Chemical Geology, 1: 93–135.

Hongn, F.D., Tubia, J.M., Aranguren, A., Mon, R. and Vegas, N., 2005. Intrusión del Granito de Tastil en areniscas Eopaleozoicas (Angosto de La Quesera, Cordillera Oriental, Salta). Actas 16° Congreso Geológico Argentino 1, La Plata, pp. 509–514.

Hoppe, A., Schobbenhaus, C. and Walde, D.H.G., 1987. Precambrian iron formation in Brazil. In: Appel, P.W.U. and LaBerge, G.L. (Eds.), Precambrian iron-formations. Athenas, pp. 347–390.

Horstmann, U.E., Ahrendt, H., Clauer, N. and Porada, H., 1992. Damaran metamorphism and its record in the Nama Group, Geocongress '92. Geological Society of South Africa, Bloemfontein, pp. 197–199.

Hou, X.-G., Aldridge, R.J., Bergström, J., Siveter, D.J. and Feng, X.-H., 2004. The Cambrian fossils of Chengjiang, China: the flowering of early animal life. Blackwell, Oxford.

Hough, M.L., Shields, GA., Evins, L.Z., Strauss, H., Henderson, R.A. and Mackenzie, S., 2006. A major sulphur isotope event at c. 510 Ma: a possible anoxia–extinction–volcanism connection during the Early–Middle Cambrian transition? Terra Nova, 18: 257–263.

Howell, D.G., 1989. Tectonics of suspect terranes. Mountain building and continental growth. Chapman and Hall, London, 232 pp.

Hua, H., Chen, Z., Yuan, X., Zhang, L. and Xiao, S., 2005. Skeletogenesis and asexual reproduction in the earliest biomineralizing animal *Cloudina*. Geology, 33: 277–280.

Hua, H., Pratt, B.R. and Zhang, L., 2003. Borings in *Cloudina* shells: Complex predator – prey dynamics in the terminal Neoproterozoic. Palaios, 18: 454–459.

Hughes, M., 1987. The Tsumeb orebody, Namibia, and related dolostone-hosted base metal ore deposits of Central Africa. Unpublished PhD Thesis, University of the Witwatersrand, Johannesburg, 448 pp.

Hughes, N.C., Peng, S., Bhargava, O.N., Ahluwalia, A.D., Walia, S., Mirow, P.M. and Parcha, S.K., 2005. Cambrian biostratigraphy of the Tal Group, Lesser Himalaya, India, and early Tsanglangpuan (Late early Cambrian) trilobites from the Nigali Dhar syncline. Geological Magazine, 142: 57–80.

Humphrey, F.L. and Allard, G.O., 1968. The Propriá geosyncline, a newly recognized Precambrian tectonic province in the Brazilian shield. Abstracts XXIII International Geological Congress, Prague, 4, pp. 123–139.

Hunter, D.R. and Reid, D.L., 1987. Mafic dyke swarms in southern Africa. In: Halls, H.C. and Fahrig, W.F. (Eds.), Mafic Dyke swarms. Geological Association of Canada, Special Paper, pp. 445–456.

Huntley, J.W., Xiao, S. and Kowalewski, M., 2006. 1.3 Billion years of acritarch history: an empirical morphospace approach. Precambrian Research, 144: 52–68.

Hurtgen, M.T., Arthur, M.A. and Halverson, G.P., 2005. Neoproterozoic sulfur isotopes, the evolution of microbial sulfur species, and the burial efficiency of sulfide as sedimentary sulfide. Geology, 33: 41–44.

Hurtgen, M.T., Halverson, G.P., Arthur, M.A. and Hoffman, P.F., 2006. Sulfur cycling in the aftermath of a Neoproterozoic (Marinoan) snowball glaciation: evidence for a syn-glacial sulfidic deep ocean. Earth and Planetary Science Letters, 245: 551–570.

Hurtgen, M.T., Pruss, S. and Knoll, A.H., 2009. Evaluating the relationship between the carbon and sulfur cycles in the later Cambrian ocean: an example from the Port au Port Group, western Newfoundland, Canada. Earth and Planetary Science Letters, 281: 288–297.

Hyde, W.T., Crowley, T.J., Baum, S.K. and Peltier, W.R., 2000. Neoproterozoic "snowball Earth" simulations with a coupled climate/ice-sheet model. Nature, 405: 425–429.

Inda, H.A.V. and Barbosa, J.F., 1978. Texto explicativo para o mapa geológico do Estado da Bahia, escala 1:1.000.000. Publ. Secretaria de Minas e Energia, Salvador, Brasil, 137 pp.

Iñiguez, A.M., del Valle, A., Poiré, D.G., Spalletti, L.A. and Zalba, P.E., 1989. Cuenca Precámbrica/Paleozoica inferior de Tandilia, Provincia de Buenos Aires. In: Chebli, G. and Spalletti, L. (Eds.), Cuencas Sedimentarias Argentinas. INSUGEO, Serie Correlación Geológica 6, pp. 245–263.

Iñiguez,, A.M. and Zalba, P.E., and Zalba, 1974. Nuevo nivel de arcilitas en la zona de Cerro Negro, Partido de Olavarría, Provincia de Buenos Aires. LEMIT, Serie II, 264: 95–100.

Innes, J. and Chaplin, R.C., 1986. Ore bodies of the Kombat mine, South West Africa/Namibia. In: Anhaeusser, C.R. and Maske, S. (Eds.), Mineral deposits of Southern Africa. Geological Society of South Africa 2, Johannesburg, pp. 1789–1805.

Ireland, R., Flötmann, T., Fanning, C.M., Gibson, G.M. and Preiss, W.V., 1998. Development of the early Paleozoic Pacific margin of Gondwana from detrital zircon ages across the Delamerian orogen. Geology, 26: 243–246.

Isozaki, Y., 1997. Permo-Triassic boundary superanoxia and stratified superocean: records from lost deep sea. Science, 276: 235–238.

Ivanov, A.I., Lifshits, V.I., Perevalov, O.V., Strakhova, T.M., Yablonovskiy, B.V, Gryzer, M.I., Il'inskaya, K.H.G. and Golovenok, V.K., 1995. Precambrian of Patom Highland. Nedra, Moscow, 353 pp. (in Russian).

Ivantsov, A.Y. and Fedonkin, M.A., 2002. Conulariid-like fossil from the Vendian of Russia: a metazoan clade across the Proterozoic/Palaeozoic boundary. Palaeontology, 45: 1219–1229.

Iyer, S.S., Babinski, M., Krouse, H.R. and Chamale, F., Jr., 1995. Highly $^{13}C$ enriched carbonate and organic matter in the Neoproterozoic sediments of the Bambui Group, Brazil. Precambrian Research, 73: 271–282.

Jacob, R.E., 1974. Geology and metamorphic petrology of part of the Damara Orogen along the lower Swakop River, South West Africa. Precambrian Research Unit, University of Cape Town, Bulletin, 17: 1–185.

Jacob, R.E., Moore, J.M. and Armstrong, R.A., 2000. Zircon and titanite age determinations from igneous rocks in the Karibib district, Namibia: implications for Navachab vein-style gold mineralization. Geological Survey of Namibia Communications, 12: 157–166.

Jacobs, J., Pisarevsky, S., Thomas, R.J. and Becker, T., 2008. The Kalahari Craton during the assembly and dispersal of Rodinia. Precambrian Research, 160: 142–158.

Jacobsen, S.B. and Kaufman, A.J., 1999. The Sr, C and O isotopic evolution of Neoproterozoic seawater. Chemical Geology, 161: 37–57.

Jain, A.K., Klootwijk, C.T. and Goswami, K.C., 1981. Late Palaeozoic diamictites of the Garhwal Lesser Himalaya. In: Hambrey, M.J. and Harland, W.B. (Eds.), Earth's pre-Pleistocene glacial record. Cambridge University Press, Cambridge, pp. 294–307.

Janikian, L., Almeida, R.P., Trindade, R.I.F., Fragoso-Cesar, A.R.S., D'Agrella Filho, M.S., Dantas, E.L. and Tohver, E., 2008. The continental Record of Ediacaran volcano-sedimentary successions in southern Brazil and its global implications. Terra Nova, 20: 259–266.

Pelechaty, S., Kaufman, A.J. and Grotzinger, J.P., 1996. Evaluation of $\delta^{13}C$ isotope stratigraphy for intrabasinal correlation: data from Vendian strata of the Olenek uplift and Kharaulakh Mountains, Siberian platform, Russia. The Bulletin of the Geological Society of America, 108: 992–1003.

Peltier, W.R., Liu, Y. and Crowley, J.W., 2007. Snowball Earth prevention by dissolved organic carbon remineralization. Nature, 450: 813–818.

Peralta, S.H., 2000. Quebrada de Zonda Field Trip: the Cambrian carbonate sequence, litho and biostratigraphic features. Eastern Precordillera, San Juan, Argentina. In: Aceñolaza, G.F. and Peralta, S.H. (Eds.), Cambrian from the Southern Edge. INSUGEO, Tucumán, pp. 21–28.

Peres, G.G., Alkmim, F.F. and Jordt-Evangelista, H., 2004. The southern Araçuaí belt and the Dom Silvério Group: geologic architecture and tectonic significance. Anais Academia Brasileira Ciências, 76: 771–790.

Peryt, T.M., Hoppe, A., Bechstadt, T., Koster, J., Pierre, C. and Richter, D.K., 1990. Late Proterozoic aragonitic cement crusts, Bambui Group, Minas Gerais, Brazil. Sedimentology, 37: 279–286.

Peterson, K.J. and Butterfield, N.J., 2005. Origin of the Eumetazoa: Testing ecological predictions of molecular clocks against the Proterozoic fossil record. Proceedings of the National Academy of Sciences USA, 102: 9547–9552.

Peterson, K.J., Cotton, J.A., Gehling, J.G. and Pisani, D., 2008. The Ediacaran emergence of bilaterians: congruence between the genetic and the geological fossil records. Philosophical Transactions of the Royal Society B, 363: 1435–1443.

Peterson, K.J., Summons, R.E. and Donoghue, P.C.J., 2007. Molecular palaeobiology. Palaeontology, 50: 775–809.

Peterson, K.J., Waggoner, B.M. and Hagadorn, J.W., 2003. A fungal analog for Newfoundland Ediacaran fossils. Integrative and Comparative Biology, 43: 127–136.

Petrov, G.A., Maslov, A.V. and Ronkin, Y.L., 2005. Pre-Palaeozoic magmatic complexes of Kvarkush-Kamennogorsk anticlinorium (Middle Ural): new data on chemistry and geodynamic. Lithosphere, 4: 42–69 (in Russian).

Petrovich, R., 2001. Mechanisms of fossilization of the soft-bodied and lightly armored faunas of the Burgess Shale and of some other classical localities. American Journal of Science, 301: 683–726.

Philipp, R.L., Nardi, L.V.S. and Bitencourt, M.F., 2000. O Batólito Pelotas no Rio Grande do Sul. In: Holz, M. and De Ros, L.F. (Eds.), Geologia do Rio Grande do Sul, Porto Alegre, pp. 133–160.

Piacentini, T., Boggiani, P.C., Kazuo Yamamoto, J., Freitas, B.T. and Campanha, G.A.C., 2007. Formação ferrífera associada à sedimentaç ão glaciogênica da Formação Puga (Marinoano) na Serra da Bodoquena, MS. Revista Brasileira de Geociências, 37: 530–541.

Pick, F.R. and Lean, D.R.S., 1987. The role of macronutrients (C, N, P) in controlling cyanobacterial dominance in temperate lakes. New Zealand Journal of Marine and Freshwater Research, 21: 425–434.

Pierrehumbert, R.T., 2004. High levels of atmospheric carbon dioxide necessary for the termination of global glaciation. Nature, 429: 646–649.

Pimentel, M.M., Dardenne, M.A., Fuck, R.A., Viana, M.G., Junges, S.L., Fischel, D.P., Seer, H.J. and Dantas, E.L., 2001. Nd isotopes and the provenance of detrital sediments of the Neoproterozoic Brasília Belt, central Brazil. Journal of South American Earth Sciences, 14: 571–585.

Pimentel, M.M. and Fuck, R.A., 1992. Neoproterozoic crustal accretion in central Brazil. Geology, 20: 375–379.

Pimentel, M.M., Fuck, R.A. and Alvarenga, C.J.S.de, 1996. Post-Brasiliano (Pan-African) high-K granitic magmatism in central Brazil: the role of late Precambrian-early Paleozoic extension. Precambrian Research, 80: 217–238.

Pimentel, M.M., Fuck, R.A. and Fischel, D.P., 1999. Estudo isotópico Sm–Nd regional da porção central da Faixa Brasília: implicações para idade e origem dos granulitos do Complexo Anápolis-Itauçu e sedimentos do Grupo Araxá. Revista Brasileira de Geociências, 29: 271–276.

Pimentel, M.M., Fuck, R.A. and Gioia, S.M., 2000. The Neoproterozoic Goiás Magmatic Arc: a review and new Sm–Nd isotopic data. Revista Brasileira de Geociências, 30: 35–39.

Pinho, F.E.C., Sial, A.N. and Figueiredo, M.F., 2003. Contribution to the Neoproterozoic C- and O-isotopic record: carbonate rocks from the Paraguay Belt, Mato Grosso, Brazil. IV South American Symposium on Isotope Geology, Salvador, Brazil, Short Papers, 1: 386–389.

Piper, J.D.A., 1982. The Precambrian palaeomagnetic record: the case for the Proterozoic Supercontinent. Earth and Planetary Science Letters, 59: 61–89.

Piper, J.D.A., 2000. The Neoproterozoic Supercontinent: Rodinia or Palaeopangaea? Earth and Planetary Science Letters, 176: 131–146.

Pirajno, F., Petzel, V.F.W. and Jacob, R.E., 1987. Geology and alteration-mineralization of the Brandberg West Sn-W deposit, Damara Orogen, South West Africa/Namibia. South African Journal of Geology, 90: 266–269.

Pisarevsky, S.A., Murphy, J.B., Cawood, P.A. and Collins, A.S., 2008. Late Neoproterozoic and Early Cambrian palaeogeography: models and problems. In: Pankhurst, R. J., Trouw, R.A.J., de Brito Neves, B.B. and De Wit, M.J. (Eds.), West Gondwana: Pre-Cenozoic correlations across the South Atlantic Region. Geological Society, London, Special Publications 294, pp. 9–31.

Pitcher, WS., 1995. Origin and nature of granite. Blackie Academic and Professional, New York, 321 pp.

Piuzana, D., Pimentel, M.M., Fuck, R.A. and Armstrong, R., 2003. Neoproterozoic magmatism and high-grade metamorphism in the Brasília Belt, central Brazil: regional implications of SHRIMP U–Pb and Sm–Nd geochronological studies. Precambrian Research, 125: 245–273.

Poidevin, J.-L., 2007. Stratigraphie isotopique du strontium et datation des formations carbonatées et glaciogénique néoprotérozoiques due Nord et de l'Ouest due craton du Congo, C.R. Geoscience, 339: 259–273.

Poiré, D.G., 1987. Mineralogía y sedimentología de la Formación Sierras Bayas en el núcleo septentrional de las sierras homónimas, Partido de Olavarría, Provincia de Buenos Aires. PhD Thesis, Facultad de Ciencias Naturales y Museo, Universidad Nacional de La Plata, pp. 1–271.

Poiré, D.G., 1989. Stromatolites of the Sierras Bayas Group, Upper Proterozoic of Olavarría, Sierras Septentrionales, Argentina. Stromatolite Newsletter, 14: 58–61.

Poiré, D.G., 1990. Ciclos estromatolíticos y cuerpos monoestromatolíticos de la Formación Villa Mónica, Precámbrico de Olavarría: su significado sedimentológico. 3 Reunión Argentina de Sedimentología, Actas, pp. 223–228.

Poiré, D.G., 1993. Estratigrafía del Precámbrico sedimentario de Olavarría, Sierras Bayas, provincia de Buenos Aires, Argentina. XII Congreso Geológico Argentino y II Congreso Exploración de Hidrocarburos Actas II, Mendoza, pp. 1–11.

Poiré, D.G., 2002. The precambrian/lower paleozoic sedimentary cover of Tandilia System, Argentina. In: Gaucher, C. and Poiré, D.G. (Eds.), II International Colloquium Vendian-Cambrian of W. Gondwana, field trip guide. Facultad de Ciencias-UNESCO, Montevideo, pp. 55–66.

Poiré, D.G., 2004. Sedimentary history of the Neoproterozoic of Olavarría, Tandilia System, Argentina: new evidence from their sedimentary sequences and unconformities-A "snowball Earth" or a "phantom" glacial? 1st Symposium on Neoproterozoic-Early Paleozoic Events in SW-Gondwana, Extended Abstracts, pp. 46–48.

Poiré, D.G, Canessa, N.D., García Repeto F. and Canalicchio, F.M., 2006. Digitate columnar stromatolites from Neoproterozoic dolostones of Gibraltar Formation, Mina Verdún Group, Uruguay. V South American Symposium on Isotope Geology, Short Papers, Punta del Este, pp. 284–286.

Poiré, D.G. and del Valle, A., 1996. Trazas fósiles en barras submareales de la Formación Balcarce (Ordovícico), Cabo Corrientes, Mar del Plata, Argentina. Asociación Paleontológica Argentina, Publicación Especial 4, pp. 89–102.

Poiré, D.G., del Valle, A. and Regalía, G.M., 1984. Trazas fósiles en cuarcitas de la Formación Sierras Bayas (Precámbrico) y su comparación con las de la Formación Balcarce (Cambro-Ordovícico), Sierras Septentrionales de la provincia de Buenos Aires. 9 Congreso Geológico Argentino, Actas 4, pp. 249–266.

Poiré, D.G. and Gaucher, C., 2007. Lithostratigraphy and correlations of two Neoproterozoic basins from the Río de la Plata Craton, SW-Gondwana. III Symposium on Neoproterozoic-Early Palaeozoic Events in southwestern Gondwana, Programme and Short Papers, Stellenbosch, pp. 23–27.

Poiré, D.G., Gómez Peral, L., Bertolino, S. and Canalicchio, J.M., 2005a. Los niveles con pirofilita de la Formación Villa Mónica, Precámbrico de Olavarría, Sistema de Tandilia, Argentina. XVI Congreso Geológico Argentino, Actas II, pp. 863–866.

Poiré, D.G., González, P.D., Canalicchio, J.M. and García Repetto, F., 2003a. Litoestratigrafía y estromatolitos de la sucesión sedimentaria Precámbrica de la cantera Mina Verdún, Minas, Uruguay. Revista Sociedad Uruguaya de Geología, 3, Publicación Especial 1, pp. 108–123.

Poiré, D.G., González, P.D., Canalicchio, J.M., García Repetto, F. and Canessa, N.D., 2005b. Estratigrafía del Grupo Mina Verdún, Proterozoico de Minas, Uruguay. Latin American Journal of Sedimentology and Basin Analysis, 12: 125–143.

Poiré, D.G. and Spalletti, L.A., 2005. La cubierta sedimentaria precámbrica-paleozoica inferior del Sistema de Tandilia. In: de Barrio, R.E., Etcheverry, R.O., Caballé, M.F. and Llambías, E. (Eds.), Geología y Recursos Minerales de la Provincia de Buenos Aires. Relatorio del XVI Congreso Geológico Argentino, La Plata, pp. 51–68.

Poiré, D.G. and Spalletti, L.A., 2008. The transition from Upper Ordovician ice house to Lower Silurian green house in the Tandilia System, Rio de la Plata Craton, Argentina. Palaeozoic Climates International Congress, Abstract, Lille.

Poire, D.G., Spalletti, L.A. and del Valle, A., 2003b. The Cambrian-Ordovician siliciclastic platform of the Balcarce Formation (Tandilia System, Argentina): facies, trace fossils, palaeoenvironments and sequence stratigraphy. Geologica Acta, 1: 41–60.

Pokrovskii, B.G., 1996. Boundary of Proterozoic and Paleozoic: isotopic anomalies in sections of Siberian Platform and global changes of natural environments. Lithology and Mineral Resources, 4: 376–392.

Pokrovskii, B.G., Melezhik, V.A. and Bujakaite, M.I., 2006. Carbon, oxygen, strontium, and sulfur isotopic compositions in late Precambrian rocks of the Patom Complex, central Siberia: Communication 1. Results, isotope stratigraphy, and dating problems. Lithology and Mineral Resources, 41: 450–474.

Pokrovsky, B.G., Melezhik, B.A. and Buyakaite, M.I., 2006. Geochemistry of isotopes C, O, Sr and S, chemostratigraphy and environments of sedimentation of Late Precambrian deposits of Patom trough. Lithology and Mineral Resources, 5: 505–530.

Popp, B.N., Trull, T., Kenig, F., Wakeham, S.G., Rust, T.M., Tilbrook, B., Griffiths, F.B., Wright, S.W., Marchant, H.J., Bidigare, R.R. and Laws, E.A., 1999. Controls on the carbon isotopic composition of Southern Ocean phytoplankton. Global Biogeochemical Cycles, 13: 827–843.

Porada, H., 1979. The Damara-Ribeira orogen of the Pan-African/Brasiliano cycle in Namibia (South West Africa) and Brazil as interpreted in terms of continental collision. Tectonophysics, 57: 237–265.

Porada, H., 1989. Pan-African rifting and orogenesis in Southern to Equatorial Africa and Eastern Brazil. Precambrian Research, 44: 103–136.

Porada, H. and Berhorst, V., 2000. Towards a new understanding of the Neoproterozoic-Early Palaeozoic Lufilian and northern Zambezi Belts in Zambia and the Democratic Republic of Congo. Journal of African Earth Sciences, 30: 727–771.

Porada, H. and Bouougri, E., 2008. Neoproterozoic trace fossils vs. microbial mat structures: examples from the Tandilia Belt of Argentina. Gondwana Research, 13: 480–487.

Porada, H., Ghergut, J. and Bouougri, E.H., 2008. Kinneyia-type wrinkle structures – critical review and model of formation. Palaios, 23: 65–77.

Porter, S.M., 2004. The fossil record of early eukaryotic diversification. Paleontological Society Papers, 10: 35–50.

Porter, S.M., 2007. Seawater chemistry and early carbonate biomineralization. Science, 316: 1302.

Porter, S.M. and Knoll, A.H., 2000. Testate amoebae in the Neoproterozoic Era: evidence from vase-shaped microfossils in the Chuar Group, Grand Canyon. Paleobiology, 26: 360–385.

Porter, S.M., Knoll, A.H. and Affaton, P., 2004. Chemostratigraphy of Neoproterozoic cap carbonates from the Volta Basin, West Africa. Precambrian Research, 130: 199–212.

Porter, S.M., Meisterfeld, R. and Knoll, A.H., 2003. Vase shaped microfossils from the Neoproterozoic Chuar Group, Grand Canyon: a classification guided by modern testate amoebae. Journal of Paleontology, 77: 409–429.

Porto, J.C., Fernández, R. and Carrión, M.H., 1990. Calizas y dolomías de la Formación Puncoviscana *s.l.* In: Aceñolaza, F.G., Miller, H. and Toselli, A.J. (Ed.), El Ciclo Pampeano en el Noroeste Argentino. Serie Correlación Geológica 4, pp. 37–52.

Postnikov, A.A. and Terleev, A.A., 2004. Stratigraphy of Neoproterozoic of the Altay-Sayan folded region. Geology and Geophysics, 45: 295–309.

Pothe de Baldis, E.D., Baldis, B.A. and Cuomo, J., 1983. Los fósiles precámbricos de la Formación Sierras Bayas (Olavarría) y su importancia intercontinental. Asociación Geológica Argentina Revista, 38: 73–83.

Powell, C.Mc.A., 1995. Are Neoproterozoic glacial deposits preserved on the margins of Laurentia related to the fragmentation of two supercontinents? Comment and Reply Geology, 23: 1053–1054.

Powell, C.McA., Jones, D.L., Pisarevsky, S. and Wingate, M.T.D., 2001. Palaeomagnetic constraints on the position of the Kalahari craton in Rodinia. Precambrian Research, 110: 33–46.

Praekelt, H.E., 1996. The geology of the Cango Caves and Environs. In: Grobbelaar, J.U., Smith, V.R., Praekelt, H.E., Du Preez, P.J. and Brink, J.S. (Eds.), Scientific study aimed at establishing a management for the Cango Caves. Report Oudtshoorn Municipality, Oudtshoorn, pp. 22–33.

Prave, A.R., 1996. Tale of three cratons: tectonostratigraphic anatomy of the Damara orogen in northwestern Namibia and the assembly of Gondwana. Geology, 24: 1115–1118.

Prazeres Filho, H.J., 2005. Caracterização Geológica e Petrogenética do Batólito Granítico Três Córregos (SP-PR): Geoquímica Isotópica (Nd-Sr-Pb), Idades (ID-TIMS/SHRIMP) e $\delta^{18}O$ em Zircão. Doctoral thesis, Intituto de Geociências da Universidade de S. Paulo, S. Paulo, 207 p.

Prazeres Filho, H.J., Harara, O.M., Basei, M.A.S., Passarelli, C.R. and Siga, O., Jr., 2003. Litogeoquímica, Geocronologia U-Pb e Geologia Isotópica (Sr-Nd-Pb) das rochas graníticas dos batólitos Cunhaporanga e Três Córregos, na porção sul do cinturão Ribeira, Estado do Paraná. Geologia USP: Série Científica, 3: 51–70.

Preciozzi, F., Masquelin, H. and Basei, M.A.S., 1999. The Namaqua/Grenville terrane of eastern Uruguay. II South American Symposium on Isotope Geology, Actas, Córdoba, Argentina, pp. 338–340.

Preciozzi, F., Sanchez Betucci, L. and Masquelin, H., 1993. Geología de la porción sur del Cinturón Cuchilla Dionisio. I Simposio Internacional del Neoproterozoico-Cámbrico de la cuenca del Plata. Guía de Excursión, La Paloma, Rocha, pp. 1–39.

Preciozzi, F., Spoturno, J., Heinzen, W. and Rossi, P., 1985. Carta Geológica del Uruguay a escala 1/500.000. DINAMIGE, Montevideo, Uruguay.

Preiss, W.V., 1987. The Adelaide geosyncline. Geological Survey of South Australia Bulletin, 53: 438 pp.

Preiss, W.V., 2000. The Adelaide geosynclines of South Australia and its significance in Neoproterozoic continental reconstruction. Precambrian Research, 100: 21–63.

Puffer, J.N., 2002. A late Neoproterozoic eastern Laurentian superplume: location, size, chemical composition and environmental impact. American Journal of Science, 302: 1–27.

Puhan, D., 1983. Temperature and pressure of metamorphism in the central Damara Orogen. In: Miller, R.M. (Ed.), Evolution of the Damara Orogen of South West Africa/Namibia. Geological Society of South Africa, Special Publication, Marshalltown, pp. 219–223.

Quartino, B. and Quartino G., 1996. Las Sierras Australes de Santiago del Estero. Caracteres diferenciales e interpretación. Actas 13 Congreso Geológico Argentino 1, pp. 563–574.

Queiroga, G.N., Pedrosa-Soares, A.C., Noce, C.M., Alkmim, F.F., Pimentel, M.M., Dantas, E., Martins, M., Castañeda, C., Suita, M.T.F. and Prichard, H., 2007. Age of the Ribeirão da Folha ophiolite, Araçuaí orogen: the U–Pb zircon dating of a plagiogranite. Geonomos, 15: 61–65.

Queiroga, G.N., Pedrosa-Soares, A.C., Quéméneur, J. and Castañeda, C., 2006. A unidade metassedimentar do ofiolito de Ribeirão da Folha, Orógeno Araçuaí, Minas Gerais: petrografia, geotermobarometria e calcografia. Geonomos, 14: 9–12.

Quintas, M.C.L., 1995. O embasamento da Bacia do Paraná: reconstrução geofísica de seu arcabouço. Doctoral thesis, Instituto de Astronomia, Geofísica e Ciências Atmosféricas, Universidade de São Paulo, S. Paulo, 218 p.

Quintas, M.C.L., Mantovani, M.S.M. and Zalán, V.P., 1997. Contribution to the study of the mechanical evolution of the Paraná Basin. B Geoci. Petrobras, Rio de Janeiro, 11 (1/2): 48–73.

Quintas, M.C.L., Mantovani, M.S.M. and Zalán, P.V., 1999. Uma contribuição para o estudo da evolução mecânica da Bacia do Paraná. Revista Brasileira de Geociências, 29: 217–226.

Raith, J.G., Cornell, D.H., Frimmel, H.E. and de Beer, C.H., 2003. New insights into the geology of the Namaqua Tectonic Province, South Africa, from ion probe dating of detrital and metamorphic zircon. The Journal of Geology, 111: 347–366.

Ramdohr, P., 1938. Die Khan-Grube bei Arandis, Südwestafrika. Zeitschrift fur Praktische Geologie, 46: 41–50.

Ramos, V.A., 1988. Late Proterozoic–Early Paleozoic of South America – a collisional history. Episodes, 11: 168–174.

Ramos, V.A., 1999. Rasgos estructurales del territorio argentino. 1. Evolución tectónica de Argentina. Geología Argentina, Anales, 29: 715–784.

Ramos, V.A., 2008. The basement of the Central Andes: the Arequipa and related terranes. Annual Review of Earth and Planetary Sciences, 36: 289–324.

Ramos, V.A., Cristallini, E.O. and Pérez, D.J., 2002. The Pampean flat-slab of the Central Andes. Journal of South American Earth Sciences, 15: 59–78.

Ramos, V., Vujovich, G., Kay, S. and McDounough, J., 1993. La Orogénesis de Greenville en las Sierras Pampeanas Occidentales: La Sierra de Pié de Palo y su integraciòn al continente Proterozoico. Actas 13° Congreso Geológico Argentino, 3: 343–357.

Ramos, V., Vujovich, G., Mahlburg Kay, S. and McDonough, M., 1993. La orogénesis de Grenville en las Sierras Pampeanas occidentals: la Sierra Pie de Palo y su integración al supercontinente Proterozoico. XII Congreso Geológico Argentino y II Congreso de Exploración de Hidrocarburos, Actas III, pp. 343–358.

Ransome, I.G.D., 1992. The geochemistry, kinematics and geodynamics of the Gannakouriep dyke swarm. Unpublished MSc Thesis, University of Cape Town, Cape Town, 182 pp.

Rapalini, A.E., 2006. New Late Proterozoic paleomagnetic pole for the Río de la Plata craton: implications for Gondwana. Precambrian Research, 147: 223–233.

Rapalini, A.E., Poiré, D.G., Trindade, R.I. and Richarte, D., 2008. Geochronologic and geodynamic implications of palaeomagnetic results from the Sierras Bayas Group, Río de la Plata craton (Argentina). VI South American Symposium on Isotope Geology, Bariloche (CD-ROM).

Rapalini, A.E. and Sanchez Bettucci, L., 2008. Widespread remagnetization of late Proterozoic sedimentary units of Uruguay and the apparent polar wander path for the Rio de La Plata craton. Geophysical Journal International, 174: 55–74.

Rapela, C.W., Casquet, C., Pankhurst, R.J., Baldo, E.G., Fanning, C.M., Galindo, C., Dahlquist, J. and González-Casado, J.M., 2008. The final assembly of Gondwana: the evidence from the Sierras Pampeanas. 33rd International Geological Congress, Oslo (CD-ROM).

Rapela, C.W., Dalla Salda, L.H. and Cingolani, C.A., 1974. Un intrusivo básico ordovícico en la Formación La Tinta (Sierra de los Barrientos, Provincia de Buenos Aires). Revista de la Asociación Geológica Argentina, 29: 319–331.

Rapela, C.W., Pankhurst, R.J., Casquet, C., Baldo, E.G., Saavedra, J. and Galindo, C., 1998a. Early evolution of the Proto-Andean margin of South America. Geology, 26: 707–710.

Rapela, C.W., Pankhurst, R., Casquet, C., Baldo, E., Saavedra, J., Galindo, C. and Fanning, C.M., 1998b. The Pampean orogeny of the southern proto-Andes: Cambrian continental collision in the Sierras de Córdoba. In: Pankhurst, R. and Rapela, C.W. (Ed.), The Proto-Andean margin of Gondwana. Geological Society, Special Publication 142, pp. 181–217.

Rapela, C.W., Pankurst, R., Casquet, C., Fanning, C., Baldo, E., González Casado, J., Galindo, C. and Dahlquist, J., 2007. The Rio de la Plata Craton and the assembly of SW Gondwana. Earth-Science Reviews, 83: 49–82.

Rapela, C.W., Pankhurst, R.J., Fanning, C.M. and Grecco, L.E., 2003. Basement evolution of the Sierra de la Ventana Fold Belt: new evidence for Cambrian continental rifting along the southern margin of Gondwana. Journal of the Geological Society of London, 160: 613–628.

Raub, T.D. and Evans, D.A., 2006. Magnetic reversals in basal Ediacaran cap carbonates: a critical review. EOS Transactions AGU, 87 (36), Joint Assembly Supplement, Abstract GP41B-02.

Reid, D.L., 1982. Age relationships within the Vioolsdrif batholith, lower Orange River region. II. Two stage emplacement history and the extent of Kibaran overprinting. Transactions of the Geological Society of South Africa, 85: 105–110.

Reid, D.L., 1991. Alkaline rocks in the Kuboos-Bremen igneous province, southern Namibia: the Kanabeam multiple ring complex. Communications of the Geological Survey of Namibia, 7: 3–13.

Reid, D.L., 1997. Sm-Nd age and REE geochemistry of Proterozoic arc-related igneous rocks in the Richtersveld Subprovince, Namaqua Mobile Belt, Southern Africa. Journal of African Earth Sciences, 24: 621–633.

Reid, D.L., Ransome, I.G.D., Onstott, T.C. and Adams, C.J., 1991. Time of emplacement and metamorphism of Late Precambrian mafic dykes associated with the Pan-African Gariep orogeny, Southern Africa: implications for the age of the Nama Group. Journal of African Earth Sciences, 13: 531–541.

Renne, P.R., Ernesto, M., Pacca, I.G., Coe, R.S., Glen, J.M., Prévot, M. and Perrin, M., 1992. The age of Paraná flood volcanism, rifting of Gondwanaland, and the Jurassic–Cretaceous Boundary. Science, 258: 975–979.

Reuning, E., 1937. Die Goldfelder von Ondundu, Südwestafrika. Geologische Rundschau, 28: 229–239.

Reynolds, C.S., Jaworski, G.H.M., Cmiech, H.A. and Leedale, G.F., 1981. On the annual cycle of the blue-green alga Microcystis aeruginosa Kütz. Emend. Elenkin. Philosophical Transactions of the Royal Society of London, Series B, Biological Science, 293: 419–477.

Riccomini, C., Nogueira, A.C.R. and Sial, A.N., 2007. Carbon and oxygen isotope geochemistry of Ediacaran outer platform carbonates, Paraguay Belt, central Brazil. Anais da academia Brasileira de Ciências, 79: 519–527.

Richards, T.E., 1986. Geological characteristics of rare-metal pegmatites of the Uis Type in the Damara Orogen, South West Africa/Namibia. In: Anhaeusser, C.R. and Maske, S. (Eds.), Mineral deposits of Southern Africa. Geological Society of South Africa, Johannesburg, Vol. 2, pp. 1845–1862.

Ridgwell, A.J., Kennedy, M.J. and Caldeira, K., 2003. Carbonate deposition, climate stability, and Neoproterozoic ice ages. Science, 302: 859–862.

Ridgwell, A. and Zeebe, R.A., 2005. The role of the global carbonate cycle in regulation and evolution of the Earth system. Earth and Planetary Science Letters, 234: 299–315.

Riding, R., 1994. Evolution of algal and cyanobacterial calcification. In: Bengtson, S. (Ed.), Early life on earth, Nobel Symposium, Vol. 84, Columbia University Press, New York, pp. 426–438.

Rieu, R., Allen, P.A., Cozzi, A., Kosler, J. and Bussy, F., 2007. A composite stratigraphy for the Neoproterozoic Huqf Supergroup of Oman: integrating new litho-, chemo- and chronostratigraphic data of the Mirbat area, southern Oman. Journal of Geological Society, London, 164: 997–1009.

Rigobello, A.E., Branquinho, J.A. Dantas, M.G.S., Oliveira, T.F. and Neves Filho, W., 1988. Mina de zinco de Vazante. In: Schobbenhaus, C. and Coelho, C.E.S. (Coords.), Principais Depósitos Minerais do Brasil, DNPM, Brasília, Vol. 3, pp. 101–110.

Ring, U., Kröner, A., Buchwaldt, R., Toulkeridis, T. and Layer, P.W., 2002. Shear-zone patterns and eclogite-facies metamorphism in the Mozambique belt of northern Malawi, east-central Africa: implications for the assembly of Gondwana. Precambrian Research, 116: 19–56.

Rino, S., Kon, Y., Sato, W., Maruyama, S., Santosh, M. and Zhao, D., 2008. The Grenvillian and Pan-African orogens: world's largest orogenies through geologic time, and their implications on the origin of superplume. Gondwana Research, 14: 51–72.

Ripperdan, R.L., 2002. Stratigraphic variation in marine carbonate carbon isotope ratios. In: Valley, J.W. and Cole, D.R. (Eds.), Stable isotope geochemistry. Reviews in Mineralogy and Geochemistry, 43: 637–662.

Ritsema, J., van Heijst, H.J. and Woodhouse, J.H., 2004. Global transition zone tomography. Journal of Geophysical Research, 109 (BO2302): 1–14.

Ritter, U., 1980. The Precambrian evolution of the eastern Richtersveld. Precambrian Research Unit, University of Cape Town, Bulletin, 26: 276 pp.

Rivalenti, G., Mazzucchelli, M., Molesini, M., Petrini, R., Girardi, V.A.V., Bossi, J. and Campal, N., 1995. Petrology of Late Proterozoic mafic dikes in the Nico Pérez region, central Uruguay. Mineralogy and Petrology, 55: 239–263.

Robb, L.J., Armstrong, R.A. and Waters, D.J., 1999. The history of granulite-facies metamorphism and crustal growth from single zircon U-Pb geochronology: Namaqualand, South Africa. Journal of Petrology, 40: 1747–1770.

Rocha-Campos, A.C., Brito-Neves, B.B., Babinski, M., Santos, P.R. and Romano, A.W., 2007. Laminito Moema: Unidade Neoporterozóica de provável origem glagiogênica no centro-leste do Estado de Minas Gerais. Resumos X Simpósio de Geologia do Sudeste, Sociedade Brasileira de Geologia, Núcleo Minas Gerais, Diamantina, p. 90.

Rodrigues, J.B., 2007. Estudos de proveniência de sedimentos da região Centro-Oeste da Faixa Brasília-uma abordagem geocronológica. Written Qualifying Examination, University of Brasília, 50 pp.

Rodrigues, J.B. and Pimentel, M.M., 2008. Zircões detríticos da Formação Sete Lagoas, Grupo Bambuí: idades e implicações tectônicas. 44° Congresso Brasileiro Geologia, Anais, Sociedade Brasileira de Geologia, Curitiba, p. 106.

Roesener, H. and Schreuder, C.P., 1992. Iron, the mineral resources of Namibia. Geological Survey of Namibia, Windhoek, pp. 2.4-1–2.4-11.

Romeiro-Silva, P.C. and Zalan, P.V., 2005. Contribuição da sísmica de reflexão na determinação do limite oeste do Cráton do São Francisco. Anais II Simpósio sobre o Cráton do São Francisco, Sociedade Brasileira de Geologia, Salvador, pp. 44–47.

Zalba, P.E., Manassero, M., Laverret, E.M., Beaufort, D., Meunier, A., Morosi, M. and Segovia, L., 2007. Middle Permian telodiagenetic processes in Neoproterozoic sequences, Tandilia System, Argentina. Journal of Sedimentary Research, 77: 525–538.

Zalba, P.E., Poiré, D.G., Andreis, R.R. and Iñiguez Rodriguez, A.M., 1992. Precambrian and Lower Paleozoic records and paleosurfaces of the Tandilia System, Buenos Aires Province, Argentina. In: Schmitt, J.M. and Gall, Q. (Eds.), Mineralogical and Geochemical Records of Paleoweathering. Memoire des Sciences de la Terre 18, pp. 93–113.

Zang, W.L., 1992. Sinian and Early Cambrian floras and biostratigraphy on the South China Platform. Palaeontographica Abteilung A, 224: 75–119.

Zang, W.L. and Walter, M.R., 1992. Late Proterozoic and Cambrian microfossils and biostratigraphy, Amadeus Basin, Central Australia. Memoirs of the Association of Australasian Paleontologists, 12: 1–132.

Zhadnova, T.P., 1961. Stratigraphy of north-east part of Patom Highland. Trudy Tsentral'nogo nauchno-issledovatel'skogo geologorasvedochnogo institute, No. 38, pp. 103–123 (in Russian).

Zhang, Q.-R., Li, X., Feng, L., Huang, J. and Song, B., 2008a. A new age constraint on the onset of the Neoproterozoic glaciation in the Yangtze Platform, South China. Journal of Geology, 116: 423–429.

Zhang, S., Jiang, G. and Han, Y., 2008b. The age of the Nantuo Formation and Nantuo glaciation in South China. Terra Nova, 20: 289–294.

Zhang, S., Jiang, G., Zhang, J., Song, B., Kennedy, M.J. and Christie-Blick, N., 2005. U-Pb sensitive high-resolution ion microprobe ages from the Doushantuo Formation in South China: constraints on late Neoproterozoic glaciations. Geology, 33: 473–476.

Zhang, X., Shu, D., Li, Y. and Han, J., 2001. New sites of Chengjiang fossils: crucial windows on the Cambrian explosion. Journal of the Geological Society of London, 158: 211–218.

Zhang, Y., Yin, L., Xiao, S. and Knoll, A.H., 1998. Permineralized fossils from the terminal Proterozoic Doushantuo Formation, South China. Journal of Paleontology, Memoir, 50: 1–52.

Zheng, Z., Li, Y., Lu, S. and Li, H., 1994. Lithology, sedimentology and genesis of the Zhengmuguan Formation of Ningxia, China. In: Deynoux, M., Miller, J.M.G., Domack, E.W., Eyles, N., Fairchild, I.J.G. and Young, G.M. (Eds.), Earth's glacial record. Cambridge University Press, Cambridge, pp. 101–108.

Zhou, C., Tucker, R., Xiao, S., Peng, Z., Yuan, X. and Chen, Z., 2004a. New constraints on the ages of Neoproterozoic glaciations in south China. Geology, 32: 437–440.

Zhou, C., Xie, G., McFadden, K., Xiao, S. and Yuan, X., 2007. The diversification and extinction of Doushantuo-Pertatataka acritarchs in South China: causes and biostratigraphic significance. Geological Journal, 42: 229–262.

Zhou, G., Tucker, R., Xiao, S., Yuan, X. and Chen, Z., 2004b. New constraints on the ages of Neoproterozoic glaciations in south China. Geology, 32: 437–440.

Zhou, G. and Xiao, S., 2007. Ediacaran $\delta^{13}C$ chemostratigraphy of South China. Chemical Geology, 237: 89–108.

Zhu, M., Gehling, J.G., Xiao, S., Zhao, Y. and Droser, M.L., 2008. Eight-armed Ediacara fossil preserved in contrasting taphonomic windows from China and Australia. Geology, 36: 867–870.

Zhu, M., Strauss, H. and Shields, G.A., 2007. From snowball earth to the Cambrian bioradiation: calibration of Ediacaran-Cambrian earth history in South China. Palaeogeography, Palaeoclimatology, Palaeoecology, 254: 1–6.

Zhuravlev, A.Y., 2001. Biotic diversity and structure during the Neoproterozoic-Ordovician transition. In: Zhuravlev, A.Y. and Riding, R. (Eds.), The Ecology of Cambrian Radiation. Columbia University Press, New York, pp. 173–199.

Zhuravlev, A.Y. and Wood, R.A., 1996. Anoxia as the cause of the mid-early Cambrian (Botomian) extinction event. Geology, 24: 311–314.

Zhuravlev, A.Y. and Wood, R.A., 2008. Eve of biomineralization: controls on skeletal mineralogy. Geology, 36: 923–926.

Zimmermann, U., 2005. Provenance studies of very low-to low-grade metasedimentary rocks of the Puncoviscana Complex, northwest Argentina. In: Vaughan, A., Leat, P. and Pankhurst, R. (Eds.), Special Publication Geological Society of London 246, pp. 381–416.

Zimmermann, U. and Van Staden, A., 2002. Neoproterozoic to Pre-Ordovician very-low to low-grade metasedimentary rocks from Siján (Sierra de Ambato) and Campo Volcán (Puna) in northwestern Argentina. Actas 15° Congreso Geológico Argentino 2, pp. 229–234.

# Subject Index